Radiation and Homeostasis

Radiation and Homeostasis

Proceedings of the International Symposium of Radiation and Homeostasis, held in Kyoto, Japan, 13-16 July 2001

Editors:

Tsutomu Sugahara
Health Research Foundation
Kyoto
Japan

Osamu Nikaido
Dept. of Molecular and Cellular Biology
Kanazawa University
Kanazawa
Japan

Ohtsura Niwa
Radiation Biology Center
Kyoto University
Kyoto
Japan

2002

ELSEVIER
Amsterdam – Boston – London – New York – Oxford – Paris – San Diego – San Francisco – Singapore – Sydney – Tokyo

ELSEVIER SCIENCE B.V.
Sara Burgerhartstraat 25
P.O. Box 211, 1000 AE Amsterdam, The Netherlands

© 2002 Elsevier Science B.V. All rights reserved.

This work is protected under copyright by Elsevier Science, and the following terms and conditions apply to its use:

Photocopying
Single photocopies of single chapters may be made for personal use as allowed by national copyright laws. Permission of the Publisher and payment of a fee is required for all other photocopying, including multiple or systematic copying, copying for advertising or promotional purposes, resale, and all forms of document delivery. Special rates are available for educational institutions that wish to make photocopies for non-profit educational classroom use.

Permissions may be sought directly from Elsevier Science via their homepage (http://www.elsevier.com) by selecting "Customer support" and then "Permissions". Alternatively you can send an e-mail to: permissions@elsevier.co.uk, or fax to: (+44) 1865 853333.
In the USA, users may clear permissions and make payments through the Copyright Clearance Center, Inc., 222 Rosewood Drive, Danvers, MA 01923, USA; phone: (+1) (978) 7508400, fax: (+1) (978) 7504744, and in the UK through the Copyright Licensing Agency Rapid Clearance Service (CLARCS), 90 Tottenham Court Road, London W1P 0LP, UK; phone: (+44) 207 631 5555; fax: (+44) 207 631 5500. Other countries may have a local reprographic rights agency for payments.

Derivative Works
Tables of contents may be reproduced for internal circulation, but permission of Elsevier Science is required for external resale or distribution of such material.
Permission of the Publisher is required for all other derivative works, including compilations and translations.

Electronic Storage or Usage
Permission of the Publisher is required to store or use electronically any material contained in this work, including any chapter or part of a chapter.

Except as outlined above, no part of this work may be reproduced, stored in a retrieval system or transmitted in any form or by any means, electronic, mechanical, photocopying, recording or otherwise, without prior written permission of the Publisher. Address permissions requests to: Elsevier Science Global Rights Department, at the mail, fax and e-mail addresses noted above.

Notice
No responsibility is assumed by the Publisher for any injury and/or damage to persons or property as a matter of products liability, negligence or otherwise, or from any use or operation of any methods, products, instructions or ideas contained in the material herein. Because of rapid advances in the medical sciences, in particular, independent verification of diagnoses and drug dosages should be made.

First edition 2002

Library of Congress Cataloging-in-Publication Data
International Symposium of Radiation and Homeostasis (2001: Kyoto, Japan)
   Radiation and homeostasis: proceedings of the International Symposium of Radiation
and Homeostasis, held in Kyoto, Japan 13-16 July 2001/ editors, Tsutomu Sugahara,
Osamu Nikaido, Ohtsura Niwa.-- 1st ed.
     p. cm.-- (International congress series; 1236)
   Includes bibliographical references and index.
   ISBN 0-444-50406-0 (alk. paper)
   1. Radiation carcinogenesis--Congresses. 2. Radiation--Physiological effect--
Congresses. 3. Homeostasis--Congresses. I. Sugahara, Tsutomu. II. Nikaido, Osamu.
III. Niwa, Ohtsura. IV. Title. V. Series.

RC268.55 .I58 2002
616.99'4071--dc21

                                                           2002075451

British Library Cataloging in Publication Data
International Symposium of Radiation and Homeostasis (2001: Kyoto, Japan)
   Radiation and homeostasis: proceedings of the International Symposium of Radiation and
Homeostasis, held in Kyoto, Japan, 13-16 July 2001. - (International congress series; 1236)
   1. Radiobiology - Congresses 2. Homeostasis - Congresses 3. Radiation - Physiological
effect - Congresses 4. Radiation injuries - Epidemiology - Congresses
I. Title II. Sugahara, Tsutomu III. Nikaido, Osamu IV. Niwa, Ohtsura
612'.01448

ISBN 0444504060

International Congress Series No. 1236
ISBN: 0-444-50406-0
ISSN: 0531-5131

∞ The paper used in this publication meets the requirements of ANSI/NISO Z39.48-1992 (Permanence of Paper).
Printed in The Netherlands.

Preface

Risks associated with radiation of artificial as well as natural origin have been the central issue of investigation in the field of radiation research. Scientific basis of risk assessment had two major foundations; phenomenological analysis of radiation exposed population such as epidemiological studies of atomic bomb survivors and mechanistic analysis by laboratory investigation at molecular, cellular and animal levels.

Ever increasing safety consciousness is the inevitable consequence of modern societies and now radiation safety has to deal with extremely low level risks associated with low doses of radiation where previous studies were unable to make any prediction on the health effect on human.

Past consensus on the mechanism of radiation effect assumed that radiation-induced DNA damage is the basis of all biological effect. Since the damage linearly increases with radiation dose, so was assumed the biological effect.

Recent years have witnessed a tremendous progress in the basic understanding of radiation effects on cells and tissues. These studies now have excavated that risks are brought about not only by damage itself, but also as a consequence of damage response of cells and tissues, and the molecular mechanism of the radio responses are now emerging. Rich knowledge of radio responses now excavated the intricate biological processes in which each dose ranges of radiation is associated with unique biological responses. Thus, the classical linear non-threshold hypothesis has to be reevaluated in the light of new advances of the field.

The above is the understanding of many researchers and therefore the time is ripe to have a meeting to discuss the issue from two major disciplines of radiation research, epidemiologists and radiation biologists, and, together with policy makers, try to come up the future direction of the better understanding of the mechanism of risk development, the health hazard of radiation at low dose range, and to have scientific basis of radiation safety. This is the purpose of the International Symposium on Radiation and Homeostasis held in Kyoto International Conference Hall from July 13 to 16, 2001 supported by the Health Research Foundation commemorating its 60th anniversary. This book is the proceedings of the Symposium published with the aim to continue and strengthen the collaboration among epidemiologists, radiation biologists and risk assessors as in this Symposium.

Tsutomu Sugahara
President of ISRH 2001
Chairman, Health Research Foundation

Preface v

Contents vii

Epidemiological findings of radiation carcinogenesis

Risk estimates for fast neutrons and implications for radiation protection
A.M. Kellerer (Germany)   3

Gaps in the epidemiology of in utero radiation-exposure effects
R.E. Shore (USA)   13

Do the findings on the health effects of prolonged exposure to very high levels of natural radiation contradict current ultra-conservative radiation protection regulations?
S.M.J. Mortazavi, M. Ghiassi-Nejad (Iran) *and T. Ikushima* (Japan)   19

Where are the radon-induced lung cancer cases? Is it time for a re-evaluation of the radon problem?
A. Enflo (Sweden)   23

Investigation of cancer mortality in the Gastein Valley, an area of high-level natural radiation
J. Pohl-Rüling and W. Hofmann (Austria)   27

Prevalence of lung cancer and indoor radon in Thailand
S. Bovornkitti (Thailand)   31

ICRP evolutionary recommendations and the reluctance of the members of the public to carry out remedial work against radon in some high-level natural radiation areas
P.A. Karam (USA), *S.M.J. Mortazavi, M. Ghiassi-nejad* (Iran), *T. Ikushima* (Japan), *J.R. Cameron and A. Niroomand-rad* (USA)   35

Some correlation aspects of thyroid cancer epidemiology in Ukraine after Chernobyl accident
N.P. Dikiy, E.P. Medvedeva, N.I. Onishchenko and V.D. Zabolotny (Ukraine)   39

Radiation risks of leukemia incidence among Russian emergency workers, 1986–1997
V. Ivanov, A. Gorski, A. Tsyb and S. Khait (Russia)   43

Fallout exposure in the Semipalatinsk Nuclear Test Site area and the induction of thyroid nodules diseases
Z.H. *Zhumadilov* (Japan, Kazakhstan), C. *Land* (USA), M. *Hoshi*, A. *Kimura*, N. *Takeichi* (Japan), T. *Zhigitaev*, G. *Abisheva* (Kazakhstan) *and* K. *Kamiya* (Japan)   47

The mortality and cancer morbidity experience of workers at British Nuclear Fuels plc, 1946–1997
D. *McGeoghegan and* K. *Binks* (UK)   51

European Commission's research programme related to the health effects of ionizing radiation
C. *Desaintes and* D. *Teunen* (Belgium)   55

Does radiation cause liver cancer? Comparison of radiation effects in atomic bomb survivors and other populations
G.B. *Sharp* (Japan)   59

Radiosensitivity and expression of nucleotide excision repair genes in peripheral blood mononuclear cells of myelodysplastic syndrome patients
S. *Ban* (Japan), K. *Kuramoto* (Japan, USA), K. *Oda*, H. *Tanaka*, A. *Kimura*, G. *Suzuki and* T. *Imai* (Japan)   67

An analysis of persistent inflammation among atomic bomb survivors with respect to sex and age at the time of the bombings
K. *Neriishi and* E. *Nakashima* (Japan)   71

Basic study on the radon effects and the thermal effects in radon therapy
K. *Yamaoka*, T. *Mifune*, S. *Kojima*, S. *Mori*, K. *Shibuya*, Y. *Tanizaki and* K. *Sugita* (Japan)   75

Low dose radiation carcinogenesis

Methodological aspects of low-dose epidemiological studies
C.R. *Muirhead* (UK)   83

Recent advances of "Epidemiological study in high background radiation area in Yangjiang, China"
L.-X. *Wei* (China) *and* T. *Sugahara* (Japan)   91

Induction of myeloid leukemias in C3H/He mice with low-dose rate irradiation
T. *Furuse*, Y. *Noda and* H. *Otsu* (Japan)   101

Dose and dose rate effect in mutagenesis, teratogenesis and carcinogenesis
T. Nomura (Japan) — 105

Spontaneous tumorigenesis and mutagenesis in mice defective in the *MTH1* gene encoding 8-oxo-dGTPase
T. Tsuzuki, A. Egashira, K. Yamauchi, K. Yoshiyama and H. Maki (Japan) — 111

Spontaneous and radiation-induced tumorigenesis in *p53*-deficient mice
R. Baskar, H. Ryo, H. Nakajima, T. Hongyo, L. Li, M. Syaifudin, X.E. Si and T. Nomura (Japan) — 115

Oncogenes and tumor suppressor genes in murine tumors induced by neutron- or gamma-irradiation in utero
S. Antal, K. Lumniczky, J. Pálfalvi, E. Hidvégi, F. Schneider and G. Sáfrány (Hungary) — 119

Low dose fetal irradiation, chromosomal instability and carcinogenesis in mouse
P. Uma Devi, M. Hossain and M. Satyamitra (India) — 123

Effects of radioactive iodine (^{131}I) on the thyroid of newborn, pubertal and adult rats
Y. Nitta, M. Hoshi and K. Kamiya (Japan) — 127

Modeling carcinogenic effects of low doses of inhaled radionuclides
I. Balásházy (Hungary), W. Hofmann (Austria) and A. Dám (Hungary) — 133

Effects of a cell phone radiofrequency (860 MHz) on the latency of brain tumors in rats
B.C. Zook and S.J. Simmens (USA) — 137

Molecular analysis of radiogenic tumors: Experimental

Genetic analysis of radiation-induced thymic lymphoma
R. Kominami, Y. Saito, T. Shinbo, A. Matsuki, H. Kosugi-Okano, A. Matsuki, Y. Ochiai, Y. Kodama, Y. Wakabayashi, Y. Takahashi, Y. Mishima and O. Niwa (Japan) — 143

Genetic analysis of radiation-induced mouse hepatomas
K. Kamiya, M. Sumii, Y. Masuda, T. Ikura, N. Koike, M. Takahashi and J. Teshima (Japan) — 151

Development and molecular analysis of thymic lymphomas induced by ionizing radiation in Scid mice
T. Ogiu, H. Ishii-Ohba, S. Kobayashi, M. Nishimura, Y. Shimada, H. Tsuji, H. Ukai, F. Watanabe, F. Suzuki and T. Sado (Japan) — 157

"Second hit" of *Tsc2* gene in radiation induced renal tumors of Eker rat model
O. Hino, H. Mitani and J. Sakaurai (Japan)   163

PTCH (patched) and *XPA* genes in radiation-induced basal cell carcinomas
F.J. Burns, R.E. Shore, N. Roy, C. Loomis and P. Zhao (USA)   175

Differences of molecular alteration between radiation-induced and *N*-ethyl-*N*-nitrosourea-induced thymic lymphomas in B6C3F1 mice
S. Kakinuma, M. Nishimura, A. Kubo, J. Nagai, K. Mita, T. Ogiu, H. Majima, Y. Katsura, T. Sado and Y. Shimada (Japan)   179

Molecular analysis of radiogenic tumors: Human

Transgenic murine models of human cancer: bridging the gap from mouse to man
J.-L. Luo, B. Zielinski (Germany), W.-M. Tong (France), M. Hergenhahn (Germany), Z.-Q. Wang (France) and M. Hollstein (Germany)   185

Cancers induced by alpha particles from Thorotrast
Y. Ishikawa, I. Wada and M. Fukumoto (Japan)   191

The *p53* and *M6P/IGF2r* genes of Thorotrast- and atomic bomb-induced liver cancers: a glimpse into the mechanisms of radiation carcinogenesis
K.S. Iwamoto (Japan)   195

Molecular epidemiology of childhood thyroid cancer around Chernobyl
S. Yamashita, Y. Shibata, H. Namba, N. Takamura and V. Saenko (Japan)   201

Molecular analysis of radiation-induced thyroid carcinomas in humans
H.M. Rabes (Germany)   207

Influence of XPD variant alleles on p53 mutations in lung tumors of nonsmokers and smokers
S.-M. Hou (Sweden), A. Kannio (Finland), S. Angelini, S. Fält, F. Nyberg (Sweden) and K. Husgafvel-Pursiainen (Finland)   217

Genetic instability in Thorotrast induced liver cancers
D. Liu, H. Momoi, L. Li, Y. Ishikawa and M. Fukumoto (Japan)   221

Molecular and cellular mechanisms of radiation responses

Transmission of damage signals from irradiated to nonirradiated cells
J.B. Little, H. Nagasawa, S.M. de Toledo and E. Azzam (USA)   229

Specific gene expression by extremely low-dose ionizing radiation which related to enhance proliferation of normal human diploid cells
M. Watanabe, K. Suzuki and S. Kodama (Japan) — 237

The Yin and Yan of bystander versus adaptive response: lessons from the microbeam studies
H. Zhou, A. Xu, M. Suzuki, G. Randers-Pehrson, C.A. Waldren, E.J. Hall and T.K. Hei (USA) — 241

Deficient PCNA expression and radiation sensitivity
G.E. Woloschak, T. Paunesku and M. Protić (USA) — 249

Mutation induction by continuous low dose rate gamma irradiation in human cells
J. Kiefer, M. Kohlpoth and M. Kuntze (Germany) — 255

DNA-PK activity plays a role on radioadaptation for radiation-induced apoptosis
T. Ohnishi, A. Takahashi and K. Ohnishi (Japan) — 265

The responses of the haemopoietic system to ionizing radiation
E.G. Wright (UK) — 271

Does radiation enhance promotion of already-initiated cells in protracted high-LET carcinogenesis via a bystander effect?
S.B. Curtis, E.G. Luebeck, W.D. Hazelton and S.H. Moolgavkar (USA) — 283

A pulsed laser generated soft X-ray source for the study of gap junction communication and 'bystander' effects in irradiated cells
R.A. Meldrum, G.O. Edwards, J.K. Chipman, C.W. Wharton, S.W. Botchway, G.J. Hirst and W. Shaikh (UK) — 289

Induction of radioresistance by a nitric oxide-mediated bystander effect
H. Matsumoto, S. Hayashi, Z.-H. Jin, M. Hatashita, H. Shioura, T. Ohtsubo, R. Kitai, Y. Furusawa, O. Yukawa and E. Kano (Japan) — 295

Roles of protein kinase C in radiation-induced apoptosis signaling pathways in murine thymic lymphoma cells (3SBH5 cells)
T. Nakajima, O. Yukawa, H. Ohyama, B. Wang, I. Hayata and H. Hama-Inaba (Japan) — 299

Cellular mechanisms of radiation adaptive response in cultured glial cells
Y. Miura, K. Abe and S. Suzuki (Japan) — 303

Radiation-induced genomic instability and delayed activation of p53
K. Suzuki, S. Kodama and M. Watanabe (Japan)   309

Unstable nature of the X-irradiated human chromosome in unirradiated mouse m5S cells
S. Kodama, K. Yamauchi, T. Tamaki, A. Urushibara, S. Nakatomi, K. Suzuki, M. Oshimura and M. Watanabe (Japan)   313

Delayed cell-cycle arrest following heavy-ion exposure
S. Goto, S. Morimoto, T. Kurobe, M. Izumi, N. Fukunishi, M. Watanabe and F. Yatagai (Japan)   317

Analysis of radiation-inducible hSNK gene in human thyroid cells
Y. Shimizu-Yoshida, K. Sugiyama, T. Rogounovitch, H. Namba, V. Saenko and S. Yamashita (Japan)   319

Cellular response in normal human cells exposed to chronically low-dose radiation in heavy-ion radiation field
M. Suzuki, H. Yasuda, R. Lee, C. Ohira, H. Majima, Y. Yamaguchi, C. Yamaguchi and K. Fujitaka (Japan)   323

Inhibition of radiation-induced DNA-double strand break repair by various metal/metalloid compounds
S. Takahashi (Japan), R. Okayasu (USA), H. Sato, Y. Kubota (Japan) and J.S. Bedford (USA)   327

Induction of a large deletion in mitochondrial genome of mouse cells by X-ray irradiation
T. Ikushima, T. Andoh, T. Kaikawa and K. Hashiguchi (Japan)   331

Effects of increased telomerase activity on radiosensitive human SCID cells
Y. Arase, K. Sugita, T. Hiwasa, H. Shirasawa, K. Agematsu, H. Ito and N. Suzuki (Japan)   335

Protein synthesis, cellular defence and *hprt*-mutations induced by low-dose neutron irradiation
A. Dám, N.E. Bogdándi, I. Polonyi, M.M. Sárdy, I. Balásházy and J. Pálfalvy (Hungary)   341

Enhanced induction of mutation by X-irradiation in Werner syndrome cells
G. Kashino, S. Kodama, K. Suzuki, A. Tachibana, M. Oshimura and M. Watanabe (Japan)   347

An ESR and ESEEM study of long-lived radicals which cause mutation in irradiated mammalian cells
J. Kumagai, T. Miyazaki, T. Kumada, S. Kodama and M. Watanabe (Japan)   351

A role of long-lived radicals in radiation mutagenesis and its suppression by epigallocatechin gallate
T. Ise, S. Kodama, K. Suzuki, T. Tanaka, J. Kumagai, T. Miyazaki and M. Watanabe (Japan)   355

Low dose of wortmannin reduces radiosensitivity of cells
K. Okaichi, K. Suzuki, N. Morita, M. Ikeda, N. Matsuda, H. Takahashi, M. Watanabe and Y. Okumura (Japan)   359

Defective accumulation of p53 protein in X-irradiated human tumor cells with low proteasome activity
M. Yamauchi, K. Suzuki, S. Kodama and M. Watanabe (Japan)   363

High susceptibility and possible involvement of telonomic instability in the induction of delayed chromosome aberrations by X-irradiation in *scid* mouse cells
A. Urushibara, S. Kodama, K. Suzuki, F. Suzuki and M. Watanabe (Japan)   367

Possible role of ATM-dependent pathway in phosphorylation of p53 in senescent normal human diploid cells
M. Suzuki, K. Suzuki, S. Kodama and M. Watanabe (Japan)   371

Suppressive effects of p53 protein on heat-induced centrosomal abnormality
M. Miyakoda, K. Suzuki, S. Kodama and M. Watanabe (Japan)   375

Susceptibility of calcium-deficient hydroxyapatite-collagen composite to irradiation
M. Ohta, M. Yasuda and H. Okamura (Japan)   379

RBE values and dose rate effects on the ratio of translocation to dicentrics yields in neutrons with low-energy spectrum
K. Tanaka (Japan), N. Gajendiran and M. Mohankumar (India)   383

Antioxidants as radioprotecting agents for low-level irradiation
E.B. Burlakova, A.N. Goloshchapov, A.A. Konradov, E.M. Molochkina, Yu.A. Treshchenkova and L.N. Shishkina (Russia)   387

Tissue responses

Intestinal metaplasia induced by X-irradiation: its biological characteristics
H. Watanabe (Japan)   393

Aberrant extracellular signaling induced by ionizing radiation and its role in carcinogenesis
R.L. Henshall-Powell, C.C. Park and M.H. Barcellos-Hoff (USA)   399

Apoptosis induced in small intestinal crypts by low doses of radiation protects the epithelium from genotoxic damage
C.S. Potten (UK) — 407

Radiation-induced apoptosis and its role in tissue response
J.H. Hendry (UK) — 415

Essential role of *p53* gene in apoptotic tissue repair for radiation-induced teratogenic injury
T. Norimura, F. Kato and S. Nomoto (Japan) — 423

Susceptibility

Individual differences in chromosomal radiosensitivity: implications for radiogenic cancer
D. Scott (UK) — 433

Susceptibility loci for radiation lymphomagenesis in mice
N. Mori and M. Okumoto (Japan) — 439

Age dependence of susceptibility for long-term effects of ionizing radiation
S. Sasaki (Japan) — 447

Nutrition status and radiation-induced cancer in mice
K. Yoshida, Y. Hirabayashi, T. Sado and T. Inoue (Japan) — 455

Mutations in cervical cancer as predictive factors for radiotherapy
K.M.Y. Wani (Japan, India), N.G. Huilgol (India), T. Hongyo, H. Ryo, H. Nakajima, L.Y. Li (Japan), N. Chatterjee, C.K.K. Nair (India) and T. Nomura (Japan) — 459

Lectin staining as a predictive test for radiosensitivity of oral cancers
P. Remani, V.N. Bhattathiri, L. Bindu and M. Krishnan Nair (India) — 463

Small dose pre-irradiation induced radioresistance and longevity after challenging irradiation in splenectomized C57BL/6 mice
K. Horie, K. Kubo, H. Kondo and M. Yonezawa (Japan) — 467

Suppression of X-ray-induced apoptosis by low dose pre-irradiation in the spleen of C57BL/6 mice
M. Yonezawa, A. Takahashi, K. Ohnishi, J. Misonoh and T. Ohnishi (Japan) — 471

Effect of pre-irradiation of mice whole-body with X-ray on radiation-induced killing, induction of splenic lymphocyte apoptosis, and expression of mutated Ca^{2+} channel α_{1A} subunit
K. Sawada, A. Takahashi, T. Ohnishi, H. Sakata-Haga and Y. Fukui (Japan) — 477

Elevation of antioxidants in the kidneys of mice by low-dose irradiation and its effect on Fe^{3+}–NTA-induced kidney damage
T. Nomura, K. Yamaoka and K. Sakai (Japan) ... 481

Suppressive effect of long-term low-dose rate gamma-irradiation on chemical carcinogenesis in mice
K. Sakai, T. Iwasaki, Y. Hoshi, T. Nomura, T. Oda, K. Fujita, T. Yamada and H. Tanooka (Japan) ... 487

Possible role of elevation of glutathione in acquisition of enhanced immune function of mouse splenocytes exposed to low-dose γ-rays
S. Kojima, S. Matsumori, H. Ishida, M. Takahashi and K. Yamaoka (Japan) ... 491

Radiation protection effect on Hatakeshimeji (lyophyllum decastes sing)
Y. Gu, Y. Ukawa, S. Park, I. Suzuki, K. Bamen and T. Iwasa (Japan) ... 495

General discussion

General discussion: molecular mechanisms of radiation carcinogenesis and their implications in radiation policy
chaired by J.B. Little (USA) and T. Sugahara (Japan) ... 503

Author index ... 549

Keyword index ... 553

Epidemiological findings of radiation carcinogenesis

Risk estimates for fast neutrons and implications for radiation protection

Albrecht M. Kellerer*

Radiobiological Institute, University of Munich, Schillerstraße 42, 80336 Munich, Germany

Abstract

The risks of low neutron doses are of increasing interest with regard to exposures in aviation, in the transport of nuclear fuel, or even in an accident as that in Tokaimura. In the absence of epidemiological information, the neutron risk coefficient is currently taken to be the product of two uncertain low-dose extrapolations, nominal risk coefficient for the photons and presumed maximum relative biological effectiveness of the neutrons. An approach is presented here that avoids the low-dose extrapolations and invokes, instead, the product of the excess risk from epidemiological observations at an intermediate *reference dose* of γ-rays—here taken to be 1 Gy—and the assumed value, R_1, of the neutron RBE against this reference dose. With R_1 between 20 and 50 the solid, cancer mortality data of the A-bomb survivors provide an excess relative risk (ERR) for the neutrons between 8 and 16/Gy. This result is converted into a risk coefficient in terms of the lifetime attributable risk (LAR) relative to the neutron *effective dose* as defined by ICRP. The result is at high neutron energies somewhat in excess of the current ICRP nominal risk factor, while it agrees with ICRP for the degraded fast neutron spectra that are of major pragmatic interest. The so-called Hiroshima neutron discrepancy will not influence the neutron risk estimate because any remaining uncertainty of the neutron doses in Hiroshima is minor at the reference dose of 1 Gy. © 2002 Elsevier Science B.V. All rights reserved.

Keywords: Neutrons; RBE; Risk coefficient; Solid cancers; A-bomb survivors; Effective dose; Ambient dose equivalent

1. Introduction

Neutrons are known to be far more effective than γ-rays, particularly in small doses. However, this knowledge was derived exclusively from cell studies or animal experiments. No late effects from neutrons have been observed in man.

---

* Corresponding author. Tel.: +49-89-5996-818; fax: +49-89-5996-840.
  E-mail address: amk.sbi@Lrz.uni-muenchen.de (A.M. Kellerer).

On the basis of the former A-bomb dosimetry system, *TD65* [1], it has been argued that much of the excess cancer rate in Hiroshima was due to neutrons [2,3]. However, this assumption led to a re-examination of the A-bomb dosimetry and then to the current dosimetry system, *DS86* [4], which specified much lower neutron doses than the earlier system, TD65. It was then concluded, perhaps prematurely, that any effect contributed by the neutrons would be minor and that no direct information on the neutron risks could be derived from the A-bomb data.

2. The familiar estimation procedure

In the absence of epidemiological evidence, it was realized that the neutron risk estimates would have to be derived by combining information on the effects of γ-rays, with the radiobiological knowledge on the relative biological effectiveness of the neutrons. While there are (as will be seen) different approaches, a seemingly straightforward but somewhat problematic procedure was adopted.

In this approach, a maximum RBE of the neutrons, RBE_{max}, is inferred from the experiments and is multiplied into the risk coefficient, α, for γ-rays:

$$\alpha_n = RBE_{max} \cdot \alpha. \tag{1}$$

This procedure has the serious weakness to be based not just on one extrapolation but actually on a product of two extrapolated numbers, RBE_{max} and α. This involves considerable uncertainty, permits widely different conclusions, and thus, invites fruitless controversy.

A consideration of the numerical values can illustrate the problem. For the maximum RBE, RBE_{max}, the values that range from at least 10 to 100 have been quoted. For the photon risk coefficient, a typical lifetime attributable cancer risk estimate is 0.1/Gy. However, ICRP invokes a reduction factor of 2 and obtains 0.05/Gy [5]. Others, for example the United Nations Committee UNSCEAR [6], have considered a reduction factor between 2 and 10, which involves photon risk estimates down to 0.01/Gy. The resulting options are sufficiently broad to make the result almost meaningless. The pessimist's view is then:

$$\alpha_n = RBE_{max} \cdot \alpha = 100 \cdot 0.1/Sv = 10/Gy \tag{2}$$

while the optimist might choose:

$$\alpha_n = RBE_{max} \cdot \alpha = 10 \cdot 0.01/Sv = 0.1/Gy. \tag{3}$$

Less extreme assessments lie somewhere in between, and the *ICRP radiation-weighting factor* for the neutrons [5] reflects such a balanced view. However, it is clear that any inference will remain judgmental.

The issue would be academic if the neutrons were irrelevant to radiation protection, however, they are of increasing concern. Half of the effective dose in air travel is from neutrons, and the radiation exposure of the aircrews is a topic of growing interest, as well as the neutron exposures from the handling or transport of spent nuclear fuel. In Germany, this latter issue has in fact become a very hot political item. Extravagant claims of high neutron risks have contributed to the problem. In short, there are good reasons to ask whether there is a better way to deduce the risk factor for neutrons.

3. A more direct approach

The diagram in Fig. 1 explains an alternative and simpler procedure that can replace the familiar approach [7].

All experimental evidence suggests that the dose relation is linear for neutrons. For γ-rays, most radiobiological studies show upward curved dose relations as indicated in the diagram in terms of excess relative risk (ERR) although no curvature is seen in the solid cancer data of the A-bomb survivors. Thus, the effect of larger γ-ray doses is fairly well known, while the initial slope of the γ-ray curve is somewhat uncertain. However, there is no need to invoke the shape of the γ-ray curve at low doses. It is sufficient to consider the rather well-defined γ-ray effect at an intermediate reference dose D_1, e.g., $D_1 = 1$ Gy. One then requires the RBE of the neutrons against this fairly high reference dose. Let this RBE be termed as R_1. It can be more reliably determined in animal experiments than the elusive low-dose extrapolation RBE_{max}. Experiments on the induction of non-lethal and lethal tumors in rats suggest $R_1 = 50$ [8,9], and experiments on life shortening in mice suggest values closer to $R_1 = 20$ [10,11]. Which of these values is more representative for late neutron effects in man is uncertain, but it appears reasonable to assume that the most plausible value lies in the range 20–50.

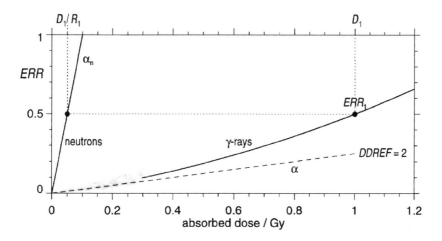

Fig. 1. Schematic diagram of the dose dependence of the excess relative risk after an acute γ-exposure and the dependence for fast neutrons that is inferred in terms of the relative biological effectiveness, R_1, of the neutrons against the reference γ-ray dose $D_1 = 1$ Gy. The grey area covers the γ-ray dependence to indicate the uncertainty at low doses. It brackets all possible dependencies from a simple linear dose relation with the reference slope $c = ERR_1/D_1$ (DDREF = 1) to a linear–quadratic dependence with the initial slope, α, six times smaller than the reference slope (DDREF = 6).

If the excess relative risk ERR_1 due to the acute γ-ray dose D_1 is known from an epidemiological study, then the risk coefficient for the neutrons, α_n, is the product of the reference slope, $c = ERR_1/D_1$, for the γ-rays and the assumed neutron RBE, R_1:

$$\alpha_n = R_1 c \qquad \text{'reference slope' for the } \gamma - \text{rays}: \; c = ERR_1/D_1. \qquad (4)$$

4. Application to the solid cancer mortality of the A-bomb survivors

The A-bomb data are the major basis of γ-ray risk estimates. In the familiar treatment, the neutrons are accounted for roughly in terms of a weighted dose, and the reference slope for the γ-rays is determined by the ERR at the *weighted dose* of 1 Gy [12–14]. The neutron risk coefficient is then readily obtained by multiplication with R_1.

In reality, the issue is somewhat more complex because it is a premature conclusion that the A-bomb data are more or less representative for γ-rays only. The neutron contribution to the absorbed dose is small, only about 11 mGy in a 1-Gy total absorbed dose [4]. However, if, for example, R_1 is 35, then 30 mGy of neutrons have the same effect as 1 Gy of γ-rays, and 11 mGy have 35% of the effect of 1 Gy of γ-rays [7,15].

A detailed analysis can quantify the neutron contribution to the observed solid cancer excess mortality of the A-bomb survivors [16]. Consider first the conventional analysis. It assigns an RBE of $w = 10$ to the neutrons and makes reference to the colon dose. Both choices are inadequate. $w = 10$ is a very low weight factor to represent the neutron RBE, and reference to the colon, which is the deepest lying organ, is a poor choice because the organ-averaged neutron dose is actually twice as large as the colon dose [7]. It is not

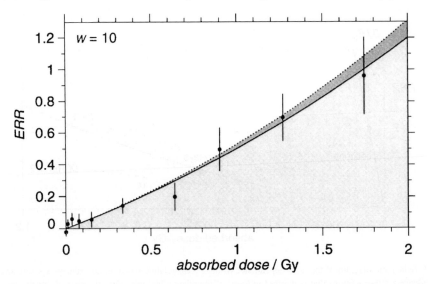

Fig. 2. Dose dependence for solid cancer mortality according to the A-bomb data (1950–1990) and the conventional treatment in terms of a weighting factor, $w = 10$, and reference to the colon dose [16]. The inferred small effect contribution by the neutrons is indicated by the narrow shaded band. The dots and standard errors represent the fit to individual dose categories.

surprising that these biased assumptions suggest a very small neutron contribution to the observed effect represented by the narrow dark shaded band in Fig. 2.

The explicit analysis refers to the organ-averaged dose and considers RBE values, R_1, between 20 and 50. With these more realistic assumptions, the effect attribution to neutrons is, of course, considerably larger. The result is given in Fig. 3. As shown for $R_1 = 35$, the neutron contribution is now fairly substantial. For $R_1 = 50$, it would be still somewhat larger, and it is important to note that the effect attributable to the γ-rays is correspondingly smaller.

Taking into account these considerations, it is now fairly simple to derive the neutron risk coefficient in its dependence on the assumed neutron RBE, R_1. Fig. 4 gives the result in terms of the ERR for solid cancer mortality [7]. The coefficient increases with the assumed neutron RBE, but this increase is somewhat less than proportional because a higher attribution to the neutrons somewhat reduces the effect attribution to 1 Gy of γ-rays. This is also represented in Fig. 4.

The main conclusion is that the uncertainty of the neutron risk coefficient has reasonably narrowed down to about a factor of 2, which is a notable improvement in comparison to the factor of 100 that was mentioned at the outset. In fact, if one considers the doubtful reduction factor DDREF for the photons, one might be tempted to say that the risk coefficient for the neutrons is less uncertain than the risk coefficient for the photons.

One widely discussed uncertainty with regard to neutrons is, of course, the possibility of a revision of the Hiroshima neutron doses. Since the publication of DS86, there have been indications that DS86 underestimates the neutron doses at larger distances, i.e., at low

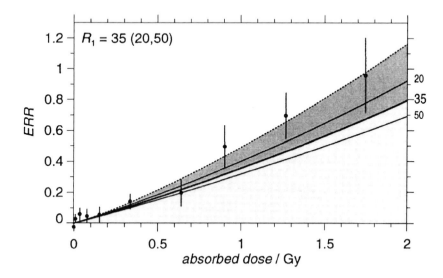

Fig. 3. Dose dependence for solid cancer mortality according to the A-bomb data (1950–1990) and the explicit analysis in terms of an assumed RBE of the neutrons, R_1, against 1 Gy of γ-rays and reference to the organ-averaged dose [16]. The inferred larger effect contribution by the neutrons is indicated by the shaded band for $R_1 = 35$ (the two other solid lines indicate the cases $R_1 = 20$ or 50).

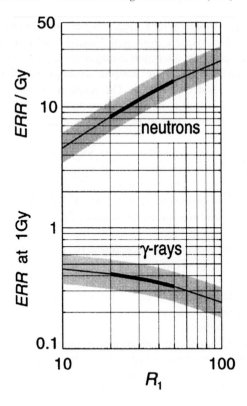

Fig. 4. The risk factor, α_n, for fast neutrons in terms of the excess relative risk, ERR/Gy, for solid cancer mortality [7]. The excess relative risk (ERR) is here taken to be an average for the age at exposure and the attained age model (see text). The lower curve represents the inferred reference slope, i.e., ERR from 1 Gy γ-rays. The values are given in their dependence on the assumed neutron RBE, R_1, versus the acute γ-ray dose 1 Gy. The grey band represents the 95% confidence interval in the fit to the solid cancer mortality data.

doses. An NAS Committee [15] has published an interim report that suggests—on the basis of the determination of fast neutron fluence through Ni-63 measurements in Hiroshima copper samples [17]—that the current dosimetry revision will lead to much smaller changes in the neutron doses than had been indicated by the thermal neutron activation measurements [18]. Some correction may be required at low doses, but at a total dose of 1 Gy—and this is the reference dose in the present analysis—the DS86 doses are not going to change much. The present results are therefore essentially independent of the current dosimetry revision.

5. Comparison to the ICRP nominal risk coefficient

A central issue in the current discussion on neutron risk—with considerable impact on the practice of radiation protection—is, of course, the validity of the ICRP nominal risk coefficient if applied to neutrons. The ICRP risk coefficient [5] is expressed not in terms of the excess relative risk, ERR, but in terms of the lifetime attributable risk. One must therefore translate the ERR in the diagram for the risk coefficient into the lifetime attributable risk, LAR. In line with the conventions adopted by ICRP, one obtains for a working population [19]:

$$LAR = 0.11 ERR. \tag{5}$$

The values ERR/Gy = 8 and ERR/Gy = 16 that result (according to Fig. 4) to the neutron RBE of $R_1 = 20$ and $R_1 = 50$ correspond then to LAR/Gy = 0.9 and LAR/Gy = 1.8.

ICRP gives the nominal solid cancer fatality risk coefficient 0.036/Sv for the occupational exposure [5]. Its radiation-weighting factor for neutrons has a maximal value around $w_R = 22$ for fission energy neutrons [5]. Naively, one would use this factor to convert the risk estimate in terms of Sv to the risk estimate in terms of Gy, and this would provide the value:

$$LAR/Gy = 22 \cdot LAR/Sv = 22 \cdot 0.036 = 0.8. \tag{6}$$

The values 0.9–1.8/Gy that result for the plausible range of neutron RBEs would then exceed the ICRP nominal risk coefficient.

However, this numerical exercise would be erroneous. ICRP defines the neutron effective dose not as the product of the weighting factor, w_R, and the neutron absorbed dose, i.e., the neutron tissue kerma. Instead, it applies the weighting factor to the mixture of the neutron and the γ-ray dose that is caused by the neutrons incident on the human body [5]. This makes a substantial difference because at the most effective and most prevalent neutron energies, there is a considerable γ-ray fraction [20] as shown in the diagram of Fig. 5.

The ICRP radiation-weighting factor must, therefore, not be seen as representing a true neutron RBE. It represents an RBE of a *mixed* γ-ray and neutron radiation. Once this difference is realized, it is fairly straightforward to derive the risk estimate for the unit neutron effective dose by summing the risk estimates for the γ-ray and for the neutron component.

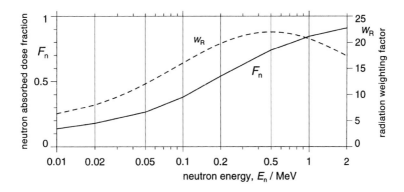

Fig. 5. The genuine neutron fraction, F_n, of the organ-weighted absorbed dose from a (anterior–posterior) neutron exposure (left ordinate–solid curve [20]) and the radiation-weighting factor, w_R, (right ordinate–dashed curve [5]) in their dependence on neutron energy, E_n.

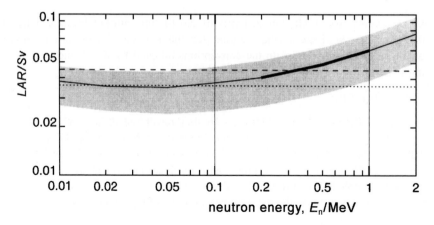

Fig. 6. Lifetime solid cancer fatality risk (LAR/Sv) relative to the neutron effective dose for a working population. The values are given in their dependence on neutron energy, E_n. The solid curve results with $R_1 = 35$. The grey band represents the possible values that result for R_1 between 20 and 50. The assumed RBE values between 20 and 50 refer to the energy range 0.2–1 MeV where the neutrons are most effective. Outside this region, the values LAR/Sv need to be seen as conservative. For a population of all ages, roughly the same values are obtained in terms of the attained age model, while the age at exposure model provides values that are higher by a factor of 1.6. The current ICRP nominal risk coefficients are included for comparison as a dotted line (0.036/Sv, working population) and a dashed line (0.045/Sv, population of all ages).

Fig. 6 gives the result in terms of the shaded band. The values correspond to the assumed neutron RBE values between 20 and 50, which are realistic for the energy range 0.2–1 MeV where the neutrons are most effective. Outside this energy range, the RBE values are lower, which makes the results conservative where the solid line is drawn lightly. It is seen that the present analysis provides a risk estimate for solid cancer mortality that slightly exceeds the ICRP nominal risk coefficient. The estimate agrees with the ICRP

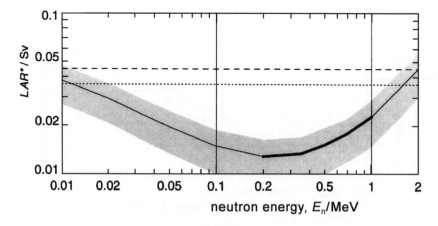

Fig. 7. Lifetime attributable risk (LAR*/Sv) relative to the ambient dose equivalent of the neutrons (H^*) in its dependence on neutron energy, E_n. The diagram is—apart from the difference between LAR/Sv and LAR*/Sv—analogous to Fig. 6.

value for the fairly low energies that are of particular importance in moderated neutron spectra since they are typical for an occupational exposure or for a population exposure due to an accident as that in Tokaimura.

The numerical risk estimates ought to be seen as guidelines and not as precise predictions. Any of the differences between the results that are obtained here and the current ICRP nominal risk coefficient are thus not very significant. An additional factor makes the difference even less essential. This is the fact that the neutron effective dose is, in practice, monitored in terms of the *ambient dose equivalent*, H^* [21]. This *operational quantity* overestimates the effective dose at most neutron energies [22]. The ICRP risk coefficient is therefore (as shown in Fig. 7) strongly conservative against the risk coefficient of the neutrons relative to H^*.

6. Conclusion

The neutron risk coefficient that is derived from typical neutron RBE values and from the reliably known effect of larger doses on the A-bomb survivors is in remarkable agreement with the current ICRP nominal risk coefficient. Its numerical value is independent of the still somewhat controversial issue of the reduction factor DDREF for the photons, and it is also unlikely to be substantially affected by the outcome of the current revision of the A-bomb dosimetry. This conclusion is important since it may reduce some of the apprehension that has recently arisen with regard to neutron exposures from the handling and transport of nuclear fuel or from exposure in aviation.

There are of course implications on risk modelling for γ-rays. As has been suggested here, if the neutrons have had a higher contribution to the effects among the A-bomb survivors than previously assumed, the effect attribution to the γ-rays is somewhat reduced. However, this is a separate issue and needs separate consideration.

References

[1] R.C. Milton, T. Shohoji, Tentative 1965 radiation dose (TD65) estimation for atomic bomb survivors, ABCC Report TR, Radiation Research Effects Foundation, Hiroshima, 1968, pp. 1–68.
[2] H.H. Rossi, A.M. Kellerer, The validity of risk estimates of leukemia incidence based on Japanese data, Radiat. Res. 58 (1974) 131–140.
[3] H.H. Rossi, C.W. Mays, Leukemia risk from neutrons, Health Phys. 34 (1978) 353–360.
[4] W.C. Roesch (Ed.), US–Japan Joint Reassessment of Atomic Bomb Radiation Dosimetry in Hiroshima and Nagasaki, vols. 1 and 2 RERF, Hiroshima, 1987.
[5] International Commission on Radiological Protection, The 1990 recommendations of the International Commission on Radiological Protection, ICRP Publication 60, Ann. ICRP 21 (1–3) Pergamon, Oxford, 1991.
[6] UNSCEAR, Sources and Effects of Ionizing Radiation, Report to the General Assembly, United Nations, New York, 1988, p. 39.
[7] A.M. Kellerer, L. Walsh, Risk estimation for fast neutrons with regard to solid cancer (2001), submitted for publication.
[8] J. Lafuma, D. Chmelevsky, J. Chameaud, M. Morin, R. Masse, A.M. Kellerer, Lung carcinomas in Sprague–Dawley rats after exposure to low doses of radon daughters, fission neutrons, or γ-rays, Radiat. Res. 118 (1989) 230–245.

[9] C. Wolf, J. Lafuma, R. Masse, M. Morin, A.M. Kellerer, Neutron RBE for tumors with high lethality in Sprague–Dawley rats, Radiat. Res., 2000.
[10] B.A. Carnes, D. Grahn, J.F. Thomson, Dose–response modeling of life in a retrospective analysis of the combined data from the Janus program at Argonne National Laboratory, Radiat. Res. 119 (1989) 39–56.
[11] V. Covelli, M. Coppola, V. Di Majo, S. Rebessi, B. Bassini, Tumor induction and life shortening in BC3F1 female mice at low doses of fast neutrons and X-rays, Radiat. Res. 113 (1989) 362–374.
[12] D.A. Pierce, Y. Shimizu, D.L. Preston, M. Vaeth, K. Mabuchi, Studies of the mortality of atomic bomb survivors: report 12. Part I. Cancer, Radiat. Res. 146 (1996) 1–27.
[13] D.E. Thompson, K. Mabuchi, E. Ron, M. Soda, M. Tokunaga, S. Ochikubo, S. Sugimoto, T. Ikeda, M. Terasaki, S. Izumi, D.L. Preston, Cancer incidence in atomic bomb survivors: Part II. Solid tumors, 1958–1987, Radiat. Res. 137 (1994) 17–67.
[14] UNSCEAR, Sources and Effects of Ionizing Radiation, Annex I, Epidemiological Evaluation of Radiation-Induced Cancer, vol. 2, United Nations, New York, 2000, pp. 297–450.
[15] NAS/NRC, Status of the dosimetry for the Radiation Effects Research Foundation (DS86), Report of the Committee on Dosimetry for RERF, National Academy Press, Washington, DC, USA, 2001 (in press, available at http://www.nap.edu/catalog/10103.html).
[16] A.M. Kellerer, L. Walsh, Risk coefficient for γ-rays with regard to solid cancer, Technical Report, SBI-LMU-216/6.2001.
[17] W. Rühm, K. Knie, G. Rugel, A.A. Marchetti, T. Faestermann, C. Wallner, J.E. McAninch, T. Straume, G. Korschinek, Accelerator mass spectrometry of ^{63}NI at the Munich tandem laboratory for estimating fast neutron fluences from the Hiroshima atomic bomb, Health Phys. 79 (2000) 358–364.
[18] T. Straume, S.D. Egbert, W.A. Woolson, R.C. Finkle, P.W. Kubik, H.E. Grove, P. Sharma, M. Hoshi, Neutron discrepancies in the DS86 dosimetry system, Health Phys. 63 (1992) 421–426.
[19] A.M. Kellerer, L. Walsh, On the Conversion of Solid Cancer ERR into Lifetime Attributable Risk, Technical Report, SBI-LMU-214/4.2001.
[20] G. Dietze, B.R.L. Siebert, Photon and neutron dose contributions and mean quality factors in phantom of different size irradiated by monoenergetic neutrons, Radiat. Res. 140 (1994) 130–133.
[21] International Commission on Radiation Units and Measurements (ICRU), Quantities and Units in Radiation Protection Dosimetry, Report 51, 1993.
[22] ICRP Publication 74, Conversion Coefficients for Use in Radiological Protection Against External Radiation, vol. 26, no. 3/4, Pergamon, 1996.

Gaps in the epidemiology of in utero radiation-exposure effects

Roy F. Shore

Department of Environmental Medicine, New York University School of Medicine, 650 First Avenue, New York, NY 10016-3240, USA

Abstract

Several gaps exist in our knowledge about in utero radiation exposure and cancer induction. More data are needed on the magnitude of risk from in utero exposure, especially since the case-control studies tend to show stronger associations than the cohort studies do. There is also uncertainty as to whether the relative risk (RR) for various cancer types is similar following prenatal irradiation. The lifetime cancer risk from prenatal radiation exposure is undefined; the atomic bomb study suggests that for ages 17–46 the RR at 1 Gy is about 3-fold, but this is based on <10 cancer deaths. Clearly more information is needed on the magnitude of risk during the adult years. The carcinogenic risk from radiation exposure to the embryo or first trimester fetus is very poorly defined and is important because of the potential for radiation exposures in early pregnancy. Regarding neurological and mental effects of prenatal radiation exposures, an important question is whether there is a dose–effect threshold for intellectual impairment by ionizing radiation. This has been examined in the Japanese atomic bomb in utero cohort for severe mental retardation, but not for intelligence in general. There is also residual uncertainty about the effects of irradiation on brain development during the first 7 weeks postconception. No human information is available on the impact of dose fractionation or dose rate, or of high-LET radiations, upon mental-deficit outcomes. Data on the impact of prenatal irradiation upon neurological or cognitive functions other than intelligence are sparse. © 2002 Elsevier Science B.V. All rights reserved.

Keywords: Prenatal radiation; Diagnostic radiation; Childhood cancer; Mental retardation; Intelligence

1. Carcinogenic effects of in utero irradiation

Even though there have been several dozen studies of the carcinogenic effects of prenatal irradiation, there is still much uncertainty about the magnitude of effects. The uncertainty stems from a series of limitations in the existing studies:

- Most of the available case-control studies have the potential for selection bias, and especially recall bias. For instance, in the large Oxford Study of Childhood Cancers

(OSCC), compared to control mothers, mothers of cases more frequently reported X-rays before marriage (RR=1.22), between marriage and conception (RR=1.16), and non-abdominal X-rays during pregnancy (RR=1.17) [1]. All of these findings suggest reporting bias.

• The studies of medical prenatal radiation have limited information and substantial uncertainties as to fetal doses. Individual doses are probably highly uncertain based on the retrospective reconstructions of "average" doses across medical practices, although the dose assessment errors would be approximately random and would therefore still tend to yield meaningful dose–response analyses. Nevertheless, the risk estimates could be biased upward or downward because average doses per film, particularly for obstetric examinations in the early calendar-year eras and for non-obstetric X-ray procedures, were not well characterized.

• There is questionable biological plausibility of some of the findings in the OSCC, for instance, the putative elevated risk for embryonal tumors (e.g., neuroblastoma, Wilms tumor) after irradiation in the last few weeks of gestation, in spite of the fact that embryonal tumors are believed to originate in the early weeks after conception.

• The cohort studies of prenatal medical irradiation have found essentially no radiation effect (the RR for total cancers was 1.03, 95% confidence interval (CI): 0.7, 1.4) and hence nominally disagree with the OSCC study results, but the numbers of cancers available in these studies were very small (total of 48 cancers in six studies) [2–7].

• The Japanese atomic bomb study found virtually no excess childhood cancers in those irradiated in utero—only one case was observed when about five to six would have been expected based on the risk estimate from the OSCC.

In addition to uncertainties, there are also several gaps in our knowledge concerning in utero radiation exposure and cancer induction.

One question is whether the relative risk (RR) for prenatal irradiation is equivalent for leukemia and solid cancers as the OSCC reported. The estimated RR of about 1.4 for both leukemia and solid cancers in the OSCC study translates into an excess RR per Gy of over 30-fold for each. By comparison, the atomic bomb estimates of RR at 1 Gy after irradiation during childhood are about 18 for leukemia but only 3 for solid cancers [8]. Similarly, in case-control studies of prenatal irradiation other than the OSCC, the RR estimates tend to be larger for leukemia than for solid cancers (Table 1), albeit the difference is not quite statistically significant ($p=0.07$).

The lifetime cancer risk from prenatal radiation exposure is undefined. Doll and Wakeford [9] estimated the excess absolute risk of cancer before age 15 as 6% per Gy,

Table 1
Combined relative risk estimates of childhood cancer from medical in utero irradiation: case-control and cohort studies that examined both leukemia and solid cancers

Studies included in analysis	Leukemia: RR (95% CI)[a]	Solid cancers: RR (95% CI)[a]
All case-control studies, including OSCC	1.43 (1.31, 1.56)	1.43 (1.27, 1.61)[b]
All case-control studies, except OSCC	1.49 (1.24, 1.80)	1.25 (1.04, 1.52)

[a] The log RRs of the respective studies were combined by the method in Ref. [34]. Statistical heterogeneity among studies in the risk estimates was not found, except as noted.

[b] Confidence interval incorporates variance due to heterogeneity among study risk estimates [34].

based largely on the OSCC, and others have derived similar estimates [10]. This is equivalent to over a 30-fold relative risk at 1 Gy, which would amount to a very large lifetime risk if this relative risk continues unabated throughout adulthood. The only available data on this point are from the atomic bomb study of prenatally exposed, which indicates that for ages 17–46 years, the relative risk at 1 Gy is about 3-fold, similar to that among those irradiated in childhood—but this estimate is very weak, being based on only the seven cancer deaths for whom doses could be estimated in the irradiated group [11]. Clearly more information is needed as to the magnitude of risk during the adult years after prenatal exposure.

The carcinogenic risk from radiation exposure to the embryo or first trimester fetus is very poorly defined, although two studies suggest, based on sparse data, that it may be two to three times as great as from fetal irradiation during the third trimester [11,12]. For example, 6 out of the 10 cancer deaths in the atomic bomb study were irradiated during the first trimester. Any new quantitative information on this issue would be valuable from a radiation protection viewpoint, given the potential for radiation exposures to occur inadvertently in early pregnancy.

2. Mental and neurological effects of in utero irradiation

There are a number of unanswered epidemiological research questions that arise from the existing data on the neurological and mental effects of prenatal radiation exposures. Probably the most important question is whether there is a dose–effect threshold for intellectual impairment by ionizing radiation. The Japanese atomic bomb data are reasonably persuasive that there is a dose threshold for severe mental retardation, although there is substantial uncertainty as to the dose at which the threshold occurs. Specifically, for irradiation during 8–15 weeks postconception, the threshold is estimated as 0.55 (CI = 0.31, 0.61) Gy [13], while for 16–25 weeks postconception it is estimated as 0.87 (CI = 0.28, 1.07) Gy. It is noteworthy that the lower confidence bound on the estimates of a dose threshold for mental retardation for irradiation at either 8–15 or 16–25 weeks postconception is about 0.3 Gy, suggesting that this may be a prudent value to use in considering radiation protection.

A threshold for mental retardation is not surprising, in that the radiation dose would presumably have to derange a large number of cells to produce severe mental effects. However, many fewer cells might have to be affected to produce subtle mental effects upon intelligence, in which case there might be either a low dose threshold or none at all. Inspection of the intelligence quotient (IQ) data suggests there may be a threshold in the vicinity of 0.1 Gy [14], although the confidence interval on any estimated threshold might well encompass zero dose. Most of the analyses that have been performed upon the IQ data have assumed a linear dose–response relationship [14,15]. When a linear-quadratic dose–IQ association was evaluated, a statistically significant quadratic component with upward curvature was found for those exposed at 16–25 weeks postconception but not for those exposed at 8–15 weeks [16]. However, the lack of statistical upward-quadratic curvature in this case may be partly due to the relative alignment of the higher-dose data points, so it does not necessarily rule out a threshold. Because the atomic bomb IQ data have not been tested for a dose threshold, there is currently no definitive answer to this question. Inspection of the school performance data also suggests the possibility of a threshold at a low dose [17], but a statistical test for a threshold has not been performed for this endpoint either.

There is residual uncertainty about the effects of irradiation during the first 8 weeks postconception on brain development. The atomic bomb study did not find such effects, but instead found a substantial deficit of births in the subset irradiated at 0–7 weeks postconception. However, Dekaban [18] reported that 6 out of 8 children in their series who received high radiation exposures at 2–7 weeks postconception suffered mental retardation and other neurological effects. There may have been selection bias in the Dekaban study, but nevertheless it at least raises the question of whether there may be some early in utero mental effects, especially at high doses to the fetus. At the other age extreme, Ron et al. [19] reported deficits in intelligence and scholastic aptitude following postnatal irradiation at the mean age of about 7 years.

Another question is the magnitude of intelligence deficit associated with in utero irradiation. Applying a linear model to the data of the Nagasaki atomic bomb in utero cohort with the mentally retarded cases excluded, the decline in IQ scores was 21 (CI = 9, 33) IQ points per Gy in the 8–15-week cohort based on the most widely applied IQ test (Koga test) in the series, but only 4 (CI = − 5, 13) IQ points per Gy for an alternate intelligence test (Tanaka-B) given to the same children [14]. The difference between the two tests in the dose-related decline in intelligence was statistically significant. This finding raises additional uncertainty regarding the magnitude of radiation effects upon intelligence and suggests the possibility that the effect may be less than has been cited conventionally (namely, 20–30 IQ points per Gy). (However, it should be noted that among those exposed during 16–25 weeks postconception the difference between the two tests was appreciably less. The regression coefficients for a dose-related decline in IQ at exposure ages of 16–25 weeks were 18 and 13 IQ points for the Koga and Tanaka-B tests, respectively.) Schull et al. [14] have suggested that the Nagasaki data comparing the two tests are not robust because of the restricted dose range, i.e., the small numbers of subjects with > 50 cGy, so one should not place great stock in the putative difference. The difference does, however, highlight the uncertainty in the magnitude of radiation-associated deficits in intelligence.

The limited information on human prenatal irradiation and mental effects with respect to the impact of dose fractionation or dose protraction has come mainly from Russia and the Chernobyl experience. With regard to mental retardation, two studies of children of Mayak workers have been negative [20,21]. A study of children born along the Techa River during 1950–1953, the period of the greatest exposures, was also negative with regard to mental retardation [22]. A Ukrainian study [23] purported to show an increase in mental retardation among those living in Chernobyl fallout areas. A study in Belarus showed no correlation between IQ and thyroid dose from Chernobyl but did suggest a lower IQ in those from exposed areas than from unexposed areas [24]. There are also a few reports pertaining to neurological or cognitive function. A study of adolescents whose mothers worked at Mayak during pregnancy found an elevation of "mild pyramidal symptoms" among them [25]. Children born along the Techa River did not have an elevated prevalence of neurological syndromes or of deficits on tests of balance or orientation-search reaction time [26]. Interpreting these studies is problematic for the most part because of scanty information about study methods and data quality.

Very little information is available on the effects of prenatal radiation upon neurological processes and cognitive functions other than intelligence. Schull et al. [27,28] have shown distinctive neuroanatomical features associated with cell-migration abnormalities in the brains of several children who were mentally retarded after exposure to large doses from the atomic bomb. In order to determine the generality of these findings, it would be of interest to know if there are qualitatively similar findings among those with moderate doses who are without frank mental retardation but who may have suffered intellectual impairment from the radiation exposure. In the atomic bomb study, there were fairly weak and imprecise data on seizure disorders in relation to prenatal radiation dose; if the mentally retarded subjects were excluded these associations were not statistically significant [29]. Another atomic bomb study evaluated gross motor function (grip strength) and fine motor coordination in relation to in utero radiation dose; unless the mentally retarded were included, there was no significant association [30].

For comparison, following postnatal brain doses on the order of 1.4 Gy for treatment of *Tinea capitis* at a mean age of about 7 years, EEG changes [31], deviations on psychological tests [19,32] and some suggestion of psychiatric disorders [19,32,33] have been found. Comparable variables have not been examined following prenatal irradiation.

Acknowledgements

The bibliographic and secretarial help of Mrs. Denise Heimowitz and Mrs. LaVerne Yee is gratefully acknowledged. This work was supported in part by Center Grant ES-00260 from the National Institute of Environmental Health Sciences.

References

[1] A. Stewart, J. Webb, D. Hewitt, A survey of childhood malignancies, Br. Med. J. 1 (1958) 1495–1508.
[2] W.M. Court-Brown, R. Doll, R. Hill, Incidence of leukaemia after exposure to diagnostic radiation in utero, Br. Med. J. 2 (1960) 1539–1545.
[3] J. Golding, M. Paterson, L. Kinlen, Factors associated with childhood cancer in a national cohort study, Br. J. Cancer 62 (1990) 304–308.
[4] M.L. Griem, P. Meier, G. Dobben, Analysis of the morbidity and mortality of children irradiated in fetal life, Radiology 88 (1967) 347–349.
[5] E.L. Diamond, H. Schmerler, A. Lilienfeld, The relationship of intra-uterine radiation to subsequent mortality and development of leukemia in children: a prospective study, Am. J. Epidemiol. 97 (1973) 283–313.
[6] P. Shiono, C. Chung, N. Myrianthopoulos, Preconception radiation, intrauterine diagnostic radiation, and childhood neoplasia, J. Natl. Cancer Inst. 65 (1980) 681–686.
[7] R. Murray, P. Heckel, L. Hempelmann, Leukemia in children exposed to ionizing radiation, New Engl. J. Med. 261 (1959) 585–589.
[8] Y. Yoshimoto, R. Delongchamp, K. Mabuchi, In utero exposed atomic bomb survivors: cancer risk update, Lancet 344 (1994) 345–346.
[9] R. Doll, R. Wakeford, Risk of childhood cancer from fetal irradiation, Br. J. Radiol. 70 (1997) 130–139.
[10] C. Muirhead, G. Kneale, Prenatal irradiation and childhood cancer, J. Radiol. Prot. 9 (1989) 209–212.
[11] R.R. Delongchamp, K. Mabuchi, Y. Yoshimoto, D. Preston, Cancer mortality among atomic bomb survivors exposed in utero or as young children, October 1950–May 1992, Radiat. Res. 147 (1997) 385–395.

[12] E. Gilman, G. Kneale, E. Knox, A. Stewart, Pregnancy X-rays and childhood cancers: effects of exposure age and radiation dose, J. Radiol. Prot. 8 (1988) 3–8.
[13] M. Otake, W.J. Schull, S. Lee, Threshold for radiation-related severe mental retardation in prenatally exposed A-bomb survivors: a re-analysis, Int. J. Radiat. Biol. 70 (1996) 755–763.
[14] W.J. Schull, M. Otake, H. Yoshimaru. Effect on intelligence test score of prenatal exposure to ionizing radiation in Hiroshima and Nagasaki: a comparison of the T65DR and DS86 dosimetry systems. RERF (Radiation Effects Research Foundation), Hiroshima, Japan (1988). Report No.: RERF TR/3–88.
[15] W.J. Schull, M. Otake, Effects on Intelligence of Prenatal Exposure to Ionizing Radiation, Radiation Effects Research Foundation, Hiroshima, 1986.
[16] UNSCEAR, Radiation effects on the developing human brain (Appendix H), Sources and Effects of Ionizing Radiation (United Nations Scientific Committee on the Effects of Atomic Radiation), United Nations, New York, 1993, pp. 805–867.
[17] M. Otake, W.J. Schull, Y. Fujikoshi, H. Yoshimaru. Effect on school performance of prenatal exposure to ionizing radiation in Hiroshima: a comparison of the T65DR and DS86 dosimetry systems. RERF (Radiation Effects Research Foundation), Hiroshima, Japan. Report No.: RERF TR (1988) 2–88.
[18] A.S. Dekaban, Abnormalities in children exposed to X-radiation during various stages of gestation: tentative timetable of radiation injury to the human fetus, Part 1, J. Nucl. Med. 9 (1968) 471–477.
[19] E. Ron, B. Modan, S. Floro, I. Harkedar, R. Gurewitz, Mental function following scalp irradiation during childhood, Am. J. Epidemiol. 116 (1982) 149–160.
[20] N.V. Patrusheva, Z.I. Voronin, N.E. Voronina, N.E. Melnikova, A.I. Golubaya, Morbidity among children exposed to gamma-radiation in utero, Bull. Radiat. Med. 2 (1976) 46–51 (Russian).
[21] L.A. Buldakov, Y.P. Ovcharenko, Y.P. SBP, et al., On the status of offspring born to women exposed to a combined gamm-radiation and Pu-239, Bull. Radiat. Med. 1 (1981) 32–36 (Russian).
[22] A.V. Akleyev, M.F. Kisselyov. Medical–Biological and Ecological Impacts of Radioactive Contamination of the Techa River (Russian). Moscow (2000).
[23] A.I. Nyagu, K.N. Loganovsky, T.K. Loganovskaja, Psychophysiologic after effects of prenatal irradiation, Int. J. Psychophysiol. 30 (1998) 303–311.
[24] Y. Kolominsky, S. Igumnov, V. Drozdovitch, The psychological development of children from Belarus exposed in the prenatal period to radiation from the Chernobyl atomic power plant, J. Child Psychol. Psychiatry 40 (1999) 299–305.
[25] N.V. Patrusheva, V.N. Doshchenko, Z.B. Tokarskaya, V.S. Vedeneyeva, On late effects of prenatal exposure in humans, Bull. Radiat. Med. 1 (1973) 78–83 (Russian).
[26] A.V. Akleyev, V.P. Yakovleva, V.A. Savostin, Late clinical effects of prenatal exposure, Probl. Radiat. Saf. 1 (1997) 47–50 (Russian).
[27] W.J. Schull, Ionising radiation and the developing human brain, Ann. ICRP 22 (1991) 95–118.
[28] W.J. Schull, H. Nishitani, K. Hasuo, T. Kobayashi, I. Goto, M. Otake. Brain abnormalities among the mentally retarded prenatally exposed atomic bomb survivors. Technical report. RERF (Radiation Effects Research Foundation), Hiroshima, Japan. Report No.: RERF TR (1991) 13–91.
[29] K. Dunn, H. Yoshimaru, M. Otake, J. Annegers, W. Schull, Prenatal exposure to ionizing radiation and subsequent development of seizures, Am. J. Epidemiol. 131 (1990) 114–123.
[30] H. Yoshimaru, M. Otake, W.J. Schull, S. Funamoto, Further observations on abnormal brain development caused by prenatal A-bomb exposure to ionizing radiation, Int. J. Radiat. Biol. 67 (1995) 359–371.
[31] I. Yaar, E. Ron, B. Modan, Y. Rinott, M. Yaar, M. Modan, Long-lasting cerebral functional changes following moderate dose x-radiation treatment to the scalp in childhood: an electroencephalographic power spectral study, J. Neurol., Neurosurg. Psychiatry 45 (1982) 166–169.
[32] S. Omran, R. Shore, A. Markoff, A. Friedhoff, A. Albert, H. Barr, W. Dahlstrom, B. Pasternack, Follow-up study of patients treated by X-ray epilation for *Tinea capitis*: psychiatric and psychometric evaluation, Am. J. Public Health 68 (1978) 561–567.
[33] R.E. Albert, R. Shore, N. Harley, A. Omran, Follow-up studies of patients treated by X-ray epilation for *Tinea capitis*, in: F. Burns, A. Upton, G. Silini (Eds.), Radiation Carcinogenesis and DNA Alterations, Plenum, New York, 1986, pp. 1–25.
[34] A. Whitehead, J. Whitehead, A general parametric approach to the meta-analysis of randomized clinical trials, Stat. Med. 10 (1991) 1665–1677.

Do the findings on the health effects of prolonged exposure to very high levels of natural radiation contradict current ultra-conservative radiation protection regulations?

S.M.J. Mortazavi [a,*], M. Ghiassi-Nejad [a,b], T. Ikushima [c]

[a]*National Radiation Protection Department (NRPD), Iranian Nuclear Regulatory Authority (INRA), P.O. Box 14155-4494, Tehran, Iran*
[b]*Biophysics Department, School of Science, Tarbiat Modarres University, Tehran, Iran*
[c]*Biology Division, Kyoto University of Education, Kyoto 612-8522, Kyoto, Japan*

Abstract

Inhabited areas with high levels of natural radiation are found in Yangjiang, China, Kerala, India and Guarapari, Brazil. Ramsar, a northern coastal city in Iran, has some areas with one of the highest levels of natural radiation studied so far. The effective dose equivalents in very high background radiation areas (VHBRAs) of Ramsar in particular in Talesh Mahalleh are a few times higher than the dose limits for radiation workers. Inhabitants who live in this area receive annual doses as high as 132 mSv from external terrestrial sources. The basic aim of this paper is to answer the question on whether the findings on the health effects of prolonged exposure to very high levels of natural radiation contradict current ultra-conservative radiation protection regulations. In spite of the fact that at present there is no considerable radio-epidemiological data regarding the incidence of cancer in the inhabitants of VHBRAs of Ramsar, some of the local physicians strongly believe that the population living in these areas does not reveal increased solid cancer or leukemia incidences. As the majority of the inhabitants of Ramsar have lived there for many generations, we started a study to assess whether they developed a radio-adaptive response to high levels of natural radiation. Our results indicated that the frequency of chromosome aberrations (CA) in the lymphocytes of the inhabitants of VHBRAs after exposure to a challenge dose of 1.5 Gy Gamma rays was significantly lower than that of the inhabitants of a normal background radiation area (NBRA). No statistically significant difference was found in the background frequencies of chromosome aberrations between

---

* Corresponding author.
 E-mail address: ikushima@kyokyo-u.ac.jp (T. Ikushima).

0531-5131/02 © 2002 Elsevier Science B.V. All rights reserved.
PII: S0531-5131(02)00291-1

the inhabitants of VHBRAs and a neighboring NBRA. Futility of any urgent limiting regulations is discussed. © 2002 Elsevier Science B.V. All rights reserved.

Keywords: Natural radiation; High background radiation area; Radiation protection

1. Introduction

As the biological effects of low doses of radiation are not fully understood, the current radiation protection recommendations are based on the predictions of an assumption on the linear, no-threshold (LNT) relationship between radiation dose and the carcinogenic effects.

Considering the LNT theory as a scientific fact, there is a general belief that even low levels of radiation, as well as exposures to natural sources are harmful. People in some areas of Ramsar, a city in northern Iran, receive an annual radiation dose from background radiation of up to 260 mSv year^{-1}, which is substantially higher than the 20 mSv year^{-1} that is permitted for Iranian radiation workers. The maximum radon levels in some regions of Ramsar are up to 3700 Bq/m^3. The people who live in these high radiation areas are of considerable interest, because they and their ancestors have been exposed to abnormally high radiation levels over many generations. If a radiation dose of a few hundred mSv per year is detrimental to health causing genetic abnormalities or an increased risk of cancer, it should be evident in the residents of Ramsar.

2. Natural radiation and cancer

The absorbed dose rates in some very high background radiation areas (VHBRAs) of Ramsar is approximately 55–200 times higher than that of normal background radiation areas (NBRAs) [1]. Considering the UNSCEAR-93 report and assuming the linear dose–effect relationship from high exposure levels down to environmental levels, 3–8% of all cancers are caused by the current levels of ionizing radiation [2,3]. If this estimation were true, all of the inhabitants of such an area with extraordinary elevated levels of natural radiation would have died of cancer. On the other hand, NRPB in the UK has reported that living for a life time in a house where radon is at the action level of 200 Bq/m^3 carries a 3–5% risk of fatal lung cancer [4]. It should be noted that radon levels in some regions of Ramsar are up to 3700 Bq/m^3. Interestingly, the preliminary results of our studies on the inhabitants of these areas showed no observable detrimental effect. In spite of the fact that there is not yet solid epidemiological information, most local physicians in Ramsar believe that there is no increase in the incidence rate of solid cancers or leukemia.

3. Induction of radio-adaptive response

The majority of the inhabitants of VHBRAs of Ramsar lived there for many generations, so we performed an experiment to assess whether they developed a radio-adaptive response to high levels of natural radiation. Our results indicated that the frequency of chromosome aberrations in the lymphocytes of the inhabitants of VHBRs after exposure to a challenge dose of 1.5 Gy gamma rays was significantly lower than that of the inhabitants of a natural background radiation area [5].

4. Conclusion

It can be concluded that the health effects of prolonged exposure to very high levels of natural radiation may contradict current ultra-conservative radiation protection regulations. More research on the health effects of prolonged exposure to natural radiation including either hematological or immunological aspects is needed. Due to the lack of any observable detrimental effect, no urgent limiting regulations are needed.

References

[1] M. Sohrabi, M. Bolourchi, M. Beitollahi, J. Amidi, Natural radioactivity of the soil samples, in: L. Wei, T. Sugahara, Z. Tao (Eds.), Proceedings of the 4th International Conference on High Levels of Natural Radiation, Elsevier, Amsterdam, 1997, pp. 129–132.
[2] T. Jung, W. Burkart, Assessment of risks from combined exposure to radiation and other agents at environmental levels, in: L. Wei, T. Sugahara, Z. Tao (Eds.), Proceedings of the 4th International Conference on High Levels of Natural Radiation, Elsevier, Amsterdam, 1997, pp. 167–178.
[3] United Nations Scientific Committee on the Effects of atomic Radiation, Sources and effects of ionizing radiation, 1993.
[4] NRPB, Health risks for radon, Environmental Radon Newsletter, Issue 25, 2000.
[5] S.M.J. Mortazavi, M. Ghiassi Nejad, M. Beitollahi, Very high background radiation areas (VHBRAs) of Ramsar: do we need any regulations to protect the inhabitants? Proceedings of the 34th Midyear Meeting, Radiation Safety and ALARA Considerations for the 21st Century, CA, USA, (2001) 177–182.

Where are the radon-induced lung cancer cases? Is it time for a re-evaluation of the radon problem?

Anita Enflo [*],[1]

Skiljevägen 27, SE-182 56 Danderyd, Sweden

Abstract

Most of the lung cancer cases occur among elderly smokers. Smoking is generally considered to be the main cause of lung cancer. Radon is considered to be the second largest cause of lung cancer. There are very few lung cancer cases among children and non-smokers. The radon-induced lung cancer seems to occur mostly among the smoking population. The strong correlation between smoking and radon exposure should be a challenge for radiobiologists, to find the underlying mechanisms behind radiation-induced cancer. The most cost effective method to prevent even the radon-induced lung cancer incidences seems to be, to reduce the smoking habits, which to a great extent can be made individually on a voluntary basis. © 2002 Elsevier Science B.V. All rights reserved.

Keywords: Radon; Domestic radon; Lung cancer; Risk estimation; Smoking

1. Introduction

Sweden belongs to a group of countries in the world with the highest radon levels in its dwellings. According to the Swedish authorities [1,2], it is estimated that the mean radon concentration is 140 Bq/m^3 in detached houses and 75 Bq/m^3 in apartments. In a housing stock of about 4.1 million dwellings, about 12% of the dwellings are estimated to have radon levels of more than 200 Bq/m^3. In 4% of the dwellings, the radon levels are estimated to be more than 400 Bq/m^3 and in 1% even more than 800 Bq/m^3. In addition, the total mean value is estimated to be 108 Bq/m^3, which should be compared with the corresponding values for some other countries such as Finland 128 Bq/m^3, Norway 73 Bq/

---

[*] Tel.: +46-8-753-24-54; fax: +46-8-753-46-89.
 E-mail address: anita.enflo@swipnet.se (A. Enflo).
[1] On leave from the Swedish Radiation Protection Institute.

m^3, France 62 Bq/m^3, Denmark 53 Bq/m^3, Germany 50 Bq/m^3, USA 46 Bq/m^3, Canada 34 Bq/m^3, Great Britain 20 Bq/m^3 and Japan 16 Bq/m^3 [3]. Thus, comparatively more radon-induced lung cancer cases should be expected in Sweden than in other countries. Consequently, extensive studies have been performed in Sweden about this issue, and data from Sweden as well as data from many other countries have been the basis for the very detailed investigations of the effects of radon on personal health conditions [4,5]. At present, the risk estimations are based on epidemiological studies. Epidemiological studies are, however, based on models, which can be questioned in certain aspects.

2. The present study. Age dependence of the lung cancer incidences

The present study deals with the age dependence of primary lung cancer incidences and is based on statistical data on all primary lung cancer incidences in Sweden. The numbers are taken from the Swedish Cancer Registry [6], which since 1958 has registered all cancer incidences in Sweden. This has been possible because of the structure of the Swedish health care system, which is public. Moreover, the diagnostic methods are, especially during later times well developed, so the risk of errors in diagnosis should be small.

Fig. 1 shows the age dependence of all primary lung cancer incidences in Sweden diagnosed in 1998. Similar results can be obtained for the other years [7]. Of the total 2847 primary lung cancer incidences in the Swedish population of about 8.8 million people, more than 78% occur at an age older than 60 years (and about 70% older than 65 years). Less than 1% of the lung cancer incidences occurred at an age younger than 35 years. Lung cancer in children is extremely rare, about one per year. Most of the lung cancer incidences occur among heavy smokers. In fact, there are so few non-smoking lung cancer

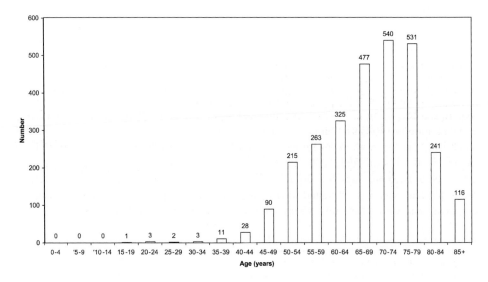

Fig. 1. Total number of primary lung cancer incidences in Sweden 1998.

incidences that epidemiological studies have been difficult to perform. Furthermore, more than half of the Swedish population is non-smoking.

3. Results and discussion

Children are exposed to domestic radon as much or even more than adults. As children are supposed to be more sensitive to radiation than adults, the risk of lung cancer from radon should be greater for children than for adults. However, lung cancer is very rare among the younger population, and in case there is a risk for children, the latent period seems to be very long maybe 40–60 years. If this is the case, the epidemiological studies performed up until now [8,9] with follow up times of less than 35 years, do not cover the whole exposure history. A follow up time of 35 years was chosen to safely cover a latent period of about 20 years, as was found from data from miners. The childhood exposure, which should be most important, is not taken into account. However, as the number of non-smoking lung cancer cases is low even at higher ages, this indicates that the risk from childhood exposure is low both for short and long latent periods. Children are not smokers. Finally if a person starts smoking at about 15–20 years old, assuming a latent period of 20 years, the first lung cancer incidences will occur at around the age of 35 years old, as is the case.

Thus, most of the lung cancer incidences, even the radon induced, seem to occur among elderly smokers. Childhood exposure to domestic radon, at the levels occurring in Swedish houses seems to have little effect on the development of lung cancer. On the other hand, smoking seems to be the main factor even for radon-induced lung cancer. This indicates that it is time for a re-evaluation of the radon problem.

References

[1] SOU 2001:7, Radon, Betänkande av Radonutredningen 2000, ISBN 91-38-21392-3. Fritzes Offentliga Publikationer. Stockholm, Sweden (In Swedish), 2001.
[2] G.-A. Swedjemark, A. Mäkitalo, Distribution of radon in houses as a basis for radiological protection, Radiation Protection Dosimetry 36 (1991) 125–128.
[3] UNSCEAR, United nations scientific committee on the effects of atomic radiation, Report to the General Assembly, with Scientific Annexes, United Nations, New York, 2000.
[4] National Research Council, Health Effects of Exposure to Radon, BEIR VI, National Academy Press, Washington DC, ISBN 0-309-05645-4.
[5] ICRP 65, Protection against radon-222 at home and at work, Recommendations of the International Commission on Radiological Protection, Pergamon, Oxford, 1994.
[6] Swedish National Board of Health and Welfare, Cancer Incidences in Sweden 1984-1999, www.sos.se/sos/statisti.html. 106 30 Stockholm.
[7] A. Enflo, Where are the radon induced lung cancer cases? in: J. Inaba, Y. Nakamura (Eds.), Comparative Evaluation on Environmental Toxicants, Kodansha Scientific, Tokyo, 1998 ISBN 4-906464-06-8, pp. 291–292, Supplement, pp. 151–156. ISBN 4-938987-05-8.
[8] G. Pershagen, G. Åkerblom, O. Axelson, B. Clavensjö, L. Damber, G. Desai, A. Enflo, F. Lagarde, H. Mellander, M. Svartengren, G.-A. Swedjemark, Residential radon and lung cancer in Sweden, New England Journal of Medicine 330 (1994) 159–164.
[9] F. Lagarde, G. Axelsson, L. Damber, H. Mellander, F. Nyberg, G. Pershagen, Residential radon and lung cancer among never-smokers in Sweden, Epidemiology 12 (2001) 396–404.

Investigation of cancer mortality in the Gastein Valley, an area of high-level natural radiation

Johanna Pohl-Rüling*, Werner Hofmann

Institute of Physics and Biophysics, University of Salzburg, Hellbrunner Str. 34, A-5020 Salzburg, Austria

Abstract

The Gastein Valley at the northern slope of the Alps contains high natural radiation areas. In the center of the town Badgastein, 19 springs originate, which supply 5×10^6 l of hot water with a mean radon content of 1.48 MBq/l. The mean radon concentrations in the atmosphere are between 30 and 100 Bq/m^3 outdoors, and between 185 and 400 Bq/m^3 indoors. Also, the mean annual radiation burden of the inhabitants range from 6 to 16 mSv. A geologically similar spa with elevated natural radiation levels, Bad Hofgastein, is situated about 10 km north of Badgastein, receiving its thermal water via pipeline from Badgastein. Its atmospheric radon content (and hence the radiation doses received by its inhabitants) are approximately 30% of those measured in Badgastein. The percentage of all cancer deaths relative to all deaths in Badgastein during the years 1947–2000 was 21%, the same value as in Bad Hofgastein despite the annual doses being three times higher. Furthermore, the number of all cancer cases, especially lung cancer, in Badgastein is lower than the expected value for the province of Salzburg, despite the annual doses being four times higher. © 2002 Elsevier Science B.V. All rights reserved.

Keywords: Radon; Lung cancer; High background area

1. Introduction

The town of Badgastein, a well-known radon spa, is situated in the 37-km-long Gastein Valley, part of the Hohe Tauern (Eastern Alps), at a mean elevation of 1000 m above sea level. At a distance of about 10 km to the north, the sister spa Bad Hofgastein is located.

The rocky ground consists of various central gneisses with mean uranium and thorium contents of 24 and 41 ppm, respectively. The pre-Mesozoic gneisses from a 15-km-long

---

* Corresponding author. Tel.: +43-662-8044-5709; fax: +43-662-8044-5709.
E-mail addresses: johanna.pohl_rueling@sbg.ac.at, physik@mh.sbg.ac.at (J. Pohl-Rüling).

and 400-m-thick cover showed higher uranium and thorium contents, with mean values of 45 ppm uranium (reaching 240 ppm at some places of muscovite gneisses) and 80 ppm thorium.

2. Natural radiation exposure

The sources of 19 thermal springs lie in a small area of about 25 000 m^2 in the center of Badgastein. They pour out more than 5×10^6 l of hot water per day with a temperature of about 48 °C and a mean radon content of 1.48 kBq/l. That means that 7.4×10^6 Bq radon is supplied to the environment per day. Almost 80% of this amount diffuses into the atmosphere. The radon content of the drinking water from the cold springs is 74 Bq/l. Therefore, the main radiation burden for the inhabitants of Badgastein is caused by the radon concentration in the air, varying between 185 and 400 Bq/m^3 indoors and between 30 and 199 Bq/m^3 outdoors. For comparison, the mean annual radiation burdens of the inhabitants of the sister town Bad Hofgastein are approximately 30% of those of the Badgastein inhabitants.

The average values of the external γ-ray dose rates, measured by radiophotoluminescence dosimeters, were 1.7 mSv/year in Bad Hofgastein and at the periphery of Badgastein, and 2.1 mSv/year in the center of Badgastein.

3. Structural chromosome aberrations

The induction of chromosomal aberrations in blood lymphocytes is a sensitive effect of ionizing radiation at low doses, where a linear shape of the dose–response curve has been disputed theoretically and experimentally. According to the results of our in vivo experiments, with detailed dose assessments for every person participating, the frequency of chromosomal aberrations initially rises sharply with the dose at the border of the normal environmental burden between 2 and 10 mGy/year [1]. Additional doses of up to about 50 mGy/year (delivered either acutely or accumulated over 1 year) change the dose–response curve to a plateau or perhaps even to a decrease. This can be explained by the existence of repair enzymes, stimulated by damages to the DNA at a given dose. Even higher doses produce a linear increase of the dose effect curve up to the range between 100 and 300 mGy, where further steps occur.

The nonlinearity of the dose effect and the occurrence of repair enzymes at low-dose irradiations are supported by theoretical considerations, as a result of "adaptive repair" and/or by radiation-induced events at the gene level of the DNA, as, e.g. the stimulation of the "GT-resistant" HPRT-gene.

4. A biophysical model of radiation carcinogenesis

In the stochastic state-vector model of radiation carcinogenesis, a cell must pass through seven states to produce a tumor [2]. The individual transitions are related to the formation of double-stranded DNA breaks, repair of breaks, interactions (translocations) between breaks, fixation of breaks, cellular inactivation, stimulated mitosis and promotion through loss of intercellular communication. Each of these transitions from the normal state 0 to the tumor state 7 is simulated stochastically in time through Monte Carlo sampling of competing events.

To simulate lung cancer incidence in the human lungs, additional biological mechanisms operating specifically under in vivo irradiation conditions have to be considered, such as the differentiation of cells in an organized cell structure. Fair agreement between epidemiological data on uranium miners and model predictions has been observed. An interesting feature of the model is that it predicts a nonlinear dose–response curve at low exposure levels.

5. Cancer mortalities and risk estimation

The cancer mortalities of the Badgastein and Bad Hofgastein areas in the years from 1947 to 2000 were collected and compared with corresponding data for the province of Salzburg. Two observations could be made:

1. The percentage of all cancer deaths to all deaths in Bagastein is 21%, the same value as in Bad Hofgastein despite the annual doses being three times higher.
2. The number of all cancer cases, especially lung cancer, in Badgastein is lower than the expected value for the province of Salzburg, despite the annual doses being four times higher.

These epidemiological results suggest that the total cancer risk in high natural radiation areas may not be elevated, or is even reduced, if compared to the normal natural radiation background levels.

References

[1] J. Pohl-Rüling, P. Fischer, The dose–effect relationship of chromosome aberrations to alpha- and gamma irradiation in a population subjected to an increased burden of natural radioactivity, Radiat. Res. 80 (1979) 61–81.
[2] D.J. Crawford-Brown, W. Hofmann, Extension of a generalized state-vector model of radiation carcinogenesis to consideration of dose rate, Math. Biosci. 115 (1993) 123–144.

Prevalence of lung cancer and indoor radon in Thailand

Somchai Bovornkitti

Department of Medicine, Faculty of Medicine Siriraj Hospital, Mahidol University, Bangkok 10700, Thailand

1. Background

In Thailand, lung cancer ranks as the second most common cause of death after liver cancer [1]. When categorized by region of the country, it is more prevalent in the northern part, and is especially high in the Chiang Mai Province where the annual incidence of lung cancer in the period of 1988–1991 was 49.8 and 37.4 per 100,000 for men and women, respectively [2]. The incidence in women at 37.4 per 100,000 of the population is among the highest in the world [3].

Although cigarette smoking is known to be the leading cause of lung cancer, such an association has not been shown in Chiang Mai [3,4]. Another leading cause of lung cancer, second only to tobacco smoking is indoor radon; it accounts for 5000–20,000 deaths a year in the United States [5]. To substantiate the relationship between lung cancer and radon in Thailand, a nationwide survey of indoor radon initiated in 1993 has disclosed the ubiquity of radon in the country and showed an interesting trend of high levels in the northern provinces [6]. Based upon the above data and the fact that there had not hitherto been any study to substantiate the association between indoor radon and lung cancer prevalence in the Chiang Mai Province, a case-control study was undertaken in the Saraphi district where the incidence of lung cancer is the highest in the country [7]. The findings suggest that indoor radon in conjunction with smoking is a risk factor for lung cancer cases in the Saraphi district.

2. A study of lung cancer prevalence vs. indoor radon in Chiang Mai

A case-control study was carried out during the period June–November 2000. The case group comprised of 224 lung cancer patients (both living and dead), and residents of the Saraphi district, diagnosed in the years 1993 through 1998. The control groups included 201 age- and sex-matched noncancer residents of the Chom Thong district (which enjoyed the lowest lung cancer incidence in the province despite their geographic and demographic closeness) and 227 age-, sex- and location-matched cancer-free healthy residents in the Saraphi district.

All subjects were interviewed regarding the profile, type and characteristics of their homes and whether they had the habit of smoking. For information on the deceased, next of kin were interviewed.

The indoor radon gas level was measured in all the subjects' houses, in the area they normally occupied, using the activated charcoal method [8].

Binary logistic regression was applied to the data in determining the association between the risk factors and lung cancer, and the mathematical models for lung cancer risk prediction.

2.1. Relevant findings

The case group had a history of lung cancer in their close cousins with the rate being 8.3% compared to none in the Chom Thong controls and 4% in the Saraphi controls. Almost half (46.6%) of the case group had smoked but quit after a duration of 35.6 ± 17.2 years. For the Saraphi control group, 40.3% had smoked for a duration of 25.8 ± 22.7 years, whereas 51.2% of the Chom Thong controls were current smokers for 27.0 ± 19.6 years. More than half (55.8%) of the case group was passive smokers; for the Saraphi controls, the figure was 55.0%, and for the Com Thong controls, 40.3%. The principal smoking materials (a locally made cheroot called *khiyo*) contain shredded tobacco leaves, tamarind shell and bark from the koy tree, wrapped in dry young banana leaves.

The radon concentrations in the houses of the case group ranged from 6.61 to 150.34 Bq m^{-3} (the arithmetic mean being 26.41 ± 14.66 Bq m^{-3}), while in the houses of the Chom Thong controls, the concentrations ranged from 4.84 to 79.53 Bq m^{-3} (the arithmetic mean being 18.32 ± 10.16 Bq m^{-3}), and in the houses of the Saraphi controls, the concentrations ranged from 0.04 to 88.07 Bq m^{-3} (the arithmetic mean being 16.68 ± 8.90 Bq.m^{-3}). The geometric means were 23.80 Bq m^{-3} for the case group, 16.28 Bq m^{-3} for the Chom Thong controls and 14.46 Bq m^{-3} for the Saraphi controls, respectively.

When the cases were compared with the controls in Chom Thong and Saraphi separately using binary logistic regression analysis, the risk factors were related statistically to lung cancer, indoor radon and smoking.

2.2. Conclusions

The results of the study were taken as confirming the multifactorial causes of lung cancer and showing smoking as being the causative factor. Furthermore, it was concluded that the relatively high indoor radon concentrations were an enhancing factor in causing lung cancer.

References

[1] Health Information Division, Bureau of Health Policy and Planning. Cancer deaths and cancer mortality rates (per 100,000) in Thailand categorized by gender 1995–1997. Available from the URL: http://www.moph.go.th/ops/bhpp/Ne.htm (April 2, 2000).
[2] K. Nakachi, P. Limtrakul, P. Sonklin, O. Sonklin, C. Tor Jarern, S. Lipigorngoson, et al., Risk factors for lung cancers among northern Thai women: epidemiological, nutritional, serological, and bacteriological surveys of residents in high- and low-incidence areas, Jpn. J. Cancer Res. 90 (1999) 1187–1195.

[3] V. Vatanasapt, N. Martin, H. Sriplung, K. Chindavijak, S. Sontipong, H. Sriamporn, et al., Cancer incidence in Thailand, 1988–1991, Cancer Epidemiol., Biomarkers Prev. 4 (1995) 475–483.
[4] S. Simarak, U.W. de Jong, N. Breslow, C.J. Dahl, K. Ruckphaopunt, P. Scheelings, et al., Cancer of the oral cavity, pharynx/larynx and lung in North Thailand: case control study and analysis of cigar smoke, Br. J. Cancer 36 (1977) 130–140.
[5] D. Bodansky, Overview of the indoor radon problem, in: D. Bodansky, M.A. Robkin, D.R. Stadler (Eds.), Indoor Radon and its Hazards, Univ. of Washington Press, Seattle, 1987, pp. 1–16.
[6] P. Polpong, S. Bovornkitti, Indoor radon in Thailand, Radiat. Res. 1999.
[7] P. Wiwatanadate, R. Voravong, T. Mahawana, M. Wiwatanadate, T. Sirisomboon, N. Ngamlur, et al., Lung cancer prevalence and indoor radon in Saraphi District, Chiang Mai, Thailand, Intern. Med. 17 (2001) 26–32.
[8] P. Polpong, S. Bovornkitti, Technique for measuring indoor radon, Siriraj Hosp. Gaz. 49 (1997) 944–947.

ICRP evolutionary recommendations and the reluctance of the members of the public to carry out remedial work against radon in some high-level natural radiation areas

P.A. Karam [a,*], S.M.J. Mortazavi [b,c], M. Ghiassi-Nejad [b,d], T. Ikushima [e], J.R. Cameron [f], A. Niroomand-rad [g]

[a] *Department of Environmental Medicine, University of Rochester, 601 Elmwood Ave. Box HPH, Rochester, NY 14642, USA*
[b] *National Radiation Protection Department (NRPD), Iranian Nuclear Regulatory Authority (INRA), P.O. Box 14155-4494, Tehran, Iran*
[c] *Medical Physics Department, School of Medicine, Rafsanjan University of Medical Sciences (RUMS), Rafsanjan, Iran*
[d] *Biophysics Department, Tarbiat Modares University (TMU), Tehran, Iran*
[e] *Biology Division, Kyoto University of Education, Kyoto 612-8522, Kyoto, Japan*
[f] *Departments of Medical Physics, Radiology and Physics, University of Wisconsin, Madison, WI, USA*
[g] *Department of Radiation Medicine, Georgetown University, LL Bles Building, 3800 Reservoir Road NY, Washington, DC, 20007-2197 USA*

Abstract

International and national advisory bodies have issued guidelines regarding acceptable levels of exposure to ionizing radiation and radon. These guidelines are intended to help protect the public against the ill effects of exposure to radiation. However, these guidelines are also far lower than the amount of radiation to which residents of Ramsar, Iran are exposed to natural sources. Under these guidelines, the good public health policy suggests that the residents should be relocated to an area of lower radiation levels. However, these residents seem perfectly healthy and, in fact, preliminary studies show that the residents of Ramsar's HBRA have no increase in the background levels of chromosomal aberrations. Furthermore, the residents exhibit fewer induced aberrations when subjected to a 1.5 Gy challenge dose, compared to the residents of nearby areas with normal background radiation levels. These findings suggest that it may be possible to relax the guidelines set

---

[*] Corresponding author. Tel.: +1-716-275-1473; fax: +1-716-273-2236.
E-mail address: andrew_karam@urmc.rochester.edu (P.A. Karam).

by these advisory bodies regarding occupational and non-occupational exposure to ionizing radiation. © 2002 Elsevier Science B.V. All rights reserved.

Keywords: Background radiation; Adaptive response; LNT; ICRP; Ramsar

1. Introduction

In the ICRP-39 Report [1], an average annual radiation dose limit of 1 mSv year^{-1} was recommended for the general public from artificially produced radiation. This document also provided guidelines, without a specific radiation limit, for exposure to natural sources of radiation. Since 1994, the ICRP-39 Report has been superseded by Publications 65 and 82 [2,3]. Publication 65 recommends an action level of 200 Bq/m^{-3} (corresponding to an annual dose of 3 mSv year^{-1}) for the radon in dwellings, while Publication 82 extends similar action levels to other situations of prolonged exposure, and provides default reference levels for the intervention. ICRP emphasizes that these levels must be interpreted with extreme caution, and stresses that the values usually are of an upper bound nature. It suggests that intervention is almost always justifiable if the total annual dose is about 100 mSv, while for the total annual doses below 10 mSv, intervention is unlikely to be justified in terms of radiological protection only. In addition, some radiation professionals may not be aware of the change in the ICRP recommendations from Publication 39.

2. Discussion

People in some areas of Ramsar, a city in northern Iran, receive an annual effective radiation dose from background radiation of up to 260 mSv year^{-1}, which is substantially higher than the 20 mSv year^{-1} that is permitted for Iranian radiation workers. The maximum radon levels in some regions of Ramsar are up to 3700 Bq/m^3 [4]. The people who live in these high radiation areas are of considerable interest because they and their ancestors have been exposed to abnormally high radiation levels over many generations. If a radiation dose of a few hundred mSv per year is detrimental to health, causes genetic abnormalities, or an increased risk of cancer, it should be evident in the residents of Ramsar. The preliminary results of our studies of the people living in high background radiation areas of Ramsar show no observable detrimental effect. Furthermore, when an in vitro challenge dose of 1.5 Gy of gamma rays was administered to the lymphocytes, they showed a significantly reduced frequency for chromosome aberrations of the people living in high background areas, compared to those in normal background areas, suggesting enhanced repair of induced DNA damage resulting from exposure to elevated levels of background radiation [4]. These results suggest that exposure to high levels of natural radiation is not harmful and does not require remedial or corrective actions.

Because of the expense of remedial actions and the long history of high background radiation levels, it is not practical to ask the inhabitants of high-level natural radiation areas (HLNRAs) of Ramsar to carry out radon remedial actions. They do not know any harmful effects because they are living in their paternal houses. Inhabitants of the HLNRAs of

Ramsar are often unaware of radon and its possible health effects. Although there is no solid epidemiological information yet, most local physicians in Ramsar believe that there is no increase in the incidence rates of cancer or leukemia in their area, and the life span of the HLNRA residents also appears no different than that of the residents of a nearby normal background radiation area (NBRA). Thus, we suggest that governments should adopt public health measures and policies that are cost-effective in risk reduction by considering the financial, social and psychological impact on their citizens. The lack of data showing the risks makes radiological remediation efforts difficult to justify, from the standpoint of improving public health.

References

[1] ICRP, Publication 35, General Principles of Monitoring for Radiation Protection of Workers, Pergamon, New York, 1982.
[2] ICRP, Publication 69, Age-dependent Doses to Members of the Public from Intake of Radionuclides: Part 3. Ingestion Dose Coefficients, Pergamon, New York, 1985.
[3] ICRP, Publication 82, Protection of the Public in Situations of Prolonged Radiation Exposure, Pergamon, New York, 1999.
[4] M. Ghiassi-Nejad, S.M.J. Mortazavi, J.R. Cameron, A. Niroomand-rad, P.A. Karam, Very High background radiation areas of Ramsar, Iran: preliminary biological studies. Health Physics 82 (2002) 87–93.

Some correlation aspects of thyroid cancer epidemiology in Ukraine after Chernobyl accident

N.P. Dikiy*, E.P. Medvedeva, N.I. Onishchenko, V.D. Zabolotny

National Scientific Center "Kharkov Institute of Physics and Technology", Academic St. 1, Kharkov 61108, Ukraine

Abstract

The correlation analysis of the thyroid cancer incident adjusted peoples age, the contamination of air, square of territory, fall-out, and others over 25 administrative regions of the Ukraine and Crimea have been considered. © 2002 Elsevier Science B.V. All rights reserved.

PACS: 87.50.Gi
Keywords: Radiation; Thyroid cancer; Correlation

Estimations of the cancer risk due to irradiation, for which the main population in the Ukraine was exposed to as a result of the accident at the Chernobyl Nuclear Power Plant (ChNPP), determines the heightened interest to the problem of the low-dose influence and epidemiology situation. The radioactive pollution of Ukrainian territory by ^{131}I was varied in 2–5 Ci/km^2. The main contribution to the inner radiation dose was caused by various isotopes of iodine. In atmosphere during radioactive clouds propagating both desorption and absorption of gas components of radioiodine at the surface of any aerosol particles not only of Chernobyl genesis. Such aerosols were transported in the atmosphere and, subsequently, were found many thousands of kilometers from their place of origin. The thyroid gland cancer (TGC) incident growth should be considered only when taking into account the influence on the human organism on the combination of factors of physical, chemical, and biological nature.

There are some publications devoted to the influence of radiation on TGC induction [1,2]. These works are concentrated on studying both the extreme high- and low-dose influence on the TGC origin. In this work, the influence of the low-dose in combination

---

* Corresponding author.
  E-mail address: onish@kipt.kharkov.ua (N.I. Onishchenko).

with the factors of physical, chemical, and biological nature on the TGC morbidity in the Ukraine are taken into consideration. The correlation analysis of the various statistical data including the thyroid cancer incident of the people aged between 1 and 80 years old, and the extent of the contaminated air with common and specific pollution, population density, square of territory, fall-out, and others over 25 administrative regions of the Ukraine and Crimea have been performed. The correlation coefficients (CC) averaged over the various regions and study parameters have been obtained.

In Fig. 1, the dependencies of the thyroid cancer incident of the Ukrainian people upon the contamination of the air are represented. Each point corresponds to the pollution and morbidity in each of the 25 regions of the Ukraine. One can see here the complicated nonlinear character of this dependency. The obtained data apparently indicate the adaptive possibilities of the human organism, under complex synergetic action of a combination of factors both of a radiation and nonradiation nature with the radiation effect prevailing.

In Fig. 2, the correlation coefficients(CC) of the TGC incident ratio and contaminants in the air over the 25 regions are represented for males in dependence upon age—for the regions with more pollution (a) and less (b) 6 tons/km^2 per year. Appreciable correlation is remarked for males which are 12 years old (CC = 0.67), who live in regions with more pollution than 6 tons/km^2 per year, and 37 years and from 50 to 80 years (CC = 0.6, 0.65 and 0.7, correspondingly) in regions with less pollution. For other age groups, the CC are negligible. This is evidence that the TGC morbidity for males obviously has multifactor genesis including an immunology factor, level of life, pernicious habits, and others.

There are similar dependencies for the TGC morbidity among females. Substantial CC = 0.6 and 0.56 correspond to the thyroid cancer incident ratio for females of 12 and 57

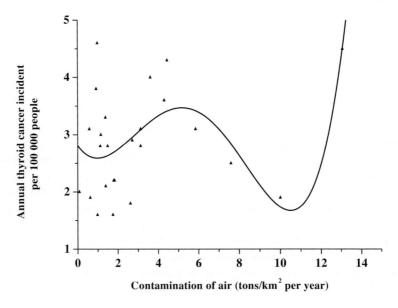

Fig. 1. The dependence of the thyroid cancer incident upon the contamination of air taken over 25 regions in the Ukraine.

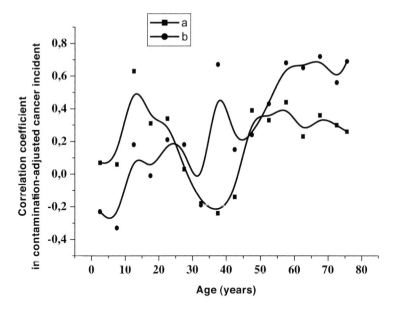

Fig. 2. The dependence of the correlation coefficient upon the age of males.

years in the regions with more pollution, and CC = 0.5 and 0.7 for 25 and 80 years with less air pollution.

Significant CC for correlation of the TGC morbidity of children aged 12 years old at high air pollution in the basin are in agreement with data of Ref. [3]; this concludes the high risk of radiogenetic cancer for children aged between 0 and 4 years at the time of the incident at the ChNPP. Noticeable CC that correspond to older males and females are apparently caused by intense testing of nuclear weapons in the years 1961–1962.

Thus, the statistical methods of available data analysis give the possibility to estimate the extent of the TGC morbidity of the people, who live in the regions with an unfavorable radiation background, and with varying degrees of air pollution. Furthermore, data analysis reveals the complex (nonlinear) character of the dose–effect dependency, conditioned by the complexity of biological processes.

References

[1] E. Ron, et al., Thyroid cancer after exposure to external radiation a pooled analysis of seven studies, Radiat. Res. 141 (1995) 259–277.
[2] R.R. Jones, R. Sounhwood, Radiation and Health: The Biological Effect of Low-Level Exposure to Ionizing Radiation, Wiley, Chichester, 1987, 281 pp.
[3] V.K. Ivanov, A.I. Gorsky, A.F. Tsyb, M.A. Maksyutov, E.M. Rastopchin, Dynamics of thyroid cancer incidence in Russia following the Chernobyl accident, Radiol. Environ. Bwphys. 38 (1999) 192–204.

Radiation risks of leukemia incidence among Russian emergency workers, 1986–1997

V. Ivanov*, A. Gorski, A. Tsyb, S. Khait

Medical Radiological Research Centre, Obninsk, Russia

The study presents an analysis of the influence of the dose and dose rate factor on leukemia incidence in the cohort of emergency workers in six regions of Russia (North-west, Volgo-Vyatsky, Povolzhsky, Central-Chernozem, North-Caucasus and Urals) in the period from 1986 to 1997.

Data supplied from these regions are characterized by reliability and high percentage of annual check-ups for emergency workers (about 86%).

A total of 70,699 emergency workers with documented dose (among them 52 leukemia cases) were considered in the radiation risk analysis.

The 8 cases diagnosed within 2 years after exposure and 12 diagnosed of Chronic Lymphocytic Leukemia (CLL) were excluded from the study. The number of follow-up person–years in this cohort was 743,845.

1. General description of the study cohort

The Russian National Medical Dosimetric Registry (RNMDR) which contains individual information about Emergency Workers was set up in 1992.

The analysis is based on individual information (from RNMDR) on date of birth, date of arrival at the controlled zone and date of departure, date of the last check-up, date of diagnosis (in case of disease) and dose ascertained in documents.

Main characteristics of the emergency workers cohort are presented in Table 1.

The structure of the leukemia incidence is given in Table 2.

2. Statistical method

Risk coefficients were estimated by the method of maximum likelihood, assuming that the number of cases is a non-stationary Poisson series of events.

---

* Corresponding author.

Table 1
Main characteristics of the emergency workers cohort

	Non-cases	Cases
Number	70,677	32[a]
Mean age at exposure	34.7	34.0
Mean dose (Gy)	0.112	0.136
Mean dose rate (Gy/day)	0.0025	0.0034
Mean days in exposure zone	86.7	95.3

[a] The 8 cases diagnosed 2 years after exposure and 12 diagnosed of Chronic Lymphocytic Leukemia (CLL) were excluded from the study.

The risk estimates were made using an external (relative spontaneous incidence in Russia as a whole) and an internal control group.

In calculations using the external control group the risk model takes the form:

$$\lambda_{i,k} = \hat{\lambda}^0_{i,k} \cdot f \cdot (1 + \text{ERR}_{1Sv} \cdot d_i)$$

where $\lambda_{i,k}$ is the intensity of the event series for person i at the k-th time interval; $\hat{\lambda}^0_{i,k}$ is the spontaneous incidence rate in Russia corresponding to the attained age of the i-th cohort member at the k-th time interval; d_i is the external radiation dose for the i-th cohort member; ERR_{1Sv} is the excess relative risk per unit dose (to be estimated parameter); f is the standardized incidence ratio (SIR) for unexposed members of the cohort.

The selected risk model has the advantage that it estimates both the dose–response and the difference in spontaneous incidence in the followed-up cohort and the referent Russian population.

The spontaneous incidence rate for each person was equal to the corresponding national rate for the attained age at a given moment of time. We believe that individual information is preferable for risk estimation because it makes it possible to minimize the influence of subjective factors and loss of information in data grouping and stratification.

Thus, it is only the relative age distribution of spontaneous mortality rates that is used for risk estimation, and this is a more robust characteristic than the absolute distribution. The value of f was assumed to be the same for all age groups.

Table 2
Structure of leukemia incidence among emergency workers

Leukemia type (ICD-9)	Number of cases	%
All acute leukemias	11	22
Acute lymphocytic leukemia (204.0)	2	4
Acute myeloid leukemia (205.0)	5	10
Other acute leukemias (206.0–208.0)	4	8
All chronic leukemias	41	78
Chronic lymphocytic leukemia (204.1)	12	23
Chronic myeloid leukemia (205.1)	20	38
Other	9	17
Total (204–208)	52	100

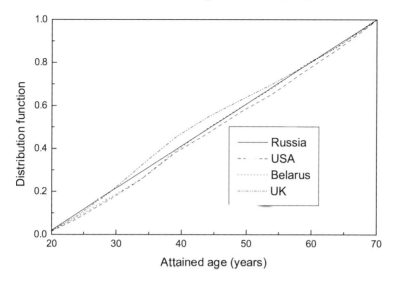

Fig. 1. Distribution function of leukemia incidence (males) by attained age for different countries.

A question arises about the reliability of data on age distribution of spontaneous incidence rates. It is reasonable to assume that the distribution of spontaneous incidence rates for leukemia by age is rather a conservative value and is weakly dependent on geographic and ethnic attributes.

To verify this hypothesis and analyze the quality of Russian data within the above hypothesis, we used the age functions of distributions of leukemia incidence rates from the world's largest cancer registries. The distribution function f_j was calculated from the relation:

$$f_{j,i} = \frac{\sum_{k=u_{\min}}^{u_j} \frac{\lambda_{k,i}}{\lambda_k^0}}{\sum_{k=u_{\min}}^{u_{\max}} \frac{\lambda_{k,i}}{\lambda_k^0}} \quad (7)$$

$u_{\max} - u_{\min}$ is the considered age range; λ_k^0 is the age distribution of the incidence rate in the registry selected as control; i is the registry index.

Table 3
Leukemia incidence (ICD 9 204–208) (EWs with documented dose)

Period	Person–years	Number of cases		SIR	95%CI	ERR/Gy	95%CI
		Obs.	Exp.				
1986–1990	288,933	16	6.1	2.63	1.51, 4.28	NA	NA
1991–1997	454,912	36	12.6	2.85	1.99, 3.94	1.02	−1.06, 4.01
1986–1997	743,845	52	18.7	2.78	2.07, 3.64	1.43	−0.66, 4.24

Table 4
Leukemia incidence (ICD 9 204–208 excluding CLL) among EWs with documented dose

Period	Number of cases		SIR	95%CI	ERR/Gy	95%CI
	Obs.	Exp.				
1986–1990	13	4.9	2.55	1.41, 4.54	NA	NA
1991–1997	27	9.5	2.84	1.88, 4.14	2.94	−0.19, 7.35
1986–1997	40	14.4	2.78	1.99, 3.79	4.59	1.15, 9.16

Fig. 1 shows the distribution function of normalized incidence rates by age at diagnosis for registries of US (white), UK, Belarus and Russia. The distribution is normalized by corresponding rates of the cancer registry selected as control (data for Russia in general). The distribution function was derived for males with an attained age of 20–70 years.

The results presented in Fig. 1 confirm the assumption about the similarity of relative distributions of age incidence rates for leukemia in different countries and the reliability of this characteristic in the Russian data was used as control.

When risk coefficients were estimated using the internal control, data were stratified by attained age and calendar time, and the spontaneous mortality was determined from the balance of the observed and expected number of cases in a given stratum.

The risk model is written as:

$$\lambda_{i,k} = \lambda_{i,k}^{EW} \cdot (1 + \text{ERR}_{1Sv} \cdot d_i)$$

where: $\lambda_{i,k}^{EW}$—is the spontaneous incidence rate among emergency workers in the stratum by attained age and calendar time in which the i-th person under study falls.

The spontaneous incidence in the stratum by attained age j, at time k was taken to be as follows:

$$\lambda_{j,k}^{EW} = \frac{n_{j,k}}{\sum_{i \subset j} \text{PY}_{i,k} \cdot (1 + \text{ERR}_{1Sv} \cdot d_i)}$$

The 95% likelihood intervals were determined from the profile likelihood function (Tables 3 and 4).

3. Conclusion

Radiation risk coefficients for leukemia incidence in the cohort of emergency workers have been estimated.

The SIR for unexposed emergency workers is equal to 1.4 (95% CI: 0.9, 1.9). The value of the excess relative risk ERR $[\text{Gy}]^{-1}$ for leukemia (excluding CLL) using external control is 4.6 (95% CI: 1.2, 9.2). Using internal control ERR $[\text{Gy}]^{-1}$ is 6.2 (95%CI: −0.1, 31.6).

The attributable risk associated with exposure (excluding CLL) is 33%.

The effect of the dose rate in the considered range of doses and dose rates is not statistically significant.

Fallout exposure in the Semipalatinsk Nuclear Test Site area and the induction of thyroid nodules diseases

Z.H. Zhumadilov [a,b,*], C. Land [c], M. Hoshi [a], A. Kimura [a], N. Takeichi [d], T. Zhigitaev [b], G. Abisheva [b], K. Kamiya [a]

[a] *Research Institute for Radiation Biology and Medicine, Hiroshima University, 1-2-3 Kasumi, Minami-ku, Hiroshima 734-8551, Japan*
[b] *Semipalatinsk State Medical Academy, 103, Abaya Str. 490050 Semipalatinsk, Kazakhstan*
[c] *Division of Cancer Epidemiology and Genetics, National Cancer Institute, National Institute of Health, Bethesda, USA*
[d] *Takeichi Thyroid Medical Clinic, Hiroshima, Japan*

Abstract

From 1949 through 1989, nuclear weapons testing carried out by the former Soviet Union at the Semipalatinsk Nuclear Test Site (SNTS), resulted in substantial levels of radiation exposure of settlements and their residents near the test site area. The results of dose reconstruction indicate a dose range from 2.0 mSv to 2.5 Sv. To investigate the possible relationship between radiation exposure and the thyroid gland abnormalities prevalence, we conducted a thyroid screening study of 3000 exposed and non-exposed current residents of the eight villages near the test site, all the residents were of similar ages (<20) at the time of the fallout and follow-up study of 311 inhabitants as well. Prevalent nodules were identified by ultrasound, malignancy was determined by fine needle aspiration biopsy and cytopathology, and circulating levels of thyroid hormones were assayed from blood samples. The nodular thyroid gland was identified in 920 of the study participants, of whom 506 were recommended for fine-needle aspiration biopsy. Furthermore, 65 cases had evidence of prior thyroid surgery. The thyroid nodule prevalence was 36% among exposed and 26% among the non-exposed residents, and increased significantly with increasing estimated age-modified gamma-ray doses (ERR=5.2 per 1 Gy, CI 95%=2.7–8.5). The cytopathology review identified 30 papillary carcinomas in 27 cases and 10 follicular neoplasms in 10 cases. The high frequency of the thyroid

---

[*] Corresponding author. Hiroshima University, 1-2-3 Kasumi, Minami-ku, Hiroshima 734-8551, Japan. Tel.: +81-082-257-5895; fax: +81-082-256-7108.
E-mail addresses: zhuma@hiroshima-u.ac.jp, zhuma@relcom.kz (Z.H. Zhumadilov), mhoshi@hiroshima-u.ac.jp (M. Hoshi), kimura@hiroshima-u.ac.jp (A. Kimura), kkamiya@hiroshima-u.ac.jp (K. Kamiya), landc@mail.nih.gov (C. Land).

dysfunction and thyroiditis among the exposed has been documented. This study revealed an apparently strong association between the fallout exposure and thyroid disease prevalence, and suggests more detailed molecular research of the thyroid cancer induction should be carried out.
© 2002 Elsevier Science B.V. All rights reserved.

Keywords: Radiogenic neoplasm; Thyroid; Fallout; Screening

1. Introduction

From 1949 through 1989, the former Soviet Union carried out 458 nuclear explosions at the Semipalatinsk Nuclear Test Site (SNTS), including 118 atmospheric tests in 1949–1962 and 340 underground blasts in 1963–1989. The total yield of atmospheric tests is nearly 6.6 Mt. Nuclear weapons testing resulted in substantial levels of radiation exposure of the settlements and their residents near the test site area. The results of dose reconstruction indicate a dose range from 2.0 mSv to 2.5 Sv. It was determined by experts from recently declassified records, that mainly 22 aboveground and 1 underground officially registered tests, resulted in substantial exposures to the inhabitants of this region, and these tests are thought to have been responsible for about 95% of the collective radiation dose from fallout to the residents of areas adjoining the SNTS [1,2]. It is well known that the thyroid gland is one of the highly radiosensitive organs. The possible relationships between radiation exposure and the thyroid gland abnormalities prevalence around the Semipalatinsk Nuclear Test Site area are not well established.

2. Materials/methods

Cohort members (2009) from six heavily exposed and two lightly exposed villages, and another 989 current residents of a similar age, who had not been present at the time of the major fallout deposition have been screened for thyroid diseases. All the screened subjects from both the exposed group and non-exposed group were <20 years at the time of the major fallout events. A study cohort of 10 000 exposed and 10 000 non-exposed residents of downwind settlements has been defined and followed since the 1960s, by the Kazakh Research Institute for Radiation Medicine and Ecology. The TSH hormone level was determined by using DELFIA® Neonatal hTSH kit during our fieldwork. This kit is intended for the quantitative determination of thyrotropin in blood specimens dried on filter paper. Urinary iodine excretion levels were assessed in 311 inhabitants from both exposed and non-exposed villages during a subsequent follow-up study. The urinary iodine levels were expressed as µI/g creatinine. Thyroid function tests included measurements of TSH, free T_3 and free T_4, thyroglobulin, and thyroid autoantibodies. Prevalent nodules, including thyroid cancer were identified by ultrasound and fine needle aspiration biopsy, cytopathology results and in some cases after thyroid surgery by morphology.

3. Results

The nodular thyroid gland was identified in 920 of the screened inhabitants, of whom 506 were recommended for fine-needle aspiration biopsy. Preliminary findings have showed that nodule prevalence was 18% among men and 39% among women, and increased by 3.5% per year of age from 40 to 70. This prevalence increased significantly with the increasing estimated age-modified gamma-ray dose (ERR = 5.2 per 1 Gy, CI 95% = 2.7–8.5). Papillary carcinoma prevalence was positive, but there was a non-significant increase of risk with the age-modified dose ($p = 0.16$). The cytopathology review identified 30 papillary carcinomas and 10 follicular neoplasms. We have revealed that the thyroid nodules increased significantly with the increasing estimated average gamma-ray dose, among persons presumably exposed, however, it was nearly three-fold higher among men ($ERR_{1\ Gy} = 4.4$, $p < 0.001$) than among women ($ERR_{1\ Gy} = 1.7$, $p = 0.002$). The follow-up study revealed the prevalence of thyroid dysfunction and chronic thyroiditis, including autoimmune thyroiditis among the exposed population compared to the non-exposed inhabitants. It should be noted that the TSH hormone level assessment using the DELFIA® Neonatal hTSH kit during our field work was easy to do, but not as informative as other TSH assessment tests, and the test does not characterize the TSH/FT4 discordances. The follow-up study revealed that the exposed people needed more precise clinical, laboratory and instrumental examinations. Thirty-three percent of the examined inhabitants had clinical or sonographic signs of thyroiditis, more frequently atrophic or Hashimoto's thyroiditis, without any changes in laboratory analysis. At the same time, among inhabitants from the exposed village of Sarzhal with the normal thyroidal echogenecity nearly 50%, various abnormalities at least in one or two laboratory tests of the circulating levels of thyroid hormones and autoantibodies were found. Especially designed screening programs and considerations are needed for this population at risk. The thyroid volume was greater among men, than that among women. The frequency distribution of thyroid volume among Russian ethnic groups as compared to Kazakh ethnic groups shifted toward a larger volume. The level of urine iodine was relatively low among all participants from both the exposed and non-exposed villages, which reflected mild and moderate iodine deficiency, but we did not find any severe iodine deficiency level in this area. The goitrogenic effect of iodine deficiency in this area needs further detailed research in order to clarify the findings. In addition, thyroid surgery revealed that an occult thyroid papillary carcinoma was with relatively benign behavior and no lymph node metastasis.

4. Discussion

It has long been recognized that thyroid nodules and thyroid cancer can be induced after radiation exposure, especially during childhood [3]. Our results show an apparently strong association between the fallout exposure and thyroid disease prevalence, and suggest more detailed molecular research of thyroid cancer induction near the Semipalatinsk Nuclear Test site area is needed. At the same time, it should be noted that the complex nature of radiation injuries from fallout, required to determine an attributable effect of gamma radiation and internally deposited ^{131}I, in order to clarify the dose–response relationships for radiation-induced thyroid nodules and thyroid cancer.

Acknowledgements

This work was supported in part by the International Award number KN2-434 of the CRDF, USA and by a Grant-in-Aid #11694282 and #09044312 for scientific research from the Ministry of Education, Science, Sports and Culture, Japan. The authors thank other members of the thyroid examination team, including Drs. J. Stracener, L. Crooks, M. Hartshorne, B. Gusev, P. Woodward, J. M. Bauman, Ph. Wiest, N. Lukyanov, and Ms. Nanci Olsen.

References

[1] R.W. Nugent, Zh.Sh. Zhumadilov, B.I. Gusev, M. Hoshi, Health Effects of Radiation Associated with Nuclear Weapons Testing at the Semipalatinsk Test Site, 1st edn., Nakamoto Sogo Printing, New York, 2000.
[2] Z.S. Zhumadilov, B.I. Gusev, J. Takada, M. Hoshi, A. Kimura, N. Hayakawa, N. Takeichi, Thyroid abnormalities trend over time in Northeastern Regions of Kazakhstan, adjacent to the Semipalatinsk Nuclear Test Site: a case review of pathological findings for 7271 patients, J. Radiat. Res. 41 (2000) 55–59.
[3] E. Ron, J.H. Lubin, R.E. Shore, K. Mabuchi, B. Modan, L.M. Pottern, A.B. Schneider, M.A. Tucker, J.D. Boice Jr., Thyroid cancer after exposure to external radiation: a pooled analysis of seven studies, Radiat. Res. 141 (1995) 259–277.

The mortality and cancer morbidity experience of workers at British Nuclear Fuels plc, 1946–1997

D. McGeoghegan*, K. Binks

Westlakes Scientific Consulting Ltd., The Princess Royal Building, Westlakes Science and Technology Park, Moor Row, Cumbria CA24 3LN, UK

Abstract

We present the preliminary results of the mortality and morbidity experience of the BNFL cohort of workers. These 62 141 employees have accumulated 1 535 730 person-years of experience to the end of 1997. The low level of untraced workers, 0.67%, is indicative of the quality of the data set. The cohort has 39 882 radiation workers who have received 2185 person-sieverts of external whole-body radiation dose. Both the radiation and non-radiation workers showed the 'healthy worker' effect, although the all causes mortality rate for radiation workers was significantly lower than that for the non-radiation workers (RR = 0.90, $p < 0.001$). There were, however, excesses of deaths amongst the radiation workers due to cancers of the pleura, connective tissue and penis as well as due to cirrhosis, when compared with non-radiation workers. For mortality, statistically significant associations with external whole-body radiation were found for all causes of death, cancer of the larynx, testicular cancer, thyroid cancer, multiple myeloma, leukaemia excluding chronic lymphatic leukaemia (CLL), mental disorders, circulatory system diseases and ischaemic heart disease. For cancer morbidity, significant associations were found with external whole-body radiation and all malignant neoplasms, leukaemia excluding CLL, non-melanoma skin cancer, testicular cancer, multiple myeloma, non-Hodgkin lymphoma and Hodgkin's disease. The excess relative risk due to all cancers excluding leukaemia was 0.48 Sv^{-1} (0.14–0.86) for mortality and 0.63 Sv^{-1} (0.34–0.94) for morbidity, when the dose was lagged 10 years. For leukaemia excluding CLL, the excess relative risk was 1.59 Sv^{-1} (−0.29–4.84) for mortality and 2.75 Sv^{-1} (0.17–7.63) for morbidity, when the dose was lagged 2 years. © 2002 Elsevier Science B.V. All rights reserved.

Keywords: Radiation; Nuclear workers; Cohort; Mortality; Cancer incidence

---

* Corresponding author. Tel.: +44-1946-514038; fax: +44-1946-514038.
 E-mail address: david@westlakes.ac.uk (D. McGeoghegan).

1. Introduction

This preliminary analysis presents the mortality and morbidity experience of the BNFL cohort with respect to exposure to external whole-body radiation dose, up to the end of 1997.

2. Study population

The BNFL cohort consists of 62141 employees of BNFL and the United Kingdom Atomic Energy Authority ever employed at the Springfields, Sellafield, Capenhurst and Chapelcross sites before the 1st January 1998 and contains 1535730 person-years of experience. The mean follow-up period was 24.7 years. Table 1 shows the vital status of the cohort. The low level of untraced workers, 0.67%, is indicative of the quality of the data set.

Up to the end of 1997, the collective dose resulting from external whole-body radiation received by the 39882 BNFL radiation workers was 2185 person-sieverts. The mean cumulative dose was 54.7 mSv. For comparison, the annual mean dose received by the UK population is estimated to be 2.6 mSv. The mean follow-up period for this cohort was about 25 years so this is equivalent to a mean cumulative dose of 65 mSv.

3. Results

Both the radiation and non-radiation workers showed the 'healthy worker' effect, although the all causes mortality rate for radiation workers was significantly lower than that for the non-radiation workers ($RR = 0.90$, $p < 0.001$). A statistically significant excess was noted for pleural cancer mortality for the radiation workers ($SMR = 213$, $p < 0.001$, Obs. = 26) when compared with the population of England and Wales and when compared with the non-radiation workers ($RR = 3.65$, $p = 0.02$). A statistically significant excess of non-melanoma skin cancer registrations ($SRR = 115$, $p = 0.01$, Obs. = 351) was noted amongst the radiation workers when compared with the population of England and Wales though not when compared to non-radiation workers ($RR = 1.10$, $p = 0.42$). Statistically

Table 1
Vital status on 31/12/97

	Radiation workers		Non-radiation workers		Total
	Male	Female	Male	Female	
Alive	27552	3004	9034	5806	45396
Deaths	8039	155	5216	1200	14610
Emigrated	951	53	540	175	1719
Untraced	114	14	209	79	416
Total	36656	3226	14999	7260	62141
Person-years	86120	57828	425276	191006	1535730

significant excesses were also noted for radiation workers for deaths from cancer of connective tissue (RR = 16.28, $p = 0.005$) and cirrhosis (RR = 2.20, $p = 0.01$) when compared with non-radiation workers, but not when compared with the population of England and Wales. The rate ratio for penile cancers was also raised when compared with non-radiation workers. For mortality, the rate ratio was undefined, as there were no cases amongst the non-radiation workers and four cases amongst the radiation workers; the rate ratio, however, was significantly different to unity ($p = 0.01$). For morbidity, the rate ratio was also significantly different to unity (RR = 7.80, $p = 0.04$) based on two cases amongst the non-radiation workers and six cases amongst the radiation workers.

Statistically significant associations with external radiation dose were noted for all cause mortality when the dose was lagged by 15 and 20 years ($z = 2.291$ $p < 0.05$, $z = 2.955$ $p < 0.01$, Obs. = 8194); mortality due to cancer of the larynx when the dose was unlagged and lagged by 2 years ($z = 1.754$ $p < 0.05$, $z = 1.811$ $p < 0.05$, Obs. = 16); mortality due to thyroid cancer when the dose was lagged 10, 15 and 20 years ($z = 2.062$ $p < 0.05$, $z = 2.523$ $p < 0.05$, $z = 3.785$ $p < 0.01$, Obs. = 7); mortality due to multiple myeloma when the dose was lagged 20 years ($z = 1.992$ $p < 0.05$, Obs. = 24); mortality due to testicular cancer when the dose was lagged 2 years ($z = 1.721$ $p < 0.05$, Obs. = 8) and mortality due to leukaemia excluding chronic lymphatic leukaemia (CLL) when the dose was unlagged and lagged 2 years ($z = 1.948$ $p < 0.05$, $z = 1.811$ $p < 0.05$, Obs. = 60).

For cancer morbidity, statistically significant trends with external radiation dose were noted for all malignant cancers when the dose was unlagged and lagged 15 years ($z = 1.673$ $p < 0.05$, $z = 1.765$ $p < 0.05$, Obs. = 3176); for all malignant cancers excluding leukaemias when the dose was lagged 15 years ($z = 1.750$ $p < 0.05$, Obs. = 3116); morbidity due to non-melanoma skin cancer when the dose was unlagged and lagged 2, 10, 15 and 20 years ($z = 2.342$ $p < 0.01$, $z = 2.322$ $p < 0.01$, $z = 2.354$ $p < 0.01$, $z = 2.622$ $p < 0.01$, $z = 2.949$ $p < 0.01$, Obs. = 623); morbidity due to leukaemia excluding CLL when the dose was unlagged and lagged by 2 and 15 years ($z = 2.201$ $p < 0.05$, $z = 2.228$ $p < 0.05$, $z = 1.763$ $p < 0.05$, Obs. = 37); morbidity due to multiple myeloma when the dose was lagged 15 years ($z = 1.812$ $p < 0.05$, Obs. = 22); morbidity due to non-Hodgkin lymphoma when the dose was lagged 20 years ($z = 1.650$ $p < 0.05$, Obs. = 73); morbidity due to Hodgkin's disease when the dose was lagged 20 years ($z = 1.850$ $p < 0.05$, Obs. = 30) and testicular cancer morbidity when the dose was lagged by 20 years ($z = 1.861$ $p < 0.05$, Obs. = 35).

For non-cancer mortality, radiation workers had a significantly lower mortality rate for circulatory system diseases (RR = 0.89, $p < 0.001$), particularly ischaemic heart disease (RR = 0.91, $p = 0.01$), when compared with the non-radiation workers. A statistically significant trend with external radiation dose was noted when the dose was lagged 15 and 20 years for circulatory system diseases ($z = 2.735$ $p < 0.01$, $z = 3.509$ $p < 0.001$, Obs. = 4120) driven by ischaemic heart disease ($z = 2.393$ $p < 0.01$, $z = 2.871$ $p < 0.01$, Obs. = 2826), and for deaths due to mental disorders when the dose was lagged 10 and 20 years.

The excess relative risk due to all cancers excluding leukaemia was 0.48 Sv^{-1} (0.14–0.86) for mortality and 0.63 Sv^{-1} (0.34–0.94) for morbidity, when the dose was lagged 10 years. For leukaemia excluding CLL, the excess relative risk was 1.59 Sv^{-1} (−0.29–4.84) for mortality and 2.75 Sv^{-1} (0.17–7.63) for morbidity, when the dose was lagged 2 years.

4. Conclusion

The workers at BNFL show the expected 'healthy worker' effect. There were, however, excesses of deaths amongst the radiation workers due to cancers of the pleura, connective tissue and penis as well as due to cirrhosis, when compared with non-radiation workers. For mortality, statistically significant associations with external whole-body radiation were found for all causes of death, cancer of the larynx, testis and thyroid, multiple myeloma, leukaemia excluding CLL, mental disorders, circulatory system diseases and ischaemic heart disease. For morbidity, significant associations were found with external whole-body radiation and all malignant neoplasms, leukaemia excluding CLL, non-melanoma skin cancer, testicular cancer, multiple myeloma, non-Hodgkin lymphoma and Hodgkin's disease. A statistically significant excess relative risk due to all cancers excluding leukaemia was found for the male workers when the dose was lagged 10 years; for mortality, it was 0.48 Sv^{-1} (0.14–0.86); for cancer morbidity, it was 0.63 Sv^{-1} (0.34–0.94). Some of these findings are likely to be a consequence of multiple hypothesis testing.

Acknowledgements

The authors thank the employees of BNFL for their support and co-operation in this study. The authors also thank the staff of the Office for National Statistics in Southport and the General Register Office in Edinburgh, who have established the tracing status of the cohort, as well as the staff of the department of Occupational Health and Medical Statistics at Westlakes Scientific Consulting.

European Commission's research programme related to the health effects of ionizing radiation

Christian Desaintes*, Diederik Teunen

European Commission, DG Research, Nuclear Fission and Radiation Protection J4, MO75 5/01, B-1049 Brussels, Belgium

Abstract

Within the European Commission's EURATOM 5th Research Framework Programme, the sub-area on Radiation Protection and Health is currently (July 2001) funding 14 shared-cost projects with a mean number of seven participants. The aim of the programme is to better understand and quantify the risk of exposure to ionizing radiation (IR) at low doses and low-dose rates, i.e., typically occurring at work places, around nuclear installations and in diagnostic medicine. It combines epidemiological, basic biological and mechanistic modeling studies to gain more insight into the risks of developing a cancer after exposure to IR. The basic radiobiological studies focus on: (1) understanding the mechanisms of chromosomal aberrations, gene mutations, genomic instability and cancer, as well as their interrelation; (2) identifying genes involved in these end points and assessing their relevance for individual susceptibility to IR. A summary of each project is available at http://www.cordis.lu/fp5-euratom/src/projects_generic.htm. © 2002 Elsevier Science B.V. All rights reserved.

Keywords: Carcinogenesis; DNA repair; Epidemiology; Genetic susceptibility; Mechanistic modeling

1. Epidemiology

Four epidemiological studies are being carried out with the support of our programme. The emphasis is put on the risk of cancer after protracted exposures at a low dose in: the Mayak workers and the population of the southern Urals (in particular, the Techa river

---

* Corresponding author.
E-mail address: christian.desaintes@cec.eu.int (C. Desaintes).

0531-5131/02 © 2002 Elsevier Science B.V. All rights reserved.
PII: S0531-5131(01)00766-X

offspring cohort), the uranium miners occupationally exposed to low concentrations of radon, populations exposed to radon in their dwellings, and workers of the nuclear industry.

2. Basic radiation biology

Our programme also supports basic radiation biological research with the aim to better understand the mechanisms of radiation-induced chromosomal aberrations, mutations, genomic instability, bystander effects and cancer. Moreover, the programme supports the identification of the genes involved in those end points, and assessing their relevance for individual susceptibility to ionizing radiation (IR).

2.1. Repair of DNA damage and mechanisms of chromosomal aberrations, mutations and genomic instability

The programme is currently focusing on double strand breaks (DSB). Induction of the initial DSB in DNA is being studied as a function of the dose and radiation quality. Investigations are also being carried out into whether there is a preferential location of these breaks on specific chromosomes or sub-chromosomal regions, and the influence of cell metabolism, nuclear architecture and chromatin structure.

The role of the two main repair pathways (Homologous Recombination Repair, or Non-Homologous End Joining) for the DSB types of damage are being studied as a function of dose, quality of the radiation, cell type, cell metabolism, cell cycle and other cellular processes, such as replication and transcription. In addition, the relative contribution of each repair pathway usage on the generation of chromosomal aberrations, mutations and genomic instability, and the role of specialised chromosomal structures, such as telomeres and centromeres, are being investigated.

2.2. Mechanisms and genes involved in radiation carcinogenesis

The examination of tumours arising in transgenic mice deficient in the different repair pathways after irradiation is aimed at establishing the role of the various repair pathways and repair genes in cancer development and correlating it with chromosome aberration, gene mutation and genomic instability. Such studies are now being extended to determine the role of repair genes in human radiation susceptibility to carcinogenesis.

In addition to the genes of high penetrance, subtle differences (polymorphism for example) inherited in multiple genes of low penetrance might account for a large part of the genetic variation of radiation susceptibility within a population. In order to identify these genes, molecular genetic approaches are being used in mice to identify the susceptibility genes involved in radiation-induced cancers of the lung, bone, thyroid, breast, skin, as well as leukaemia and lymphoma.

These animal studies are being completed by investigations of cancer in humans. A large collection of thyroid tissue has been collected from individuals who were children at the time of the Chernobyl fallout in Russia, Belarus and the Ukraine. These tissues are being used to identify genes differentially expressed or altered in thyroid cancers, in order to get an insight into the mechanism of radiation-induced thyroid cancer and to use this information to develop tools for prognostic or for clinical therapy.

3. Modeling of radiation carcinogenesis

Results from the basic biological and epidemiological studies are being used to establish mechanistic models for the induction of primary damage to cells and organs, for repair of that damage and for multi-step carcinogenesis, in particular for the induction of lung, thyroid and bone cancers. This approach is expected to lead to a better assessment of the risk.

Does radiation cause liver cancer? Comparison of radiation effects in atomic bomb survivors and other populations

Gerald B. Sharp*

Department of Epidemiology, Radiation Effects Research Foundation, 5-2 Hijiyama Park, Minami-ku, Hiroshima 732-0815, Japan

Abstract

Recent studies of atomic bomb survivors have shown that risks of liver cancer are significantly increased by radiation exposure. This contrasts with mortality studies of other radiation-exposed populations, which generally have not shown a significant radiation effect for this cancer. Because the liver is a frequent site to which other tumors metastasize, liver cancer is one of the most difficult cancers to correctly diagnose. Studies of liver cancer in A-bomb survivors and other populations have documented high percentages of tumors metastasized to the liver being incorrectly diagnosed as liver tumors. In addition, many deaths due to liver cancer have been incorrectly attributed to cirrhosis or chronic hepatitis. Studies of incident or pathology-confirmed liver cancer cases in A-bomb survivors have found higher radiation risk estimates for liver cancer than mortality studies in this cohort. Most studies of radiation and liver cancer in other radiation-exposed cohorts have been mortality studies, and thus, would include many misclassified liver cancers. Liver cancer was consistently associated with radiation exposure in studies of four cohorts exposed to Thorotrast, a previously used radiology contrast agent. However, the histologic subtypes of liver cancer and type of radiation exposure (external rather than internal) differ from those experienced by the A-bomb survivors. Liver cancer in atomic bomb survivors is primarily hepatocellular carcinoma (HCC), rather than the cholangiocarcinoma and hemangiosarcoma subtypes more associated with Thorotrast exposure. A recent case control study of the joint effects of radiation and viral hepatitis in the etiology of HCC, showed that A-bomb radiation had a significantly stronger effect among subjects who were infected with the hepatitis C virus (HCV). No significant interaction between hepatitis B viral infections and radiation in the etiology of this disease was found. We compared incidence and mortality studies of liver cancer conducted in a wide variety of radiation-exposed populations, in terms of their radiation risk estimates for liver cancer, the background level of liver

---

* Tel.: +81-82-261-1937; fax: +81-82-262-9768.
E-mail address: sharp@rerf.or.jp (G.B. Sharp).

cancer in the cohort, HCV prevalence in the population from which the cohort was drawn, and other factors. The differences between the radiation risk estimates for atomic bomb survivors and other cohorts may reflect liver cancer diagnosis errors. Varying risk estimates may also reflect differences in the prevalence of HCV in radiation-exposed cohorts. © 2002 Elsevier Science B.V. All rights reserved.

Keywords: Hepatocellular carcinoma; Radiation; Hepatitis C virus; Liver cancer

1. Introduction

Previous studies of cancer risks in atomic bomb survivors have shown that exposure to low-LET ionizing radiation significantly increases mortality rates [1] and incident rates of liver cancer [2,3], which in this cohort is primarily HCC [4]. Although, most other mortality studies of liver cancer conducted in populations where HCV infections are less prevalent have not found radiation to be a significant risk factor for liver cancer [5–12], radiation risk estimates were significantly elevated in the combined study of underground, radon-exposed miners in Europe and North America [13].

A significant radiation effect was also found in the study of Mayak nuclear workers exposed to both external gamma radiation and internally deposited plutonium [14]. Like Thorotrast-related liver cancers, liver cancers in the Mayak workers were more likely to be hemangiosarcomas (24% of liver cancers with known histologic types) [14] than were liver cancers among A-bomb survivors (84% HCC, 15% cholangiocarcinoma, and 0% hemangiosarcoma) [4]. Radioactive thorium dioxide from injected Thorotrast migrates to the connective tissue and results in greater radiation exposures of the nearby liver bile duct and vascular cells than of the hepatic cells. In contrast, subjects in the atomic bomb survivor and the other non-Thorotrast cohorts received mostly penetrating gamma rays, resulting in whole-body exposures involving not only bile duct and vascular cells in the liver and hepatic cells, but also bone marrow and hematopoietic cells. Liver cancers in the German, Danish, and Japanese cohorts exposed to Thorotrast were fairly equally divided between HCC, cholangiocarcinoma, and hemangiosarcoma (reviewed in Ref. [4]). Of the liver cancers in the Portuguese Thorotrast cohort, 61% were hemangiosarcomas, 32% cholangiocarcinoma, and just 6% HCC (reviewed in Ref. [4]).

A large number of consistently positive studies have established HBV and HCV as strong risk factors for hepatocellular carcinoma [15]. In a nested case control study of HCC among atomic bomb survivors, we found super-multiplicative interaction between radiation and HCV in the etiology of HCC not accompanied by cirrhosis ($p=0.039$). However, no interaction between radiation and HBV ($p=0.91$) was noted in the etiology of HCC regardless of cirrhosis status (data not shown).

The goal of the present study was to compare studies of liver cancer and radiation in order to determine if discrepancies in the results of these studies might be related to differing exposures of these cohorts to risk factors for liver cancer. Because of our recent finding of interaction between radiation and HCV in the etiology of HCC, we particularly wanted to determine how liver cancer risks varied with prevalences of HCV in these cohorts.

2. Materials and methods

Analysis was limited to mortality studies of radiation and liver cancer. When several reports for the same cohort study were published, we surveyed the latest report available. Study characteristics recorded include (1) time span of subject follow-up, (2) country of residence of the subjects used in the radiation risk calculations, (3) nature of the radiation exposure of the cohort, (4) mean radiation exposure for exposed subjects, (5) total number of subjects, (6) number of subjects with any exposure to radiation, (7) total liver cancer deaths during the follow-up period, (8) radiation risk estimate, and (9) confidence interval (CI) for the risk estimate. These factors for each study were compared to the prevalence of HCV in the area from which study subjects were drawn. The HCV prevalences were based on a review by Wasley and Alter [16], which surveyed studies calculating HCV prevalences based on enzyme immunoassays confirmed by supplemental testing. China's estimate was based on two studies of blood donors [17,18]; we did not find a study, which reported HCV prevalence for the region of Russia containing the Mayak facility. The HCV prevalences for A-bomb survivors were based on a clinical study of 6121 subjects [19] and an autopsy study of 897 A-bomb survivors who died before 1988 from diseases other than primary liver cancer (data not shown). We also included a study of Osaka blood donors aged 55–64 years [20].

2.1. Assessment of background level of liver cancer in cohorts

Ideally, the background level would be calculated as liver cancer deaths per 100,000 person-years of follow-up, but person-year totals were not available in most of the reports. We roughly calculated person-years as the total number of persons included in each study multiplied by the maximum duration of each study, assuming that each study had no other deaths or loss to the follow-up. The number of liver cancer deaths was divided by this number and the product multiplied by 100,000. Our method overestimated person-years and, correspondingly, underestimated liver cancer deaths per 100,000 person-years.

3. Results

The table lists the studies of radiation and liver cancer analyzed for this study. HCV prevalences were substantially higher in the A-bomb survivor cohort (7.8–8.9%) than in most other areas where cohort studies were conducted. HCV prevalences in the UK, where many of the mortality studies were conducted, were particularly low, ranging from 0.01% to 0.1%, with HCV prevalences ranging from 0.2% to 0.5% in North America and Western Europe [16].

Background liver cancer levels were low in most of the cohorts studied, ranging from 0.2 in the Canadian worker study [7] to 1.6 in the combined analysis of underground miners in North America, France, Sweden, UK, and the Czech Republic [13]. Background levels were substantially higher than this in the Chinese iron ore miner cohort, where hepatitis B was highly prevalent, in the peptic ulcer patient cohort [8] and in the uterine bleeding cohort [12]; these studies were conducted in areas of low HCV prevalence and

Table 1
Studies of radiation and liver cancer by type of radiation exposure and background prevalence of hepatitis C virus

Reference (first author)	Population	Location	Time of study	Total subjects (number radiation exposed)	Total liver cancer deaths (background level)[a]	Mean radiation exposure level	Result
Studies of chronic radiation exposure in areas of low[b] hepatitis C virus prevalence							
Cardis [5]	nuclear workers	US, UK, and Canada	1944–1988	95,673 (all)	33 (0.8)	0.040 Sv (2% > 0.400 Sv)	trend test p: 0.495
Chen [6]	underground iron ore miners	China	1970–1982	6444 (all)	17 (19.7)	0.021–0.030 Sv annually	SMR: 0.8; 95% CI: 0.4–1.2
Darby [13]	underground miners, mainly of uranium	US, Canada, UK, Sweden, Czech Republic, and Western Europe	1941–1990	64,209 (all)	50 (1.6)	155 working level months (WLM)	O/E: 1.73; 95% CI: 1.29–2.28; p for trend: 0.81
Studies of chronic radiation exposure in areas of unknown hepatitis C virus prevalence							
Gilbert [14]	Mayak nuclear workers exposed to plutonium	Russia	1948–1994	11,000 (2207)	60 (11.8)	0.31–1.74 Gy	SMR: 1.8 (1.4–2.3)
Studies of acute radiation exposure in areas of low[b] hepatitis C virus prevalence							
Ashmore [7]	workers with radiation exposures	Canada	1951–1987	206,620 (28,917)	16 (0.2)	0.006 Sv	ERR/Sv: males: −0.9; 90% CI: −22.3–20.5
Griem [8]	peptic ulcer patients receiving radiotherapy	US	1937–1985	3609 (1831)	20 (11.5)	4.61 Gy (mean liver dose)	adjusted RR: 0.79; 95% CI: 0.3–2.1
Boice [9]	cervical cancer patients receiving radiotherapy	US, Canada, UK, and Northern Europe	1960–1983	182,040 (82,616)	21 (0.5)	1.5 Gy (mean liver dose)	O/E: 1.0; p>0.05

Study	Description	Country	Years	Cohort (cases)	Deaths	Dose	Risk estimate
Darby [10]	participants in UK nuclear tests	UK	1952–1991	43,691 (21,358)	17 (1.0)	0.008 Sv	RR: 2.46; 90% CI: 0.92–6.92
Weiss [11]	ankylosing spondylitis patients receiving radiotherapy	UK	1935–1992	15,577 (14,556)	11 (1.2)	2.13 Gy (mean liver dose)	RR: 0.81; 95% CI: 0.40–1.44
Inskip [12]	women receiving radiotherapy for uterine bleeding	US	1925–1984[c]	4153 (all)	15 (6.4)	0.21 Gy	SMR: 0.7
Studies of acute radiation exposure in areas of high[d] hepatitis C virus prevalence							
Pierce [1]	atomic bomb survivors	Hiroshima and Nagasaki, Japan	1950–1990	86,572 (50,113 0.005 Sv)	753[e] (21.7)	0.20 Sv	ERR/Sv males: 0.52; 90% CI: 0.22–0.91; females: 0.11; 90% CI: 0.08–0.59

[a] Liver cancer deaths divided by [total subjects multiplied by maximum duration of study] multiplied by 100,000, our rough approximation of the mortality rate of liver cancer in each cohort, assuming no other deaths or loss to follow-up.

[b] Prevalence of hepatitis C virus based on review by Wasley and Alter [16]: UK and Scandinavia: 0.01–0.1%; North America and Western Europe: 0.2–0.5%; and China: 0.3–1.2% [17,18].

[c] 504 of the 4153 women were followed until 1966.

[d] HCV prevalence was 7.8% among hepatocellular carcinoma cases in a recent case control study and 8.9% in a clinical study of A-bomb survivors [19]; prevalence was 8.5% in Osaka blood donors age 55–64 years [20].

[e] Includes deaths with underlying cause coded as liver cancer, primary or liver cancer not specified as primary or secondary.

showed no significant radiation risks for liver cancer. Background liver cancer rates were also elevated in two studies reporting significantly increased liver cancer risks for radiation: the Mayak workers, for whom HCV prevalences were unknown, and A-bomb survivors, the group with the highest HCV prevalence rates [1]. The study of the underground miners, with both a low background rate of liver cancer and a low HCV prevalence, also reported a significantly increased risk of liver cancer (O/E ratio: 1.73; 95% CI: 1.29–2.28).

4. Discussion

There are several possible explanations for the wide variation in radiation risk estimates for liver cancer that we describe. These findings are primarily based on mortality studies. Liver cancer is one of the most difficult tumors to correctly diagnose, because the liver is a frequent site to which other tumors metastasize and because death from liver cancer is often incorrectly attributed to cirrhosis or chronic hepatitis. We found that from 1958 to 1987, about half of the A-bomb survivors whose death certificates recorded liver cancer as the cause of death, died from other causes, and that about two-thirds of the persons recorded as dying from cirrhosis or chronic hepatitis were found on pathology review to have died from primary liver cancer [21]. The A-bomb survivor studies of incident cases of primary liver cancer [2,3], which corrected for some of these diagnosis errors, reported slightly higher ERR per Sv risk estimates than the A-bomb survivor mortality study included in the table.

Because HCV prevalence estimates for the UK, Europe, and China are primarily based on blood donor studies, they are likely to be underestimated in comparison to the A-bomb survivors and US prevalence estimates. However, the high HCV prevalence in the A-bomb survivor cohort is consistent with a 1994 Osaka, Japan blood donor study that reported a HCV prevalence of 8.5% for donors aged 55–64 years [20].

Our background liver cancer index underestimates liver cancer death rates but is useful for comparing background liver cancer mortality rates in the cohorts. For example, we estimated the background level for the Mayak study at 11.8 liver cancer deaths per 100,000 per year (Table 1), which compares to an estimate of 14.0 liver cancers per 100,000 person-years that can be calculated from person-year information provided in this report [14]. It is not clear if the differences in background liver cancer rates can explain the differences in radiation risk estimates between studies. In the A-bomb survivor [1] and the Mayak worker [14] studies, which reported significant risks of liver cancer for radiation, background levels were high. In contrast, however, the Chinese iron ore miner [6], peptic ulcer patient [8], and uterine bleeding [12] studies also had high background levels of liver cancer, but found no increased risk of liver cancer for radiation. Two studies by Darby et al. [10,13], conducted in areas with low background rates, disclosed point estimates consistent with a radiation effect, but in both studies the authors concluded that the increase was unconnected with radiation exposure. Those investigators attributed their finding of a statistically significant, but not dose-dependent, radiation effect in the study of underground miners to high alcohol consumption and the possible confusion of primary with secondary liver cancers [13], a criticism that could also be leveled against all mortality studies of liver cancer deaths occurring before the mid-1980s.

The finding of interaction between HCV and radiation in the etiology of HCC, suggests that studies of radiation and liver cancer that do not adjust for HCV infections overestimate radiation risks when HCV levels are high. Thus, radiation risk estimates for the A-bomb survivors might be inflated by the high prevalence of HCV in this cohort. Because HCV prevalences appear to be low in the Chinese miner [6], peptic ulcer [8], and uterine bleeding [12] cohorts, where background liver cancer rates were high, their lower HCV prevalences may explain why these studies did not find elevated radiation risks. Radiation risk estimates were significantly elevated in the Mayak worker study [14], where background liver cancer levels were high and the HCV prevalence in the cohort could not be determined. Furthermore, the nature of the radiation exposure (internal deposition and external, rather than just external) differed from the other cohorts

Acknowledgements

This study was supported by contract NCI-4893-8-001 from the US National Institutes of Health. RERF is a private nonprofit foundation funded equally by the Japanese Ministry of Health and Welfare and the US Department of Energy through the US National Academy of Sciences.

References

[1] D.A. Pierce, Y. Shimizu, D.L. Preston, et al, Studies of the mortality of atomic bomb survivors. Report 12, Part I. Cancer: 1950–1990, Radiat. Res. 146 (1996) 1–27.
[2] J.B. Cologne, S. Tokuoka, G.W. Beebe, et al., Effects of radiation on incidence of primary liver cancer among atomic bomb survivors, Radiat. Res. 152 (1999) 364–373.
[3] D.E. Thompson, K. Mabuchi, E. Ron, et al., Cancer incidence in atomic bomb survivors. Part II: solid tumors, 1958–1987, Radiat. Res. 137S (1994) 17–67.
[4] T. Fukuhara, G.B. Sharp, T. Mizuno, et al., Liver cancer in atomic-bomb survivors: histological characteristics and relationships to radiation and hepatitis B and C viruses, J. Radiat. Res. (Tokyo) 42 (2001) 117–130.
[5] E. Cardis, E.S. Gilbert, L. Carpenter, et al., Effects of low doses and low dose rates of external ionizing radiation: cancer mortality among nuclear industry workers in three countries, Radiat. Res. 142 (1995) 117–132.
[6] S.Y. Chen, R.B. Hayes, S.R. Liang, et al., Mortality experience of haematite mine workers in China, Br. J. Ind. Med. 47 (1990) 175–181.
[7] J.P. Ashmore, D. Krewski, J.M. Zielinski, et al., First analysis of mortality and occupational radiation exposure based on the National Dose Registry of Canada, Am. J. Epidemiol. 148 (1998) 564–574.
[8] M.L. Griem, R.A. Kleinerman, J.D. Boice Jr., et al., Cancer following radiotherapy for peptic ulcer, J. Natl. Cancer Inst. 86 (1994) 842–849.
[9] J.D. Boice Jr., N.E. Day, A. Andersen, et al., Second cancers following radiation treatment for cervical cancer. An international collaboration among cancer registries, J. Natl. Cancer Inst. 74 (1985) 955–975.
[10] S.C. Darby, G.M. Kendall, T.P. Fell, et al., Further follow up of mortality and incidence of cancer in men from the United Kingdom who participated in the United Kingdom's atmospheric nuclear weapon tests and experimental programmes, BMJ 307 (1993) 1530–1535.
[11] H.A. Weiss, S.C. Darby, R. Doll, Cancer mortality following X-ray treatment for ankylosing spondylitis, Int. J. Cancer 59 (1994) 327–338.
[12] P.D. Inskip, R.R. Monson, J.K. Wagoner, et al., Cancer mortality following radium treatment for uterine bleeding, Radiat. Res. 123 (1990) 331–344.
[13] S.C. Darby, E. Whitley, G.R. Howe, et al., Radon and cancers other than lung cancer in underground miners: a collaborative analysis of 11 studies, J. Natl. Cancer Inst. 87 (1995) 378–384.

[14] E.S. Gilbert, N.A. Koshurnikova, M. Sokolnikov, et al., Liver cancers in Mayak workers, Radiat. Res. 154 (2000) 246–252.
[15] F. Donato, P. Boffetta, M. Puoti, A meta-analysis of epidemiological studies on the combined effect of hepatitis B and C virus infections in causing hepatocellular carcinoma, Int. J. Cancer 75 (1998) 347–354.
[16] A. Wasley, M.J. Alter, Epidemiology of hepatitis C: geographic differences and temporal trends, Semin. Liver Dis. 20 (2000) 1–16.
[17] Y.Y. Zhang, L.S. Guo, L.J. Hao, B.G. Hansson, A. Widell, E. Nordenfelt, Antibodies to hepatitis C virus and hepatitis C virus RNA in Chinese blood donors determined by ELISA, recombinant immunoblot assay and polymerase chain reaction, Chin. Med. J. 106 (1993) 171–174.
[18] Y. Wang, Q.M. Tao, H.Y. Zhao, et al., Hepatitis C virus RNA and antibodies among blood donors in Beijing, J. Hepatol. 21 (1994) 634–640.
[19] S. Fujiwara, S. Kusumi, J.B. Cologne, et al., Prevalence of anti-hepatitis C virus antibody and chronic liver disease among atomic bomb survivors, Radiat. Res. 154 (2000) 12–19.
[20] H. Tanaka, T. Hiyama, H. Tsukuma, et al., Prevalence of second generation antibody to hepatitis C virus among voluntary blood donors in Osaka, Japan, Cancer Causes Control 5 (1994) 409–413.
[21] G.B. Sharp, J.B. Cologne, T. Fukuhara, et al., Temporal changes in liver cancer incidence rates in Japan: accounting for death certificate inaccuracies and improving diagnostic techniques, Int. J. Cancer 93 (2001) 751–758.

Radiosensitivity and expression of nucleotide excision repair genes in peripheral blood mononuclear cells of myelodysplastic syndrome patients

Sadayuki Ban [a,*], Ken Kuramoto [b,c], Kenji Oda [d], Hideo Tanaka [e], Akiro Kimura [e], Gen Suzuki [b], Takashi Imai [f]

[a] *Department of Radiobiology, Radiation Effects Research Foundation, Hiroshima 732-0815, Japan*
[b] *Department of Clinical Studies, Radiation Effects Research Foundation, Hiroshima 732-0815, Japan*
[c] *Hematology Branch, NIH, Bethesda, MD 20892, USA*
[d] *Department of Internal Medicine, Hiroshima City Hospital, Hiroshima 730-8518, Japan*
[e] *Department of Hematology and Oncology, Research Institute for Radiation Biology and Medicine, Hiroshima University, Hiroshima 734-8553, Japan*
[f] *Frontier Research Center, National Institute of Radiological Sciences, Chiba 263-8555, Japan*

Abstract

Myelodysplastic syndrome (MDS) is the only other radiogenic blood disease apart from leukemia. Clinically, MDS involves dysplastic hematopoisis and an increased risk of leukemic transformation. We compared the micronucleus (MN) frequency in the peripheral T lymphocytes of atomic bomb survivors with MDS and normal individuals. The spontaneous and X-ray-induced MN frequencies were significantly higher in MDS patients than in normal individuals. To explain the cause of unusual radiosensitivity, we measured the expression levels of nucleotide excision repair (NER) genes in peripheral blood mononuclear cells using an RT-PCR method. The reduction of NER genes was expressed in only 1 of the 10 patients with mild symptoms, but in 4 of the 11 patients with severe symptoms. Our data suggest that chromosomal instability and DNA repair defects may be involved in the pathophysiology of disease progression. © 2002 Elsevier Science B.V. All rights reserved.

Keywords: MDS; Radiosensitivity; NER gene; RT-PCR; Micronuclei

---

* Corresponding author. Present address: Frontier Research Center, National Institute of Radiological Sciences, Anagawa 4-9-1, Inage, Chiba 263-8555, Japan.
  E-mail address: s_ban@nirs.go.jp (S. Ban).

1. Introduction

A preliminary epidemiological study demonstrated that myelodysplastic syndrome (MDS) has an excess relative risk per sievert in 13 of the atomic bomb survivors [1]. MDS is the only other radiogenic blood disease apart from leukemia. Clinically, MDS involves hematopoietic dysfunction, which often precedes the development of acute myelogenous leukemia [2]. Because it is uncertain whether MDS pathogenesis affects lymphoid progenitor cells as well as myeloid progenitor cells, we investigated the micronucleus (MN) frequency in the peripheral T lymphocytes of atomic bomb survivors with MDS and normal individuals.

2. Materials and methods

2.1. Blood samples

Peripheral blood was obtained from 5 normal individuals and 23 atomic bomb survivors with MDS: 11 with refractory anemia (RA), 1 with RA with ring sideroblasts (RARS), 5 with RA with excess blasts (RAEB) and 6 with RA with excess blasts in transformation (RAEB-T).

2.2. Micronucleus assay

A method previously used in our laboratory was used [3].

2.3. Total RNA extraction

The total RNA was extracted from peripheral blood mononuclear cells according to the instruction of Trizol (Gibco BRL).

2.4. RT-PCR

The mRNAs corresponding to the human nucleotide excision repair genes (NER: ERCC1, ERCC3, ERCC5 and XPC) were amplified by the RT-PCR method with primer sets which were reported by Cheng et al. [4].

3. Results and discussion

MDS is a clonal disorder of the hematopoietic stem cells. The multistep pathogenesis may be involved in the onset of MDS and the clonal evolution. The initial genetic event of this pathogenesis, which might occur in the hematopoietic stem cell, is unknown. The aneuploidy has long been identified as one of the most common features of cancer cells and is widely believed to stem from an imbalance of chromosomal segregation. Enumeration of micronuclei is technically faster and statistically more precise in assessing the magnitude of chromosomal instability than a full comprehensive analysis of aneuploidy [5,6]. The spontaneous MN frequencies were significantly higher in MDS patients than in normal individuals. The spontaneous frequencies increased with the increase of severity of the MDS clinical subtypes. These results suggest that the control of chromosomal stability is impaired in pluripotent hematopoietic stem cells of MDS patients.

Next, we found that X-ray-induced MN frequencies were also higher in MDS patients than in normal individuals. Interestingly, radiation sensitivity also increased along with the severity of MDS symptoms. Because many of the patients in this study had not been exposed to chemo- or radiotherapy, their unusual radiosensitivities may be related to their chromosomal or genomic instability. To explain the cause of the unusual radiosensitivity, we measured the expression levels of four nucleotide excision repair (NER) genes (ERCC1, ERCC3, ERCC5 and XPC) in peripheral blood mononuclear cells using an RT-PCR method. One gene (ERCC5) was expressed at reduced levels in only 1 of the 10 patients with mild symptoms. A reduction in the NER genes was expressed in 4 out of the 11 patients with severe symptoms. Our results suggest that DNA repair defects may be involved in the pathophysiology of disease progression.

References

[1] K. Oda, A. Kimura, T. Matsuo, M. Tomonaga, K. Kodama, K. Mabuchi, Increased relative risk of myelodysplastic syndrome in atomic bomb survivors, J. Nagasaki Med. Assoc. 73 (Suppl.) (1988) 174–179 (in Japanese).
[2] H.P. Koeffler, Myelodysplastic syndromes (preleukemia), Semin. Hematol. 23 (1986) 284–299.
[3] S. Ban, J.B. Cologne, S. Fujita, A.A. Awa, Radiosensitivity of atomic bomb survivors as determined with a micronucleus assay, Radiat. Res. 134 (1993) 170–178.
[4] L. Cheng, Y. Guan, L. Li, R.J. Legerski, J. Einspahr, J. Bangert, D.S. Alberts, Q. Wei, Expression in normal human tissues of five nucleotide excision repair genes measured simultaneously by multiple reverse transcription-polymerase chain reaction, Cancer Epidemiol., Biomarkers Prev. 8 (1999) 801–807.
[5] S. Bonassi, M. Fenech, C. Lando, Y.-P. Lin, M. Ceppi, W.P. Chang, N. Holland, M. Kirsch-Volders, W. Zeiger, S. Ban, et al., Human micronucleus project: international database comparison for results with cytokinesis-block micronucleus assay in human lymphocytes, Environ. Mol. Mutagen. 37 (2001) 31–45.
[6] S. Ban, Y. Shinohara Hirai, Y. Moritaku, J.B. Cologne, D.G. MacPhee, Chromosomal instability in BRCA1- or BRCA2-defective human cancer cells detected by spontaneous micronucleus assay, Mutat. Res. 474 (2001) 15–23.

An analysis of persistent inflammation among atomic bomb survivors with respect to sex and age at the time of the bombings

Kazuo Neriishi[a],*, Eiji Nakashima[b]

[a]*Department of Clinical Studies, Radiation Effects Research Foundation, 5-2 Hijiyama Park, 732-0815 Hiroshima, Japan*
[b]*Department of Statistics, Radiation Effects Research Foundation, 5-2 Hijiyama Park, 732-0815 Hiroshima, Japan*

Abstract

We have so far reported that atomic bomb (A-bomb) survivors have persistent subclinical inflammation. This study reports on the analysis of persistent inflammation among A-bomb survivors with respect to sex and age at the time of the bombings. Among 6258 A-bomb survivors undergoing seven inflammatory tests, standardized inflammatory scores, used as inflammation indices for each person, were regressed for city, age at examination, age at the time of the bombings, inflammatory diseases, smoking, and radiation dose, in order to analyze the effects of radiation dose. The results indicate that the inflammation scores increased significantly for both males and females younger than 20 years old at the time of bombings, while the scores increased significantly only for the females that were 20 years old or older at the time of the bombings. As evidence has been observed, on the one hand in the radiosensitive young generation, and on the other hand in the population with a high risk of radiation-induced disorders in the estrogen receptor organs, such as thyroid cancer, breast cancer, ovarian cancer and uterine myoma, a close association between the persistent inflammation and the radiation damage mechanism(s) is suggested. © 2002 Elsevier Science B.V. All rights reserved.

Keywords: Inflammation; Atomic bomb; Sex; Age

1. Background

We have so far reported that the atomic bomb (A-bomb) survivors have persistent subclinical inflammation [1]. This study reports on the characteristics in the analysis of per-

* Corresponding author. Tel.: +81-82-261-3131; fax: +81-82-263-7279.
E-mail address: neriishi@rerf.or.jp (K. Neriishi).

sistent inflammation among A-bomb survivors with respect to sex and age at the time of the bombings.

2. Subjects

The subjects are 6258 individuals participating in the Adult Health Study who, between 1988 and 1992, underwent seven inflammatory tests, i.e., white blood cell count, neutrophil count, erythrocyte sedimentation rate, corrected erythrocyte sedimentation rate, alpha 1 globulin, alpha 2 globulin, and sialic acid, and for whom we have a radiation dose (DS86) and smoking information. From among the subjects, 2923 persons with inflammatory diseases that could have influenced inflammation test levels were selected using ICD codes.

3. Statistical methods

Standardized inflammatory scores were obtained using the following formula below:

$$\text{Score} = \sum_{i=1}^{7} \alpha_i \frac{X_i - \overline{X_i}}{\text{SD}(X_i)}$$

where $X_1, X_2, X_3, X_4, X_5, X_6$, and X_7 represent the white blood cell count, neutrophil count, erythrocyte sedimentation rate, corrected erythrocyte sedimentation rate, alpha 1 globulin, alpha 2 globulin, and sialic acid, respectively. Using α_i of each test at the maximum variation (deviation) of the score, the standardized scores, utilized as inflammation indices for each

Table 1
Standardized inflammation scores at 1 Gy (DS86) by age at the time of the bombings in regression analysis of A-bomb survivors

	Age at the time of the bombings				Total
	0–9	10–19	20–29	30–	
Male					
Number of subjects	615	923	284	361	2183
Inflammation scores					
Mean	0.15	0.13	0.00	−0.05	
SD	0.05	0.04	0.08	0.09	
p value	0.0028	0.008	N.S.	N.S.	
Female					
Number of subjects	710	1354	1136	875	4075
Inflammation scores					
Mean	0.13	0.16	0.15	0.12	
SD	0.05	0.04	0.04	0.06	
p value	0.0076	0.0001	0.0002	0.0339	

person, were regressed for city, age at examination, age at the time of the bombings, inflammatory diseases, smoking, and radiation dose, in order to analyze the effects of the radiation dose.

4. Results

The results indicate that the inflammation scores increased significantly for both males and females younger than 20 years old at the time of the bombings ($p < 0.005$ in males and $p < 0.01$ in females), while the scores increased significantly only for the females that were 20 years old or older at the time of the bombings ($p > 0.5$ in males and $p < 0.05$ in females) (Table 1).

5. Discussion

The higher risk of radiation-induced diseases among young males and females is an established biological fact, as younger individuals have higher radiosensitivity [2]. The results in this study of the significant increase in the inflammation scores for young males and females younger than 20 years old at the time of bombings may suggest the same mechanism.

As for the significant increase in the inflammation scores only for females that were 20 years old or older at the time of the bombings, the evidence reminds us that in A-bomb survivors of the same generation, the cancer incidence in the estrogen receptor organs, such as breast, thyroid, parathyroid, and ovary, increases [2]. Interestingly, several reports indicate that radiation-induced cancers are frequently observed in the estrogen receptor organs without hormonal effects. In fact, Huang et al. [3] reported that radiation-induced ovarian and breast cancers are preferentially observed in hormone receptor negative tumors, while nonradiation-induced breast cancers are strongly associated with chronic estrogen exposure [4]. Also, Karlsson et al. [5] reported that radiation-induced thyroid cancers are not associated with the estrogen and progesterone receptors. In our study, the females that were 20 years old or older at the time of the bombings are no longer affected by estrogen at the time of the inflammation tests because they are in the postmenopausal stage. This suggests that radiation-induced persistent inflammation may contribute as an epigenetic and/or bystander effect, and in the development of radiation-induced disorders in the estrogen receptor organs even in postmenopausal A-bomb survivors.

Our observations in this study may therefore provide important clues for further understanding the mechanism(s) involved in radiation damage.

References

[1] K. Neriishi, E. Nakashima, R.R. Delongchamp, Persistent subclinical inflammation among A-bomb survivors, Int. J. Radiat. Biol. 77 (4) (2001 Apr) 475–482.
[2] D.A. Pierce, Y. Shimizu, D.L. Preston, M. Vaeth, K. Mabuchi, Studies of the mortality of atomic bomb survivors. Report 12, Part I. Cancer: 1950–1990, Radiat. Res. 146 (1) (1996 Jul) 1–27.
[3] W.Y. Huang, B. Newman, R.C. Millikan, M.J. Schell, B.S. Hulka, M.G. Moorman, Hormone-related factors

and risk of breast cancer in relation to estrogen receptor and progesterone receptor status, Am. J. Epidemiol. 151 (7) (2000 Apr 1) 703–714.
[4] M. Clemons, P. Goss, Estrogen and the risk of breast cancer, N. Engl. J. Med. 344 (4) (2001 Jan 25) 276–285.
[5] M.G. Karlsson, L. Hardell, A. Hallquist, No association between immunohistochemical expression of p53, c-erbB-2, Ki-67, estrogen and progesterone receptors in female papillary thyroid cancer and ionizing radiation, Cancer Lett. 120 (2) (1997 Dec 9) 173–177.

Basic study on the radon effects and the thermal effects in radon therapy

Kiyonori Yamaoka [a,*], Takashi Mifune [b], Shuji Kojima [c], Shuji Mori [d], Koichi Shibuya [a], Yoshiro Tanizaki [b], Katsuhiko Sugita [a]

[a] *Medical Radioscience, Okayama University Medical School, 2-5-1 Shikata-cho, Okayama 700-8558, Japan*
[b] *Misasa Branch Hospital, Okayama University Medical School, 827 Yamada, Misasa-cho, Tohaku-gun, Totori 682-0193, Japan*
[c] *Department of Pharmaceutical Sciences, Science University of Tokyo, 2669 Yamazaki, Noda Chiba 278-0022, Japan*
[d] *Clinical Biology, Okayama University Medical School, 2-5-1 Shikata-cho, Okayama 700-8558, Japan*

Abstract

Because most of the diseases to which radon therapy is applied are related with activated oxygen, in this study, the effects of the radioactivity of radon and the thermal effects were compared under condition with the same chemical effects using the activity of superoxide dismutase (SOD), which is an oxidation inhibitor, and lipid peroxide and low-density lipoprotein (LDL)-cholesterol, which are closely involved in arteriosclerosis, as the parameters. The results were as follows, the SOD activity was significantly increased, and the lipid peroxide and LDL-cholesterol levels were significantly decreased on days 6 and 7. The results were about twofold larger in the radon group than in the thermo group. This suggests that the antioxidation function was more enhanced by radon therapy than by thermo therapy, and it was suggested that radon therapy may prevent the causes of lifestyle-related diseases such as arteriosclerosis. These findings are important in understanding the mechanism of diseases to which radon therapy can be performed, and most of which are called activated oxygen-related diseases. © 2002 Elsevier Science B.V. All rights reserved.

Keywords: Radon therapy; Thermal effect; Superoxide dismutase; Lipid peroxide; Low-density lipoprotein cholesterol

1. Background

Therapy using radon, which mainly emits α-rays, is performed for various diseases such as osteoarthritis and asthma [1]. Several attempts have been made to clarify its mechanism;

---

* Corresponding author. Tel/fax: +81-86-235-6852.
 E-mail address: yamaoka@md.Okayama-u.ac.jp (K. Yamaoka).

however, there have only been a few studies on radon therapy in humans. Because most of the diseases to which radon therapy is applied are related with activated oxygen, the effects of the radioactivity of radon and the thermal effects were compared under the condition with the same chemical effects using the activity of superoxide dismutase (SOD), which is an oxidation inhibitor, and lipid peroxide and low-density lipoprotein (LDL)-cholesterol, which cause lifestyle-related diseases. Furthermore, they are closely involved in arteriosclerosis, as the parameters, in order to clarify the mechanism of the diseases to which radon therapy can be performed.

2. Methods

The subjects were six females in their 40s–50s, who were divided into two groups. The radon group took a hot bath with a high concentration of radon (room temperature, 36 °C; radon concentration, 2080 Bq/m^3, about 40-fold higher than that in the local sauna baths [2]) in the Misasa Branch of the Hospital of the Okayama University, Medical School. The thermo group went to a local sauna bath in the region (room temperature, 48 °C). On days 1, 2, 3, 6 and 7, inhalation through the nose was performed for 40 min once a day under a condition of high humidity.

Blood samples were collected after inhalation, and the blood samples taken before the start of inhalation were regarded as the control. In order to prevent the effects of a change in climate, the subjects were residents in the region. Inhalation through the nose was used where the uptake of radon is the most efficient.

The SOD activity was measured using the nitro blue tetrazolium (NBT) method [3], lipid peroxide by the thiobarbituric acid (TBA) method [4], and LDL-cholesterol by the enzyme method [5].

The data values are reported as the mean ± the standard error of the mean (S.E.M.). Statistical analysis was carried out using the Student's unpaired t-test to show the significance of the difference between the pairs of means.

3. Results

3.1. Sod activities

On day 1, the SOD activity significantly increased in the radon group by about 10% compared to the control (3.8 ± 0.2 U/ml), while in the thermo group, there was no significant change. On days 2 and 3, the SOD activity was slightly lower than that on the previous day in of the both groups. On days 6 and 7, the SOD activity had significantly increased by about 35% in the radon group and by about 15% in the thermo group compared to the control (Fig. 1a).

3.2. Lipid peroxide levels

On day 2, the lipid peroxide level had significantly decreased in the radon group by about 10% compared to the control (4.0 ± 0.2 nmol/ml), while in the thermo group, there

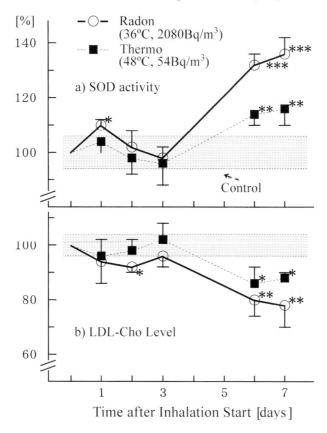

Fig. 1. The time-dependent changes in (a) SOD activities and (b) LDL-cholesterol levels in the blood of humans at each radon inhalation or thermal treatment. Each value represents the mean ± S.E.M. The number of persons per experimental point is 3. *$P<0.05$, **$P<0.01$, ***$P<0.001$ by the t-test, radon or the thermal group value vs. the control group value.

was no significant change. On day 3, the lipid peroxide level was slightly higher than that on the previous day in both of the groups. On days 6 and 7, the lipid peroxide level had significantly decreased by about 25% in the radon group and by about 10% in the thermo group compared to the control.

3.3. LDL-cholesterol levels

On day 2, the LDL-cholesterol level had significantly decreased in the radon group by about 10% compared to the control (1.10 ± 0.05 mg/ml), while in the thermo group, there was no significant change. On day 3, the LDL-cholesterol level was slightly higher than that on the previous day in both of the groups. On days 6 and 7, the LDL-cholesterol level had significantly decreased by about 20% in the radon group and by about 12% in the thermo group compared to the control (Fig. 1b).

4. Discussion

Radon is an inert gas and, as such, does not react with any chemical component of the body. Upon entry through the lungs or through the skin, it reaches the blood stream and is then distributed throughout the body. Being rather lipid soluble, radon tends to accumulate in organs rich in fat, such as the endosecretory glands, and also nerve fibers, which are surrounded and protected by a lipid-containing layer. The retention time in the body is short, 50% having disappeared after only 15–30 min. It is during this short period, however, while radon is in contact with the tissue that it launches the beneficial effects. Radon-222 (^{222}Rn) is a source of α-rays, and it can only travel a distance of about 20 µm through body tissues. The relatively large transfer of energy, which is associated with the absorption of α-particles, causes a series of complicated reactions within the tissues. As yet, the molecular processes involved are still poorly understood. It is safe, however, to assume that radiolytic radicals are released and these, in turn, stimulate detoxification processes and also may stimulate such processes as cell metabolism and energy conversion within mitochondria, as well as biosynthesis of enzymes and other proteins or bioactive peptides [1].

It has been known that the activity of SOD, a scavenger of superoxide radicals, is increased in cultured cells [6] and in various organs of rats [7] and rabbits [8] by exposure to radon. In the present study, similar results were obtained with human blood. In the thermo group, the increase in SOD activity was due to the heat shock protein (HSP), which was induced by the high temperature of 48 °C [9]. The slightly lower SOD activity and higher lipid peroxide and LDL-cholesterol levels around day 3 than on the previous day were considered to be symptoms equivalent to slight "yuatari", which is the effect of taking a hot bath for too long, accompanied with inhalation for 3 days. After the interruption of inhalation on days 4 and 5, the SOD activity had significantly increased, and the lipid peroxide and LDL-cholesterol levels had significantly decreased on days 6 and 7. In addition, the level of lipid peroxide was reduced by radon inhalation in the rabbits [8]. These results agree with the studies showing that radon therapy is more effective if performed every 2 days than every day. The results on days 6 and 7 were about twofold larger in the radon group than in the thermo group. This suggests that the anti-oxidation function was more enhanced by radon therapy than by thermo therapy and that radon therapy may prevent the causes of lifestyle-related diseases such as arteriosclerosis. These findings are important in understanding the mechanism of diseases to which radon therapy can be performed and most of which are called activated oxygen-related diseases.

In the previous study, we performed a low-dose γ-ray or X-ray irradiation to model mice with diabetes [10] and hypertension [11], for which radon hot-spring therapy is performed, and found alleviation of the symptoms of both diseases and enhancement of the anti-oxidation function was demonstrated. We also obtained similar results from mice with liver [12,13] or brain disorders [14]. Furthermore, we found that the level of β-endorphin with morphine-like activity and α-atrial natriuretic polypeptide (α-ANP) with the vasodilation activity were increased in the blood by radon inhalation in rabbits [15,16]. These results recorded from the animal experiments support the data recorded during the human experiments.

References

[1] P. Deetjen, Radon in der Kurmedizin, I.S.M.H Verlag Geretsried, 1997, pp. 32–38.
[2] M. Mifune, Spa Sci. 31 (1981) 79–93.
[3] C. Beauchamp, I. Fridovich, Anal. Biochem. 44 (1971) 276–287.
[4] H. Ookawa, S. Ooishi, K. Yagi, Anal. Biochem. 95 (1979) 351–358.
[5] C.C. Allain, Clin. Chem. 20 (1974) 470–475.
[6] H. Frick, W. Pfaller, Z. Phys. Med., Balneol., Med. Klimatol. 17 (1988) 23–33.
[7] J. Ma, H. Yonehara, M. Ikebuchi, T. Aoyama, J. Radiat. Res. 37 (1996) 12–19.
[8] K. Yamaoka, Y. Komoto, I. Suzuka, R. Edamatsu, A. Mori, Arch. Biochem. Biophys. 302 (1993) 37–41.
[9] M.A. Hass, D. Massaro, J. Biol. Chem. 26 (1988) 776–781.
[10] M. Takahashi, S. Kojima, K. Yamaoka, E. Niki, Radiat. Res. 154 (2000) 680–685.
[11] K. Yamaoka, K. Ishii, Physiol. Chem. Phys. Med. NMR 27 (1995) 161–165.
[12] K. Yamaoka, S. Kojima, T. Nomura, Free Radical Res. 32 (2000) 213–221.
[13] T. Nomura, K. Yamaoka, Free Radical Biol. Med. 27 (1999) 1324–1333.
[14] S. Kojima, O. Matsuki, T. Nomura, K. Yamaoka, M. Takahashi, E. Niki, Free Radical Biol. Med. 26 (1999) 388–395.
[15] K. Yamaoka, K. Ishii, T. Ito, Y. Komoto, I. Suzuka, R. Edamatsu, A. Mori, Neurosciences 20 (1994) 17–22.
[16] K. Yamaoka, Y. Komoto, Physiol. Chem. Phys. Med. NMR 28 (1996) 1–5.

Low dose radiation carcinogenesis

Low dose radiation carcinogenesis

Methodological aspects of low-dose epidemiological studies

Colin R. Muirhead*

National Radiological Protection Board, Chilton, Didcot, Oxon OX11 0RQ, UK

Abstract

Epidemiological studies of the survivors of the atomic bombings of Hiroshima and Nagasaki and of groups exposed to radiation for medical reasons play an important role in formulating risk estimates for radiation-induced cancer. These studies are particularly informative about the effects of acute or fractionated high-dose exposures. However, much of the interest in radiological protection concerns protracted low-dose exposures. The above epidemiological studies contain groups who received acute or fractionated low-dose exposures, while studies of natural, occupational and environmental radiation have the potential to provide information on the effects of protracted low-dose exposures. However, methodological aspects common to these studies play an important role in interpreting their findings. First, whether bias (systematic errors in the design or conduct of the study) or confounding (the effect of uncontrolled factors) might have given rise to spurious findings. Secondly, whether the study has sufficient statistical precision to estimate any radiation effect. These issues are particularly relevant to low-dose studies, in which attempts to detect and quantify a small effect of radiation could be affected adversely by small biases, residual confounding, and/or limitations on the study size. These methodological issues are illustrated using examples from epidemiological studies of natural, occupational and environmental exposures. Attention is also given to areas where biological considerations can assist the quantification of low-dose risks. © 2002 Elsevier Science B.V. All rights reserved.

Keywords: Epidemiology; Radiation; Cancer; Low doses

1. Introduction

Much is known about the risks of cancer following acute or fractionated exposures to high doses of radiation from epidemiological studies of the survivors of the atomic

---

* Tel.: +44-1235-822808; fax: +44-1235-833891.
 E-mail address: colin.muirhead@nrpb.org.uk (C.R. Muirhead).

bombings of Hiroshima and Nagasaki and from studies of groups exposed for medical reasons, for example, as treatment for an earlier cancer or for other diseases [1]. Information on acute or fractionated low-dose exposures can also be derived from such studies. Furthermore, studies of groups exposed to radiation either occupationally, from elevated levels of radon in homes or as a consequence of incidents in the former Soviet Union such as that at Chernobyl and near the Techa River, might be informative about risks from protected exposures to low or moderate doses [1]. However, the interpretation of low-dose epidemiological studies is not always straightforward, owing to possible difficulties from bias in the study design, confounding factors or low statistical power.

These potential methodological difficulties are described in this paper, with examples given from studies of natural, occupational and environmental exposures. In addition, some instances where such studies can be informative will be illustrated. Finally, some thoughts will be provided on how biological considerations can assist in quantifying low-dose risks.

2. Methodological aspects and examples

2.1. Bias and confounding

Bias can be defined as any process at any stage of inference that tends to produce results or conclusions that differ systematically from the truth. Since epidemiology is observational rather than experimental in nature, bias tends to be of more concern here than in randomised controlled trials. In particular, even relatively small biases could distort findings from epidemiological studies of low-dose exposures where the magnitude of any effect would be predicted to be small. Consequently, care is needed to minimise the potential for bias.

Bias can arise in various ways in epidemiology [1,2]. For example, when comparing disease rates in an exposed and an unexposed group, systematic differences between the groups in the collection of data can lead to bias. An example concerns a study of Russian workers involved in the cleanup of the Chernobyl plant that reported a raised risk of leukaemia incidence relative to the general Russian population [3]. However, there appears to have been over-ascertainment and misdiagnosis of some leukaemias among the workers, who received frequent medical examinations, whereas leukaemias in the general population are likely to have been under-reported [4]. It is noticeable that, amongst these workers, there was no relation between leukaemia risk and either radiation dose or the type of work conducted around Chernobyl [5], so suggesting that the comparison with the national population was subject to bias.

Even when data for exposed and unexposed groups are collected in a similar manner, there may be problems associated with a lack of comparability between the two groups. For example, it has been found in many studies that persons selected into and retained in employment tend to have better health than the general population. This "healthy worker effect" can arise among both radiation workers and other workers. For example, among nuclear industry workforces in the United Kingdom, overall mortality was less than national rates for both radiation workers and non-radiation workers [6]. Such comparisons with national mortality rates can be useful to gauging the overall health of workers.

However, to investigate effects of radiation exposure, it is preferable to compare workers with different levels of dose, as indicated in Section 2.2 below.

Another potential source of bias, which is sometimes considered separately from other sources, is confounding. A confounding factor is correlated both with the disease under study and with the exposure of primary interest. While various factors other than radiation can affect cancer rates, such factors will not confound studies of radiation and cancer risks unless they are also correlated with radiation dose. For example, older persons exposed occupationally to radiation over many years usually would have a larger cumulative dose than would young radiation workers. Since cancer risks increase with increasing age, then age could confound analyses of cancer in relation to cumulative dose. Fortunately, it is standard practice to adjust for age in epidemiological analyses, so this should not pose difficulties in general.

In low-dose studies generally, it may be important to take account of a factor that as a strong effect on cancer risks, even if there is no strong prior reason to expect that is also correlated with radiation dose, since a fairly small level of confounding could affect the analysis. For example, in studying lung cancer risk and residential radon exposure, it is highly desirable to take account of smoking habits. In some studies of this topic that collected data at the individual level, analyses have been conducted both of the individual-level data and of the same data aggregated over geographical areas, and have tended to yield different results (e.g. Refs. [7,8]). Since it is known that aggregation of individual-level data can lead to bias [9], and because data on radiation and other factors measured at the individual level—while possibly subject to uncertainty—are likely to be more relevant (e.g. in terms of the age ranges and time periods) than aggregated data obtained from other sources [10], then the findings from individual-level studies of residential radon and lung cancer should be more reliable than those from studies based on aggregated data (e.g. Ref. [11]).

2.2. Statistical power

An important aspect of an epidemiological study is its statistical power, that is, the probability that it will detect a raised risk of given magnitude with a specific degree of confidence. Statistical power would normally be calculated before initiating a study; for example, to evaluate the probability of detecting a doubling of risk, say, using a significance test at the 5% level. In contrast, once a study has been conducted, its precision can be gauged by the width of the confidence interval for the estimated radiation effect.

The limited range and distribution of doses is an inherent restriction on low-dose studies, compared with the Japanese atomic bomb survivors, for example, whose doses ranged from 0 up to several sieverts. Nevertheless, it may be possible to increase the power of a study of natural radiation by studying regions with a comparatively wide range of exposures. In particular, levels of residential radon can vary markedly within a small geographical area, in contrast to background gamma radiation. In an occupational context, a study of radiation workers who received doses over many years should be more informative than a study restricted to workers with a short period of exposure, whose cumulative doses would generally be lower than in the preceding group. Notwithstanding the limitations on dose distributions in low-dose studies, power can be increased by raising the number of study subjects and, in particular, the number of diseased cases. For a case-control study, in which radiation exposures are compared for persons with the disease of interest (cases) and persons selected from the same source population who do not have this disease (controls), efforts could similarly be made to increase the numbers of cases and controls. For a cohort study, in which a cohort of individuals is followed to determine their subsequent disease incidence or mortality, this might involve increasing the cohort size or the period of follow-up, or focussing on a disease with a relatively high baseline rate. An example concerns the UK National Registry for Radiation Workers (NRRW) [12], which, with about 125 000 radiation workers, in one of the largest such studies worldwide. As of the end of 1992, about 13 000 of these workers had died. Although the confidence interval of the dose response for leukaemia and other cancers at low doses is wider for the NRRW than for the Life Span Study (LSS) of Japanese A-bomb survivors, reflecting the wider dose range and longer follow-up in the LSS, it is noteworthy that the NRRW findings are inconsistent with risks more than four times those estimated from the LSS [12].

Another means of increasing statistical power is to combine results from different studies of the same topic. There are two approaches that are usually adopted. One consists of a meta-analysis, in which published summary measures of the relationship between dose and risk are combined. For example, Lubin and Boice [13] analysed results from eight case-control studies of residential radon and lung cancer in this way. While meta-analyses are often relatively easy to perform, they can sometimes be limited by the degree of data available and by differences in the way in which data from individual studies were collected and analysed. It is often preferable to undertake a combined analysis, in which the original study investigators participate in a pooled analysis of individual-level data from each of the available studies. This approach allows data from the studies to be analysed in parallel and may identify any lack of comparability between the studies, before pooling the data. Several combined analyses of low-dose epidemiological studies are currently in progress. In particular, data from European studies of residential radon and lung cancer are currently being combined, and a similar pooling is taking place for studies in North America [14]. Furthermore, following an earlier study of data from three countries [15], the International Agency for Research on Cancer is coordinating an international collaborative study of cancer risk in radiation workers in the nuclear industry, which will include data for more than half a million workers worldwide, including the above data from the NRRW.

2.3. Dose assessment

There has been a substantial amount of work in recent years to evaluate the impact of errors in dosimetry on dose-response analyses [16]. For example, random errors in dose estimates would tend to bias dose-response trends towards zero. Statistical methods have been developed to correct for such biases, although it is not possible to recover the loss of statistical power that is also associated with these random errors. Alternative physical and biologically based methods of exposure assessment have been considered during the past decade, including electromagnetic paramagnetic resonance of tooth enamel and the fluorescent in situ hybridisation (FISH) technique for chromosome stable translocation analysis; the latter is considered further under Section 3.1 below.

3. Input from biological considerations

The previous section focussed on the difficulties that can arise in designing and interpreting low-dose epidemiological studies. However, it should be emphasised that these studies provide direct information on the health of humans exposed to ionising radiation. In the following, the extent to which biological considerations can add to information available from epidemiological studies is discussed.

3.1. Molecular epidemiology

Traditionally, radiation epidemiology has evaluated the relationship between clinically observed disease and a physically based estimate of dose. Advances in recent years have allowed the inclusion of biologically based measurements at the molecular level, which might be used to estimate exposure, early biological response or host characteristics that influence susceptibility [17]. For example, the FISH technique for chromosome stable translocation analysis has been applied to Chernobyl cleanup workers [18]; however, it is currently difficult to evaluate individual doses of less than 100–200 mSv using this technique, so limiting its use in low-dose epidemiology. Chromosome aberrations are also used sometimes as a biomarker of disease, although the interpretation of analyses of these aberrations is not always straightforward. For example, levels of unstable chromosome aberrations in a high background area of China were raised relative to a low background area, whereas cancer rates were generally similar in the two areas [19]. Several studies of post-Chernobyl childhood thyroid tumours show oncogenic rearrangements that might be characteristic to radiation exposure, but alternatively, might be markers of tumour phenotype [1]. In such studies, particular attention needs to be given to the probability that radiation might have induced individual tumours. More generally, low-dose studies usually would not have sufficient statistical power to identify genetic susceptibility. Consequently, further developments in the identification and analysis of biological measures would be desirable.

3.2. Experimental studies and mechanisms of carcinogenesis

Experimental studies provide information on the effects of dose and dose rate on cancer and life span in animals and on cells in culture (e.g. Refs. [20,21]). While these studies are particularly informative about the effects of low-dose rate, for which human data are more sparse, epidemiological studies are better suited for obtaining quantitative cancer risks in humans, because of their direct relevance. Perhaps of greater importance than the experimental findings per se is the information that is becoming available from increased understanding of the fundamental nature of the carcinogenic process. Neoplasia is seen as a complex, multistage process that can be subdivided into neoplastic initiation, promotion, conversion and progression. While radiation mutation might affect all of these stages, neoplastic initiation is likely to be of greatest importance for low-dose radiation, and would suggest the absence of a dose threshold for risk [21]. These considerations do not provide direct quantitative estimates of risks, but can assist in developing models that may be fitted to epidemiological or experimental data; for example, the two-mutation model of Moolgavkar and Venzon [22]. The ability to distinguish between different types of mechanistic model is limited at low doses, although there is more opportunity to do so using high-dose data [1]. Advances in modelling the physical and biological aspects of radiation tumorigenesis should be valuable in estimating risks at low doses and low-dose rates.

4. Concluding remarks

Epidemiological studies of low doses can be difficult to perform well. Nevertheless, well-designed studies can provide additional direct information on risks following low-dose radiation exposures, which can supplement data from high-dose studies and from experimental studies. Further developments in understanding and modelling mechanisms of carcinogenesis should aid the estimation of risks at low doses.

Acknowledgements

This paper draws in part on an evaluation conducted for the UNSCEAR 2000 report [1]. The author thanks Dr. Elaine Ron for her extensive input in preparing Annex I of that report.

References

[1] United Nations Scientific Committee on the Effects of Atomic Radiation, Sources and Effects of Ionizing Radiation. Vol. II: Effects. 2000 Report to the General Assembly, with Scientific Annexes, United Nations, New York, 2000.
[2] B. MacMahon, D. Trichopoulous, Epidemiology, Principles and Methods, 2nd edn., Little, Brown and Company, Boston, 1996.
[3] V.K. Ivanov, A.F. Tsyb, A.I. Gorsky, et al., Leukaemia and thyroid cancer in emergency workers of the Chernobyl accident: estimation of radiation risks (1986–1995), Radiat. Environ. Biophys. 36 (1997) 9–16.
[4] V.K. Ivanov, A.F. Tsyb, A.P. Konogorov, et al., Case-control analysis of leukaemia among Chernobyl accident emergency workers residing in the Russian Federation, 1986–1993, J. Radiol. Prot. 17 (1997) 137–157.
[5] J.D. Boice Jr., Leukemia, Chernobyl and epidemiology, J. Radiol. Prot. 17 (1997) 127–133.
[6] L. Carpenter, C. Higgins, A. Douglas, et al., Combined analysis of mortality in three United Kingdom nuclear industry workforces, 1946–1988, Radiat. Res. 138 (1994) 224–238.
[7] F. Lagarde, G. Pershagen, Parallel analyses of individual and ecologic data on residential radon, cofactors, and lung cancer in Sweden, Am. J. Epidemiol. 149 (1999) 268–274.
[8] S.C. Darby, H. Deo, R. Doll, et al., A parallel analysis of individual and ecological data on residential radon and lung cancer in south-west England, J. R. Stat. Soc. A 164 (2001) 193–203.
[9] S. Greenland, J. Robins, Invited commentary: ecologic studies—biases, misconceptions, and counter-examples, Am. J. Epidemiol. 139 (1994) 747–760.
[10] B.J. Smith, R.W. Field, C.F. Lynch, Residential ^{222}Rn exposure and lung cancer: testing the linear no-threshold theory with ecologic data, Health Phys. 75 (1998) 11–17.
[11] B.L. Cohen, Test of the linear no-threshold theory of radiation carcinogenesis for inhaled radon decay products, Health Phys. 68 (1995) 157–174.
[12] C.R. Muirhead, A.A. Goodill, R.G.E. Haylock, et al., Occupational radiation exposure and mortality: second analysis of the National Registry for Radiation Workers, J. Radiol. Prot. 19 (1999) 3–26.

[13] J.H. Lubin, J.D. Boice Jr., Lung cancer risk from residential radon: meta-analysis of eight epidemiologic studies, J. Natl. Cancer Inst. 89 (1997) 49–57.
[14] N. Hunter, Report: residential radon and lung cancer, Radiol. Prot. Bull. 224 (2000) 24–26 (Available at http://www.nrpb.org.uk/Bulletin/Erpb224.htm).
[15] E. Cardis, E.S. Gilbert, L. Carpenter, et al., Effects of low doses and low dose rates of external ionizing radiation: cancer mortality among nuclear industry workers in three countries, Radiat. Res. 142 (1995) 117–132.
[16] E. Ron, O. Hoffman (Eds.), Uncertainties in Radiation Dosimetry and their Impact on Dose-Response Analyses, National Cancer Institute, 1999, NIH Publication No. 99-4541.
[17] A.J. McMichael, Invited commentary—"Molecular epidemiology": new pathway or new travelling companion? Am. J. Epidemiol. 140 (1994) 1–11.
[18] L.G. Littlefield, A.F. McFee, S.I. Salomaa, et al., Do recorded doses overestimate true doses received by Chernobyl cleanup workers? Results of cytogenetic analysis of Estonian workers by fluorescence in situ hybridization, Radiat. Res. 150 (1998) 237–249.
[19] L. Wei, T. Sugahara, Recent advances of "Epidemiological study in high background radiation area in Yangjiang, China" (In these proceedings).
[20] United Nations Scientific Committee on the Effects of Atomic Radiation, Sources and Effects of Ionizing Radiation. 1993 Report to the General Assembly, with Scientific Annexes, United Nations, New York, 1993.
[21] NRPB, Risk of radiation-induced cancer at low doses and low dose rates for radiation protection purposes, Doc. NRPB 6 (1) (1995) 1–77.
[22] S.H. Moolgavkar, D.J. Venzon, Two-event models for carcinogenesis: incidence curves for childhood and adult tumours, Math. Biosci. 47 (1979) 55–77.

Recent advances of "Epidemiological study in high background radiation area in Yangjiang, China"

Lu-Xin Wei [a],*, Tsutomu Sugahara [b]

[a] *Department of Bio-Medicine, Laboratory of Industrial Hygiene, Ministry of Health, No. 2 Xinkang Street, Deshengmenwai, Beijing 100088, China*
[b] *Health Research Foundation, Pasteur Building 5F, 103 5 Tanaka Monzen cho, Sakyo-ku, Kyoto 606-8225, Japan*

1. Introduction

The High Background Radiation Research Group (HBRRG, China) started the Health Survey in the High Background Radiation Areas (HBRA) in Yangjiang, China, in 1972. Since then, the HBRRG had conducted the epidemiological study consecutively, until 1990. The research reports and articles concerning the progress and results recording the investigation in the period from 1972 to 1990 had been published. [1–3].

In 1991, Japanese scientists of radiation research recognized the importance of the work, after a joint feasibility study with revised protocols, a China–Japan cooperative project of "Epidemiological study on the population in the high background radiation area of Yangjiang, China" involving both Chinese and Japanese scientists began in 1992, which currently is still in progress. In this article, we give a brief account on the advances since the cooperation, including the results obtained up to the year 2000.

The purposes of the Project: (1) to identify whether there exists an association between the health effects and low-dose ionizing radiation (high background radiation), (2) if there is an association, to quantify the magnitude of health risk. In the period from 1991 to 2000, the health effects we studied were cancer mortality and the frequency of chromosome aberrations.

To realize the above-mentioned purposes, we have conducted the following:

1. to have a fixed cohort study to improve the precision of individual dose estimation and the quality of observation on cancer mortality;
2. as a complement to the cancer mortality study, the frequencies of unstable type and stable type chromosome aberrations were studied;

---

* Corresponding author. Tel.: +86-106-238-9930; fax: +86-106-238-8008.

0531-5131/02 © 2002 Elsevier Science B.V. All rights reserved.
PII: S0531-5131(02)00308-4

3. to have a long-term observation on cancer mortality of large size population to raise the statistical power;
4. according to the estimated individual doses (annual doses) based on the environmental measurement and personal dose monitoring a classification of dose-groups was made to facilitate the analysis of the association between low-dose exposure and cancer mortality; [4,5]
5. to study the risk factors of cancer induction other than ionizing radiation to facilitate the analysis of confounding factors.

2. Main results up to the end of year 2000

2.1. Radiation measurement and dose assessment

Based on the data of environmental measurements of gamma exposure and the occupancy factors, annual effective doses for cohort members of HBRA and control area (CA) were estimated, thus a classification of dose groups was made [4,5]. Table 1 shows the classification.

In the last 4 years, the Task Group on Radiation Measurements and Dose Assessment has conducted the renewal of dose estimation from internal irradiation, especially that of Rn-222, Rn-220 and their decay products. Because the previous measurements were made some 15 to 20 years ago, now the methodology of measurements, especially the measurements of cumulative exposure to radon, thoron and their decay products, has been much improved. Table 2 shows the results from integrating measurements of concentrations of Rn-222 and Rn-220 (indoor air). Measurements were proceeded in two periods: (1) from September 1997 through December 1998; (2) from September 1999 through November 2000.

The concentrations of Rn-222 and Rn-220 in outdoor airs are less than that in indoor air. In HBRA, the average value is 17.30 Bq m^{-3}, in CA is 11.7 Bq m^{-3}.

It is interesting to note that the indoor concentrations of radon in HBRA were not as high as expected (Table 2). This is likely due to the geographical location of the investigated areas, which is between the Equator and the Tropic of Cancer. It has been stated in the UNSCEAR Report [6] that "Concentrations of ^{222}Rn and its progeny are

Table 1
Classification of dose groups for the cohort members of investigation (only for external irradiation)

Dose group	Annual effective dose (10^{-5} Sv a^{-1}) (for external dose only)		No. of hamlets	No. of people
	Range	Mean		
High	224.10–308.04	246.07	124	23,718
Medial	198.07–224.09	210.19	135	28,803
Low	125.29–198.06	183.31	125	26,093
Control	50.43–95.67	67.92	142	27,903

Table 2
Concentrations of Rn-222 and Rn-220 of indoor air in houses of different dose-groups[a]

Range of external radiation dose (10^{-5} Sv a^{-1})	No. of hamlets selected		No. of homes measured		Concentration[b] (Bq m^{-3})	
	Rn-222	Rn-220	Rn-222	Rn-220	Rn-222	Rn-220
High group (224.10–308.04)	55	19	63	42	57.95±25.76	140.09±80.86
Medial group (198.07–224.09)	53	15	65	38	48.23±23.67	79.15±40.24
Low group (125.29–198.06)	46	15	55	38	39.41±14.78	62.92±41.58
Control group (50.43–95.67)	41	2	52	24	18.14±5.57	12.4±3.3

[a] Four dose-groups are classified based on the measurements of environmental external radiation in 526 hamlets of the investigated areas.

[b] The Rn–Tn cup monitors were fixed in the rooms; the weather of the area is warm the whole year such that the windows are always open.

usually higher in indoor air than in outdoor air; exceptions are in tropical areas, where ^{222}Rn concentrations in well-ventilated dwellings are essentially the same as in outdoor air." In our investigated areas, air concentrations of radon in the control area were similar between indoors and outdoors. This is consistent with the notion of the UNSCEAR Report. However, the indoor concentrations were still significantly higher than that of the outdoors in HBRA [7]. Based on the data obtained from the integrating and discrete measurements of ^{222}Rn, ^{220}Rn and their decay products, the Task Group estimated the effective doses of ^{222}Rn, ^{220}Rn and their decay products to the cohort members of HBRA and CA (Table 3).

With these estimations, we are able to do the preliminary estimations of annual effective doses resulting from natural radiation sources in different dose-groups of HBRA and CA (Table 4). There is little difference in the values between the new approach and the existing estimation. The cause of the difference is the change of estimation models. The ^{222}Rn, ^{220}Rn and their decay products are the main contributors to the effective dose of

Table 3
Estimates of the effective doses of Rn-222, Rn-220 and their decay products to the cohort members (HBRA, CA)[a]

Estimate on	Annual effective dose (Sv a^{-1})					
	HBRA			CA		
	Indoor	Outdoor	Subtotal	Indoor	Outdoor	Subtotal
Rn-222	51.72	7.73	59.45	18.87	5.23	24.10
EEC, Rn-222	1259.42	216.89	1476.31	619.21	177.15	796.36
Rn-220	64.19	2.69	66.88	8.36	2.34	10.70
EEC, Rn-220	1596.77	83.05	1679.82	161.88	34.69	196.57
Total	2972.10	310.36	3282.46	808.32	219.41	1027.73

[a] Based on the model UNSCEAR-2000.

Table 4
Preliminary estimations of annual effective doses resulting from natural radiation sources in different dose-groups of HBRA and CA

Component of exposure	Annual effective dose in mSv a^{-1}				CA
	HBRA				
	Low	Medial	High	Average	Average
External exposure	1.84	2.10	2.46	2.12	0.69
Internal exposure[a]	3.22	3.72	4.40	3.76	1.30
Total	5.06	5.82	6.86	5.88	1.99

[a] Internal exposure includes exposures from ingestion and inhalation of natural radionuclides. Estimation of the effective doses of Rn-222, Rn-220 and their decay products was based on the UNSCEAR-2000.

internal exposure, other components of internal exposure are trace natural radionuclides in food, drinking water, etc. Because the sample sizes for Rn-222 and Rn-220 measurement were not large, the estimates are, thus, still preliminary.

Table 5

(A) Results (1979–1998) of cancer mortalities in HBRA and CA

	HBRA	Control	Total
No. of subjects	89,694	35,385	125,079
Person-years	1,464,929	528,010	1,992,939
No. of deaths (all causes)	8905	3539	12,444
No. of cancer deaths	855	347	1202
Death rate (1/1000)	6.08	6.70	6.24
Cancer rate (1/100,000)	58.36	65.71	60.31

(B) Adjusted RRs (95% CI) with sex and age group for site-specific cancer in HBRA and CA (1979–1998)

Site of cancer	CA		HBRA	
	No. of deaths	RR	No. of deaths	RR (95% CI)
All cancers	347	1	855	1.00 (0.89–1.14)
Leukemia	13	1	36	1.03 (0.56–2.02)
All cancers except leukemia	334	1	819	1.00 (0.88–1.14)
Nasopharynx	66	1	153	0.94 (0.71–1.26)
Esophagus	5	1	31	2.61 (1.11–7.66)
Stomach	37	1	81	0.90 (0.61–1.34)
Colon	7	1	12	0.70 (0.28–1.89)
Rectum	4	1	13	1.40 (0.49–4.97)
Liver	100	1	218	0.89 (0.70–1.13)
Pancreas	4	1	17	1.69 (0.62–5.87)
Lungs	38	1	81	0.87 (0.60–1.30)
Bone	5	1	12	0.99 (0.36–3.11)
Skin	7	1	28	1.74 (0.80–4.33)
Female breast	8	1	12	0.65 (0.27–1.66)
Cervix uterus	1	1	9	4.01 (0.75–74.02)
CNS[a]	7	1	24	1.32 (0.60–3.33)
Thyroid	2	1	5	1.09 (0.23–7.60)
Lymphoma	6	1	23	1.48 (0.64–4.01)

[a] CNS: Brain and central nervous system.

Table 6
Relative risk (RR) of solid cancer mortality by dose category (1979–98)

Dose (mSv)	No. of cases	RR	95% CI
0–99	173	1 (referent)	–
100–199	307	0.79	0.60 to 1.04
200–299	260	0.95	0.75 to 1.21
300–399	311	0.78	0.57 to 1.06
≥400	102	0.66	0.45 to 0.98

2.2. Cancer mortality study data from 1979 through 1995 were published [8,9]

During the Phase III, the follow-up of cohort study continued, data from 1979 through 1998 were collected, statistical analysis of these data has been conducted (Table 5). The data (1979 through 1998) of nearly 2 million person-years' observation show that the mortality rate of all cancers in HBRA was lower than that in CA, but not statistically significant. Table 5B shows that the differences of site-specific cancer mortality rates between HBRA and CA were not statistically significant except for esophagus cancer. However, if we compare the relative risks (RRs) for the three dose-groups in HBRA (High, Medial and Low groups), there is no trend of increase of RR with increase of radiation dose.

For detecting the association between the cancer mortality and low dose ionizing radiation, an analysis of RRs in various dose-ranges was conducted which also shows that the RRs did not increase with doses (Table 6).

It is interesting to note the RRs estimated for cohort members aged 40–69 years, i.e., in the age of high prevalence. The analysis shows that in the cohort members in HBRA even in the age of higher incidence and even with more accumulated dose, no increase of cancer mortality was found (Table 7). Table 8 shows the excess relative risk (ERR) per sievert (Sv) for major cancer sites, for all ages and both sexes (1979 through 1998). The values of ERR per sievert are negative, but the 95% confidence intervals of any single cancer site are wide.

For the purpose of detecting the improvement of risk estimation for cancer mortality, we calculated the statistical power of detecting the difference in cancer mortalities between the HBRA and CA. The calculation shows that if the RR is 1.15, the statistical power is 74.7%, if the RR is 1.20, the power will be 91.9%. This estimate is for all the solid cancers. For the site-specific cancers, including leukemia, the power was not so high.

Table 7
Relative risk (RR) of cancer mortality, estimated for cohort members of 40–69 years old (1979–1998)

Area and subcohort	No. of cases	RR and 95% CI (bounds)	Accumulated dose mSv[a]
Control	222	1.00	108.0±20.1
HBRA	550	1.01 (0.86–1.18)	298±55.8
Low	210	1.14 (0.94–1.38)	
Medial	186	0.97 (0.79–1.17)	
High	154	0.91 (0.74–1.12)	

[a] External irradiation only.

Table 8
ERR per dose in sievert for major cancer sites, all ages, and both sexes (1979–1998)

Site of cancer	No. of cases	ERR[a] per Sv	95% CI
All solid cancer	1153	−0.14	−0.65 to 0.57
Liver	318	−0.80	−1.44 to 0.30
Nasopharynx	219	−0.63	−1.50 to 1.13
Stomach	118	−0.41	−1.39 to 2.00
Lung	119	−0.40	−1.38 to 2.04

Model: assumed linear relationship and stratified by period, sex and age group.
[a] Excess relative risk.

With the increase of statistical power, the confidence intervals of RR became narrower.

For detecting the confounding factors in the HBRA and in the CA, since 1997, we have conducted case-control studies on nasopharyngeal cancer, lung cancer and leukemia in the HBRA and CA [10]. Upon these investigations, we have found that there were several risk factors associated with the above-mentioned cancers.

However, we have not found any association between the long-term higher background radiation exposure and the above-mentioned cancers. Our cross-sectional survey did not demonstrate any significant difference in risk factors between the HBRA and CA.

2.3. Cytogenetic study

Since the joint research started, there have been two steps taken to study the frequencies of chromosomal aberrations. Firstly, the Task Group on Cytogenetic Study worked at the analysis on unstable chromosomal aberrations [11]; secondly, they began the analysis on stable chromosomal aberrations in 1997 [12]. The subjects (blood samples' donors) of unstable aberration study were chosen from three generations in each family. Twenty-two

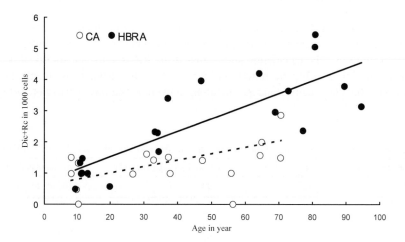

Fig. 1. Age responsible in the yield of Dic+Rc in HBRA and CA.

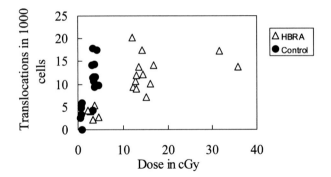

Fig. 2. Frequencies of translocation vs. dose in the lymphocytes of residents in HBRA and in the control areas.

subjects were from eight households living in HBRA, and 17 subjects were from five households in the CA. All subjects were healthy inhabitants in HBRA and in CA. They had no history of significant exposure to X-rays. The environmental external radiation levels were measured for every home of the subjects, and cumulative doses were estimated. For details of the method and results of analysis, please see Ref. [11]. A total of 55,595 cells for the HBRA subjects and 45,799 cells for the CA subjects was analyzed. The analyses showed that the frequency of unstable aberrations (dicentrics and rings) of peripheral lymphocytes was age-dependent in both HBRA and CA inhabitants; however, an obvious trend of increase with age appeared in the HBRA group, but not in the CA group (Fig. 1). The cumulative dose increased with the age of an individual. There was a difference of more than three times in the slopes of the age–response relationship. Therefore, it was considered that the increase of Dic+Rc was mainly attributable to the increase in the cumulative dose. The frequencies also increased in proportion to the cumulative doses. However, the study on stable chromosome aberrations (+translocations) obtained a different result (Figs. 2 and 3) [12]. Five children (10.8–13.5 years) and 13 aged males (53.2–89.5 years) in the HBRA, and six children (10.3–13.8 years) and 11 aged males (55.3–70.5 years) in the CA were the subjects of this study. The accumulated

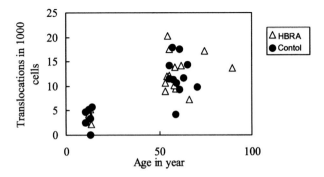

Fig. 3. Frequencies of translocation vs. age in the lymphocytes in HBRA and in the control areas.

doses (air kerma) in aged men ranged from 120.0 to 358.9 mGy in HBRA and from 31.5 to 45.8 mGy in the CA, respectively. The accumulated doses in children ranged from 22.6 to 45.3 mGy in HBRA and from 5.8 to 9.7 mGy in the CA. 37,722 cells (HBRA) and 35,378 cells (CA) of aged people, and 32,296 cells (HBRA)and 27,437 cells (CA) of children were analyzed. Translocations were determined using whole-chromosome painting probes for chromosomes 1, 2 and 4 (representing 22.44% of the human genome). The frequencies per 1000 cells in children were 2.7–5.3 in HBRA and 0–5.8 in CA, the corresponding frequencies per 1000 cells in aged people were 7.2–20.3 in HBRA and 4.3–17.9 in CA. Children had less number of translocations and less variation compared with the aged men. The frequencies of translocations among elder individuals are widely varied. The Task group could not see the effect of the high background radiation on the frequency of translocation.

3. Conclusions and perspectives

(1) The joint research once again verified that no increase of cancer risk is associated with the high levels of natural radiation in HBRA of Yangjiang when compared with the nearby control area, where the radiation level is similar to the world average radiation background. The ratio of environmental radiation levels in HBRA and CA is about three (average). This result is further supported by a relative risk (RR) analysis for a subgroup of HBRA subjects aged 40 to 69 years, who received larger cumulative dose (298.1±55.8 mSv). The RR of solid cancer is 1.01 (95% CI: 0.86–1.18).

(2) The results of our cancer mortality study may provide a possible upper bound of ERR per sievert for solid cancers in worst-case analysis.

(3) The statistical power (precision) is effective to identify the relative risk of all cancer for HBRA inhabitants (if the RR is 1.2), but less effective for most single cancers.

(4) There are paradoxes observed in HBRA studies: the increase in unstable chromosome aberrations (dicentrics and rings) in peripheral lymphocytes was observed in the HBRA, but no increase in stable (translocation) aberrations. The increase in unstable aberrations was almost parallel to dose rate, i.e., about three times, but no corresponding increase was observed in cancer mortality. This means that in spite of significant observation of DNA double-strand break (dsb), no increase in cancer mortality was observed, contrary to the expectation based on the LNT (linear, no threshold) model. A new hypothesis to explain this paradox is necessary.

(5) The advantages of fixed cohort study are prominent in radiation epidemiology; however, there is a weakness in the study of children's diseases, because the newborn baby's name does not enter the cohort, and the original children become older year by year. It is necessary to collect the data of children specifically.

(6) Our case-control studies were not complete because the cases were not many. We will continue the case-control study on lung cancer.

(7) It seems important to do some health survey for assessing the impact of radiation-induced mutations on the frequencies of multifactorial diseases in the population of HBRA. Because 90% of the families of HBRA have lived there for more than six generations.

References

[1] High-Background Radiation Research Group, China, Health survey in high background radiation areas in China, Science 209 (1980) 877–880.
[2] L. Wei, Y. Zha, Z. Tao, W. He, D. Chen, R. Yuan, Cancer mortality in high-background radiation areas of Yangjiang, China, Radiat. Biol. Res. Commun. 23 (1988) 209–220.
[3] L. Wei (Ed.), High Background Radiation Research in Yangjiang China, Atomic Energy Press, Beijing, China, 1996.
[4] Y.L. Yuan, H. Morishima, H. Shen, T. Koga, Q. Sun, K. Tatsumi, Y. Zha, S. Nakai, L. Wei, T. Sugahara, Recent advances in dosimetry investigation in the high background radiation area in Yangjiang, China, in: L. Wei, T. Sugahara, Z. Tao (Eds.), High Levels of Natural Radiation — Radiation Dose and Health Effects, 1997, pp. 223–233.
[5] H. Morishima, T. Koga, K. Tatsumi, S. Nakai, T. Sagahara, Y. Yuan, Q. Sun, L. Wei, Study of the indirect method of personal dose assessment for the inhabitants in HBRA of China, in: L. Wei, T. Sugahara, Z. Tao (Eds.), High Levels of Natural Radiation — Radiation Dose and Health Effects, 1997, pp. 235–240.
[6] United Nations Scientific Committee on the Effects of Atomic Radiation (UNSCEAR) 1993 REPORT to the General Assembly, with scientific Annexes: sources and effects of ionizing radiation, 1993, pp. 34–61, 74 UN, New York.
[7] Y.-L. Yuan, T. Morishima, T. Shen, T. Koga, L.-X. Wei, T. Sugahara, Measurements of Rn-222 and their decay products in the environmental air of the high background radiation areas in Yangjiang, China, J. Radiat. Res. 41 (2000) 25–30, Supplement.
[8] Z.-F. Tao, Y.R. Zha, S. Akiba, Q.-F. Sun, J.M. Zou, J. Li, Y.-S. Liu, H. Kato, T. Sugahara, L.-X. Wei, Cancer mortality in the high background radiation areas of Yangjiang, China during the period between 1979 and 1995, J. Radiat. Res. 41 (2000) 31–41, Supplement.
[9] Q.-F. Sun, S. Akiba, Z.F. Tao, Y.-L. Yuan, J.-M. Zou, H. Morishima, H. Kato, Y.-R. Zha, T. Sugahara, L.-X. Wei, Excess relative risk of solid cancer mortality after prolonged exposure to naturally occurring high background radiation in Yangjiang, China, J. Radiat. Res. 41 (2000) 43–52, Supplement.
[10] J.-M. Zou, Q.-F. Sun, S. Akiba, Y.-L. Yuan, Y.-R. Zha, Z.-F. Tao, L.-X. Wei, T. Sugahara, A case-control study of nasopharyngeal carcinoma in the high background radiation areas of Yangjiang, China, J. Radiat. Res. 41 (2000) 53–62, Supplement.
[11] T. Jiang, I. Hayata, C.-Y. Wang, S. Nakai, S.-Y. Yao, Y.-L. Yuan, L.-L. Dai, Q.-J. Liu, D.-Q. Chen, L.-X. Wei, T. Sugahara, Dose–effect relationship of dicentric and ring chromosomes in lymphocytes of individuals living in the high background radiation areas in China, J. Radiat. Res. 41 (2000) 63–68, Supplement.
[12] I. Hayata, C.-Y. Wang, W. Zhang, D.-Q. Chen, M. Minamihisamatsu, H. Morishima, Y.-L. Yuan, L.-X. Wei, T. Sugahara, Chromosome translocation in residents of the high background radiation areas in southern China, J. Radiat. Res. 41 (2000) 69–74, Supplement.

Induction of myeloid leukemias in C3H/He mice with low-dose rate irradiation

Takeshi Furuse*, Yuko Noda, Hiroshi Otsu

National Institute of Radiological Sciences, 4-9-1 Anagawa, Inage-ku, Chiba 263-8555, Japan

Abstract

C3H/He male mice were exposed to whole body gamma-ray irradiation at 8 weeks of age. Radiation at a high-dose rate of 88.2 cGyears/min with doses of 0.125–5.0 Gyears, a medium dose rate of 9.56 cGyears/min with doses of 1.0–5.0 Gyears, and low-dose rates of 0.0298 cGyear/min with doses of 1.0–10 Gyears, 0.0067 cGyear/min with doses of 1.0–10.0 Gyears or 0.0016 cGyear/min with doses of 1.0–4.0 Gyears were delivered from ^{137}Cs sources. The mice in the low-dose rate groups were irradiated continuously for 22 h daily during a period of 3 to 200 days. All the mice were maintained for their entire life span and were pathologically examined after their death. Myeloid leukemia developed significantly more frequently in the irradiated groups with doses over 1 Gyear than in the unirradiated groups. The maximum values were 23.5% in the high-dose rate group, 11% in the medium dose rate group, 7.3%, 7.2%, 6.3% in the low-dose rate groups, and 0.99% in the unirradiated control group. These dose–effect curves had their highest values on each curve at about 3 Gyears. We obtained the dose and dose rate effectiveness factor (DDREF) values of 2.96 by linear fittings for their dose–response curves of the dose ranges in which leukemia incidences were increasing. © 2002 Elsevier Science B.V. All rights reserved.

Keywords: Radiation carcinogenesis; Myeloid leukemia; Low-dose and low-dose rate exposure; Dose and dose rate effectiveness factor (DDREF)

1. Background

Quantitative information on the risk of cancer in human populations exposed to ionizing radiation comes largely from information available from populations that have been exposed to intermediate and high doses and dose rates of radiation. For protection against radiation, however, the assessment of the risk from environmental and occupational ex-

---

* Corresponding author.

posure to radiation must be known for exposures of low doses delivered at low-dose rates. It was concluded that the human data were not adequate to demonstrate conclusively that a dose rate effect either does or does not exist. In view of the complexity and wide spectrum of tumorigenic responses to radiation found in experimental animals and the lack of information on the detailed mechanisms of such responses in animals or man, more specific reduction factors for either individual tumor types or total tumor incidence have not been given [1–4].

2. Methods

2.1. Animals

Male C3H/He mice 8 weeks old [5], bred at our institute, were used. They were maintained under SPF conditions for high-dose rate irradiation and conventional conditions after low-dose rate irradiation in our special irradiating facilities. All the mice were observed during their lives and processed for routine histopathological examinations.

2.2. Irradiation

88.2 cGyears/min was delivered from a Cs-137 source of 115TBq equipped in Gamma-Cell 40. Second, radiation at a medium dose rate of 9.58 cGyears/min was delivered from a Cs-137 source. Third, the low-dose rate radiation at three dose rate levels, L-A, L-B and L-C, were delivered from a Cs-137 source of 0.375 TBq, which is installed at the center of our radiation room. This irradiation was performed for 22 h every day by pulling the radiation source up out of its shielding case or pushing it down into the case automatically. Mice in the cages on the racks close to the source were irradiated at a dose rate of 0.0298 cGyear/min. Those on the racks further away from the source were irradiated at a dose rate of 0.0016 cGyear/min, and those on the racks between them were irradiated at a dose rate of 0.0067 cGyear/min. These dosimetries were performed as described in a previous paper [6].

Myeloid leukemias were diagnosed pathologically using the criteria that Seki et al. [5] adopted.

3. Results

In this experiment, myeloid leukemia was the only disease that could be analyzed for a relationship between the radiation dose and the incidence of the disease. These incidences were expressed as crude incidences and age-adjusted incidences [7], which were adjusted for the differences in the distribution of ages at death among the various treatment groups. We used the latter to analyze the dose dependency of radiation leukemogenesis and to estimate the dose and dose rate effectiveness factor (DDREF) [4].

Myeloid leukemias occurred at an incidence of 0.99% in the control group. In the high-dose rate group (group H), the incidence gradually increased depending on the radiation dose, reached a maximum at an incidence of 23.57% with a dose of 3 Gyears, and then decreased, although the doses continued to increase. In this group, a significantly higher incidence was observed at a dose of 0.25 Gyear than at the values at the neighboring doses. In the medium dose rate (1/100 of the high-dose rate) group, the maximum incidence was 11.05% with a dose of 3 Gyears and the incidence of myeloid leukemia was obviously lower than that of group H. All the incidences among the low-dose rate (1/300–1/5000 of the medium dose rate) groups at doses of 1–4 Gyears showed a slight, but significant, increase. In order to apply the calculation method of DREF [2] to our data, we described linear regression lines for these dose–response curves of radiation induced leukemias, and the following formulas were obtained:

group H (882 mGyears/min): $Y=(-2.28\pm1.76)+(8.21\pm1.2)X$;
group L-A (0.298 mGyear/min): $Y=(0.54\pm0.83)+(1.16\pm0.45)X$;
group L-C (0.016 mGyear/min): $Y=(-0.96\pm1.72)+(1.80\pm1.17)X$.

These formulas can be described generally as:

$$Y = (\alpha_1 \pm \sigma_1) + (\alpha_2 \pm \alpha_2)X$$

where α_1 and α_2 are coefficients with 95% confidence limits of σ_1 and σ_2, respectively. We calculated 95% confidence limits for the DDREF values as ratios of two linear regression coefficients $\alpha_2(H)$ for the high-dose rate group and $\alpha_2(L)$ for a low-dose rate group as follows:

$$\text{DDREF} = \alpha_2(H)/\alpha_2(L)$$

DDREFs and 95% confidence limits were estimated from the above data as 7.03 ± 2.95 (4.08–9.98) for the LA group, and 4.55 ± 4.65 (0–9.20) for the LC group.

4. Conclusion

Animal experiments provide the best indication of the extent to which lowering the dose rate of low LET radiation, even at intermediate or high total doses, can reduce the effectiveness of radiation in inducing cancer or other cellular damage. Furthermore, the experiments provide guidance on the extrapolation of the risks at high doses and high-dose rates compared to the low doses and low-dose rates generally of concern in radiation protection. The ratio of the slope α_{l1} of the non-threshold linear fit to the high-dose and dose rate data to the slope α_{Exp} of the linear experimentally determined the curve for the low-dose rate data, and has been termed a dose rate effectiveness factor (DREF). From our data for the high-dose rate radiation, the DDREFs were estimated as 2.96 ± 3.21 (0–6.17) for $D=1$ Gyear, and 4.92 ± 6.42 (0–11.34) for $D=2$ Gyears.

References

[1] ICRP, Recommendations of the ICRP, B. 4. Stochastic effects: carcinogenesis, ICRP Publication 60, Ann. ICRP, vol. 21 (1–3), New York, 1991, pp. 106–114.

[2] NCRP, 9, Tumorigenesis in experimental laboratory animals, Influence of Dose and Its Distribution in Time on Dose–Response Relationship for Low-LET Radiations. NCRP Report No. 64, National Council on Radiation Protection and Measurements, Washington D.C. (1980) pp. 108–131.
[3] NAS, Health Effects of Exposure to Low Levels of Ionizing Radiation, BEIR Report. National Academy of Science, National Academy Press, Washington, DC, 1990, p. 23.
[4] UNSCEAR, Annexes, term and units of radiation exposure, Sources, Effects and Risks of Ionizing Radiation, UNSCEAR 1988 Report to the General Assembly, United Nations, New York, 1988, pp. 407–408.
[5] M. Seki, K. Yoshida, M. Nishimura, et al., Radiation-induced myeloid leukemia in C3H/He mice and the effect of prednisolone acetate on leukemogenesis, Radiat. Res. 127 (1991) 146–149.
[6] A. Shiragai, F. Sato, N. Kawashima, et al., Absorbed dose estimates in a prolonged cesium 137 gamma irradiation facility for mice, J. Radiat. Res. 21 (1980) 118–125.
[7] R.L. Ullrich, J.B. Storer, Influence of γirradiation on the development of neoplastic disease in mice: I. Reticular tissue tumours, Radiat. Res. 80 (1979) 303–316.

Dose and dose rate effect in mutagenesis, teratogenesis and carcinogenesis

Taisei Nomura*

Department of Radiation Biology and Medical Genetics, Graduate School of Medicine, Osaka University, B4, 2-2 Yamada-oka, Suita, Osaka 565-0871, Japan

Abstract

In utero exposure to X-rays, ^{60}Co γ-rays, ^{252}Cf neutron, and ^{3}H water resulted in the linear increase of in vivo somatic mutation in PTHTF$_1$ mice, while tumors were induced in the offspring after the postnatal treatment with tumor promoter. Apparent dose rate effects were observed in X-ray-induced mutation, malformation and tumor. In the young adult mice exposed to 0.4–6.8 Gy of γ-rays at the dose rates from 0.04 to 1189 mGy/min at the same age, a large and significant reduction of leukemia was observed by the low dose rate exposure, while dose rate effects were not detected in solid tumors at high doses. However, a significant reduction in the incidences of solid tumors was observed at low dose (0.4 Gy) and low dose rate (0.04 mGy/min) exposure. Double-strand break repair deficient SCID (severe combined immunodeficient) mice were extremely sensitive to radiation, and no dose rate effects were observed in γ-ray-induced mortality, malformation and leukemia. The double-strand break repair plays an essential role in homeostasis of the organism. In the normal human tissues maintained in the improved SCID mice, daily UV-light B exposure-induced mutation, keratosis and skin cancer. However, cancer has not been induced in the human thyroid gland below 60 Gy of X-rays, although *p53* and *c-kit* mutations were detected at lower doses.
© 2002 Elsevier Science B.V. All rights reserved.

Keywords: Low dose; Dose rate effect; SCID; Mouse; Human organ/tissue

1. Introduction

Dose and dose rate effectiveness is one of the most important subject to estimate the risk of radiation exposure to human beings, because humans have been exposed continuously to low dose and low dose rate radiations. However, a majority of human data come from human

---

* Tel.: +81-6-6879-3811; fax: +81-6-6879-3819.
E-mail address: tnomura@radbio.med.osaka-u.ac.jp (T. Nomura).

populations accidentally exposed to high doses of radiations at high dose rates such as atomic bomb, medical treatments, etc., and there has been no sufficient information about low dose rate exposure. In this article, analyses were made for dose and dose rate effects in radiation-induced cancer, malformation, mutation and mortality by using specific in vivo mouse and human systems to detect these defects [1–6].

2. Mutagenesis, tetatogenesis, and carcinogenesis

2.1. Dose response

Two mouse strains PT and HT, which have eight different coat color recessive genes, were used for the detection of cancer, malformation, mutation, and cell killing in the same individual. Details of experimental procedures were given previously [1,2].

When 0.1–1.5 Gy of X-rays were given at high dose rate (540 mGy/min) on the 10.5th day of gestation, a linear increase of mutant spots was observed up to the dose of 1.2 Gy. Congenital malformations and midventral white spots, indicating the killing of pigment cells, did not show any increases at low doses, but a large increase was observed at high doses [1,2]. ^{60}Co γ-rays given at dose rate of 570 mGy/min also induced mutation, and there were no differences in mutation frequency between X-rays and ^{60}Co γ-rays [7]. However, ^{252}Cf and ^{3}H water induced very high frequency of mutation, and RBEs of ^{252}Cf and ^{3}H water were 6.6 and 2.5, respectively, in comparison with reference dose of X-rays [7,8]. Consequently, RBEs of neutron and β-rays for somatic mutation were similar to those of spermatogonial mutation.

Though in utero X-ray, ^{60}Co γ-ray and ^{252}Cf exposure alone did not show any increase of tumors, the postnatal treatment (6 weeks after birth) with a tumor promoter, 12-O-

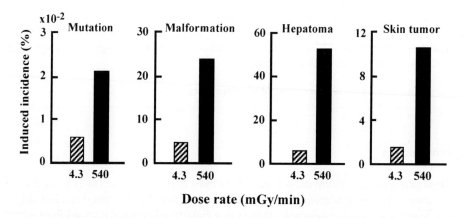

Fig. 1. Dose rate effectiveness in X-ray induced somatic mutation, malformation and tumors in PTHTF$_1$ mice. X-ray dose is 1.03 Gy at dose rate of 4.3 and 540 mGy/min, and 218 and 192 mice were used for high and low dose rate exposure, respectively [4,7]. Hepatomas and skin tumors were induced by in utero X-ray exposure and postnatal treatment with TPA [9].

tetradecanoylphorbol-13-acetate (TPA) increased incidences of hepatomas and skin tumors in proportion with in utero doses of radiations [9]. RBEs of ^{252}Cf were 6.1 for hepatoma and 9.1 for skin tumor. The values were similar to those of somatic mutation [7]. Post-irradiation process seems to be more important in radiation carcinogenesis in humans who have a high probability of exposure to various carcinogenic/promoting agents in their diet and environment in contrast to mice reared under specified condition. Initial events by in utero irradiation must be memorized in the cell and cause tumors by the post-treatment with promoting agent [9,10].

2.2. Dose rate effects

As shown in Fig. 1, one third of somatic mutations were induced at low dose rate of X-rays (4.3 mGy/min) against high dose rate (540 mGy/min) exposure [9], showing similar level of dose rate effect as to the germ line mutation. Furthermore, significant reduction of congenital malformation (1/5) was observed by low dose rate exposure (4.3 mGy/min). When in utero X-ray exposure was given at low dose rate (4.3 mGy/min), incidences of hepatomas and skin tumors decreased to 1/6 to 1/8 of those by the high dose rate exposure [9]. Consequently, there are apparent dose rate effects in mutagenesis, teratogenesis and carcinogenesis.

3. Dose rate effects in leukemia and solid tumors

Most of previous works on dose rate effects were carried out by the comparison between single and protracted exposure. Animals were irradiated at different ages and some animals died or were euthanized shortly after the last irradiation, indicating large differences in the effective doses of radiation between single and protracted exposure, though total doses were equal. To examine the real dose rate effect, young adult (6 weeks old) C3H/HeJ and C57BL/6J mice were exposed to 1.7, 0.5 or 0.1 Gy of ^{60}Co or ^{137}Cs γ-rays 4 times at 7 days' interval from 6 to 10 weeks of age at dose rates of 1189, 570, 50, 10, 3, 1, 0.2, 0.04 mGy/min. To avoid age difference in radiation sensitivity, mouse age at risk and all experimental conditions except dose rate are same in all experimental groups.

In the high dose experiments (6.8 and 2.0 Gy), high incidences of leukemia were induced in both strains and incidences decreased with decreasing dose rates, reducing to 1/20 and 1/2, respectively at 0.2 mGy/min, although there were no substantial differences in total incidences of solid tumors between high and low dose rates. Induced rate of solid tumors by the high dose rate exposure was about $11-12 \times 10^{-2}$ per Sv, and our mouse experiments suggest DDREF value of 2 for leukemia or 1.2 for solid tumors, values being similar to those in human [11]. However, the spectra of induced tumors were different among species, human populations and mouse strains. Very high incidence of tumors in tumor susceptible organs (e.g., ovary in C3H mice) at high doses is critical and often shadows or conceals dose rate effects of other organs. Consequently, lower dose experiments are necessary for the definite analysis of dose rate effects. In fact, significant reduction (1/2–1/4) in the incidences of solid tumors was observed in C3H/HeJ mice at low dose (0.4 Gy) and low dose rate (0.04 mGy/min) exposure, although no reduction

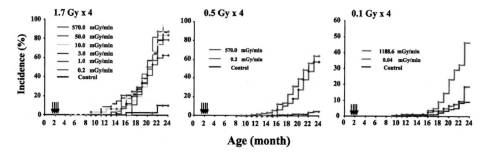

Fig. 2. Cumulative incidence of ovarian tumors in C3H/HeJ mice (1106 females) exposed to 6.8 and 2.0 Gy of ^{60}Co γ-rays and 0.4 Gy of ^{137}Cs γ-rays at various dose rates.

was observed in high dose experiments (6.8 and 2 Gy). Fig. 2 shows a typical example. Thus, significant dose rate effects are detectable only at low doses.

4. Dose rate effects and DNA repair deficiency

In C.B17 wild type mice, a large and significant reduction of congenital malformation (1/30) was observed by low dose rate (0.37 mGy/min) exposure against high dose rate (1150 mGy/min) exposure (Fig. 3), as in the case of previous experiments (Fig. 1). In C.B17-*scid* mice, however, the same dose of ^{137}Cs γ-rays at low dose rate on the same gestational period induced extremely high incidences of fetal deaths and congenital malformations, and no dose rate effects were observed. This indicates that potent dose rate effect in teratogenesis is caused by the repair of γ-ray induced double-strand breaks in most parts, although there were some dose rate effects in fetal deaths [7]. It was the case in

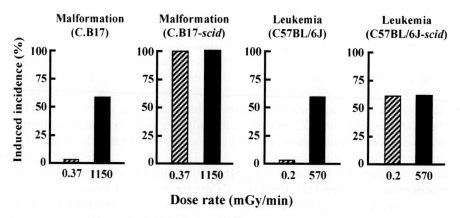

Fig. 3. Dose rate effectiveness in γ-ray induced malformation and leukemia in double-strand break repair deficient mice. Pregnant C.B17 and C.B17-*scid* mice (361 C.B17 and 63 C.B17-*scid* embryos) were exposed to 1.2 Gy of ^{137}Cs γ-rays at dose rates of 0.37 and 1150 mGy/min on the 8.5–10.5th day of gestation. Six weeks old C57BL/6J (296) and C57BL/6J-*scid* (111) mice were exposed to 6.8 and 2.0 Gy of ^{60}Co γ-rays, respectively, at high (1189 mGy/min) and low (0.2 mGy/min) dose rates.

radiation-induced leukemia in young adult C57BL/6J-*scid* and C57BL/6J mice. In C57BL/6J wild type mice, a large reduction of leukemia deaths was observed by the low dose rate (0.2 mGy/min) exposure against high dose rate (570 mGy/min) exposure, when 1.7 Gy of ^{60}Co γ-rays were given 4 times to C57BL/6J mice. When 0.5 Gy of ^{60}Co γ-rays were given 4 times to C57BL/6J mice, no substantial increases of leukemia deaths were observed. However, C57BL/6J-*scid* mice died of leukemia before 8 months of age with the same dose, and no dose rate effects were observed (Fig. 3). Consequently, double-strand break repair deficient SCID mice were extremely sensitive to radiation for malformation and leukemia induction, and no recovery of malformation and leukemia was observed by the low dose rate exposure. Double-strand break repair plays an essential part in homeostasis of the organism, especially in radiation-induced mutagenesis, teratogenesis and carcinogenesis.

5. Carcinogenesis and Mutagenesis in Human and Tissues

Morphology and function of human organs and tissues are well maintained in the improved SCID mice for several years [5,6]. Normal human skin and thyroid tissue maintained in the improved SCID mice were exposed to daily doses of UV-light B and weekly doses of X-rays or ^{137}Cs γ-rays for periods of approximately 2 years. We have succeeded for the first time in inducing cancer and solar (actinic) keratosis in human skin by UVB. Of 18 normal skins exposed to doses of 73–180 J/cm^2, 14 actinic keratoses (77.8%) and three squamous cell carcinomas (16.7%) developed, whereas neither actinic keratosis nor cancer was observed in 15 human skins not exposed to UVB [5]. Lowest doses for keratosis and skin cancer induction were 30 and 130 J/cm^2, respectively. Among *p53* mutations at various sites, mutation at codon 242 was specifically observed in both skin cancers and actinic keratoses. Later, same mutation at codon 242 was reported in the skin cancer induced iatrogenically by PUVA (psoralen + UVA) therapy of psoriasis patients [12].

Consecutive X-irradiation of normal thyroid gland from the head and neck cancer patients (60 years old) has not induced thyroid cancer below the total doses of 60 Gy, but yielded two *p53* and four *c-kit* mutations in 13 transplanted tissues with total doses more than 24 Gy, while no mutations were detected in the tissues unexposed or exposed to X-ray doses less than 20 Gy. Human thyroid glands from young hyperthyroidism patients (20 years old) showed higher radiosensitivity to ^{137}Cs γ-rays, yielding two *p53* and three *c-kit* mutations in 11 transplanted tissues exposed to cumulative doses of 9–27 Gy (1189 mGy/min), but only one *p53* mutation in eight tissues exposed to the low dose rate (0.23 mGy/min) γ-rays. Cells with such mutations can survive and expand clonally in human thyroid tissues like coat color spots in mice.

Acknowledgements

I thank H. Nakajima, T. Hongyo, K. Sutoh, L.Y. Li, M. Syaifudin, and M. Maeda for their help to prepare the article. The works reviewed here were fully supported by grants from the Ministry of Education, Science and Culture, "Research for the Future" of JSPS, and the Nuclear Safety Research Association.

References

[1] T. Nomura, Comparative inhibiting effects of methyl-xanthines on urethan-induced tumors, malformations, and presumed somatic mutations in mice, Cancer Res. 43 (1983) 1342–1346.
[2] T. Nomura, Quantitative studies on mutagenesis, teratogenesis, and carcinogenesis in mice, in: Y. Tazima, S. Kuroda, Y. Kuroda (Eds.), Problem of Threshold in Chemical Mutagenesis, Environmental Mutagen Society of Japan, Shizuoka, 1984, pp. 27–34.
[3] K.A. Biedermann, J. Sun, A.J. Giaccia, L.M. Tosto, J.M. Brown, Scid mutation in mice confers hypersensitivity to ionizing radiation and a deficiency in DNA double-strand break repair, Proc. Natl. Acad. Sci. U. S. A. 88 (1991) 1394–1397.
[4] T. Tanaka, T. Yamagami, Y. Oka, T. Nomura, H. Sugiyama, The scid mutation in mice causes defects in the repair system for both double-strand DNA breaks and DNA cross-links, Mutat. Res. 288 (1993) 277–280.
[5] T. Nomura, H. Nakajima, T. Hongyo, E. Taniguchi, K. Fukuda, Y.L. Li, M. Kurooka, K. Sutoh, T.M. Hande, T. Kawaguchi, M. Ueda, H. Takatera, Induction of cancer, actinic keratosis and specific p53 mutations by ultraviolet light B in human skin maintained in SCID mice, Cancer Res. 57 (1997) 2081–2084.
[6] K. Fukuda, H. Nakajima, E. Taniguchi, K. Sutoh, H. Wang, P.M. Hande, L.Y. Li, M. Kurooka, K. Mori, T. Hongyo, T. Kubo, T. Nomura, Morphology and function of human benign tumors and normal thyroid tissues maintained in severe combined immunodeficient mice, Cancer Lett. 132 (1998) 153–158.
[7] T. Nomura, Dose rate effectiveness and repair in radiation-induced mutagenesis, teratogenesis, and carcinogenesis in mice, in: R.N. Sharan (Ed.), Trends in Radiation and Cancer Biology, Forschungzentrum Julich, Germany, 1998, pp. 149–155.
[8] T. Nomura, O. Yamamoto, In vivo somatic mutation in mice induced by tritiated water, in: S. Okada (Ed.), Proceedings of the Third Japan–US Workshop on Tritium Radiobiology and Health Physics, Institute of Plasma Physics Nagoya University, Nagoya, 1989, pp. 230–233.
[9] T. Nomura, H. Nakajima, T. Hatanaka, M. Kinuta, T. Hongyo, Embryonic mutation as a possible cause of in utero carcinogenesis in mice revealed by postnatal treatment with 12-O-tetradecanoylphobol-13-acetate, Cancer Res. 50 (1990) 2135–2138.
[10] T. Nomura, Induction of persistent hypersensitivity to lung tumorigenesis by in utero X-irradiation in mice, Environ. Mutagen. 6 (1984) 33–40.
[11] 1990 Recommendations of the International Commission of Radiological Protection, ICRP publication 60, Ann. ICRP 21 (1991) 1–3.
[12] A.J. Nataraj, P. Wolf, L. Cerroni, H.N. Ananthaswamy, p53 mutation in squamous cell carcinomas from psoriasis patients treated with psoralen+UVA (PUVA), J. Invest. Dermatol. 109 (2) (1997) 238–243.

Spontaneous tumorigenesis and mutagenesis in mice defective in the *MTH1* gene encoding 8-oxo-dGTPase

Teruhisa Tsuzuki [a,*], Akinori Egashira [a], Kazumi Yamauchi [a], Kaoru Yoshiyama [a,b], Hisaji Maki [b]

[a]*Department of Medical Biophysics and Radiation Biology, Faculty of Medical Sciences, Graduate School, Kyushu University, Fukuoka 812-8582, Japan*
[b]*Department of Molecular Biology, Graduate School of Biological Sciences, Nara Institute of Science and Technology, Ikoma 630-0101, Japan*

Abstract

Oxygen radicals, which can be produced through normal cellular metabolism as well as X-ray irradiation, are thought to play an important role in mutagenesis and tumorigenesis. Among various classes of oxidative DNA damage, 8-oxo-7,8-dihydroguanine (8-oxoG) is most important because of its abundance and mutagenicity. The *MTH1* gene encodes an enzyme that hydrolyzes 8-oxo-dGTP to monophosphate in the nucleotide pool, thereby preventing occurrence of mutations. When examined 18 months after birth, a greater number of tumors had formed in the lungs, livers and stomachs of *MTH1*-deficient mice, as compared with wild-type mice. Next, we analysed the mutation frequency and their spectra in each *MTH1*-deficient or proficient mice at the age of 4 and 24 weeks. The mutation frequency on the *rpsL* transgene in the spleen samples was determined. However, the spontaneous mutation frequency observed in the spleen samples from the $MTH1^{-/-}$ mice showed no apparent increase compared to the value of the one in the $MTH1^{+/+}$ mice. Furthermore, the site distribution of the mutations that occurred on the *rpsL* gene was slightly different between these two *MTH1* genotypes. © 2002 Elsevier Science B.V. All rights reserved.

Keywords: Oxidative DNA damage; 8-Oxoguanine; Nucleotide sanitization; Spontaneous mutation; Gene targeting

1. Introduction

There are many oxidative lesions in endogenous mammalian DNA. The spectrum of the lesions exceeds 100 different types of which 8-oxoG is one of the most abundant and appears

---

[*] Corresponding author. Tel.: +81-92-642-6141; fax: +81-92-642-6145.
E-mail address: tsuzuki@med.kyushu-u.ac.jp (T. Tsuzuki).

to play crucial roles in carcinogenesis and in aging [1,2]. 8-OxoG can pair with both cytosine and adenine during DNA synthesis and, as a result, G:C to T:A transversions are induced [3]. Oxidation of guanine also occurs in the cellular nucleotide pool. 8-Oxo-dGTP, when formed, is a potent mutagenic substrate for DNA synthesis; it is equally incorporated opposite adenine and cytosine in DNA [4], resulting in both A:T to C:G and G:C to T:A transversions. In *Escherichia coli*, mutations owing to misincorporation of 8-oxo-dGTP can be prevented by the *mutT* gene product, which hydrolyzes 8-oxo-dGTP to 8-oxo-dGMP [4]. Such enzyme activity similar to those of the *E. coli* MutT protein exists in mammalian cells. Based on partial amino acid sequences obtained from a purified preparation of human 8-oxo-d-GTPase, cDNA and the gene for the human enzyme were isolated and named *MTH1* (*mutT* homolog-1) [5,6]. It is interesting to note that the *E. coli* and mammalian enzymes have some different spectra in their substrate specificity. Human MTH1 hydrolyzes 2-hydrodeoxyadenosine triphosphate but acts only slightly on 8-oxoguanosine triphosphate, a ribonucleic counterpart of 8-oxo-dGTP. In contrast, the *E. coli* MutT exhibits opposite actions on these nucleotides [7]. To elucidate the roles of MTH1 with 8-oxo-dGTPase activity in vivo, we generated a mouse line carrying a mutant *MTH1* allele created by targeted gene disruption.

2. Susceptibility to tumorigenesis and mutagenesis

Oxidative damage is a major source of DNA damage in living organisms and has long been suggested to play a role in tumorigenesis, although the proof of a direct link is not clear to date [8]. Since mutagenic substrate, 8-oxo-dGTP, is sanitized from the nucleotide pool by MTH1, we examined the susceptibility of the gene-targeted mice to spontaneous tumorigenesis.

MTH1 homozygous mutant mice were found to have a normal physical appearance. Systematic pathological examination revealed a statistically significant difference in the incidence of tumors in $MTH1^{+/+}$ and $MTH1^{-/-}$ mice. Around the time of 18 months after birth, many tumors were found in the lungs and livers of the *MTH1*-deficient mice, but there were a few in the $MTH1^{+/+}$ mice. The elevated incidence of tumor formation in the liver of $MTH1^{-/-}$ mice was well correlated with the highest content of the MTH1 protein in this organ of the wild-type mice [9]. In summary, more tumors were formed in the three internal organs: lungs, liver and stomach of the $MTH1^{-/-}$ mice than in those of the wild-type mice. When statistical analysis was carried out, comparing the total number of mice bearing tumors between the $MTH1^{-/-}$ and $MTH1^{+/+}$ groups, there was a significant difference between these two groups: 34 out of 93 (36%) $MTH1^{-/-}$ mice compared to 10 out of 90 (11%) $MTH1^{+/+}$ mice ($P<0.001$) [14].

The study of mutation in vivo is facilitated by the use of transgenic mice, where mutational responses can be measured in virtually any tissue as a function of age, sex and external stresses. The mutational target in HITEC transgenic mice is the exceptionally well-characterized *rpsL* gene from *E. coli*. The *rpsL* gene is highly sensitive to base substitution and frameshift mutations, as well as deletions and insertions, making the transgene an ideal choice for the recovery of spontaneous and induced mutations [10]. Currently, spleen DNA samples from three 24-week-old wild type and null mutant animals, hemizygous for the *rpsL* transgene, were analysed by the HITEC mutational assay to investigate the frequency and specificity of mutation in the transgene in vivo. Consistent with the previous report [10], the mutation frequencies for the $rpsL/MTH1^{+/+}$ transgenic mice were approximately 3.7×10^{-6}. The spontaneous mutation frequency observed in spleen samples from the $MTH1^{-/-}$ mice showed no apparent increase compared to the value of the one in $MTH1^{+/+}$ mice (less than twofold). However, the site distribution of the mutations that occurred on the $rpsL$ gene seemed slightly different between the two $MTH1$ genotypes in the spleen samples. In the $MTH1^{-/-}$ mice, there are 1-basepair frameshift mutations at the mononucleotide; however, repeats of those were not found in $MTH1^{+/+}$ mice. It is interesting to note that an approximately twofold higher mutation rate was observed in two independently isolated $MTH1^{-/-}$ cells compared with the value of $MTH1^{+/+}$ cells when mutation rates toward 6-thioguanine resistance were determined by the fluctuation test [14]

To prevent occurrence of cancers, organisms are equipped with multiple defense mechanisms including DNA repair systems, apoptotic cell death and immune systems. Indeed, in *MSH2*-deficient mice, where one of the mismatch repair enzymes is absent [11], and *p53*-deficient mice, where appropriate apoptotic cell death is disturbed, tumors frequently occur during early life. Though it is generally conceived that blockage of DNA replication by DNA damage somehow acts as a signal for inducing apoptosis, some types of DNA lesions which do not prevent progression of the replication fork also cause apoptosis. Accumulation of 8-oxoG paired with adenine may be recognized by the mismatch protein complex [12]. If cells carrying 8-oxoG were eliminated through apoptosis, the incidence of mutation and tumors in the $MTH1^{-/-}$ mice might be reduced. Moreover, another pathway, such as the nucleotide excision repair pathway, may also participate in repairing the mutagenic lesion, 8-oxoG [13]. More definitive studies designed to evaluate the role of this oxidation-induced DNA damage defense system in the intact animal will be benefited from double or triple mutant mice, which are defective in other pathways of DNA repair.

Acknowledgements

We thank Prof. M. Sekiguchi, the principal investigator of this project, and his colleagues at the Medical Institute of Bioregulation, Kyushu University. We would also like to thank Prof. M. Katsuki's group and Prof. Ishikawa's group, at the University of Tokyo, for their assistance and pertinent advice. This work was supported by grants-in-aid from the Ministry of Education, Culture, Sports, Science and Technology, and the Ministry of Public Welfare and Labor.

References

[1] B.N. Ames, L.S. Gold, Endogenous mutagens and the causes of aging and cancer, Mutat. Res. 250 (1991) 3–16.
[2] H. Kasai, P.F. Crain, Y. Kuchino, S. Nishimura, A. Ootsuyama, H. Tanooka, Formation of 8-hydroxyguanine moiety in cellular DNA by agents producing oxygen radicals and evidence for its repair, Carcinogenesis 7 (1986) 1849–1851.
[3] S. Shibutani, M. Takeshita, A.P. Grollman, Insertion of specific bases during DNA synthesis past the oxidation-damaged base 8-oxoG, Nature 349 (1991) 431–434.

[4] H. Maki, M. Sekiguchi, MutT protein specifically hydrolyses a potent mutagenic substrate for DNA synthesis, Nature 355 (1992) 273–275.
[5] K. Sakumi, M. Furuichi, T. Tsuzuki, T. Kakuma, S. Kawabata, H. Maki, M. Sekiguchi, Cloning and expression of cDNA for a human enzyme that hydrolyzes 8-oxo-dGTP, a mutagenic substrate for DNA synthesis, J. Biol. Chem. 268 (1993) 23524–23530.
[6] M. Furuichi, M.C. Yoshida, H. Oda, T. Tajiri, Y. Nakabeppu, T. Tsuzuki, M. Sekiguchi, Genomic structure and chromosome location of the human *mutT* homologue gene *MTH1* encoding 8-oxo-dGTPase for prevention of A:T to C:G transversion, Genomics 24 (1994) 485–490.
[7] K. Fujikawa, H. Kamiya, H. Yakushiji, Y. Fujii, Y. Nakabeppu, H. Kasai, The oxidized forms of dATP are substrate for the human MutT homologue, the MTH1 protein, J. Biol. Chem. 274 (1999) 18201–18205.
[8] A.L. Jackson, L.A. Loeb, On the origin of multiple mutations in human cancers, Semin. Cancer Biol. 8 (1998) 421–429.
[9] K. Kakuma, J. Nishida, T. Tsuzuki, M. Sekiguchi, Mouse MTH1 protein with 8-oxo-7,8-dihydro-2′-deoxyguanosine 5′-triphosphatase activity that prevents transversion mutation, J. Biol. Chem. 270 (1995) 25942–25948.
[10] Y. Shioyama, Y. Gondo, K. Nakao, M. Katsuki, Different mutation frequencies and spectra among organs by *N*-methyl-*N*-nitrosourea in *rpsL* (*strA*) transgenic mice, Jpn. J. Cancer Res. 91 (2000) 482–491.
[11] A.H. Reitmair, R. Schmits, A. Ewel, B. Bapat, M. Redston, A. Mitri, P. Waterhouse, H.W. Mittrucker, A. Wakeham, B. Liu, A. Thomason, H. Griesser, S. Gallinger, W.G. Ballhausen, R. Fishel, T.W. Mak, *MSH2*-deficient mice are viable and susceptible to lymphoid tumors, Nat. Genet. 11 (1995) 64–70.
[12] T.L. DeWeese, J.M. Shipman, N.A. Larrier, N.M. Buckley, L.R. Kidd, J.D. Groopman, R.G. Cutler, H. te Riele, W.G. Nelson, Mouse embryonic stem cells carrying one or two defective *Msh2* alleles respond abnormally to oxidative stress inflicted by low-level radiation, Proc. Natl. Acad. Sci. U. S. A. 95 (1998) 11915–11920.
[13] J.T. Reardon, T. Bessho, H.C. Kung, P.H. Bolton, A. Sancar, In vitro repair of oxidative DNA damage by human nucleotide excision repair system: possible explanation for neurodegeneration in xeroderma pigmentosum patients, Proc. Natl. Acad. Sci. U. S. A. 94 (1997) 9463–9468.
[14] T. Tsuzuki, A. Egashira, H. Igarashi, T. Iwakuma, Y. Nakatsuru, Y. Tominaga, H. Kawate, K. Nakao, K. Nakamura, F. Ide, S. Kura, Y. Nakabeppu, M. Katsuki, T. Ishikawa, M. Sekiguchi, Spontaneous tumorigenesis in mice defective in the *MTH1* gene encoding 8-oxo-dGTPase, Proc. Natl. Acad. Sci. U. S. A. 98 (2001) 11456–11461.

Spontaneous and radiation-induced tumorigenesis in *p53*-deficient mice

Rajamanickam Baskar, Haruko Ryo, Hiroo Nakajima,
Tadashi Hongyo, Liya Li, Mukh Syaifudin, Xiao E Si,
Taisei Nomura*

Department of Radiation Biology and Medical Genetics, Graduate School of Medicine, Osaka University, B-4, 2-2 Yamada-oka, Suita, Osaka, 565-0871, Japan

Abstract

To investigate the spontaneous and radiation induced tumorigenesis, tumor development and molecular changes were examined in *p53* gene-deficient mice. All of the null ($p53^{-/-}$) animals died of leukemia, sarcoma, etc. before 12 months of age, and predominantly of lymphocytic leukemia at the early age of 3–5 months, while heterozygous ($p53^{+/-}$) and wild type ($p53^{+/+}$) animals survived more than 30 and 40 months, respectively. Half of the heterozygotes developed sarcomas (osteosarcomas, rhabdomyosarcomas, hemangiomas), and some developed tumors in the lung and liver; however, a lower incidence of leukemia was found. The onset of the tumor was late in their lives in the heterozygotes, though the tumors grew rapidly. In the wild type mice, leukemia and sarcomas were rarely observed. The ^{137}Cs γ-irradiation shortened the lifespan of the heterozygotes with an earlier tumor development, though this effect was not evident in the null mice. Among the 54 spontaneous tumors studied in the heterozygotes, 12 (22.2%) showed loss of the wild type *p53* allele (loss of heterozygosity, LOH) and one functional mutation was found at exon 6 of the *p53* gene. All the LOH was found in leukemia and sarcomas, except in one breast carcinoma. © 2002 Elsevier Science B.V. All rights reserved.

Keywords: p53 knockout; γ-radiation; Tumorigenesis; LOH; Mutation

1. Introduction

Normal cells have a number of intrinsic mechanisms that involve molecular 'gatekeepers' to protect them from rapid uncontrolled proliferation [1]. Tumor formation and

---

* Corresponding author. Tel.: +81-6-6879-3811; fax: +81-6-6879-3819.
 E-mail address: tnomura@radbio.med.osaka-u.ac.jp (T. Nomura).

growth are characterized by uncontrolled cellular proliferation, and it can occur when essential regulatory proteins are altered. In human cancers, one of the tumor suppressor genes closely related to ionizing radiation is *p53*. It determines the growth arrest and/or cell death following irradiation [2]. The gene *p53* appears to have a pivotal function in protecting humans from cancer, because this gene is mutated in more than 50% of human cancers [3].

The tumorigenic effects of ionizing radiation in both experimental animals and humans have been well studied. However, there is no direct evidence to demonstrate the changes in carcinogenesis or mutagenesis of *p53*-deficient tissues by radiation. To investigate the spontaneous and radiation initiated tumorigenesis, the onset and latency of tumors and their molecular changes were examined in the *p53* gene-deficient mice.

2. Materials and methods

2.1. Animals

Mice lacking one allele of the *p53* gene (C57BL/6-*p53*$^{+/-}$) [4] were provided by Prof. M. Katsuki, University of Tokyo. The mice were maintained by brother–sister inbreeding in complete barrier conditions with light from 4:00 to 18:00 at 23 ± 1 °C and 50–70% humidity, with an autoclaved mouse diet CRF-1 (Charles River Japan, Kanagawa, Japan) and acidified, chlorinated and filtrated (by Millipore) water. The animal experiments were carried out in the barrier section of the Institute of Experimental Animal Sciences following the Osaka University Guidelines for Animal Experimentation.

2.2. Identification of genotypes

The genotypes of the individual mice were determined by polymerase chain reaction (PCR, GeneAmp PCR System 9700, Perkin-Elmer, Foster City, CA, USA) from the tail DNA, using the vector specific primers and *p53* primers covering exons 6 and 7. We designated wild types as *p53*$^{+/+}$, heterozygotes as *p53*$^{+/-}$ and knockout homozygotes as *p53*$^{-/-}$.

2.3. Irradiation

At the age of 6 weeks, the mice were exposed to 0.5 Gy of ^{137}Cs γ-rays four times at 7-day intervals at a dose rate of 1.07 Gy/min.

2.4. Examination of tumor

Mice were checked 6 days a week, and allowed to live out their lifespan or were euthanatized when moribund. All animals were fully necropsied, and the tumors and normal tissues were frozen at -80 °C for molecular analysis. The tumors and selected pathological lesions were fixed in 10% neutral buffered formaldehyde solution and submitted to histological examinations to confirm diagnosis.

2.5. LOH and mutation analysis

DNA was extracted from the frozen tissue and LOH of the *p53* gene was determined by PCR. The DNA was amplified using four specific pairs of primers for the *p53* gene exons 5, 6, 7 and 5–7. The PCR amplification of the *p53* knockout allele by vector specific primers was also carried out as a control. In addition, LOH was determined based on the signal of PCR products in agarose gel electrophoresis. A very weak or no signal of PCR products of the wild type allele amplified by all the four specific primers was classified as LOH, in comparison with the standard signal by the *p53* knockout specific primers.

PCR and single strand conformational polymorphism (SSCP) was carried out to screen exons 5, 6, 7 and 8 of the *p53* gene for the presence of mutations. Single-strand conformational polymorphism (SSCP) was carried out as described previously [5]. Briefly, SSCP bands, which showed altered mobility, were extracted from the gel and reamplified by 25 cycles of PCR to enrich the mutated alleles. Direct sequencing was performed by the dideoxy chain termination method using the Big Dye terminator cycle sequencing kit (Perkin-Elmer). The sequencing primers were the same as those for PCR. After ethanol precipitation, sequencing was done using an automated DNA sequencer (Genetic Analyzer, ABI Prism 310, Perkin-Elmer).

2.6. Statistical analysis

Statistical analysis was done using the SPSS system (SPSS, Chicago, USA).

3. Results and discussion

Large and significant differences were observed in survival periods among null ($p53^{-/-}$), heterozygous ($p53^{+/-}$) and wild type ($p53^{+/+}$) mice. All unexposed null mice died of leukemia (88/104, 84.6%), sarcomas, etc. (14/104, 13.5%) before the age of 12 months. Especially, null mice died of lymphocytic leukemia at the early age of (3–5 months), while heterozygous and wild type mice developed a low incidence of leukemia (7/88, 8.0% and 2/16, 12.5%) and many of them survived more than 30 and 40 months, respectively. The heterozygotes developed various types of tumors (61/88, 69.3%), mainly in the connective tissue, at older ages. The spectrum of the observed tumors was osteosarcomas (22.7%), other sarcomas (21.6%), cancers (9.1%), reticulum cell neoplasms (4.5%) and leukemia (8.0%). The wild-type animals developed reticulum cell neoplasms (4/16, 25%) and cancers (2/16, 12.5%), which were commonly observed in the back ground strain (C57BL/6) at the end of their lives.

^{137}Cs γ-irradiation of heterozygous mice shortened their lifespan significantly, when compared to the unirradiated heterozygotes. Most of the irradiated heterozygotes had multiple tumors in different organs. In contrast, unirradiated heterozygous animals showed tumor development at older ages and only a small number of animals had multiple tumors in different organs. However, there were no substantial differences in the survival periods of the null mice between the irradiated and unirradiated groups. In the present study, the animals survived longer than those in other studies reported [6,7], probably because our animals were maintained in the complete barrier condition. However, the spectrum of observed tumors was not different from that of the previous study [6].

We analyzed the status of the wild type allele in the tumors of unirradiated $p53^{+/-}$ mice and found that 12 out of 54 (22.2%) tumors showed LOH, while there were no LOH in 37 normal tissues examined. Among the tumors examined, LOH was found in 5 out of 8 (62.5%) leukemia and malignant lymphomas, 3 out of 13 (23.1%) osteosarcomas, 2 out of 2 (100%) fibrosarcomas, 1 out of 3 (33.3%) mammary adenocarcinomas, and 1 out of 1 (100%) neuroblastoma. Most of the LOH (5 of 12, 41.7%) were observed in leukemia and malignant lymphomas. Among the rest of the 42 tumors, which did not show any LOH, only one mutation (Tyr → Stop codon) was found in exon 6 in the reticulum cell neoplasm. The majority of the tumor samples with LOH showed a residual signal band of the wild type allele on the gel. We assume it inevitable due to the contamination of normal cells in the tumor tissue. Detailed studies of the LOH and mutations in spontaneous and radiation induced tumors are in progress.

In conclusion, the present results demonstrate that the mice lacking one or both alleles of the *p53* gene are highly susceptible to tumor induction, and irradiation of the heterozygotes enhances tumor deaths as observed in the unirradiated null mice, indicating the important role of the *p53* gene to prevent tumorigenesis.

Acknowledgements

We thank K.Y.M. Wani, M. Maeda, R. Kaba, R. Tsuboi, Y. Kitagawa and Y. Hattori for their assistance. This study was supported by grants from the Ministry of Education, Science and Culture and the Research for the Future (JSPS).

References

[1] C.J. Kemp, T. Wheldon, A. Balmain, *P53*-deficeint mice are extremely susceptible to radiation-induced tumorigenesis, Nat. Genet. 8 (1994) 66–69.
[2] A.J. Levine, *p53*, the cellular gatekeeper for growth and division, Cell 88 (1997) 323–331.
[3] M.S. Greenblatt, M. Hollstein, C.C. Harris, Mutations in the *p53* tumor suppressor gene: clues to cancer etiology and molecular pathogenesis, Cancer Res. 54 (1994) 4855–4878.
[4] Y. Gondo, K. Nakamura, K. Nakao, T. Sasaoka, K.I. Ito, M. Kimura, M. Katsuki, Gene replacement of the *p53* gene with the *lacZ* gene in mouse embryonic stem cells and mice by using two steps of homologous recombination, Biochem. Biophys. Res. Commun. 202 (2) (1994) 830–837.
[5] T. Hongyo, G.S. Buzzard, R.J. Calvert, C.M. Weghorst, 'Cold SSCP': a simple, rapid and non radioactive method for optimized single strand conformation polymorphism analysis, Nucleic Acids Res. 21 (1993) 3637–3642.
[6] M. Harvey, M.J. McArthur, C.A. Montgomery, J.S. Butel, A. Bradley, L.A. Donehower, Spontaneous and carcinogen induced tumorigenesis in *p53*-deficinet mice, Nat. Genet. 5 (1993) 225–229.
[7] S. Venkatachalam, Y.P. Shi, S.N. Jones, H. Vogel, A. Bradley, D. Pinkle, L.A. Donehower, Retention of wild-type *p53* in tumors from *p53* heterozygous mice: reduction of *p53* dosage can promote cancer formation, EMBO J. 17 (16) (1998) 4657–4667.

Oncogenes and tumor suppressor genes in murine tumors induced by neutron- or gamma-irradiation in utero

Sára Antal [a,*], K. Lumniczky [a], J. Pálfalvi [b], E. Hidvégi [a], F. Schneider [c], G. Sáfrány [a]

[a] *National Research Institute for Radiobiology and Radiohygiene, Anna St. 5, 1221 Budapest, Hungary*
[b] *Atomic Energy Research Institute, National Academy of Sciences, Budapest, Hungary*
[c] *National Institute of Oncology, Budapest, Hungary*

Abstract

Half a century after the A-bomb explosions, the cancer mortality is still increasing in prenatally exposed survivors. In this study, mice were exposed to neutron- or γ-irradiation of different embryonic stages. The tumor incidence increased to 35% after 2.0 Gy γ-irradiation on the 18th day of gestation, compared with 15% in the unirradiated controls. Most of the tumors developed in the lymphoid system, liver, lung and uterus. Some brain tumors were observed in mice irradiated in utero both with neutron and γ-rays, but none were found in the unirradiated controls. The expression of the p53 gene decreased by 2–5-fold in 60% of lung tumors. Allelic losses on chromosome 11 at the p53 locus were observed in 30–40% of the spontaneous and radiation-induced lymphoid tumors and liver adenocarcinomas. In lung tumors, about one-third of the spontaneous adenocarcinomas exhibited allelic losses at the p53 locus not associated with point mutation. This fact may suggest that the loss at one of p53 alleles is an earlier event during carcinogenesis than development of point mutations. © 2002 Elsevier Science B.V. All rights reserved.

Keywords: Tumor induction; Radiation; Oncogenes; Suppressor genes; Loss of heterozygosity

1. Introduction

Half a century after the A-bomb explosions, the cancer mortality is still increasing in prenatally exposed survivors [1]. After the Chernobyl accident, the incidence of devel-

---

* Corresponding author.
E-mail address: antal@hp.osski.hu (S. Antal).

opmental abnormalities increased in the most radionuclide-contaminated regions of Belorussia. Infants exposed in utero to ionizing radiation from the Chernobyl accident had an incidence of leukemia 2.6 times higher than in the unexposed children [2]. Recently, Shibata et al. [3] reported significantly high incidences of meningiomas among the Nagasaki atomic bomb survivors. In 1999, Shintani et al. [4] suggested that meningiomas were induced in the survivors of the atomic bomb in Hiroshima as well.

In our experiments, pregnant mice were irradiated either with fission neutrons or gamma-rays, and the tumor incidence and alterations in oncogenes in the induced tumors were studied.

2. Materials and methods

Female C57Bl/6 mice were mated with male DBA2 mice. The pregnant mice were irradiated either with high LET neutrons or with ^{60}Co–gamma-ray.

In order to study the tumor incidence, mice were killed when they were either moribund or at the end of the second year of age and autopsied, and tissues showing macroscopic alterations were examined histologically. Changes in the oncogenes, tumor suppressor genes and loss of heterozygosity were determined as described [5].

3. Results

Table 1 shows that five brain tumors (ependymoma) were found in 1080 mice after 0.1–2.0 Gy γ-irradiation in utero, and 17 brain tumors (ependymoma, meningioma) were observed in 818 mice irradiated prenatally with 0.1–2.0 Gy neutrons. No brain tumors were found in the 596 unirradiated control animals.

We have been investigating the alterations in oncogenes in tumor suppressor genes and the loss of heterozygosity in the tumors of animals irradiated in utero (Table 2). The expression of the p53 gene decreased in 60% of lung tumors. Allelic losses on chromosome 11

Table 1
Tumor incidence in mouse brain after exposure to neutron- and gamma-irradiation in utero

	No. of animals	No. of tumors	No. of brain tumors
Unirradiated controls	596	89 (15%)	0
Neutron-irradiated (Gy)			
0.1–0.3	338	142 (42.0%)	4 (1.2%)
0.5–0.6	358	138 (38.5%)	9 (2.5%)
0.7–1.5	122	54 (44.3%)	4 (3.3%)
Whole dose range (0.1–1.5)	818	334 (40.8%)	17 (2.1%)
Gamma-irradiated (Gy)			
0.1–0.2	221	33 (14.9%)	1 (0.4%)
0.3–0.5	283	82 (29.3%)	2 (0.7%)
0.6–1.0	321	98 (30.5%)	0
1.1–2.0	255	91 (35.7%)	2 (0.8%)
Whole dose range (0.1–2.0)	1080	304 (28.1%)	5 (0.5%)

Table 2
Frequencies of oncogenic alterations in tumors after irradiation in utero

Tumors	Alteration	Unirradiated mice (%)[a]	Irradiated mice (%)[a]
Lymphoid	myc expression[a]	28↑	23↑
	p53 mutations	25	13
	LOH at Acrb	40	40
	LOH at D4Mit77	40	23
Liver	Ha-ras expression	33↓	20↓
	N-ras expression	33↑	20↑
	Ha-ras mutations	33	40
	LOH at Acrb	33	40
	LOH at D4Mit77	33	0
Lung	Ha-ras expression	50↓	40↓
	p53 expression	50↓	60↓
	Ki-ras mutations	33	17
	LOH at Acrb	33	0
	LOH at D4Mit77	0	17
Uterus	LOH at D4Mit77	33	25

↓, decreased gene expression; ↑, increased gene expression.

[a] Frequencies of alterations are related to the unirradiated healthy mice.

at the p53 locus were observed in 30–40% of spontaneous and radiation-induced lymphoid tumors and in liver adenocarcinomas. In lung tumors, about one-third of the spontaneous adenocarcinomas exhibited allelic losses at the p53 locus not associated with point mutation.

4. Discussion

The cancer incidence and oncogenic changes in the tumors were investigated in mice after γ- or neutron-irradiation in utero. Both types of radiation increased the brain tumor incidence supporting the findings of Shibata et al. [3] and Shintani et al. [4] with A-bomb survivors. In our experiments, no brain tumors were observed in the unirradiated control animals (Table 1). This finding coincides with Fraser's [6] results obtained from different strains of mice.

We have already reported that mutations in the p53 gene and LOH at the D4Mit77 locus occurred slightly less frequently in radiation-induced lymphoid tumors than in spontaneous ones [5]. The incidence in higher myc expressions and the frequency of LOH at the Acrb locus were nearly unchanged (Table 2). Allelic losses on chromosome 11 at the Acrb locus have not been detected previously in spontaneous and chemically induced tumors of mice [7]. We found a nearly identical incidence of allelic losses at the Acrb locus in both spontaneous and radiation-induced liver tumors (Table 2). The allelic losses were not accompanied by point mutations in exons 5–8 of the p53 gene, which may suggest that the loss of one p53 allele may occur earlier in carcinogenesis than the point mutation. Moreover, allelic loss without point mutation might represent a random event associated with increased genetic instability of tumors. In the human lung, Ki-ras and p53 are the genes most frequently involved in oncogenesis [8]. However, we did not find p53 mutations in murine lung tumors although the expression of the p53 gene decreased (Table 2).

The point mutations were tissue-specific like: the mutations in the p53 gene were present mainly in lymphoid tumors, and Ha-ras mutations were detected in lung adenocarcinomas. The frequency of carcinogenic alterations was nearly identical in the spontaneous and radiation-induced lymphoid, liver and uterus tumors, suggesting that similar oncogenic events occur in these tissues during spontaneous and in utero radiation-induced carcinogenesis. In lung tumors, however, the frequencies of many alterations were different in radiation-induced and spontaneous tumors, suggesting that different oncogenic pathways were activated during the spontaneous and in radiation-induced lung carcinogenesis of mice.

Acknowledgements

This work was supported by a Hungarian OTKA grant (T-023952), by a Hungarian OMFB grant (94-97-47-0747) to Antal Sára, Hungarian OTKA grant (T-025333) to Hidvegi Egon and European Commission PECO grant to Sáfrány Géza (F13P-CT92-0030).

References

[1] Y. Yoshimoto, H. Kato, W.J. Schull, Risk of cancer among children exposed in utero to A-bomb radiation: 1950–84, Lancet 2 (1988) 665–669.
[2] E. Petridou, et al., Infant leukemia after in utero exposure to radiation from Chernobyl, Nature 382 (1996) 352–353.
[3] S. Shibata, N. Sadamori, M. Mine, I. Sekine, Intracranial meningiomas among Nagasaki atomic bomb survivors, Lancet 344 (1994) 1770.
[4] T. Shintani, et al., High incidence of meningioma among Hiroshima atomic bomb survivors, J. Radiat. Res. 40 (1999) 49–57.
[5] K. Lumniczky, et al., Oncogenic changes in murine lymphoid tumors induced by in utero exposure to ionizing radiation, Radiat. Oncol. Invest. 5 (1997) 158–162.
[6] H. Fraser, Brain tumors in mice, with particular reference to astrocytoma, Food. Chem. Toxic 24 (1986) 105–111.
[7] L.M. Davis, et al., Loss of heterozygosity in spontaneous and chemically induced tumors of the B6C3F1 mouse, Carcinogenesis 15 (1994) 1637–1645.
[8] J.W. McDonald, et al., P53 and K-ras in radon associated lung adenocarcinoma, Cancer Epidemiol., Biomarkers Prev. 4 (1995) 791–793.

Low dose fetal irradiation, chromosomal instability and carcinogenesis in mouse

P. Uma Devi*, M. Hossain, M. Satyamitra

Department of Radiobiology, Kasturba Medical College, T.M. A. Pai Research Centre, Manipal, Karnataka 576 119, India

Abstract

A study was carried out to see if acute low dose fetal irradiation of mouse could induce long lasting chromosomal anomalies and adult cancers. The abdomen of pregnant Swiss mice was exposed to 0.1–1 Gy of ^{60}Co gamma radiation on day 14 of gestation. The animals were killed 24 h later or allowed to deliver their young. Hemopoietic cell death and chromosomal aberrations were studied in the fetal liver. Chromosomal instability was traced in the 1 Gy exposed group from the fetal liver cells through their CFU-S to postnatal bone marrow. The F1 mice were observed for chromosomal aberrations in the bone marrow and for solid tumor incidence. Irradiation produced a significant dose-dependent increase in the hemopoietic cell death and aberrant metaphases in the fetal liver. The main types of aberrations were chromatid breaks and fragments. The aberrations persisted in the spleen colonies developed from these cells. Their frequency decreased in the postnatal bone marrow, however, showed a significant increase at 12 months of age. Some exposed animals had abnormally high leukocyte counts and after 1 Gy their bone marrow had eight times more polyploid cells than the control. Prenatal exposure also significantly increased the incidence of solid tumors, the ovaries showing the highest risk. It is concluded that a single exposure below 1 Gy of gamma radiation at the early fetal stage of these mice can induce persistent chromosomal instability in the hemopoietic cells, and significantly increase the incidence of solid tumors in adults. © 2002 Elsevier Science B.V. All rights reserved.

Keywords: Low dose fetal irradiation; Chromosomal instability; Carcinogenesis

1. Introduction

The fetal period in mammalian development is relatively resistant to radiation teratogenesis. However, exposure to doses below 1 Gy of X- and gamma radiation has been

---

* Corresponding author. Department of Research, Jawaharlal Nehru Cancer Hospital and Research Centre, Bhopal, 462001, India. Tel.: +91-8252-72100; fax: +91-8252-70062.
E-mail address: jncancer@vsnl.com (P. Uma Devi).

linked with hematological disorders and malignant changes during postnatal life. However, there is very little information on the changes in the fetal hemopoietic cells and their late consequences, or the carcinogenic effect after prenatal irradiation. Therefore, a study was undertaken to see if acute low-dose irradiation during the early fetal period of mouse can lead to (a) long-term chromosomal instability, and (b) cancer induction in the adults.

2. Experimental details

Virgin Swiss albino mice of 6–8 weeks were mated with males of the same age. The abdominal area of the pregnant females was exposed to 0, 0.1, 0.25, 0.5 or 1.0 Gy of ^{60}Co gamma radiation at a dose rate of 1.6 Gy/min on day 14 post-conception (p.c.).

Five animals from each group were killed at 24 h post-irradiation and single cell suspensions were prepared from the fetal liver. Hemopoietic cell death was studied by exogenous spleen colony assay (CFU-S) and chromosomal aberrations were scored in metaphase plates [1]. In the 1 Gy group, the CFU-S was propagated through two passages in bone marrow ablated adult mice and the chromosomal aberrations were scored in the spleen colony cells. The second passage animals were killed after 3 weeks and the chromosomal aberrations in the femur marrow were studied. The pups from five mothers exposed to 1 Gy on day 14 p.c. were killed at 1 and 3 months of age, and the bone marrow chromosomes were studied.

Twenty-four animals per group were left to deliver their young. The F1 mice were physically examined from the age of 4–18 months for any signs of tumor development externally, and in the abdominal cavity. All the animals were killed at the age of 18 months and the solid tumors in the peritoneal cavity were recorded. Peripheral blood cell counts were done up to the age of 12 months. At 12 months, the chromosomal aberrations were scored in the femur marrow of 10 animals per group.

3. Results and discussion

3.1. Chromosomal instability

The fetal liver 24 h after the irradiation showed a dose-dependent, significant decrease in the total cellularity and CFU-S at doses from 0.25 to 1 Gy. The percent aberrant metaphases showed a significant increase above the control even at 0.1 Gy. The main types of structural aberrations were chromatid breaks and fragments; the exposed animals also showed a dose-dependent increase in the number of polyploid cells [1]. A higher number of aberrant metaphases were observed consistently in the bone marrow of adult mice repopulated by the CFU-S cells of the fetal liver. Their number decreased in the bone marrow after birth, but again showed a significant increase at the age of 12 months, along with a high increase in polyploid cells [2].

In general, the in-utero exposed adult mice showed significantly low peripheral blood cell counts at doses of ≥ 0.5 Gy. However, some animals had 4–10 times the normal WBC counts at the age of 12 months, the major component affected being the neutrophils.

Persistent genomic instability is indicated to significantly influence the development of leukemia in the people exposed to the atom bomb [3]. Edwards [4] suggested that in-utero irradiation of the early fetus might induce genomic instability, leading to aberrations in the hemopoietic stem cells, which migrate to the bone marrow. Experimental evidence in support of this comes from our study. While the fetal liver cells from 14-day p.c. irradiated mice and their spleen colony forming units (CFU-S) showed a higher incidence of polyploids than in the sham-irradiated controls, such cells were absent in the adult bone marrow repopulated by the irradiated fetal liver CFU-S cells or in the young pups irradiated on day 14 p.c. However, polyploidy again appeared in the bone marrow at the age of 12 months, along with a higher number of breaks and fragments than the control. This effect was much more pronounced in the mice with high leukocyte counts, which showed a two-fold increase in the polyploids compared to their low WBC counterparts exposed to 1 Gy on day 14 p.c., and eight times that found in the control (Table 1), thus suggesting an association between the development of polyploidy in the hemopoietic cells and radiation-induced leukemia. Under the influence of promoting factors during the course of postnatal life, the radiation induced genomic instability in the fetal hemopoietic cells may develop into explicit structural and numerical chromosomal aberrations, which contribute to hematological disorders, including leukemia, at later ages.

3.2. Solid tumor incidence

Adult mice that had been exposed on day 14 p.c. showed a significantly higher incidence of solid tumors compared to the sham-treated animals. The main organs involved were the ovaries, uterus, liver and spleen and the total tumor incidence showed

Table 1
Chromosomal aberrations in mice exposed to 1 Gy on day 14 p.c.

Observation time	Metaphase scored	Aberrant metaphases (%)	Breaks and frag./cell	Polyploids (%)
Pre/postnatal				
Control[a]	2235	10.5 (0.5)	0.0054	0
Fetal liver[b]	2457	113 (4.2)	0.0078	12 (0.5)
CFU 1	550	23 (4.4)	0.0546	2 (0.4)
CFU 2	568	18 (3.5)	0.0396	1 (0.2)
3 wk BM[c]	1527	27 (1.8)	0.0195	0
1 month BM	2023	23 (1.1)	0.0089	0
3 month BM	2500	24 (0.9)	0.0100	0
Adult				
Control	2961	31.9 (1.0)	0.0203	4 (0.1)
12 month	2468	128 (5.2)	0.0858	16 (0.7)
12 month[d]	2190	174.3 (7.9)[e]	0.1010	33 (1.5)[e]

BM: bone marrow; CFU-colony forming unit (spleen) cells.
[a] Pooled mean of all sham-irradiated values.
[b] Twenty-four hours after irradiation.
[c] Adult bone marrow 3 weeks after repopulation with CFU-S from the irradiated fetal liver cells.
[d] Animals showed six to eight times the normal WBC counts.
[e] $p < 0.001$ compared to all the adult groups.

a radiation dose dependent increase. The females showed a higher incidence than the males, mainly due to ovarian tumors, which showed a significant increase when compared to the control at doses above 0.1 Gy. Tumors of the ovary also developed earlier (≤ 6 months of age) than the other tumors (9–12 months) [5]. Nitta et al. [6] had also reported a higher incidence of tumors in the female B6C3F1 hybrid mice than in the males after fetal irradiation. They found a significant increase in the tumors of the pituitary and mammary glands in the adult mice exposed on day 16.5 p.c. to 2.7 Gy of gamma radiation. In both these studies the radiation did not induce any new type of tumor, however, increased the incidence of tumors spontaneously occurring in that strain.

4. Conclusion

Low-dose gamma radiation induces chromosomal instability in the fetal hemopoietic stem cells, which is transmitted to the postnatal bone marrow and expressed as structural and numerical aberrations in the adults. There seems to be a close association between the development of chromosome polyploidy in the bone marrow and leukemia induction in the adults after fetal irradiation. Exposure to gamma radiation below 1 Gy during the early fetal period of mouse significantly increases the risk of tumor development in adults. In addition, females are more sensitive to this effect than males.

References

[1] P. Uma Devi, M. Hossain, Evaluation of the cytogenetic damage and progenitor cell survival in fetal liver of mice exposed to gamma radiation during the early fetal period, Int. J. Radiat. Biol. 76 (2000) 413–417.
[2] P. Uma Devi, M. Hossain, Induction of chromosomal instability in mouse hemopoietic cells by fetal irradiation, Mutat. Res. 456 (2000) 33–37.
[3] M. Nakanishi, K. Tanaka, T. Shintani, T. Takahashi, N. Kamada, Chromosomal instability in acute myelocytic leukemia and myelodisplastic syndrome patients among atomic bomb survivors, J. Radiat. Res. 40 (1999) 159–167.
[4] R. Edwards, Radiation roulette, New Scientist 156 (1997) 36–40.
[5] P. Uma Devi, M. Hossain, Induction of tumors in the Swiss albino mouse by low-dose fetal irradiation, Int. J. Radiat. Biol. 76 (2000) 95–99.
[6] Y. Nitta, K. Kamia, K. Yokoro, Carcinogenic effect of in utero ^{252}Cf and ^{60}Co irradiation in C57BL/6N × C3H/He (B6C3F1) mice, J. Radiat. Res. 33 (1992) 319–333.

Effects of radioactive iodine (^{131}I) on the thyroid of newborn, pubertal and adult rats

Yumiko Nitta*, Masaharu Hoshi, Kenji Kamiya

Research Institute for Radiation Biology and Medicine, Hiroshima University, 1-2-3 Kasumi, Minami, Hiroshima 734-8553, Japan

Abstract

Age is a potent modifier of thyroid cancer. The short latency for the development of thyroid cancer in the post-Chernobyl cases proposes that we need to be sure of the thyroid susceptibility to internal exposure, especially at young ages. We started a large-scale carcinogenesis project 6 years ago with the purpose to evaluate the carcinogenic potential of ^{131}I when irradiated at young ages. First, we established a method to estimate the absorbed doses in the thyroid of different age groups. Irradiation at 1 week of age caused heavier exposure than at 4 and 9 weeks of age by eight times, while damages of the thyroid tissue were more obvious in the 4-week-old groups than in the 1-week-old groups. Second, we tested the responsiveness of thyroid epithelium to radiation. Apoptosis was not detected in the 1-week-old-thyroid epithelium, however, it did appear in the 4-week-old thyroid epithelium. While the proliferating cell nuclear antigen (PCNA)-labeling index was vice versa. Third, the carcinogenesis of ^{131}I has been tested. Papillary carcinomas have developed in rats internally irradiated with ^{131}I at the age of 1 week. A very low dose rate of irradiation by ^{131}I could induce thyroid carcinomas with a short latency. © 2002 Elsevier Science B.V. All rights reserved.

Keywords: Chernobyl; Thyroid cancer; ^{131}I; Internal exposure; Rat

1. Introduction

Medical internal exposures to ^{131}I showed no age-dependency in the risk of thyroid cancer. However, the short latency for the development of thyroid cancer in children of the post-Chernobyl cases signals to us to make sure of the thyroid's susceptibility to radiation at young ages. We started a large-scale carcinogenesis project for the purpose of re-evaluating the carcinogenic potential of ^{131}I when irradiated at young ages. First, we established a method to estimate the doses of ^{131}I in the thyroid at different ages [1,2]. Second, we

---

* Corresponding author. Tel./fax: +81-82-257-5877.
E-mail address: yumiko@hiroshima-u.ac.jp (Y. Nitta).

compared the thyroid susceptibility to radiation at different ages [3]. Third, the carcinogenesis experiment is on going.

2. Materials and methods

For the internal and external irradiation experiments, female rats of the Fischer 344 strain of the 1-, 4- and 9-week-old groups at irradiation were set up. Each age group was composed of the standard diet (SD) and the iodine-deficient diet (IDD) group [1,2]. ^{131}I was injected with activities to give 3.0 Gy as the thyroid dose for the SD-rats for the internal irradiation [1,2]. The ^{60}Co-rays were applied externally to the rat thyroid with a dose of 3.0 Gy. The rats

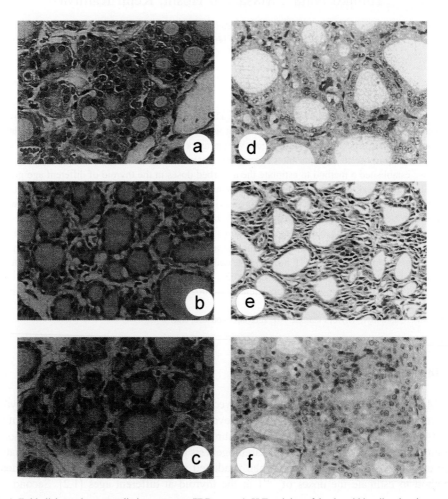

Fig. 1. Epithelial atrophy post-radiation exposure (IDD groups). H·E staining of the thyroid irradiated at the age of 1 week (left column) or at the age of 4 weeks (right column). (a) and (d) were the section just before the irradiation, (b) and (e) were those at 6-h post-irradiation, and (c) and (f) at 48-h post-irradiation.

were autopsied at 0-, 6-, 12-, 24-, 48-, 96-, 192- and 384-h post-irradiation. Using the paraffin-embedded thyroid section, the proliferation profile of the epithelium was expressed by labeling the index of the proliferating cell nuclear antigen (PCNA), and the apoptosis was detected by the TdT-mediated dUTP-biotin nick end labeling (TUNEL) method.

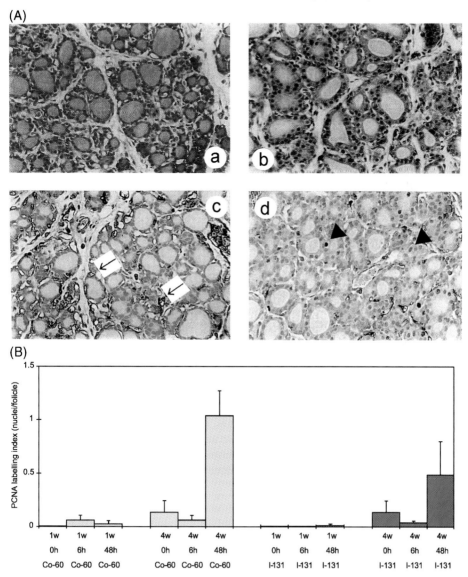

Fig. 2. (A) Epithelial responsiveness to radiation exposure (SD groups). Columns indicate the PCNA labeling index (nuclei/follicle). The bar indicates standard error. (B) Epithelial responsiveness to radiation exposure (IDD groups). H·E staining (a, b) and PCNA immunostaining (c, d). (a) and (c) were the thyroid at 48-h post-irradiation of the 1-week-old group.(b) and (d) were the thyroid at 48-h post-irradiation of the 4-week-old group. The arrow indicates PCNA positive blood vessel's endothelium. The arrowhead indicates PCNA positive foliclar epithelium.

3. Results

3.1. Kinetics of the thyroid treated with ^{131}I

The kinetics of ^{131}I in the thyroid and the whole body showed that IDD-treatment increased the exposure to ^{131}I by 4.2, 2.4 and 2.1 for the 1-, 4- and 9-week-old groups, respectively [2]. The epithelial atrophy 6 h after irradiation was prominent in the IDD-treated thyroid after the injection (Fig. 1).

3.2. Epithelial apoptosis and proliferating responsiveness post-irradiation

When ^{131}I was irradiated, no TURNEL-positive cell was found in the thyroid, however, it was found in the spleen. When ^{60}Co was irradiated, severe apoptosis was observed in the external granular cells in the cerebellum, however, not in the thyroid. When irradiated with ^{131}I or ^{60}Co, there were lots of PCNA-positive nuclei in the 4- but not in the 1-week-old thyroid (Fig. 2A, B).

4. Discussion

The 1-week-old groups were exposed more heavily than the 4- and 9-week-old groups when irradiated internally, and both internal and external irradiations caused atrophy on the thyroid epithelium 6 to 12 h later [2]. However, the atrophic cells did not show apoptosis in all the three age groups. Cell proliferation started after 48 h of irradiation in the 4-week-old groups, however, it did not start at that time in the 1-week-old groups. This attenuation of the epithelial proliferation post-irradiation in the 1-week-old groups might contribute to thyroid cancer development. One year has passed since we set up a large-scale carcinogenesis experiment for the induction of thyroid cancer by radiation. At the moment, multiple carcinomas developed in the rats whose thyroids were exposed to 3.6, 8.9 or 21.4 Gy of ^{131}I at the age of 1 week with the latency of 36 weeks. Pathologically, most of the cancers were diagnosed as papillary carcinomas. In conclusion, papillary carcinomas are found to be inducible in rats by the combined treatment of ^{131}I and an iodine deficient diet with a short latency, when the exposure to radiation is at a young age. Finally, a low dose rate of the moderate dose irradiation could induce thyroid cancer in rats.

Acknowledgements

This work was supported by a Grant-in-Aid from the Ministry of Education, Science, Sports and Culture of Japan (Y.N. 09788107).

References

[1] S. Endo, Y. Nitta, M. Ohtaki, J. Takada, V. Stepanenko, K. Komatsu, H. Tauchi, S. Matsuura, E. Iaskova, M. Hoshi, Estimation of dose absorbed fractions in rat thyroid, J. Radiat. Res. 39 (1998) 223–230.

[2] Y. Nitta, S. Endo, N. Fujimoto, K. Kamiya, M. Hoshi, Age-dependent exposure to radioactive iodine (^{131}I) in the thyroid and total body of newborn, pubertal and adult Fischer 344 rats, J. Radiat. Res. 42 (2001) 143–155.

[3] Y. Nitta, K. Kamiya, M. Hoshi, Thyroid susceptibility to radioactive iodine (^{131}I) in newborn, pubertal and adult rats, Proceedings of the 2nd International Symposium for Low Dose and Very Low Dose of Radiation on Human Health, Dublin, 27th ~ 29th, June, 2001.

Modeling carcinogenic effects of low doses of inhaled radionuclides

Imre Balásházy [a,*], Werner Hofmann [b], Annamária Dám [c]

[a] *Health Physics Department, KFKI Atomic Energy Research Institute, PO Box 49, H-1525 Budapest, Hungary*
[b] *Institute of Physics and Biophysics, University of Salzburg, Hellbrunner Str. 34, A-5020 Salzburg, Austria*
[c] *National Research Institute for Radiobiology and Radiohygiene, PO Box 101, H-1775 Budapest, Hungary*

Abstract

Clinical studies have indicated that carinal regions of the airway generation 3–5 are preferential sites of tumor development following inhalation of radon progenies. These coincide with the locations where primary hot spots of deposition have been found. However, current lung dosimetry models do not take into consideration the inhomogeneity of deposition within the airways. In this study, computed local distributions of deposited inhaled radionuclides, such as radon progenies in morphologically realistic human airway bifurcation models, are analyzed for different flow rates and particle sizes. Then, local deposition enhancement factors (EF), defined as the ratio of local to average deposition densities, are computed by scanning along the surface of the bifurcation with prespecified surface area elements. The computed enhancement factors indicate that cells located at carinal ridges may receive localized doses, which are two orders of magnitude higher than the average values. Here, the probability of multiple hits can be quite high even at low doses. If the number of multiple cellular hits plays a crucial role in lung cancer development, then the maximum enhancement factor could serve as a useful parameter for a mechanism-based risk analysis of inhaled radionuclides. © 2002 Elsevier Science B.V. All rights reserved.

Keywords: Radon inhalation; Bronchial carcinoma; Deposition enhancement factors

1. Introduction

Inhaled naturally occurring radionuclides have been implicated in causal relationships with lung carcinomas. Histological studies have noted highly localized preneoplastic and neoplastic lesions at the carinal regions of large central airway bifurcations. Primary hot spots of deposition have been found at the same locations. There is no detailed analysis in the literature for the characterization of the deposition patterns of radon progenies within the airways. The objective of the current research is to describe and quantify the local deposition distributions and to explore the biophysical consequences regarding the development of alpha radiation-induced bronchial carcinomas.

2. Method

In this paper, local distributions of deposited radon progenies in airway generation 3–4 are computed during inhalation applying morphologically realistic three-dimensional airway bifurcation models. The airflow is simulated by the FIRE computational fluid dynamics (CFD) program package at different flow rates. Trajectories of randomly selected particles are followed by considering all four characteristic deposition mechanisms: inertial impaction, gravitational sedimentation, diffusion and interception. Details of the model are published in Ref. [1]. By selecting large number of particles, deposition patterns of the inhaled aerosols can be computed. For the quantification of the distribution of deposition along an airway bifurcation, the whole surface was scanned with prespecified surface area elements, and local deposition enhancement factors, EF, were computed as the ratio of local to average deposition densities. Details of the technique are presented in Ref. [2]. Here, the maximum and the distribution of enhancement factors for unattached and attached radon progenies are analyzed in a more detailed way as earlier for a physiologically realistic bifurcation geometry at low, intermediate and high flow rates.

3. Results

The local to average deposition density values, EF, depend strongly on the patch size of scanning. Here, we have selected only one size which is about 10×10 epithelial cells, namely, 100×100 μm. The reason for this selection is that the presence of several neighboring cells seems to be necessary for the development of a tumor. Fig. 1 displays the computed enhancement factor maxima, EF_{max}, in airway generation 3–4 during inhalation in the particle size range of 1–200 nm at sleeping, light physical activity and heavy exercise breathing conditions for adults. The number of selected particles was 100 000 in each case and the inlet flow, and thus the inlet particle distribution profile as well, were parabolic. The range of the maximum enhancement factors at the examined parameter selections is between 35 and 115. In the case of diffusion-dominated deposition (i.e., 1, 10 and 20 nm), EF_{max} values do not depend very much on the flow rate. In contrast, for larger particles, EF_{max} is consistently larger at the low flow rate than those for the other two higher flow rates. This may be explained by the greater dispersion of particle trajectories at higher flow rates, leading to a more extended deposition pattern and thus smaller EF_{max}.

From a biophysical point of view, it is also important to know the location of EF_{max} within the bifurcation. In the case of the wide-curved carina, the locations of the hot spots depend especially on flow rate and particle size because of the reverse flow in the vicinity

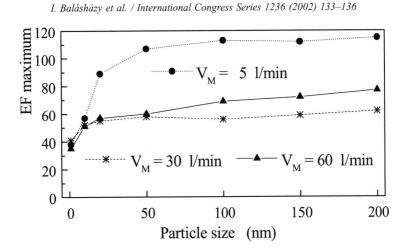

Fig. 1. The maximum enhancement factor as a function of particle size in airway generation 3–4 for three inspiratory minute volumes. Parameters: 100×100-μm patch size of scanning, physiologically realistic curved geometry, parabolic inlet velocity profiles, V_M: inspiratory minute volume, EF: enhancement factor.

of the carina. In case of submicron particles, such as radon progenies, usually, there is no deposition exactly at the peak of the carina, only a little bit over or below in the central zone or at the inner sides of the daughter airways. It is worth mentioning that deposition is strongly inhomogeneous even in the case of the lowest flow rate and smallest particle size ($EF_{max} = 35$).

Other characteristic of the inhomogeneity of the deposition patterns is the frequency distribution of the EF. Fig. 2 demonstrates this distribution for 1-and 200-nm particles at light physical activity breathing conditions and for the 100-μm patch size. In the case of 1-nm particles at 4600 patches (52%) and 200-nm particles at 7685 patches (87%), there is no deposition at all, that is, EF = 0. The plot of these values is cut by the top of the figure to

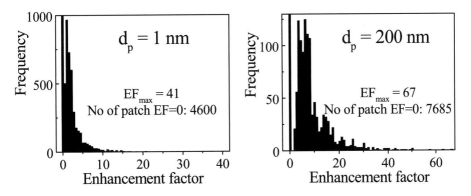

Fig. 2. Frequency distribution of enhancement factors in airway generation 3–4 for unattached and attached radon progeny particle sizes in the case of 30 l/min inspiratory minute volume. Parameters: d_p: particle diameter, others as in the previous figure.

visibly demonstrate the remainder part of the EF distributions, that is, where there is deposition. Both panels of the figure present the high degree of inhomogeneity of deposition along the surface of the bifurcation.

4. Summary and conclusion

The computed enhancement factors indicate that cells located in the vicinity of the carinal ridge or at the inner sides of the daughter branches may receive localized doses, which are two orders of magnitude higher than the average values in the case of radon progeny deposition in the large central human airways. As a result, the probability of multiple hits can be quite high even at low doses. Hence, the maximum EF may serve as a measure of the probability of multiple hits. If the number of multiple cellular hits plays a crucial role in lung cancer development, then the maximum EF could serve as a useful parameter for a mechanism-based risk analysis of inhaled radionuclides.

By applying in vitro cell transformation and inactivation probabilities for the computed data, we hope to receive useful information for the LNT hypothesis as well.

Acknowledgements

This research was supported by the Hungarian OTKA T030571 and T034564 Projects and by the CEC Contract No. MCFI-2000-01310, FIGH-CT-1999-00005 and FIGD-CT-2000-00053.

References

[1] I. Balásházy, T. Heistracher, W. Hofmann, Air flow and particle deposition patterns in bronchial airway bifurcations: the effect of different CFD models and bifurcation geometries, J. Aerosol Med. 9 (3) (1996) 287–301.
[2] I. Balásházy, W. Hofmann, Quantification of local deposition patterns of inhaled radon decay products in human bronchial airway bifurcations, Health Phys. 78 (2) (2000) 147–158.

Effects of a cell phone radiofrequency (860 MHz) on the latency of brain tumors in rats

B.C. Zook [a],*, S.J. Simmens [b]

[a] *Department of Pathology, Rm. 609, Ross Hall, The George Washington University Medical Center, 2300 Eye St., NW, Washington, DC 20037, USA*
[b] *Department of Epidemiology and Biostatistics, The George Washington University Medical Center, Washington, DC 20037, USA*

Abstract

The purpose of this study was to determine if an 860-MHz pulsed radiofrequency (PRF) shortened the latency of ethylnitrosourea (ENU)-induced brain tumors in rats. A total of 1080 Sprague–Dawley rats born of mothers given 6.3 or 10.0 mg/kg ENU intravenously on day 15 of gestation was randomized by ENU dose group into 18 groups of 60 rats (30 males and 30 females). Six groups were exposed to the PRF, 6 groups were sham exposed and 6 were cage controls. The PRF was delivered 6 h per day, 5 days per week beginning at 52 ± 3 days of age. The specific absorption rate was 1 ± 0.2 W/kg average to the brain. An equal number of rats from each exposure group was randomly selected for euthanasia and necropsy every 30 days from 828 rats between 172 and 322 days of age. A total of 252 rats were sacrificed because of the development of nervous disorders or they died spontaneously. There was no statistically significant difference between the incidence rates of brain tumors among PRF, sham or cage control groups in either the serial sacrifice or morbid groups. These results revealed no evidence that the PRF affects the incidence rate (or latency) of ENU-induced brain tumors in rats. © 2002 Elsevier Science B.V. All rights reserved.

Keywords: Radiofrequency; Cellular telephones; Microwaves; Brain cancer; Ethylnitrosourea

1. Introduction

Most animal and epidemiologic studies have suggested that there is little or no risk of brain cancer from the use of cellular telephones. Recent long-term initiation-promotion studies in rats have indicated no carcinogenic effects of cell phone radiofrequencies on

---

* Corresponding author. Tel.: +1-202-994-3391; fax: +1-202-994-2618.
  E-mail address: resbcz@gwumc.edu (B.C. Zook).

brain [1–3] or other tissues [3]. The ever-increasing use of these devices and the occasional report of physical effects induced by radiofrequencies, however, have warranted comprehensive study of the matter. The results of a large study conducted in this laboratory indicated that some rats born of mothers given ethylnitrosourea (ENU), and exposed to an 860-MHz pulsed radiofrequency (PRF), had a trend to develop fatal brain tumors with a shorter latency than controls [3]. Although the shortened latency was not statistically significant, it was deemed prudent to repeat the study, but this time euthanizing rats from each group at 30-day intervals to determine any consequence of the PRF on brain tumor latency.

2. Materials and methods

A total of 1080 Sprague–Dawley rats born of mothers given 6.3 or 10.0 mg/kg ENU intravenously on day 15 of gestation was studied. The rats were randomized by ENU dose groups into 18 groups of 60 rats (30 males and 30 females) each. Six groups (360 rats) were exposed to an 860-MHz PRF for 6 h per day, 5 days per week from 52 ± 3 days of age while restrained in Plexiglas tubes 2.0 ± 0.5 cm from the antenna. The 1.0-W average output from the antenna resulted in a specific absorption rate of 1 ± 0.2 W/kg average to the brain. Six groups of rats were sham exposed in an identical manner save no PRF and 6 groups were cage controls. The rats were euthanized and necropsied following random selection of an equal number from each group at 30-day intervals between 172 and 322 days of age. Coronal sections of brain were taken every millimeter and examined histologically. Rats that died spontaneously were similarly necropsied and studied histologically.

3. Results

Eight hundred twenty-eight (828) rats were serially sacrificed and 252 rats were sacrificed because of signs of nervous disorders or were found dead. The overall incidence of brain tumors in the serial sacrifice groups was PRF 48%, sham 52% and cage control 48%. Tables 1 and 2 depict the brain tumor incidence at 30-day intervals for the 6.3 and 10.0 mg/kg ENU dose groups. Of the 252 rats that developed nervous signs and were sacrificed or found dead, the incidence of brain tumors was PRF 45%, sham 58% and cage

Table 1
Brain tumor incidence among serially sacrificed rats exposed to 6.3 mg/kg ENU; rats irradiated with a PRF are compared to sham and cage control rats

Age at sacrifice (days)	172	202	232	262	292	322
Exposure groups						
PRF (%)	50	41	33	40	19	18
Sham (%)	50	33	32	52	32	38
Cage (%)	20	32	41	50	30	33

Table 2
Brain tumor incidence among serially sacrificed rats exposed to 10.0 mg/kg ENU; rats irradiated with a PRF are compared to sham and cage control rats

Age at sacrifice (days)	172	202	232	262	292
Exposure groups					
PRF (%)	67	63	73	58	57
Sham (%)	57	57	77	78	33
Cage (%)	73	40	58	80	60

controls 55%. A survival analysis of brain tumor incidence was performed for all 1080 rats conducted separately for the two ENU dose groups. This analysis revealed no significant difference between PRF, sham or cage control groups.

4. Discussion

The primary purpose of this study, to determine the effect of a PRF on brain tumor latency, has been accomplished. No evidence of an effect was discovered. Complimentary findings will follow; the histologic type, multiplicity, volume and malignancy of brain tumors are currently being studied. In addition, the incidence of cranial and spinal nerve and spinal cord tumors induced by ENU will be examined for possible promotion by the PRF. Finally, all relevant body tissues will be evaluated for any evidence of an effect due to the PRF.

5. Conclusion

This study revealed no evidence that the 860-MHz PRF used affects the incidence rate (or latency) of brain tumors induced by ENU in rats.

Acknowledgements

This study was supported in part by a grant from Motorola.

References

[1] W.R. Adey, C.V. Byus, C.D. Cain, R.J. Higgins, R.A. Jones, C.J. Kean, N. Kuster, A. MacMurray, R.B. Stagg, W. Haggren, Spontaneous and nitrosourea-induced primary tumors of the central nervous system in Fischer 344 rats chronically exposed to 836 MHz modulated microwaves, Radiat. Res. 152 (1999) 293–302.
[2] W.R. Adey, C.V. Byus, C.D. Cain, R.J. Higgins, R.A. Jones, C.J. Kean, N. Kuster, A. MacMurray, R.B. Stagg, G. Zimmerman, Spontaneous and nitrosourea-induced primary tumors of the central nervous system in Fischer 344 rats exposed to frequency-modulated microwave fields, Cancer Res. 60 (2000) 1857–1863.
[3] B.C. Zook, S.J. Simmens, The effects of 860 MHz radiofrequency radiation on the induction or promotion of brain tumors and other neoplasms in rats, Radiat Res. 155 (2001) 572–583.

Molecular analysis of radiogenic tumors: Experimental

Genetic analysis of radiation-induced thymic lymphoma

R. Kominami [a,b,*], Y. Saito [a], T. Shinbo [a], A. Matsuki [a],
H. Kosugi-Okano [a], A. Matsuki [a], Y. Ochiai [a], Y. Kodama [a],
Y. Wakabayashi [a], Y. Takahashi [a], Y. Mishima [a], O. Niwa [b]

[a] *Department of Gene Regulation, Niigata University Graduate School of Medical and Dental Sciences, Asahimachi 1-757, Niigata 951-8122, Japan*
[b] *Radiation Biology Center, Kyoto University, Yoshida-Konoecho, Sakyou, Kyoto, 606-8315, Japan*

Abstract

Mouse thymic lymphomas are one of the classic models of radiation-induced malignancies. However, little genetic study has been performed, although the mouse systems offer a number of useful features for genetic and physical mapping. We have carried out large-scale mapping toward the isolation of the genes involved in lymphoma development. Two different types of genes are chosen as targets for positional cloning. One is the tumor suppressor gene and the other is the susceptibility or resistance-giving gene, which predispose to the lymphoma development. One susceptibility locus was localized near *D4Mit12* on chromosome 12 by an association study for backcross and congenic mice, and three loci, probably harboring a tumor suppressor gene, were localized by allelic loss mapping on physical maps that were covered by BAC clones. The maps are invaluable to facilitate the identification of candidate tumor suppressor genes. Also, success in identification of *Ikaros* as a tumor suppressor gene is described. © 2002 Elsevier Science B.V. All rights reserved.

Keywords: Tumor suppressor gene; Tumor susceptibility; Positional cloning; Mouse thymic lymphoma; Polymorphism

1. Introduction

Mouse thymic lymphomas are one of the classic models of radiation-induced malignancies and have offered important clues to how cells become malignant [1]. Recent

---

* Corresponding author. Tel.: +81-25-227-2077; fax: +81-25-227-0757.
 E-mail address: rykomina@med-niigata-u.ac.jp (R. Kominami).

0531-5131/02 © 2002 Elsevier Science B.V. All rights reserved.
PII: S0531-5131(01)00743-9

studies show a series of genes that affect lymphoma induction in mice, and these include K-ras, p53, ATM, mismatch repair genes, and p21 [2–4]. However, few formal genetic studies have been performed, although mice systems offer a number of useful features for genetic and physical mapping. They include the availability of thousands of polymorphic markers and the ability to breed genetically informative hybrid mice that can provide an essentially unlimited number of tumors.

We have conducted large-scale genetic mapping toward the isolation of the genes involved in thymic lymphomas [5–9]. Two different types of genes are chosen as targets of positional cloning. One is the tumor suppressor gene and the other is the susceptibility or resistance-giving gene, which predispose to lymphoma development.

2. Mapping a susceptibility locus on chromosome 4

Epidemiological data suggest that cancer-susceptibility genes are of low penetrance but comprise an important hereditary genetic component affecting the cancer incidence [10]. Mapping and isolation of such genes in humans is complicated, but animal models of susceptibility have proved to be useful [10,11]. BALB/c is a susceptible strain to the development of thymic lymphoma after exposure to γ-ray radiation [12], while MSM, an inbred strain derived from Japanese wild mice, *Mus musculus molossinus*, shows resistance [5]. Crossing of the sensitive and resistant strains may provide valuable information about the number of genes that are involved, their approximate locations, and the dominant or recessive nature of the susceptibility.

We thus generated a total of 293 backcross mice from BALB/c and MSM strains, which were subjected to four times γ-irradiation of 2.5 Gy at weekly intervals at an age of 4 weeks. The mice were inspected for labored breathing, and thymic lymphomas were confirmed upon autopsy of the mice. The intersubspecific F1 hybrid and N2C backcross mice (backcrossed to BALB/c) showed incidence similar to that of BALB/c, while N2M mice (backcrossed to MSM) exhibited resistance to lymphoma development, suggesting multiple tumor susceptibility loci in the BALB/c genome. Accordingly, N2M mice were used for an association study. Genotyping was carried out with polymerase chain reaction (PCR) of 67 micro satellite marker loci [13], and the average and maximum distances between the marker loci were 24 and 32 cm, respectively. Linkage was evaluated by the χ^2 test for goodness of fit against an expected 1:1 ratio of BALB/MSM and MSM/MSM genotypes. Results revealed three marker loci with skewed genotypic ratios: at *D2Mit15* on chromosome 2, at *D4Mit12* on chromosome 4, and at *D5Mit5* on chromosome 5 (Table 1). Since the three markers were the most closely linked, susceptibility loci were likely to reside near the three marker loci. The evidence for each linkage was modest, however, especially given the concern about the use of multiple statistical tests [14]. None of the three loci provided significant linkage; their linkages were marginally suggestive.

We generated mice partially congenic for *D4Mit12* by backcrossing MSM to BALB/c five or six times according to the marker-assisted protocol [15]. The partially congenic mice were mated with MSM mice and their 78 offspring were subjected to γ-ray irradiation. Genotyping with five markers, *D4Mit9*, *D4Mit331*, *D4Mit278*, *D4Mit12*, and *D4Mit338*, revealed that 32 mice received the BALB/c chromosome from the heterozygous mice and

Table 1
The association of markers with the development of γ-ray-induced thymic lymphoma

	Lym(+)	Lym(+)	Lym(−)	Lym(−)	χ^2 value	P value
Marker locus	C/M	M/M	C/M	M/M		
D2Mit15	38	22	66	94	8.54	0.0035
D4Mit12	38	22	67	93	8.05	0.0045
D5Mit5	21	39	89	71	3.52	0.061

C/M, heterozygous for BALB/c and MSM alleles; M/M, homozygous for MSM alleles. χ^2 and P values were obtained by analysis with StatView-J 4.11 software on a Macintosh personal computer. Evaluation of linkage followed the criteria of Lander and Kruglyak [14]. Linkage was taken as suggestive in the backcross mice when P was less than 3.4×10^{-3}.

46 inherited the recombinant chromosome bearing the MSM genome, eight of which underwent recombination in the heterozygous region (Fig. 1A). The χ^2 test for the association at each of the four loci provided the lowest P value at the *D4Mit12* locus (Fig. 1B). Fig. 1C shows the cumulative lymphoma incidence of the homozygotes and the heterozygotes at the *D4Mit12* locus. The two groups showed a significant difference in the incidence and latency ($P = 0.0037$ in the Mantel–Cox test). These data confirmed the presence of a dominant susceptibility-giving allele near the *D4Mit12* locus on the BALB/c chromosome.

In general, susceptibility/resistance locus could involve factors either extrinsic or intrinsic to the tumor cell lineage. Action extrinsic is exemplified by the activity of the secretary phospholipase A2 gene, that is a modifier of the *Apc*-gene mutation in intestinal

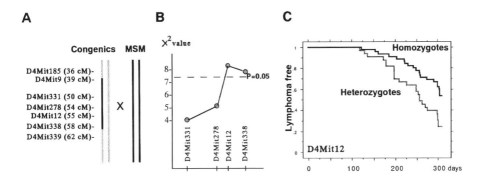

Fig. 1. (A) Schematic drawing of a part of chromosome 4. Gray lines display chromosomal regions of BALB/c and black lines represent chromosomal regions of MSM. The partially congenic mice were obtained by backcrossing five or six times according to the marker-assisted protocol. Six backcross mice used in this experiment carry at most the 25 cm interval of the MSM genome marked by the black line. The six mice were mated with MSM and their 78 progeny were subjected to irradiation. (B) χ^2 value of the Mantel–Cox test is shown. Evaluation of linkage followed the criteria of Lander and Kruglyak [14]. The statistical threshold of the congenic mouse data was corrected for multiple comparisons using the formula, $m(T) = [C + 2rGT^2]$ a (T): C (number of chromosomes) = 1, r (to account for corrected results among linked loci: for backcross with 1 df) = 1, G (genome size in organs) = 0.25, T (for the χ^2 statistic with 1 df) = 3.84, and $m(T) = 0.05$. To achieve a false positive rate of 5% or less in the analysis of markers on the congenic region, we used a $P = 0.0060$ threshold. (C) Lymphoma frequencies in the C/M heterozygotes and M/M homozygotes at *D4Mit12* of 78 hybrids that were produced by crossing partially congenic mice to MSM.

tumors [8,16–18]. The phospholipase A2 is secreted and acts outside of the tumor cells, and it probably affects the net growth rate and multiplicity of adenomas by a yet to be clarified mechanism. Furthermore, action intrinsic might involve mutational inactivation of susceptibility/resistance genes required for tumor development.

Efforts are now being made to generate more congenic strains for the *D4Mit12*. Those mice will be useful for lymphoma susceptibility testing to further narrow down the interval of the susceptibility locus. A recent advance in human and mouse genome projects will aid the identification of transcripts or candidate genes within the loci. Subsequent cloning of these genes would elucidate the underlying mechanisms of cancer susceptibility.

2.1. Allelic loss mapping

Identification of novel tumor suppressor genes is important for understanding the molecular mechanism underlying carcinogenesis, but only about 25 suppressor genes have been identified. Many human and murine cancers contain mutations in tumor suppressor genes, which often map to chromosomal regions that exhibit allelic loss or loss of heterozygosity. Allelic loss analysis is therefore a standard method to locate the regions carrying tumor suppressor genes [19,20]. We performed a genome-wide allelic loss (LOH) analysis for murine thymic lymphomas that were induced by γ-irradiation in F_1 hybrid mice, between BALB/c and MSM strains [5]. Among the 62 micro satellite loci that were examined, three loci, *D11Mit71*, *D12Mit279*, and *D16Mit122*, exhibited high frequencies of allelic loss, 40%, 65%, and 45%, respectively. This suggested that novel tumor suppressor genes may reside in these chromosomal regions. Further analysis assigned the LOH region on chromosome 11 to an interval between *D11Mit71* and *D11Mit19* where a putative tumor suppressor gene, *Ikaros*, was localized. On the other hand, no candidate tumor suppressor gene was found either near loci Tlsr4 on mouse chromosome 12 or Tlsr7 on chromosome 16. Positional cloning requires not only genetic mapping but also physical mapping that provide sequence data of candidate tumor suppressor regions.

2.2. Physical mapping of Tlsr4 and Tlsr7

Four probes of *D12Mit53*, *D12Mit233*, *D12Mit279*, and *D12Mit80* on chromosome 12 were used as starting points for screening a YAC library and 11 YAC clones were isolated (Fig. 2A). Their centromeric and telomeric ends were recovered and sequenced to yield sequence-tagged sites (STSs). Two STS markers, *Y11(II)* and *YE6(II)*, detected polymorphisms between BALB/c and MSM strains, and hence, the markers were used for fine LOH mapping of Tlsr4. Among the 251 informative lymphomas examined in total, 13 lymphomas and 5 lymphomas still retained both alleles in the *Y11(II)* and *YE6(II)* marker sites, respectively (Fig. 2A).

To further narrow down the LOH peak, we constructed a BAC contig consisting of 15 overlapping BAC clones (Fig. 2B). Four polymorphic STSs of *B19E(I)*, *MCA1*, *B2N(I)*, and *B3N(II)* were obtained by sequence analysis of the BAC DNA, followed by detecting the variation between the two strains. Thirty-three informative samples were subjected to allelic loss mapping using the four BAC STS probes together with the two YAC-end probes. The frequency of allelic loss was highest in the *B2N(I)* locus (69%). A comparison

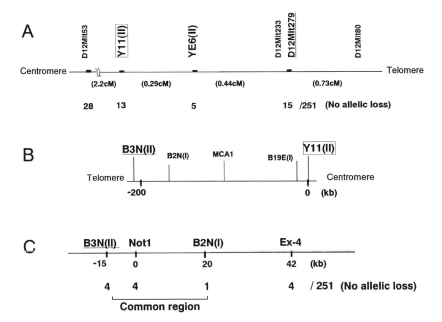

Fig. 2. (A) The Tlsr4 region exhibiting frequent allelic loss in the vicinity of *D12Mit279* on mouse chromosome 12. The position of markers is displayed above the bar and genetic distances (cM) between marker positions are shown in parentheses. Numbers below a thin line indicate lymphomas retaining both alleles in 251 informative lymphomas examined. (B) Physical map of the Tlsr4 locus by BAC contigs. The position of BAC-end and YAC-end markers is indicated above the line. MCA1 is a CA-repeat marker. (C) The four markers on the line detect polymorphisms, and were thereby used for allelic loss mapping. Physical distances (kb) between markers are shown under the line. Numbers below indicate lymphomas retaining both alleles in the 251 informative lymphomas.

of the extent of allelic loss in the six polymorphic sites showed boundaries of allelic losses relative to the flanking markers (Fig. 2C). The region between *B3N(II)* and *B2N(I)* was the only site to be consistently lost in these lymphomas, and was 35 kb long from the pulsed-field gel electrophoresis analysis of BAC DNA [7]. Random DNA sequencing of BACs covering the 35-kb region has revealed a gene that encodes a novel 884-amino acid protein comprising six zinc finger domains of the C2H2 type (Accession numbers: AB043551). Details of the analysis carried out on the gene and its relevance for lymphomagenesis will be published elsewhere.

A similar genetic and physical mapping was carried out of the vicinity of Tlsr7 on chromosome 16. Finally, we constructed a physical map of the Tlsr7 region near the *D16 Mit122* locus by scanning a total of 587 thymic lymphomas. The map consisted of 13 overlapping BAC clones, and isolation of the BAC-derived polymorphic probes led to fine mapping of the allelic losses. Eleven lymphomas showed informative breakpoints of allelic loss regions relative to the flanking markers on the map. Pulsed-field gel electrophoresis of *Not*I digests of the clones, showed that the commonly lost region was localized within an approximately 300-kb interval near *D16Mit192* [8]. These maps are invaluable to facilitate the identification of the genes in the Tlsr4 and Tlsr7 regions.

3. Identification of *Ikaros* as a tumor suppressor gene

Genetic mapping localized the LOH region on chromosome 11 within the interval between *D11Mit71* and *D11Mit19*. Interestingly, to this region, the mouse *Ikaros* gene was mapped [21]. The *Ikaros* gene encodes, by alternative splicing, a family of zinc finger proteins essential for the development of the lymphoid system [22–24]. Homozygous mice for a deletion in the *Ikaros* DNA-binding domain failed to generate mature lymphocytes, whereas heterozygous mice possessed lymphocytes with normal cell surface antigens during the first month of their lives, but underwent dramatic changes in T cell populations shortly afterwards [21,22]. The heterozygotes developed lympho-proliferative disorders and ultimately died of T-cell leukemias and lymphomas.

Fine allelic loss mapping of 191 lymphomas indicated that the critical region of allelic loss was centered at the *Ikaros* locus. Since it suggested *Ikaros* as a candidate, homozygous deletion of the gene was examined with PCR for 108 lymphomas that were selected by allelic loss screening in the region surrounding the *Ikaros* locus. The results showed homozygous deletions at the *Ikaros* gene region in 9 of the 108 lymphomas, and that some of the deleted regions may be limited within the coding regions. Subsequent Southern blot hybridization also confirmed these findings. We next examined mutations of the *Ikaros* gene. The N-terminal zinc finger domain and activation domain regions were chosen for mutation analysis [24]. Six and five point mutations were identified in the N-terminal zinc finger domain and the activation domain regions, respectively (Fig. 3). Among the eleven mutations, five resulted in amino acid-substitutions, two of which changed key amino acid residues of cysteine and histidine in the zinc finger motif. Three were mutations producing stop codons, and the other three comprising of one-base insertion or deletion resulted in frame shifts.

Radiation is known to induce DNA strand breaks, which are repaired by homologous and illegitimate recombination processes. The latter mechanism of repair frequently produces promiscuous rejoining of two DNA ends, and it is well accepted that the major type of radiation-induced mutations is a so-called large mutation, such as deletion and translocation

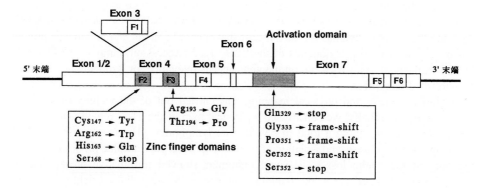

Fig. 3. Mutations of the *Ikaros* gene. Two zinc finger regions, F2 and F3, and the activation domain region were examined for small mutation, and 11 missense or nonsense mutations were detected. The number after the amino acid indicates the position, assuming that the first ATG initiation codon is 1.

[25]. Although homozygous deletion of the *Ikaros* gene in radiogenic thymic lymphomas is consistent with the mode of action of radiation, small mutations found in the present study contradict the expected spectrum of radiation-induced mutation. Therefore, the role of radiation in the development of thymic lymphomas may not simply be the direct one in which radiation induces carcinogenic mutation on the T lymphocytes, but may involve a variety of biological processes such as stimulation of DNA synthesis during thymic regeneration and delayed mutation in the irradiated target cells [26].

The mouse *Ikaros* gene has been postulated to participate in the proliferation of thymocytes and oncogenic process from studies of *Ikaros* KO mice. However, no genetic evidence has been presented implicating this gene in the development of lymphomas in *Ikaros* wild-type mice. Besides, the importance of *Ikaros* inactivation in lymphomagenesis was not addressed. Here, we have presented evidence for a role of *Ikaros* in oncogenesis of mice thymic lymphomas.

Acknowledgements

This work was supported by grants-in-aid for Research on Human Genome and Gene Therapy from the Ministry of Health and Welfare of Japan.

References

[1] H.S. Kaplan, M.B. Brown, J. Natl. Cancer Inst. 13 (1952) 185–208.
[2] T. Jacks, Annu. Rev. Genet. 30 (1996) 603–633.
[3] A.J. Giaccia, M.B. Kastan, Genes Dev. 12 (1998) 2973–2983.
[4] P.A. Jeggo, A.M. Carr, A.R. Lehmann, Trends Genet. 14 (1998) 312–316.
[5] Y. Matsumoto, S. Kosugi, T. Shinbo, D. Chou, M. Ohashi, Y. Wakabayashi, K. Sakai, M. Okumoto, N. Mori, S. Aizawa, O. Niwa, R. Kominami, Oncogene 16 (1998) 2747–2754.
[6] H. Okano, Y. Saito, T. Miyazawa, T. Shinbo, D. Chou, S. Kosugi, Y. Takahashi, S. Odani, O. Niwa, R. Kominami, Oncogene 18 (1999) 6677–6683.
[7] T. Shinbo, A. Matsuki, Y. Matsumoto, S. Kosugi, Y. Takahashi, O. Niwa, R. Kominami, Oncogene 18 (1999) 4131–4136.
[8] A. Matsuki, H. Kosugi-Okano, Y. Ochiai, S. Kosugi, T. Miyazawa, Y. Wakabayashi, K. Hatakeyama, O. Niwa, R. Ryo Kominami, Biochem. Biophys. Res. Commun. 282 (2001) 16–20.
[9] Y. Saito, Y. Ochiai, Y. Kodama, Y. Tamura, T. Togashi, H. Kosugi-Okano, T. Miyazawa, Y. Wakabayashi, K. Hatakeyama, S. Wakana, O. Niwa, R. Ryo Kominami, Oncogene, in press.
[10] A. Balmain, H. Nagase, Trends Genet. 14 (1998) 139–144.
[11] W.F. Dietrich, E.S. Lander, J.S. Smith, A.R. Moser, K.A. Gould, C. Luongo, N. Borenstein, W. Dove, Cell 75 (1993) 631–639.
[12] M. Okumoto, R. Nishikawa, S. Imai, J. Hilgers, Cancer Res. 50 (1990) 3848–3850.
[13] W.F. Dietrich, J. Miller, R. Steen, M.A. Merchant, D. Damron-Boles, Z. Husain, R. Dredge, M.J. Daly, K.A. Ingalls, T.J. O'Connor, C.A. Evans, M.M. DeAngelis, D.M. Levinson, L. Kruglyak, N. Goodman, N.G. Copeland, N.A. Jenkins, T.L. Hawkins, L. Stein, D.C. Page, E.S. Lander, Nature 380 (1996) 149–152.
[14] E. Lander, L. Kruglyak, Nat. Genet. 11 (1995) 241–247.
[15] M. Liyanage, Z. Weaver, C. Barlow, A. Coleman, D.G. Pankratz, S. Anderson, A. Wynshaw-Boris, T. Ried, Blood 96 (2000) 1940–1946.
[16] P. Markel, P. Shu, C. Ebeling, G.A. Carlson, D.L. Nagle, J.S. Smutko, K.J. Moore, Nat. Genet. 17 (1997) 280–284.

[17] L.-K. Su, K.W. Kinzler, B. Vogelstein, A.C. Preisinger, A.R. Moser, C. Luongo, K.A. Gould, W.F. Dove, Science 256 (1992) 668–670.
[18] M. Macphee, K.P. Chepenik, R.A. Liddell, K.K. Nelson, L.D. Siracusa, A.M. Buchberg, Cell 81 (1995) 957–966.
[19] R.T. Cormier, K.H. Hong, R.B. Halberg, T.L. Hawkins, P. Richardson, R. Mulherkar, W.F. Dove, E.S. Lander, Nat. Genet. 17 (1997) 88–91.
[20] A. Darvasi, Nat. Genet. 18 (1998) 19–24.
[21] A.G. Knudson, Cancer Res. 45 (1985) 1437–1443.
[22] R.A. Weinberg, Science 254 (1991) 1138–1146.
[23] S. Winandy, P. Wu, K. Georgopoulos, Cell 83 (1995) 289–299.
[24] K. Georgopoulos, D.D. Moore, B. Derfler, Science 258 (1992) 808–812.
[25] K. Georgopoulos, M. Bigby, J.-H. Wang, A. Molnar, P. Wu, S. Winandy, A. Sharpe, Cell 79 (1994) 143–156.
[26] A. Molnar, K. Georgopoulos, Mol. Cell. Biol. 14 (1994) 8292–8303.
[27] J. Thacker, Adv. Radiat. Biol. 16 (1992) 77–117.
[28] J.B. Little, H. Nagasawa, T. Pfenning, H. Vetrovs, Radiat. Res. 148 (1997) 299–307.

Genetic analysis of radiation-induced mouse hepatomas

Kenji Kamiya*, Masaharu Sumii, Yuji Masuda, Tsuyoshi Ikura, Norimichi Koike, Mamoru Takahashi, Jun Teishima

Department of Developmental Biology and Oncology, Division of Molecular Biology, Research Institute for Radiation Biology and Medicine, Hiroshima University, 1-2-3 Kasumi, Minami, Hiroshima 734-8553, Japan

Abstract

In order to understand the molecular mechanism of radiation carcinogenesis, we conducted two experiments. Firstly, we analyzed overexpressed 14 genes in radiation-induced mouse hepatomas, including novel two genes named *CRAD3*, which was a member of *cis*-retinol/androgen dehydrogenase (CRAD) family, and *Sdf2l1*, which was a member of Pmt/rt family. CRAD plays an important role in androgen metabolism, which converts inactive 3α-adiol into active dihydrotestosterone and consequently increases androgen activity. Actually, oxidative 3α-hydroxysteroid dehydrogenase activity in mouse hepatomas was found to be higher than that in normal liver at physiological 3α-adiol level. Dihydrotestosterone is well known to promote hepatocarcinogenesis. Therefore, the overexpression of *CRAD3* must modify the radiation-induced mouse hepatocarcinogenesis by increasing local dihydrotestosterone level. Secondly, we compared the mutation spectrum of β-catenin and found that there were no statistical differences in the frequency and sites of β-catenin mutation between radiation-induced and spontaneous hepatomas. Since the similarity of mutation spectra between radiation-induced and spontaneous cancers was well documented, we try to clarify the involvement of spontaneous mutations in radiation carcinogenesis. We characterized mouse *Rev1* gene which was a member of the *UmuC/DinB/XPV* gene family, played important roles in spontaneous mutations. Biochemical analysis of the mouse Rev1 protein revealed that the mouse Rev1 protein possessed not only deoxycytidyl transferase activity as human REV1 protein, but also the ability to insert a dGMP or a dTMP residue opposite template guanine and an AP site. The expression of mouse *Rev1* gene of primary embryonic fibroblasts in culture was induced by radiation exposure. These observations might

Abbreviations: CRAD, *cis*-retinol/androgen dehydrogenase; HCC, hepatocellular carcinoma; 3α-HSD, 3α-hydroxysteroid dehydrogenase; SDRs, short chain dehydrogenases/reductases; AP, apurinic/apyrimidinic; h6-mRev1, mouse Rev1 protein tagged with hexa-histidine.

* Corresponding author. Tel.: +81-82-257-5842; fax: +81-82-257-5844.

E-mail address: kkamiya@hiroshima-u.ac.jp (K. Kamiya).

be important for the study of molecular mechanism in radiation carcinogenesis. © 2002 Elsevier Science B.V. All rights reserved.

Keywords: Radiation carcinogenesis; Mutation; Hepatoma; CRAD; Rev1

1. Introduction

Although many epidemiological data have indicated that radiation could induce human cancer, molecular mechanisms of radiation carcinogenesis had not been elucidated. Rodent models are the suitable model to study the molecular mechanisms of radiation carcinogenesis. We analyzed radiation-induced mouse hepatomas to get some clues to understand the molecular mechanisms of radiation carcinogenesis from the following view points: (1) The Analysis of overexpressed genes in hepatomas, (2) the comparison of mutation spectrum of β-catenin between radiation-induced and spontaneous hepatomas.

Furthermore, we analyzed mouse *Rev1* gene, which is the member of UmuC/XPV/DinB family and is speculated to play a critical role in damage-induced and spontaneous mutations by its translation DNA replication activity, to elucidate the relationship between radiation carcinogenesis and spontaneous mutations.

2. Cloning and characterization of new genes overexpressed in radiation-induced mouse hepatomas by differential display analysis

Mouse hepatomas were induced in B6C3F1 mice by 3Gy of ^{60}Coγ-ray exposure, and mRNAs were isolated from hepatomas and normal liver in the same mice. To identify differentially expressed genes between mouse HCC and normal liver, we used a differential display technique, and subsequently Northern blot analysis. We analyzed 19 differentially expressed genes in radiation-induced mouse hepatomas. Expressions of five out of 19 genes were decreased and those of other 14 genes, including novel two genes named *CRAD3* which was a member of *cis*-retinol/androgen dehydrogenase (*CRAD*) family and *Sdf2l1* which was a member of Pmt/rt family, were increased in hepatomas [1].

2.1. Cloning of CRAD3

Nucleotide sequence analysis of the detected cDNA fragment did not match any known genes or homologues. Using this cDNA fragment as a probe, cDNA clones of *CRAD3* were isolated from mouse HCC library. The full-sized cDNA sequence of *CRAD3* was 2644 bp in length and encoded a protein of 317 aa. The deduced amino acid sequence of *CRAD3* belonged to short chain dehydrogenases/reductases (SDRs). Because of close amino acid sequence similarity between *CRAD3* and other SDRs, especially *CRAD1* or *CRAD2* [2,3], *CRAD3* occurred with a *cis*-retinol/3α-hydroxysterol dehydrogenase isozyme.

2.2. Tissue distribution of CRAD3 mRNA

The highest expression of *CRAD3* mRNA was observed in liver, and relatively high expression of *CRAD3* mRNA was seen in kidney and adrenal glands. *CRAD1* mRNA and *CRAD2* mRNA were also highly expressed in liver, kidney and adrenal glands. However, tissue distribution of these three enzymes were obviously different.

2.3. Enzyme activity of recombinant CRAD3

We examined whether *CRAD3* catalyzed 3α-hydroxysteroids or *cis*-retinols (or all-*trans*-retinol) as its substrates like *CRAD1* or *CRAD2* [2,3]. As expected, recombinant *CRAD3* could oxidize 3α-adiol and androsterone to dihydrotestosterone and androstanedione, respectively, in the presence of NAD^+ as a cofactor. The K_m values of *CRAD3* for 3α-adiol and androsterone were 0.04 and 0.1 µM, respectively. This result indicated that *CRAD3* would reveal the highest oxidative 3α-HSD activity among CRADs at a physiological concentration of 3α-adiol.

2.4. Oxidative 3α-HSD activity in mouse HCCs

We investigated oxidative 3α-HSD activities in mouse HCCs and normal livers using their homogenates. We determined their oxidative 3α-HSD activities as the amount of produced dihydrotestosterone using 3α-adiol as substrate. Oxidative 3α-HSD activity in mouse HCCs was significantly increased at 10 nM 3α-adiol as substrates [0.076±0.012 nmol/min/mg in HCCs versus 0.020±0.002 nmol/min/mg in normal livers ($p<0.001$)]. Therefore, oxidative 3α-HSD activity for 3α-adiol was thought to be higher at physiological 3α-adiol level in mouse HCC than normal liver.

2.5. The role of overexpressed CRAD3 in mouse hepatocarcinogenesis

It is well known that testosterone promotes hepatocarcinogenesis. *CRAD3* was one of the SDRs and had close similarity in amino acid sequence with *CRAD1* and *CRAD2* [2,3]. As expected, *CRAD3* as well as *CRAD1* and *CRAD2* oxidized 3α-adiol and androsterone to dihydrotestosterone and androstanedione, respectively.

Dihydrotestosterone is the most powerful ligand in androgens for the androgen receptor. Dihydrotestosterone is inactivated to 3α-adiol in a reversible reaction catalyzed by reductive 3α-HSD, while 3α-adiol is oxidized back to active androgen, dihydrotestosterone, by oxidative 3α-HSD like CRADs [2–4]. Therefore, 3α-adiol can be an important source for dihydrotestosterone synthesis by oxidative 3α-HSD. Indeed, dihydrotestosterone was detected in castrated and functionally hepatectomized rats, and 5α-reductase inhibitors did not obliterate androgen activity [5]. Moreover, even the administration of 3α-adiol, inactive androgen, stimulated the growth of prostate in vivo [6–8] and in prostate organ culture [9]. These data indicate that oxidative 3α-HSDs including CRADs take an important part in increasing local active androgen level in androgen target tissues. Therefore, overexpressed *CRAD3* may modify mouse hepatocarcinogenesis by increasing the local active androgen level.

3. Comparison of mutation spectrum of β-catenin in mouse hepatomas

It was well documented that mutation spectrum of p53 in lung cancers observed among uranium miners were similar to that observed among unirradiated people. In human and mouse HCCs, nuclear accumulation of β-catenin had been reported in nearly 50%, suggesting that defects in the Wnt signaling pathway and following overexpressions of β-catenin Tcf target genes played a crucial role in hepatocarcinogenesis. This result also indicated that the mutation of β-catenin was one of the most common genetic events in hepatocarcinogenesis. We compared the mutation spectrum of β-catenin between radiation-induced and spontaneous mouse hepatomas to get some information about the feature of genetic events in radiation-induced hepatomas.

3.1. β-catenin mutations

After RT-PCR-SSCP and sequence analyses, missense mutations of β-catenin involving the GSK-3β phosphorylation sites were confirmed in 7 of 36 HCCs (19.4%) (3 of 13 spontaneous HCCs (23.1%) and 4 of 23 radiation-induced HCCs (17.4%). There were no statistical differences in the frequency of β-catenin mutation between spontaneous and radiation-induced HCCs. Codons 33, 37, 41 and 45 are phosphorylable residues by GSK-3β and most β-catenin mutations reported in human and mouse HCCs were found at these codons or at neighbors of these codons. All missense mutations in this study were also found at phosphorylable residues or at their neighbors in the GSK-3β phosphorylation site of β-catenin.

4. New approach to radiation carcinogenesis

In *Escherichia coli* and *Saccharomyces cerevisiae*, DNA damage-induced mutagenesis is regulated by a cellular system which may be conserved from bacteria to humans [10]. In *S. cerevisiae*, members of the *RAD6* epistasis group play a critical role in this system [10–18]. The *REV1* gene belongs to the *RAD6* group and is required for damage-induced and spontaneous mutagenesis [12,15–17,19,20]. A defect in the *REV1* gene was found to decrease the translesion replication of an apurinic/apyrimidinic (AP) site, a T–T (6-4) UV photoproduct and a N-2-acetylaminofluorene (AAF)-modified guanine [21]. The *REV1* gene encodes a protein containing a BRCT domain in its N-terminus, and the protein possesses a deoxycytidyl transferase activity in a template-directed reaction [22].

It is speculated that the translesion DNA replication plays a role in carcinogenesis, because mutagenesis is an important factor during the initiation and progression of human cancer. Accumulated data also suggested that spontaneous mutation might play important role in radiation carcinogenesis. So we focus our study on *REV1* gene to elucidate molecular mechanisms of radiation carcinogenesis.

In this work, we characterized the *Rev1* gene of the mouse, a commonly used animal model for studying human diseases.

4.1. Primary structure of mouse Rev1 protein

The mouse *Rev1* cDNA encodes a putative protein of 1249 amino acid residues with a calculated molecular mass of 137 kDa. Comparison of the amino acid sequences of the human and mouse Rev1 proteins revealed an overall amino acid identity of 84% and similarity of 90%. All of the motifs found in the human REV1 protein were conserved in mouse Rev1 protein. The BRCT domain, motif I and motif VIII are specific to the Rev1 family. Motifs II, III, IV, V, VI and VII are conserved to the polymerases of the UmuC superfamily. We previously showed the minimum region required for the deoxycytidyl transferase activity of the human REV1 protein [23]. This region was highly conserved with 88% identity and 94% similarity. This supported our previous finding that the region was important for the activity [23].

4.2. Expression of the Rev1 gene in mouse tissues

Northern blot analysis detected the mouse *Rev1/Rev1s* mRNA in all tissues examined, indicating that the *Rev1* gene is ubiquitously expressed. Expression of the *Rev1* gene was relatively high in the heart, skeletal muscle and testis.

4.3. Damage response of mouse Rev1 gene expression

The expression of the yeast *Rev1* gene does not change after UV irradiation or treatment with 4-nitroquinoline-N-oxide [24]. We investigated whether the mouse *Rev1* gene was induced by DNA damage. The cells were irradiated with γ rays and mRNA was isolated for Northern blot analysis at 7 h after γ irradiation. In contrast to the yeast, the mouse *Rev1* gene was slightly induced by γ ray irradiation in a low dose range. The *Rev1* expression was increased in a dose-dependent manner until 5 Gy and then decreased. We also examined the damage responses of *Polh* (*Rad30A/XPV*), *Poli* (*Rad30B*) and *Polk* (*DinB1*) genes, which are other members of the *umuC* gene family. The profiles of the gene expressions were quite similar.

4.4. Deoxyribonucleotidyl transferase activity of the mouse Rev1 protein

Using six different primer-templates, we examined the substrate specificity of the transferase activity of mouse Rev1 protein tagged with hexa-histidine (h6-mRev1) protein in the presence of each of 0.1 mM dNTP. In this experiment, the respective primer-templates differ only at the template nucleotide immediately downstream from the annealed primer. As a G template was incubated with h6-mRev1 protein and each of the dNTPs, we surprisingly detected a one-base-extended product in the presence of not only dCTP but also dGTP and dTTP. These abilities have not been reported in the human REV1 protein [25]. This result indicates that the h6-mRev1 protein has a potential for transfer of a dGMP and a dTMP residue. The efficiencies for insertion of a dGMP and a dTMP were five-times less than that of a dCMP. In the presence of all four dNTPs, the dCMP transfer reaction predominated over the dGMP and dTMP transfer reactions. We also found abilities to insert a dGMP and a dTMP opposite the template AP site.

As other templates were tested, unexpectedly, the h6-mRev1 protein inserted a dCMP opposite not only the template AP site but also all bases examined. The insertion efficiency opposite the template AP site was slightly higher than that opposite templates G, A and U (1.5–2 fold) and six times higher than that opposite templates T and C. These activities were proportional to the enzyme concentration, and the kinetics was linear until 15 min.

These results indicated that the mouse Rev1 protein possessed the ability to insert a dGMP or a dTMP residue opposite template guanine and an AP site. This ability of Rev1 protein might be required for damage bypass replication of unidentified lesions.

References

[1] S. Fukuda, M. Sumii, Y. Masuda, M. Takahashi, N. Koike, J. Teishima, H. Yasumoto, T. Itamoto, T. Asahara, K. Dohi, K. Kamiya, Biochem. Biophys. Res. Commun. 280 (1) (2001) 407–414.
[2] X. Chai, Y. Zhai, J.L. Napoli, J. Biol. Chem. 272 (1997) 33125–33131.
[3] J. Su, X. Chai, B. Kahn, J.L. Napoli, J. Biol. Chem. 273 (1998) 17910–17916.
[4] W.H. Gough, S. VanOoteghem, T. Sint, N. Kedishvili, J. Biol. Chem. 273 (1998) 19778–19785.
[5] D.W. Russell, J.D. Wilson, Annu. Rev. Biochem. 63 (1994) 25–61.
[6] N. Bruchovsky, Endocrinology 89 (1971) 1212–1222.
[7] R.J. Moore, J.D. Wilson, Endocrinology 93 (1973) 581–592.
[8] F.M. Schultz, J.D. Wilson, Endocrinology 94 (1974) 979–986.
[9] E.E. Baulieu, I. Lasnitzki, P. Robel, Nature 219 (1968) 1155–1156.
[10] E.C. Friedberg, G.C. Walker, W. Siede, DNA Repair and Mutagenesis, American Society for Microbiology, Washington, DC, 1995.
[11] A. Datta, J.L. Schmeits, N.S. Amin, P.J. Lau, K. Myung, R.D. Kolodner, Mol. Cell 6 (2000) 593–603.
[12] C.W. Lawrence, R.B. Christensen, Genetics 82 (1976) 207–232.
[13] C.W. Lawrence, R.B. Christensen, Mol. Gen. Genet. 177 (1979) 31–38.
[14] C.W. Lawrence, R.B. Christensen, Mol. Gen. Genet. 186 (1982) 1–9.
[15] C.W. Lawrence, R.B. Christensen, J. Mol. Biol. 122 (1978) 1–21.
[16] C.W. Lawrence, J.W. Stewart, F. Sherman, R. Christensen, J. Mol. Biol. 85 (1974) 137–162.
[17] C.W. Lawrence, J.W. Stewart, F. Sherman, F.L.X. Thomas, Genetics 64 (1970) 836–837.
[18] L. Prakash, Genetics 78 (1974) 1101–1118.
[19] B.J. Glassner, L.J. Rasmussen, M.T. Najarian, L.M. Posnick, L.D. Samson, Proc. Natl. Acad. Sci. U. S. A. 95 (1998) 9997–10002.
[20] D.P. Kalinowski, F.W. Larimer, M.J. Plewa, Mutat. Res. 331 (1995) 149–159.
[21] K. Baynton, A. Bresson-Roy, R.P. Fuchs, Mol. Microbiol. 34 (1999) 124–133.
[22] J.R. Nelson, C.W. Lawrence, D.C. Hinkle, Nature 382 (1996) 729–731.
[23] Y. Masuda, M. Takahashi, N. Tsunekuni, T. Minami, M. Sumii, K. Miyagawa, K. Kamiya, J. Biol. Chem. 276 (18) (2001) 15051–15058.
[24] F.W. Larimer, J.R. Perry, A.A. Hardigree, J. Bacteriol. 171 (1989) 230–237.
[25] W. Lin, H. Xin, Y. Zhang, X. Wu, F. Yuan, Z. Wang, Nucleic Acids Res. 27 (1999) 4468–4475.

Development and molecular analysis of thymic lymphomas induced by ionizing radiation in Scid mice

Toshiaki Ogiu [a,*], Hiroko Ishii-Ohba [a], Shigeru Kobayashi [a], Mayumi Nishimura [a], Yoshiya Shimada [a], Hideo Tsuji [a], Hideki Ukai [a], Fumiaki Watanabe [a], Fumio Suzuki [b], Toshihiko Sado [c]

[a] *Low-Dose Radiation Effects Research Project, Research Center for Radiation Safety, National Institute of Radiological Sciences, Anagawa 4-9-1, Inage-ku, Chiba 263-8555, Japan*
[b] *Department of Regulatory Radiobiology, Research Institute for Radiation Biology and Medicine, Hiroshima University, Hiroshima, Japan*
[c] *Department of Health Sciences, Oita University of Nursing and Health Sciences, Oita, Japan*

Abstract

The Scid mouse is well known to be immunodeficient and also radiosensitive because of point mutation in the DNA-dependent protein kinase catalytic subunits (DNA-PKcs) gene. To analyze the effects of Scid mutation on radiation carcinogenesis, Scid (*Scid* homozygote), C.B-17 (wild-type) and their (C.B-17 × Scid) F1 hybrid (*Scid* heterozygote) were used. These strains of mice were first examined for acute effects, and then irradiated with 1–3 Gy gamma rays for carcinogenesis experiments. Scid mice are extremely susceptible to the induction of thymic lymphomas by ionizing radiation. Molecular analyses demonstrated that mutation of the *Ras* gene contributed less to the induction of thymic lymphomas, although *Notch1* mutation played a role. These results suggest a close relationship between radiosensitivity and the development of thymic lymphomas in *Scid* mice, and correlation of reduced DNA-PK activity with an increase in the yield of deletions or insertions in oncogene(s) rather than point mutations. © 2002 Elsevier Science B.V. All rights reserved.

Keywords: Scid mice; Radiation-induced thymic lymphomas; Oncogenes; *Ras*; *Notch1*

Abbreviations: DNA-PK, DNA-dependent protein kinase; ENU, *N*-ethyl-*N*-nitrosourea; IAP, Intracisternal A particle; MuLV, Murine leukemia virus.
* Corresponding author. Tel.: +81-43-251-2111; fax: +81-43-206-4138.
E-mail address: ogiu@nirs.go.jp (T. Ogiu).

1. Introduction

The Scid mouse was found in the C.B-17 strain and established as a severe combined immunodeficient mouse [1]. The functional deficiency is caused by highly repressed V(D)J recombination in T- and B-lymphocyte maturation, due to a reduced ability to rejoin coding ends of T-cell receptor genes and immunoglobulin genes. Scid mice are also hypersensitive to ionizing radiation because of a defect in rejoining radiation-induced DNA double strand breaks. Recently, it was reported that DNA-dependent protein kinase (DNA-PK) activity was depressed in Scid cells because a T to A transversion of the Tyr-4046 in the DNA-PK catalytic subunit (DNA-PKcs) gene results in the truncated protein [2].

To analyze the relationship between sensitivity to acute effects of ionizing radiation and susceptibility to the development of tumors by radiation, Scid (*Scid* homozygote), its parental strain C.B-17 (wild-type) and their hybrid (C.B-17 × Scid) F1 (*Scid* heterozygote) mice with the same C.B-17 genetic background were used for carcinogenesis experiments. Furthermore, molecular changes in the oncogenes and tumor suppressor genes in induced thymic lymphomas were analyzed to consider the mechanisms of radiation carcinogenesis.

2. Acute effects of ionizing radiation on Scid mice/cells

The acute effects of ionizing radiation on the survival of the whole body, bone marrow cells and cultured cell lines of Scid, C.B-17 and their F1 mice were examined [3].

Mice receiving various doses of gamma rays were maintained in the SPF animal facility and observed for 30 days. $LD_{50(30)}$ was 4.05 Gy in Scid mice and the values were 6.5 and 7.2 Gy, in F1 and C.B-17 mice, respectively.

Bone marrow cells were irradiated with various doses of X-rays in vitro, and the CFU-GM colonies were counted 7 days later. Survival curves for the C.B-17 and F1 cells had a shoulder region and decreased exponentially with dose, though no shoulder was observed for the dose response of Scid cells. The D_{10} values (10% survival dose) were 3.1, 2.8 and 1.0 Gy for the C.B-17, F1 and Scid cells, respectively. In addition, eight established cell lines derived from fetuses of these three mice strains also revealed different radiosensitivities.

These results showed that the Scid mutation was manifested as a partial dominant trait in a heterozygous state for sensitivity to ionizing radiation both in vivo and in vitro.

3. Development of thymic lymphomas in Scid mice

3.1. Radiation carcinogenesis in Scid mice

To examine the characteristics of radiation carcinogenesis in Scid mice, groups of Scid, C.B-17 and F1 mice were exposed to a single whole body radiation with 1–3 Gy gamma rays, and then maintained under the SPF condition. Most of the irradiated Scid mice died with thymic lymphomas at an early stage, 20–40 weeks after irradiation. Incidences of thymic lymphomas were 70–82% with a peak at 2 Gy in irradiated Scid mice, and 38% in unirradiated Scid mice. On the contrary, C.B-17 and F1 mice survived longer, and the incidences of thymic lymphomas were less than 10%.

Since thymic lymphomas were induced by 1 or 3 Gy gamma rays with a very high incidence, groups of Scid mice were further irradiated with 0.1–0.5 Gy gamma rays. Although this experiment has not been completed yet, the incidence of thymic lymphoma seemed to increase as a function of radiation dose, and a peak of thymic lymphoma development was seen before the 50th experimental week in all dose groups. The dose–effect relationship curve for thymic lymphomas for doses 0.1–3.0 Gy at the 50th experimental week seems to fit to linear quadratic or linear models.

3.2. Chemical carcinogenesis in Scid mice

We also examined carcinogenicity of *N*-ethyl-*N*-nitrosourea (ENU), a strong lymphomagenic agent in Scid mice [4]. The 400-ppm ENU solution was administered as drinking water for 2–10 weeks. A rather low incidence of thymic lymphomas was caused as compared with ionizing radiation. In other organs, no clear relationship between the total dose and tumor development was observed. This result means that in Scid mice, development of thymic lymphoma is correlated with DNA double strand breaks but not with DNA alkylation.

It is concluded that Scid mice were extremely susceptible to the induction of thymic lymphomas by ionizing radiation as compared to C.B-17 and F1 mice, and that the mutated *Scid* gene plays a crucial role in radiation-induced thymic lymphomas.

4. Change of oncogenes in radiation-induced thymic lymphomas

Possible oncogenes involved in the development of thymic lymphomas were examined.

4.1. Change of Ras oncogenes in radiation-induced thymic lymphomas

It has been reported that K-*Ras* or N-*Ras* genes were mutated in 40–85% of chemically induced murine thymic lymphomas. Therefore, *Ras* gene mutations of Scid thymic lymphomas were analyzed with the PCR-SSCP method and DNA sequence.

It turned out that the incidence of the *Ras* gene mutation was less than 18% in radiation-induced Scid thymic lymphomas. Almost all mutations were G to A transition in codons 12 or 13. This result suggests that point mutation of the *Ras* gene contributes only a little to the development of thymic lymphomas, and that mutations in other oncogenes or tumor suppressor genes are involved [5].

Analysis of the ENU-induced thymic lymphomas demonstrated that the K-*Ras* gene mutations were observed in 78% of the Scid thymic lymphomas. Observed mutations were G to A transition in codon 12 and A to C transversion in codon 61 [4].

These data indicate that the *Ras* gene mutations contribute to the induction of thymic lymphomas by ENU but do not to that by radiation in Scid mice.

4.2. Change of Notch1 gene in radiation-induced thymic lymphomas

To clarify the roles of other genes involved in thymic lymphoma development, thymic lymphoma cell lines derived from Scid mice as well as wild-type strains of mice were established and mutations of tumor-associated genes, which are murine homologues of activated oncogenes in human T-cell leukemia, were screened. As a result, frequent rearrangements of the *Notch1* gene were detected. In radiation-induced Scid thymic lymphomas, 35% exhibited DNA rearrangements of the *Notch1* gene, whereas 20% of thymic lymphomas of wild-type mice displayed abnormalities. This suggested an important role of *Notch1* abnormalities in thymic lymphoma induction by radiation.

For detailed analysis rearrangement, almost all sequences of the *Notch1* genomic DNA were first determined. Analysis showed that intragenic deletions were frequently observed in Scid thymic lymphomas and also in wild-type lymphomas, though hot spots were different in each lymphoma. In addition, insertion of intracisternal A particle (IAP) was observed in Scid thymic lymphomas and that of murine leukemia virus (MuLV) in a wild-type lymphoma.

These data suggest that the defective *Scid* gene participates in the increase of yield of deletion or insertion in the *Notch1* gene, and that dysregulated *Notch1* plays a role in murine thymic lymphomagenesis.

5. Conclusions

From these results, it will be pointed out: (1) the mutated *Scid* gene displays a partial dominance for the radiosensitivity, (2) Scid mice are extremely susceptible to induction of thymic lymphomas by ionizing radiation, (3) *Ras* mutation may only contribute a little to the development of Scid thymic lymphomas, and (4) mutated *Notch1* plays a role in Scid thymic lymphomagenesis.

It is concluded that there is a close relationship between radiosensitivity and the development of thymic lymphomas as shown in *Scid* mice. The development of thymic lymphomas in Scid mice may correlate with reduced DNA-PK that is high in the thymus, in which the activity is high in the wild-type mice, as compared with other organs, and may correlate with an increase in the yield of deletion or insertion in oncogene(s) rather than point mutation.

Acknowledgements

We are grateful to Mrs. Hara, Sasaki, Osada, Kondo, Iki and Seo and Mr. Nagai for their technical assistance. We also thank the entire staff of the Animal Facility and Technical Services of this institute. Furthermore, a Special Project Research Grant from the Science and Technology Agency, Japan and a Grant-in-Aid from the Ministry of Education, Science, Sports and Culture, Japan supported this work.

References

[1] G.C. Bosma, R.P. Custer, M.J. Bosma, A severe combined immunodeficiency mutation in the mouse, Nature 301 (2000) 527–530.

[2] R. Araki, A. Fujimori, K. Hamatani, K. Mita, T. Saito, M. Mori, R. Fukumura, M. Morimyo, M. Muto, M. Itoh, K. Tatsumi, M. Abe, Nonsense mutation at Tyr-4046 in the DNA-dependent protein kinase catalytic subunit of severe combined immune deficiency mice, Proc. Natl. Acad. Sci. U. S. A. 94 (1997) 2438–2443.
[3] S. Kobayashi, M. Nishimura, Y. Shimada, F. Suzuki, A. Matsuoka, H. Sakamoto, M. Hayashi, T. Sofuni, T. Sado, T. Ogiu, Increased sensitivity of *scid* heterozygous mice to ionizing radiation, Int. J. Radiat. Res. 72 (1997) 537–545.
[4] M. Nishimura, S. Kakinuma, S. Wakana, K. Mita, T. Sado, T. Ogiu, Y. Shimada, Reduced sensitivity to and *Ras* mutation spectrum of *N*-ethyl-*N*-nitrosourea-induced thymic lymphomas in adult C.B-17 scid mice, Mutat. Res. 486 (2001) 275–283.
[5] M. Nishimura, S. Wakana, S. Kakinuma, K. Mita, H. Ishii, S. Kobayashi, T. Ogiu, T. Sado, Y. Shimada, Low frequency of *Ras* gene mutation in spontaneous and gamma-ray-induced thymic lymphomas of scid mice, Radiat. Res. 151 (1999) 142–149.

"Second hit" of *Tsc2* gene in radiation induced renal tumors of Eker rat model

Okio Hino*, Hiroaki Mitani, Junko Sakaurai

Department of Experimental Pathology, Cancer Institute, Japanese Foundation for Cancer Research, 1-37-1 Kami-ikebukuro, Toshima-ku, Tokyo 170-8455, Japan

Abstract

Cancer is a heritable disorder of the somatic cells. The environment and heredity both operate in the origin of human cancer. Hereditary cancers should prove valuable in elucidating carcinogenesis. The Eker rat model of hereditary renal carcinoma (RC) is an example of Mendelian dominantly inherited predisposition to a specific cancer in an experimental animal. We, along with others, identified a germline mutation in the rat homologus of the human tuberous sclerosis gene (*TSC2*) as the predisposing *Eker* gene. We previously reported that a qualitative difference in the second hit of the *Tsc2* gene exists between spontaneous and ENU-induced mutations (e.g., deletion or duplication versus point mutation). In this study, we show the second hit of the *Tsc2* gene in radiation-induced RCs. © 2002 Elsevier Science B.V. All rights reserved.

Keywords: Eker rat; *Tsc2* gene; Two-hit carcinogenesis; Renal carcinogenesis; Radiation carcinogenesis

1. Introduction

Hereditary cancers in animals provide valuable experimental models for understanding the mechanisms of disease, and the development of therapeutic treatments which can be translated into human patients, as well as how environmental factors interact with cancer susceptibility genes. Such a naturally occurring hereditary cancer in rats was described by Reidar Eker in 1954 [1].

The Eker rat hereditary renal carcinoma (RC) was originally described by Reidar Eker in 1954. Reidar Eker, a pathologist and former Director of the Norwegian Cancer Center in Oslo, died in 1996 at the age of 92 years old. After 1971, Alfred G. Knudson (Fox Chase

---

* Corresponding author. Tel./fax: +81-3-5394-3815.

E-mail address: ohino@ims.u-tokyo.ac.jp (O. Hino).

Cancer Center, Philadelphia) was looking for an animal model of "two-hit carcinogenesis" that might parallel hereditary retinoblastoma in humans [2], and found it in the Eker rat. In 1983, Reidar Eker kindly sent some Eker rats to Alfred G. Knudson in Philadelphia. Progress toward genetic linkage analysis was slow because so few genetic markers were available in the rat. Then, I (Okio Hino) went to Knudson's laboratory on an American Cancer Society—Eleanor Roosevelt International Fellowship (UICC) from 1989 to 1991, and began to isolate rat genetic markers [3]. Unfortunately, these initial linkage studies were negative. When I (O.H.) returned to the Cancer Institute in Tokyo, Alfred G. Knudson sent some Eker rats to me and they were bred on a normal Long–Evans (LE) strain background at the Animal Facility of Cancer Institute since 1991. I (O.H.) have since continued investigating them in Tokyo, independently of Knudson's group in Philadelphia. Thus, Eker rat research is continued by a third generation of investigators [Eker (Oslo)→Knudson (Philadelphia)→Hino (Tokyo)].

2. The *TSC2* gene mutant (Eker) rat

The predisposing gene of the Eker rat was mapped to the proximal part of a rat chromosome 10 [4,5]. Meanwhile, Walker's group demonstrated that dimethylnitrosamine (DMN) could increase the number of renal tumors considerably [6]. We then established a method for early detection of the *Eker* gene carrier rats as early as 2 weeks of age by transplacental administration of *N*-ethyl-*N*-nitrosourea (ENU) [7], and dramatically sped up the linkage analysis [5,8]. Shortly after, we used the conserved linkage between parts of rat chromosome 10q and human chromosome 16p to show that the *Eker* mutation is tightly linked to the tuberous sclerosis (*Tsc2*) gene [8]. Finally, we and Knudson's group independently identified a germline retrotransposon insertion in the rat homologue of the human tuberous sclerosis (*TSC2*) gene, resulting in an aberrant RNA expression from the mutant allele, 40 years after the discovery of the Eker rat in Oslo [9,10]. In contrast, the 15 kb polycystic kidney disease (*Pkd1*) gene transcript, which is located adjacent to the Tsc2 gene in a tail-to-tail orientation as in the human case, showed no differences between non-carriers and carriers [9]. This suggests that the genetic alteration influence relates only to the *Tsc2* gene. Then, we constructed transgenic Eker rats with a wild-type *Tsc2* gene and ascertained that germline suppression of the Eker phenotype is possible for both embryonic lethality of the homozygote and tumor predisposition in the heterozygote (gene therapy or prevention), and finally confirmed that a tumor predisposition in the Eker rat is caused by the *Tsc2* germline mutation [11].

To the best of our knowledge, this was the first isolation of a Mendelian dominantly predisposing cancer gene in a naturally occurring animal model.

3. *Tsc2* gene

Following the cloning of the human gene by the European Chromosome 16 Tuberous Sclerosis Consortium [12], we described the entire cDNA (5375 bp without exons 25 and 31) of the rat *Tsc2* gene [13], and genomic structure of the rat *Tsc2* gene [13]. Surprisingly, there are 41 coding exons with relatively small sized introns, and the length covering all the exons is 35 kb, although there is a non-coding exon(s) in the 5′ upstream region [14]. Two alternative splicing events [involving exons 25 (129bp) and 31 (69bp)] make for a complex diversity in the Tsc2 product [13]. The deduced amino acid sequence (1743 amino acids) shows 92% identity to the human counterpart [13]. The determination of the Tsc2 gene and establishment of strong conservation between the rat and man provide clues to gene functions.

As a next step, we have checked the effect of wild-type *Tsc2* gene expression in Eker RC cells tested, using a tetracycline-responsive promoter system. Transfection and expression of an exogenous *Tsc2* gene affect cell morphorogy and growth rate [15], as well as anchorage independence in vitro and tumorigenecity in vivo [16]. This provides direct support for the classification of *Tsc2* as a tumor suppressor gene.

Although it does contain a short region as sequence homology to the ras family GTPase-activating proteins (Rap1GAP; exons 36 and 37) located downstream of the Eker insertion site (intron 30) [9,12,13], and has weak Rap1 GAP [17] and Rab5 GAP activities [18], little is known about how it operates to prevent renal carcinogenesis.

We described transcriptional activation domains [AD1 and AD2 in the carboxyl terminus of the Tsc2 product (in exons 30 and 32 and exon 41, respectively) [19]. The Eker insertion mutation (intron 30) disrupts their transcriptional activities. Examination of intragenic somatic mutations (the second hit of Knudson) in the *Tsc2* gene using PCR-single-strand conformational polymorphism (PCR-SSCP) analysis, indeed revealed a second hit at exon 40 in one case [20], and recently in a human TSC patient, a germline mutation at exon 40 was also reported [21]. Interestingly, Henry et al. [22] reported that the C-terminus of the TSC2 product modulates transcription mediated by steroid hormone receptor family members. Recently, van Slegtenhorst et al., showed that the TSC1 product (hamartin) and the TSC2 product (tuberin) associate physically in vivo [23], although its biological significance is still unknown [24].

Our transgenic rescue system [11] using various deletion mutants of the *Tsc2* gene will be useful for the analysis of functional *TSC2/Tsc2* domains, whose study is now being carried out.

4. Pathology of the Eker rat

We found that the homozygous mutant condition is lethal at around the 13th day of fetal life [25]. In heterozygotes, renal carcinomas develop from early preneoplastic lesions (phenotypically altered renal tubules, which begin to appear at two months of age) to adenomas around the age of one year; penetrance for this RC (*Tsc2*) gene is virtually complete.

Investigation of extra-renal primary tumors revealed hemangiomas/hemangiosarcomas of the spleen, leiomyomas/leiomyosarcomas of the uterus [26] and pituitary adenomas [27–29]. The Eker rat thus bears a single gene mutation with a dominant predisposition to four different tumors, although predisposition for extra-renal tumors is not as complete as with RCs [30]. Recently, brain lesions in the Eker rat, such as subependymal and subcortical hamartomas, were also reported [31]. Very recently, in the Eker rat cerebrum we identified two novel lesions, a cortical tuber and anaplastic ganglioglioma, in addition to the two types of hamartomas described above [32]. The presence of a cortical tuber is important, as tuber is epileptogenic and presumably associated with autism and other neurological symptoms of human TSC. Recently, *Gigas* (*Tsc2* gene mutant in *Drosophila*) was reported, and importantly it showed the phenotype of enlarged cells [33]. In addition, we generated *Tsc1* and *Tsc2* knock out mice [34,35].

5. Multi-step renal carcinogenesis in the Eker rat

Successive stages in the development of RCs were observed, beginning with isolated phenotypically altered renal tubule (which begin to appear at 2 months of age), characterized by partial or total replacement of the proximal tubular epithelium by large, or weakly acidophilic cells with different degrees of nuclear atypia, or by basophilic cells. These foci developed into atypical hyperplasia, then to adenomas of either eosinophilic or basophilic tubular type, and finally, to fully developed carcinomas that were predominantly of one type, but also included carcinomas of mixed type with both basophilic and eosinophilic components. These foci of atypical tubules were seen in all kidneys exhibiting tumors and are presumed to give rise to the smaller, rounded early adenomas as well as to the later, larger carcinomas; they were not observed in normal rats. Most tumors were composed of periodic acid/Schiff—negative large cells with clear, finely granular, and sometimes vacuolated, cytoplasm; some of the cells stained positively for acid mucopolysaccharides and weakly with alcian blue. Their tubular origin and histochemical pattern are compatible with the so-called chromophobe cell tumor according to the Bannasch nomenclature [36]. Some tumors were cystic and had papillary projections. Microscopic invasion of perirenal tissue was detected infrequently.

The Eker rat provides a promising model for analyzing the essential events of carcinogenesis at different stages. We previously reported that ionizing radiation induces additional tumors (large adenoma and carcinomas), with a linear dose–response relationship [25]. Loss of heterozygosity (LOH) at chromosome 10, where the predisposing *Tsc2* gene is located, was found in the renal carcinomas that developed from hybrid F1 rats carrying the Eker mutation, indicating that in heterozygotes at least two events (one inherited, one somatic) are necessary to produce large adenomas and carcinomas [18]. Using laser microscopic dissection, we detected a loss of heterozygosity of the wild-type allele even in the earliest pre-neoplastic lesions, e.g., phenotypically altered renal tubules [37], supporting the hypothesis that a second, somatic mutation (second hit) might be a rate-limiting step for renal carcinogenesis in the Eker rat model of dominantly inherited cancer; as well as indicating a tumor-suppressor function for the *Tsc2* gene. Thus, heterozygosity is not itself a sufficient condition for the development of cancer, but only one hit is enough to produce phenotypically altered renal tubules in the Eker rat. Such a lesion may initially be benign, but continued proliferation virtually ensures that other critical, though not rate limiting, events will occur. Although the initial event that triggers Eker rat renal cancer is a somatic mutation of the *Tsc2* wild-type allele, other genetic or epigenetic modifications may also contribute to tumor progression in multistep renal carcinogenesis.

Micro satellite instability was not observed in 26 Eker rat tumors [38]. Nonrandom loss of rat chromosome 5 in RC-derived cell lines is sometimes associated with homologous deletion of the interferon gene loci at rat chromosome bands 5q31–33 [25,39]. Homozygous deletion of the *Ink4* homologue on rat chromosome 5q was observed in 14 of the 24 (58%) RC-derived cell lines, *Ifna* gene in 5 out of the 24 cases (21%) and the *Ifnβ* gene in 1 of the 24 cases (4%), the order of the genes may be *Ink4–Ifna–Ifnβ* [38]. Since this locus is not linked with the predisposing inherited gene in the Eker rat, it probably represents a second tumor suppressor gene involved in late events in tumor progression.

Although inactivation of *Tsc2* results in development of renal tumors, it is not sufficient for metastatic RCs in the Eker rat. Their transformation to carcinomas may require additional rate-limiting events. To investigate the additional genetic event(s) necessary for cancer metastasis, we recently established highly metastatic cell lines from a non-metastatic RC cell line [40]. They should be useful experimental tools to investigate metastasis-promoting events in renal carcinogenesis.

6. Alterations of gene expression

Carcinogenesis looks like an opened Japanese fan because initiated cells grow in several directions and clinical tumors suggest the edge of the fan having many gene abnormalities (Fig. 1). To search for such alterations, we identified genes that were expressed more abundantly in Eker rats RC cells than in the normal kidney by representational difference analysis [41–43].

We identified the highly expressed genes in Eker RCs as *C3* gene encoding the third component of complement, *annaxin II* gene encoding the calpactine 1 heavy-chain, *Erc* (expressed in renal carcinoma) gene and *fra-1* gene encoding a transcriptional factor activator protein 1 (AP-1) [42]. We also found that other members of the AP-1 transcriptional factors were involved in the renal carcinogenesis in the Eker rat model [44]. Interestingly, AP-1 proteins were highly expressed even in the earlist preneoplatic lesions (e.g., phenotypically altered tubules) as suggested by immunohistochemistry. Moreover, we transfected antisense oligonucleotides targeting *AP-1* genes into RC cells and demonstrated that their growth was strongly inhibited [44]. Thus, the data suggest that expession of the *AP-1* genes might play a crucial role in renal carcinogenesis in the Eker rat model.

After we determined the complete primary structure of rat *Erc* cDNA, we showed that the putative rat Erc product has 56.1% identity to human MPF/mesothelin [45]. Rat *Erc* and its human homologue were localized in the chromosome 10q12–21 and 16p13.3, respectively, both of which coincided with the locus of the *Tsc2/TSC2* gene [45]. We also found that *Erc* was expressed at higher levels in primary RCs compared with the normal kidney of the Eker rat [45]. As mesothelin is a cell surface protein, it may function as a cell adhesion molecule. Our preliminary transfection data also suggest a role of the Erc product in cell adhesion and/or cell shape dynamics [45]. Furthermore, *Erc* may be related to carcinogenesis in the Eker rat model [45]. In addition, we found the *MPF*, *Erc* human homologue expression in the human renal cell carcinoma cell lines [45]. It shows the potential property of *Erc* as a tumor marker for renal cell carcinoma.

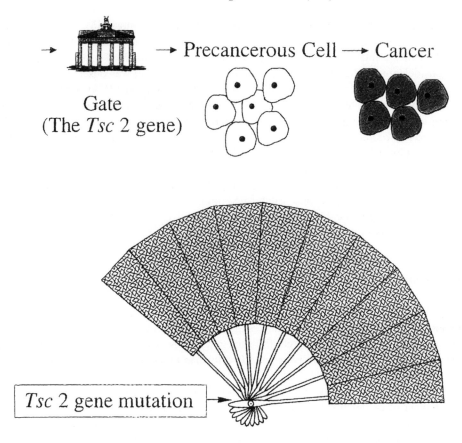

Fig. 1. Fanning the carcinogenesis. Schema of multistep renal carcinogenesis in the *Tsc2* gene mutant (Eker) rat. "Two hits" in the initial (*Tsc2*) gene set renal carcinogenesis in motion. Initiated cells grow in many directions and culminate in diverse clinical cancers at the edge [54].

Very recently, we also isolated a novel gene, which was named the "Niban", or the second in Japanese, because it is the second new gene to be found after *Erc* in our laboratory [46]. "Niban" is well expressed even in small primary rat Eker renal tumors, and also in human renal carcinoma cells, but not in normal human or rat kidney [46]. This "*Niban*" gene has the potential to be a good marker of renal tumors, and its molecular analysis might provide new insights into multistep carcinogenesis in the kidney [46].

7. Genetic alterations involved in renal carcinogenesis

A number of cancer genes have been identified by study of hereditary human cancers and shown to be involved in sporadic genesis of the same tumors. von Hippel-Lindau (*VHL*) gene mutations were detected in non-inherited, sporadic human RCs, especially of the clear cell type, at a high frequency. Recently, we searched for mutations in the *Tsc2* gene in chemically induced non-Eker rat RCs by SSCP analysis [47,48]. We simultaneously searched for mutation in the *Vhl* gene, a rat homologue of von Hippel-Lindau disease (*VHL*). Mutations in the *VHL* gene were not detected in any spontaneous RCs of the Eker rat model, nor chemically induced non-Eker rat tumors [47–49]. In contrast, *Tsc2* gene mutations were detected at a high frequency in N-ethyl-N-hydroxyethylnitrosamine (EHEN)- and diethylnitrosamine (DEN)-induced non-Eker rat primary RCs [48]. We infrequently found mutations of the *Tsc1* gene, a rat homologue of the *TSC1* gene, in chemically induced rat RCs [50]. These findings call attention to possible *TSC1* and *TSC2* gene mutation in human RCs, especially of the non-clear cell type, which are not related to the *VHL* gene. Until now, mutations of *TSC1/TSC2* genes in human RCs have not been extensively examined. Interestingly, we recently found the LOH of the *TSC2* gene locus in human lung carcinomas [51].

Moreover, we discovered a new hereditary renal carcinoma in the rat, and the rat was named the "Nihon" rat, and as a result its predisposing gene could be a novel [52].

8. ENU-induced renal carcinogenesis in the Eker rat

ENU-induced transplacental renal carcinogenesis in the rat results primarily in Wilms' tumors, not renal cell carcinomas, apparently because primitive nephroblasts are the preferred target. To further examine this specificity of tumor type, we undertook transplacental carcinogenesis in the Eker rat, which is heterozygous for a mutation that predisposes to renal cell carcinoma with high penetrance, but not to Wilms' tumor [7]. Surprisingly, RCs but no Wilms' tumors began to appear from as early as 1 week after birth [7]. Thus, the Eker rat is highly susceptible to induction of RCs (but not Wilms' tumors) by transplacental administration of ENU. The inheritance of the Eker mutation reduces the required number of carcinogenic events, as shown by the findings of renal lesions under conditions where ENU produces no tumors in controls, and determines the specificity of tumor histology even with in utero carcinogenesis; where nephroblasts are the preferred target. Nephroblasts might be the targets for development of both kinds of tumor in ENU-induced transplacental renal carcinogenesis. Our data thus support the hypothesis that the Wilms' tumor gene may control differentiation of nephroblasts (metanephric stem cells) into renal tubular stem cells, while the *Eker* gene may control terminal differentiation of these stem cells but permit differentiation of the nephroblasts to renal cells [7].

9. Qualitative differences in the second-hit among mutagens

We previously reported that in spontaneous RCs, 60% (6 of 10) showed LOH covering over 30 cM and, in contrast, 0% (0 of 9) in ENU-induced RCs using several DNA markers located on rat chromosome 10 [28]. Now, we can characterize the second hit (intragenic mutations including point mutation) in LOH-negative RCs at the DNA-sequencing level of the predisposing *Tsc2* gene using methods such as, polymerase chain reaction-single strand conformations in the *Tsc2* gene. The availability of the cDNA sequence of *Tsc2* permits comparative analysis of spontaneous and chemically induced

GGC T<u>TG TC</u>C TCA
(4bp deletion)
(Exon 5)
17 ℓ T3

AGC TG<u>G</u> CTG
G→A (Trp → Stop)
(Exon 30)
3 LT2

AGC TG<u>G</u> CTG
G→A (Trp → Stop)
(Exon 30)
3 LT4

TTG <u>AAG CAG</u>
(6bp deletion)
(Exon 18)
3 LT8

CTC C<u>G</u>C CAC
G→C (Arg → Pro)
(Exon 40)
4 LT2

AAG <u>TAC</u> AGG
(2bp deletion)
(Exon 10)
18 ℓ T3

ttatatgttagCTT TCA GCC CCG AAG ACC CTT GAG
(35bp deletion)
(Exon 20)
7 ℓ T3

Fig. 2. The second hit in the LOH-negative RCs in the irradiated group.

tumors in the Eker rat. We showed that a qualitative difference in the second hit exists between spontaneous and ENU-induced mutations (e.g., deletion or duplication versus point mutation) [20]. We further detected LOH in 4 of the 11 uterine leiomyosarcomas (36%) and in 11 out of the 31 pituitary adenomas (35%) from Eker rats, but in none of the 9 pituitary adenomas from non-carrier rats. This suggests that inactivation of the *Tsc2* gene is also a critical event in the pathogenesis of these extra-renal tumors [18]. Our data indicate that there might be different pathways for tumorigenesis of pituitary adenomas between Eker and non-carrier rats [29]. It is, however, noted that none of the 5 hemangiosarcomas of the spleen exhibited LOH, although one explanation might be contamination with appreciable normal cellular components in such tumors and/or the small number of cases investigated [29].

Very recently, Niida et al. [53] reported that the frequency of LOH of the second allele of either *TSC1* or *TSC2* is different among various types of human hamartomas.

10. Radiation-induced renal carcinogenesis in the *Tsc2* gene mutant (Eker) rat

We investigated the second hit of the *Tsc2* gene in radiation-induced RCs. Four-week-old pups (Eker/+) were lightly anesthetized and irradiated. They were positioned so that both kidneys were within the tissue volume irradiated bilaterally by 250 KvpX-ray (9 Gy for 3 min). Non-irradiated Eker rats were compared as a control. Animals were sacrificed at 12 months, and their kidneys were removed. Tumors (>2 mm in size) were counted (3.4 tumors/rat vs. 0). Thus, radiation induced early development of RCs in Eker rats.

We detected LOH in 10 out of the 35 RCs (29%) in the irradiated group. Then, we characterized the second hit in LOH-negative RCs at the DNA-sequencing level of the predisposing *Tsc2* gene using SSCP analysis, and we found extra bands in 10 out of the 35 RCs (29%). Interestingly, genomic rearrangement by Southerblot analysis was observed in one case (one out of 35). Among 10 extra bands by SSCP analysis, seven were proved at the DNA-sequencing level (Fig. 2). These data permit comparative analysis of spontaneous, chemically and radiation induced tumors in the Eker rat. Importantly, a qualitative difference in the second hit exists among spontaneous, ENU and radiation induced mutations (e.g., LOH; 60% VS. 0% VS. 29%, respectively). It is also noted that, even in radiation-induced RCs, point mutations, as well as intragenic small deletions, were found.

Thus, the mutant *Tsc2* gene should also be a valuable experimental model for the understanding of radiation carcinogenesis, as well as how radiation interacts with the cancer susceptibility gene [54].

Acknowledgements

We thank Dr. O. Niwa for the radiation experiment. This work was supported in part by Grants-in Aid for Cancer Research from the Ministry of Education, Science and Culture and the Ministry of Health and Welfare of Japan and the Organization for Pharmaceutical Safety and Research (OPSR).

References

[1] R. Eker, J. Mossige, A dominant gene for renal adenomas in the rats, Nature 189 (1961) 858–859.
[2] A.G. Knudson, Mutation and cancer: statistical study of retinoblastoma, Proc. Natl. Acad. Sci. U. S. A. 68 (1971) 820–823.
[3] O. Hino, J. Testa, K. Buetow, T. Taguch, J.-Y. Zhou, M. Bremer, A. Bruzel, R. Yeung, G. Levan, A.G. Knudson, K.D. Tartof, Universal mapping probes and origin of human chromsome 3, Proc. Natl. Acad. Sci. U. S. A. 90 (1993) 730–734.
[4] R.S. Yeung, K.H. Buetow, J.R. Testa, A.G. Knudson, Susceptibility to renal carcinoma in the Eker rat involves a tumor suppressor gene on chromosome 10, Proc. Natl. Acad. Sci. U. S. A. 90 (1993) 8038–8042.
[5] O. Hino, H. Mitani, M. Nishizawa, H. Katsuyama, E. Kobayashi, Y. Hirayama, A novel renal cell carcinoma susceptibility gene maps on chromosome 10 in the Eker rat, Jpn. J. Cancer Res. 84 (1993) 1106–1109.
[6] C. Walker, T.L. Goldsworthy, D.C. Wolf, J. Everitt, Predisposition to renal cell carcinoma due to alteration of a cancer susceptibility gene, Science 255 (1992) 1693–1695.
[7] O. Hino, H. Mitani, A.G. Knudson, Genetic predisposition to transplacentally induced renal cell carcinomas in the Eker rat, Cancer Res. 53 (1993) 5856–5858.
[8] O. Hino, T. Kobayashi, H. Tsuchiya, Y. Kikuchi, E. Kobayashi, H. Mitani, Y. Hirayama, The predisposing gene of the Eker rat inherited cancer syndrome is tightly linked to the tuberous sclerosis (Tsc2) gene, Biochem. Biophys. Res. Commun. 203 (1994) 1302–1308.
[9] T. Kobayashi, Y. Hirayama, E. Kobayashi, Y. Kubo, O. Hino, A gemline insertion in the tuberous sclerosis (Tsc2) gene gives rise to the Eker rat model of dominantly inherited cancer, Nat. Genet. 9 (1995) 70–74.
[10] R.C. Yeung, G.H. Xiao, F. Jin, W.-C. Lee, J.R. Testa, A.G. Knudson, Predisposition to renal carcinoma in the Eker rat is determined by germ-line mutation of the tuberous sclerosis 2 (TSC2) gene, Proc. Natl. Acad. Sci. U. S. A. 91 (1994) 11413–11416.
[11] T. Kobayashi, H. Mitani, R. Takahashi, M. Hirabayashi, M. Ueda, H. Tamura, O. Hino, Transgenic rescue from embryonic lethality and renal carcinogenesis in the Eker rat model by introduction of a wild-type Tsc2 gene, Proc. Natl. Acad. Sci. U. S. A. 94 (1997) 3990–3993.
[12] The European Chromosome 16 Tuberous Sclerosis Consortium: Identification and chracterization of the tuberous sclerosis gene on chromosome 16, Cell 75 (1993) 1305–1315.
[13] T. Kobayashi, M. Nishizawa, Y. Hirayama, E. Kobayashi, O. Hino, cDNA structure, alternative splicing and exon-intron organization of the predisposing tuberous sclerosis (Tsc2) gene of the Eker rat model, Nucleic Acids Res. 23 (1995) 2608–2613.
[14] T. Kobayashi, S. Urakami, J.P. Cheadle, R. Aspinwall, P. Harris, J.R. Sampson, O. Hino, Identification of a leader exon and a core promoter for the rat tuberous sclerosis 2 (Tsc2) gene and structural comparison with the human homologue, Mamm. Genome 8 (1997) 554–558.
[15] K. Orimoto, H. Tsuchiya, T. Kobayashi, T. Matsuda, O. Hino, Suppression of the neoplastic phenotype by replacement of the Tsc2 gene in Eker rat renal carcinoma cells, Biochem. Biophys. Res. Commun. 219 (1996) 70–75.
[16] F. Jin, R. Wienecke, G.H. Xiao, J.C. Maize Jr., J.E. DeClue, R.S. Yeung, Supression of tumorigenicity by the wild-type tuberous sclerosis 2 (Tsc2) gene and its C-terminal region, Proc. Natl. Acad. Sci. U. S. A. 93 (1996) 9154–9159.
[17] R. Wienecke, A. Koning, J.E. DeClue, Identification of tuberin, the tuberous sclerosis 2 product, J. Biol. Chem. 207 (1995) 16409–16414.
[18] G.H. Xiao, F. Shoarinejad, F. Jin, E.A. Golemis, R.S. Yeung, The TSC2 gene product, tuberin, functions as a Rab5 GTPase activating protein (GAP) in modulating endocytosis, J. Biol. Chem. 272 (1997) 6097–6100.
[19] H. Tsuchiya, K. Orimoto, T. Kobayashi, O. Hino, Presence of potent transcriptional activation domains in the predisposing tuberous sclerosis (Tsc2) gene product of the Eker rat model, Cancer Res. 56 (1996) 429–433.
[20] T. Kobayashi, S. Urakami, Y. Hirayama, T. Yamamoto, M. Nishizawa, T. Takahara, O. Hino, Intragenic Tsc2 somatic mutations as Knudson's second hit in spontaneous and chemically induced renal carcinomas in the Eker rat model, Jpn. J. Cancer Res. 88 (1997) 254–261.
[21] A. Kumar, C. Wolpert, R.S. Kandt, J. Segal, J. Pufky, A.D. Roses, M.A. Pericak Vance, J.R. Gilbert, De novo frame-shift mutation in the tuberin gene, Hum. Mol. Genet. 4 (1995) 1471–1472.

[22] K.W. Henry, X. Yuan, N.J. Koszewski, H. Onda, D.J. Kwiatkowski, D.J. Noonan, Tuberous sclerosis gene 2 product modulates transcription mediated by steroid hormone receptor family members, J. Biol. Chem. 273 (1998) 20535–20539.
[23] M. van Slegtenhorst, M. Nellist, B. Nagelkerken, J. Cheadle, R. Snell, A. van den Ouweland, A. Reuser, J. Sampson, D. Halley, P. van der Sluijs, Interaction between hamartin and tuberin, the TSC1 and TSC2 gene products, Hum. Mol. Genet. 7 (1998) 1053–1057.
[24] T. Fukuda, T. Kobayashi, S. Momose, H. Yasui, O. Hino, Distribution of Tsc1 protein detected by immunohistochemistry in various normal rat tissues and the renal carcinomas of Eker rat: detection of limited co-localization with Tsc1 and Tsc2 gene products in vivo, Lab. Invest. 80 (2000) 1347–1359.
[25] O. Hino, A.J.P. Klein-Szanto, J.J. Freed, J.R. Testa, D.Q. Brown, M. Vilensky, R.S. Yeung, K.D. Tartof, A.G. Knudson, Spontaneous and radiation-induced renal tumors in the Eker rat model of dominantly inherited cancer, Proc. Natl. Acad. Sci. U. S. A. 90 (1993) 327–331.
[26] J.I. Everitt, T.L. Goldsworthy, D.S. Wolf, C.L. Walker, Hereditary renal cell carcinoma in the Eker rat: a rodent familial cancer syndrome, J. Urol. 148 (1992) 1932–1936.
[27] O. Hino, H. Mitani, H. Katsuyama, Y. Kubo, A novel cancer predisposition syndrome in the Eker rat model, Cancer Lett. 83 (1994) 117–121.
[28] Y. Kubo, H. Mitani, O. Hino, Allelic loss at the predisposing gene locus in spontaneous and chemically induced renal cell carcinomas in Eker rat, Cancer Res. 54 (1994) 2633–2635.
[29] Y. Kubo, Y. Kikuchi, H. Mitani, E. Kobayashi, T. Kobayashi, O. Hino, Allelic loss at the tuberous sclerosis (Tsc2) gene locus in spontaneous uterine leiomyosarcomas and pituitary adenomas in the Eker rat model, Jpn. J. Cancer Res. 86 (1995) 828–832.
[30] O. Hino, T. Kobayashi, H. Mitani, H. Kubo, Y. Tsuchiya, Y. Kikuchi, M. Nishizawa, Y. Hirayama, The Eker rat, a model of dominantly inherited cancer syndrome, Transplant. Proc. 27 (1995) 1529–1531.
[31] R.S. Yeung, C.D. Katsetos, A. Klein-Szanto, Subependymal astrocytic hamartomas in the Eker rat model of tuberous sclerosis, Am. J. Pathol. 151 (1997) 1477–1486.
[32] M. Mizuguchi, S. Takahashi, H. Yamanouchi, Y. Nakazato, H. Mitani, O. Hino, Novel cerebral lesions in the Eker rat model of tuberous sclerosis: cortical tuber and anaplastic ganglioglioma, J. Nuropathol. Exp. Neurol. 59 (2000) 188–196.
[33] N. Ito, G.M. Rubin, Gigas, a drosophila homolog of tuberous sclerosis gene product-2, regulates the cell cycle, Cell 96 (1999) 529–539.
[34] T. Kobayashi, O. Minowa, J. Kuno, H. Mitani, O. Hino, T. Noda, Renal carcinogenesis, hepatic hemangiomatosis, and embryonic lethality caused by a germ-line Tsc2 mutation in mice, Cancer Res. 59 (1999) 1206–1211.
[35] T. Kobayashi, O. Minowa, Y. Sugitani, S. Takai, H. Mitani, E. Kobayashi, T. Noda, O. Hino, A germ-line Tsc1 mutation causes tumor development and embryonic lethality that are similar, but not identical to, those caused by Tsc2 mutation in mice, Proc. Natl. Acad. Sci. U. S. A. 98 (2001) 8762–8767.
[36] P. Bannasch, H. Zerban, Animal models and renal carcinogenesis, in: J.N. Ebel (Ed.), Tumors and Tumor-like Conditions of the Kidneys and Ureters, Churchill Livingstone, New York, 1990, pp. 1–34.
[37] Y. Kubo, F. Klimek, Y. Kikuchi, P. Bannasch, O. Hino, Early detection of Knudson's two-hits in preneoplastic renal cells of the Eker rat model by the laser microdissection procedure, Cancer Res. 55 (1995) 989–990.
[38] O. Hino, E. Kobayashi, Y. Hirayama, T. Kobayashi, Y. Kubo, H. Tsuchiya, Y. Kikuchi, H. Mitani, Molecular genetic basis of renal carcinogenesis in the Eker rat model of tuberous sclerosis (Tsc2), Mol. Carcinog. 14 (1995) 23–27.
[39] J.R. Testa, T. Taguchi, A.G. Knudson, O. Hino, Localization of the interferon—α gene cluster to rat chromosome bands 5q31–q33 by fluorescence in situ hybridization, Cytogenet. Cell Genet. 60 (1992) 247–249.
[40] T. Fukuda, Y. Hirayama, H. Mitani, H. Maeda, M. Tsutsumi, Y. Konishi, O. Hino, Generation of metastatic variants of Eker renal carcinoma cell lines for experimental investigation of renal cancer metastasis, Jpn. J. Cancer Res. 89 (1998) 1104–1108.
[41] H. Tsuchiya, Y. Tsuchiya, T. Kobayashi, Y. Kikuchi, O. Hino, Isolation of genes differentially expressed between the Yoshida sarcoma and long-survival Yoshida sarcoma variants: origin of Yoshida sarcoma revisited, Jpn. J. Cancer Res. 85 (1994) 1099–1104.

[42] O. Hino, E. Kobayashi, M. Nishizawa, Y. Kubo, T. Kobayashi, Y. Hirayama, S. Takai, Y. Kikuchi, H. Tsuchiya, K. Orimoto, K. Kajino, T. Takahara, H. Mitani, Renal carcinogenesis in the Eker rat, J. Cancer Res. Clin. Oncol. 121 (1995) 602–605.
[43] K. Orimoto, H. Tsuchiya, J. Sakurai, M. Nishizawa, O. Hino, Identification of cDNAs induced by the tumor suppressor Tsc2 gene using conditional expression system in Tsc2 mutant (Eker) rat renal carcinoma cells, Biochem. Biophys. Res. Commun. 247 (1998) 728–733.
[44] S. Urakami, H. Tsuchiya, K. Orimoto, T. Kobayashi, M. Igawa, O. Hino, Overexpression of members of the AP-1 transcriptional factor family from an early stage of renal carcinogenesis and inhibition of cell growth by AP-1 gene antisense oligonucleotides in the Tsc2 gene mutant (Eker) rat model, Biochem. Biophys. Res. Commun. 241 (1997) 24–30.
[45] Y. Yamashita, M. Yokoyama, E. Kobayashi, S. Takai, O. Hino, Mapping and determination of the cDNA sequence of the Erc gene preferentially expressed in renal cell carcinoma in the Tsc2 gene mutant (Eker) rat model, Biochem. Biophys. Res. Commun., in press.
[46] S. Majima, K. Kajino, T. Fukuda, F. Otsuka, O. Hino, "Niban" upregulated in renal carcinogenesis—cloning by the cDNA-AFLP (Amplified Fragment Length Ploymorphism) approach, Jpn. J. Cancer Res. 91 (2000) 869–874.
[47] S. Urakami, R. Tokuzen, H. Tsuda, M. Igawa, O. Hino, Somatic mutation of the tuberous sclerosis (Tsc2) tumor suppressor gene in chemically induced rat renal carcinoma cell, J. Urology 158, 275–278.
[48] N. Satake, S. Urakami, Y. Hirayama, K. Izumi, O. Hino, Biallelic mutations of theTsc2 gene in chemically induced rat renal carcinoma, Int. J. Cancer 77 (1998) 895–900.
[49] Y. Kikuchi, E. Kobayashi, M. Nishizawa, S. Hamazaki, S. Okada, O. Hino, Cloning of the rat homologue of the von Hippel-Lindau tumor suppressor gene and its non-somatic mutation in rat renal cell carcinomas, Jpn. J. Cancer Res. 86 (1995) 905–909.
[50] N. Satake, T. Kobayashi, E. Kobayashi, K. Izumi, O. Hino, Isolation and characterization of a rat homologue of the human tuberous sclerosis 1 gene (Tsc1) and analysis of its mutations in rat, Cancer Res. 59 (1999) 849–855.
[51] K. Kajino, J. Yakurai, Y. Hirayama, T. Takahara, I. Fukui, Y. Ishikawa, O. Hino, LOH of tumor suppressor TSC2 gene locus in human renal cell carcinomas and lung carcinomas, Proc. Jpn. Cancer Assoc. (1997) 279.
[52] K. Okimoto, M. Kouchi, E. Kikawa, K. Toyosawa, T. Koujitani, K. Tanaka, N. Matsuoka, J. Sakurai, O. Hino, A novel "Nihon" rat model of a Mendelian dominantly inherited renal cell carcinoma, Jpn. J. Cancer Res. 91 (2000) 1096–1099.
[53] Y. Niida, A.O. Stemmer-Rachamimov, M. Logrip, D. Tapon, R. Perez, D.J. Kwiatkowski, K. Sims, M. MacCollin, D.N. Louis, V. Ramesh, Survey of somatic mutations in tuberous sclerosis complex (TSC) hamartomas suggests different genetic mechanisms for pathogenesis of TSC lesions, Am. J. Hum. Genet. 69 (2001).
[54] O. Hino, S. Majima, T. Kobayashi, S. Honda, S. Momose, Y. Kikuchi, H. Mitani, Multistep renal carcinogenesis as gene expression disease in tumor suppressor TSC2 gene mutant model-genotype, phenotype and environment, Mutat. Res. 477 (2001) 155–164.

PTCH (patched) and XPA genes in radiation-induced basal cell carcinomas

F.J. Burns [a,*], R.E. Shore [a], N. Roy [a], C. Loomis [b], P. Zhao [a]

[a] Department of Environmental Med., NYU School of Medicine, 550 First Avenue, New York, NY 10016, USA
[b] Department of Dermatology, NYU School of Medicine, 550 First Avenue, New York, NY 10016, USA

Abstract

A study of X-irradiated tinea capitis patients has identified a small subgroup (about 1.5%) of people at high risk for developing multiple basal cell carcinomas (BCCs). Among 1680 Caucasians who received about 3 Gy of low voltage X-ray, the incidence of BCCs, 35 years after exposure, was 0.10 tumors per person. A random distribution indicates that no one should have developed three or more BCCs and yet 25 with three or more were observed. Among three selected from the latter group, one presented with five previously unreported BCCs; one presented with four; and one presented with one. Varied patterns of loss of heterozygosity (LOH) were found in the BCCs from all three patients in the region of the *PTCH* and xeroderma pigmentosum A (*XPA*) genes in chromosome region 9q22.3–31. The subject with five BCCs reported that 25 BCCs had previously been removed. Amplification primers for *XPA* were picked from intronic sequences and positive SSCP results were followed-up by sequencing the identified regions to establish the exact nature of the mutation. A large number of negative SSCP results were sequenced, but no mutations undetected by SSCP were found. Normal blood DNA of the subject with five BCCs showed a 14 base deletion in exon 6 of the *XPA* gene, strongly suggesting *XPA* heterozygosity. Biopsy samples of the BCCs showed the same 14 base deletion in all five BCCs. Allelic identification verified that the normal *XPA* allele was missing in two of two BCCs selected for testing. These data imply that a mutational inactivation of one *XPA* allele increased the susceptibility to BCC induction by ionizing radiation, possibly because the normal *XPA* allele was lost in the same deletional event that produced LOH in the *PTCH* region. These findings support the idea that the lost event associated with LOH at 9q22.3 frequently (5/5) includes both the *PTCH* and *XPA* loci. Frequently, an *XPA* allele located within about 3 megabases distal to *PTCH* is lost along with the *PTCH* gene. Overall, these results suggest how inactivation of the *XPA* gene may play an important role in susceptibility to BCC induction by increasing the UV-induced mutational rate of *PTCH* or other cancer-relevant genes (work supported by NIEHS, NCI, and EPA). © 2002 Elsevier Science B.V. All rights reserved.

Keywords: PTCH; XPA; Human; Skin; Cancer; Radiation

---

* Corresponding author. Tel.: +1-845-731-3551; fax: +1-845-351-2118.
E-mail address: burns@env.med.nyu.edu (F.J. Burns).

1. Introduction

This study was designed to develop information that might provide insight why certain X-irradiated subjects in the tinea capitis population develop basal cell carcinomas (BCCs) in their irradiated skin far in excess of random expectations [1,2]. The *PTCH* gene, situated in chromosome region 9q22.3, is the causative gene of Gorlin's syndrome, one symptom of which is an extreme sensitivity to BCC induction in UV-exposed regions of the skin. The *PTCH* gene has also has been shown to be a tumor suppressor gene with specificity for basal cell carcinomas [3–7]. The xeroderma pigmentosum A (*XPA*) gene resides in chromosome region 9q22.3, about 3 megabases distant from the *PTCH* locus, so that deletions or recombinations of large segments of DNA in this region would likely affect both genes. It is well established that inactivation of a DNA repair gene, such as *XPA*, increases a cell's vulnerability to the carcinogenic effects of UV as can be seen clearly in people with repair deficiency syndrome, xeroderma pigmentosum (XP) [8,9]. The *XPA* gene consists of 273 amino acids in 6 exons distributed over 25 kb of genomic DNA. Its product is involved in the earliest repair stage of damage recognition and DNA binding [10–12].

An opportunity to study the possible interaction between these two carcinogenic radiations was found in a population that was about 8 years old when given about 3.0 Gy of X-irradiation as part of the treatment for tinea capitis. Subsequently, subjects were exposed inadvertently to the UV component of sunlight during the course of normal life activities. By 35–40 years of follow-up, the X-irradiated group has exhibited a 3.7-fold higher skin BCC incidence in the treatment region relative to non-X-irradiated controls [1].

Patients received X-ray treatments at the NYU Dermatology Clinic between the years 1940 and 1959 [13]. Three patients were selected for a clinical examination on the basis of a prior history of multiple (> 5) skin BCCs. The patients presented with BCCs in the head or neck region were as follows: patient 1—1 BCC, patient 2—4 BCCs, and patient 3—5 BCCs.

2. Results

Amplification primers were picked from the intronic sequences described by Satokata et al. [14] and SSCP analysis was carried out as described by Hensel et al. [15]. Positive SSCP results were followed-up by sequencing the identified region to establish the exact nature of the mutation. A missense mutation was found in exon 4 of BCC #1. Identical deletion mutations involving a 14 base segment were found in exon 6 of BCCs 3A, 3B, 3C, 3D, and 3E. In addition, a missense mutation was found in exon 2 of BCC 3C. All five BCCs showing the 14 base deletion were from the same patient, and analysis of this patient's blood indicated the presence of the same deletion, indicating that this person was probably an *XPA* heterozygote. All five BCCs showed LOH in 9q22.3. Of nine BCCs tested at 9q22.3 for loss of *XPA* heterozygosity, seven exhibited mutations in the undeleted allele. One mutation was a 14 base deletion in the region of codons 256–260, found in all five BCCs of one patient. One BCC out of four from a second patient exhibited a GAT to GGT missense mutation in codon 154. One BCC only from a third patient was negative for *XPA* mutations.

3. Discussion

A working hypothesis to explain the current results is that susceptibility to BCC induction by radiation exposures (X-ray and UV) depends on the status of the *PTCH* and *XPA* genes as follows: the first and most critical event is the loss or recombination, probably caused by the X-ray, of a large segment of chromosome 9q that deletes or inactivates one *PTCH* allele, thereby converting the cell to one with Gorlin syndrome susceptibility characteristics (Fig. 1). In principle, the initial inactivation event could have been a conventional point mutation, but deletion is observed to be far more likely in both sporadic BCCs and in Gorlin syndrome BCCs. Development of a BCC requires inactivating *PTCH*'s second allele which most likely occurred as a result of UVR mutations. Some of the time, an *XPA* allele, which is about 3 megabases distal to *PTCH*, is also inactivated along with the *PTCH* gene. In this double deletion scenario, the chance of a BCC could be substantially increased by an inactivating mutation in the surviving *XPA* allele, because the affected cell then becomes DNA repair deficient with enhanced sensitivity to UV inactivation of the second *PTCH* allele, as experienced by XP patients.

Cells with deficient DNA excision repair would be expected to exhibit a much higher risk of UV damage to the second *PTCH* allele, which very likely accounts for the presence of an *XPA* heterozygote in the multiple BCC sample. A recent result indicating no associated between *XPA* and squamous cell carcinomas (SCCs) is not contradictory, because the causative gene for SCCs is not *PTCH* [10]. Overall, these results support previous findings that the *PTCH* gene plays an important role in skin BCC development

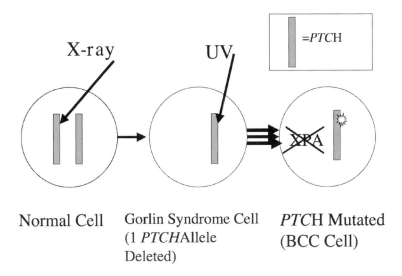

Fig. 1. Schematic showing how *PTCH* and *XPA* may contribute to BBC induction in skin exposed to X-ray and UV radiation in an *XPA* heterozygote. The first event is a deletion of one *PTCH* allele and one *XPA* allele by the X-ray. The second event is a UV-induced mutation in the second *PTCH* allele effectively producing a BCC cell. If *XPA* is completely inactivated, the probability of the second event is greatly elevated because of the abrogation of DNA excision repair.

and further indicate that the XP genes may also play an important secondary role by enhancing the likelihood of mutational inactivation of the second *PTCH* allele.

References

[1] R. Shore, R. Albert, M. Reed, N. Harley, B. Pasternack, Skin cancer incidence among children irradiated for ringworm of the scalp, Radiation Research 100 (1984) 192–204.
[2] R. Shore, Overview of radiation-induced skin cancer in humans [Review], International Journal of Radiation Biology 57 (4) (1990) 809–827.
[3] R. Gorlin, Nevoid basal cell carcinoma syndrome [Review], Dermatologic Clinics 13 (1) (1995) 113–125.
[4] A. Reis, W. Kuster, G. Linss, E. Gebel, H. Hamm, W. Fuhrmann, G. Wolff, W. Groth, G. Gustafson, M. Kuklik, Localisation of gene for the nevoid basal cell carcinoma syndrome, [letter], Lancet 339 (8793) (1992) 328–617.
[5] P. Farndon, D. Morris, C. Hardy, C. McConville, J. Weissenbach, M. Kilpatrick, A. Reis, Analysis of 133 meiosis places the genes for nevoid basal cell carcinoma (Gorlin) syndrome and fanconi anemia group C in a 26-cm interval and contributes to the fine map of 9q223, Genomics 23 (2) (1994) 486–489.
[6] M. Gailani, S. Bale, D. Leffell, J. Digiovanna, G. Peck, S. Poliak, M. Drum, B. Pastakia, O. Mcbride, R. Kase, Developmental defects in Gorlin syndrome related to a putative tumor suppressor gene on chromosome 9, Cell 69 (1) (1992) 111–117.
[7] A. Chidambaram, A. Goldstein, M. Gailani, B. Gerrard, S. Bale, J. DiGiovanna, A. Bale, M. Dean, Mutations in the human homologue of the *Drosophila* patched gene in Caucasian and African–American nevoid basal cell carcinoma syndrome patients, Cancer Research 56 (20) (1996) 4599–4601.
[8] N. Copeland, C. Hanke, J. Michalak, The molecular basis of xeroderma pigmentosum [Review] [52 refs], Dermatologic Surgery 23 (6) (1997) 447–455.
[9] S. Benhamou, A. Sarasin, Variability in nucleotide excision repair and cancer risk: a review [Review] [33 refs], Mutation Research 462 (2–3) (2000) 149–158.
[10] L. Eklund, E. Lindstrom, A. Unden, B. Lundh-Rozell, M. Stahle-Backdahl, P. Zaphiropoulos, R. Toftgard, P. Soderkvist, Mutation analysis of the human homologue of *Drosophila* patched and the xeroderma pigmentosum complementation group A genes in squamous cell carcinomas of the skin, Molecular Carcinogenesis 21 (2) (1998) 87–92.
[11] M. Ichihashi, K. Naruse, S. Harada, T. Nagano, T. Nakamura, T. Suzuki, N. Wadabayashi, S. Watanabe, Trends in nonmelanoma skin cancer in Japan, Recent Results in Cancer Research 139 (1995) 263–273.
[12] A. Halpern, J. Altman, Genetic predisposition to skin cancer [Review] [49 refs], Current Opinion in Oncology 11 (2) (1999) 132–138.
[13] R. Schulz, R. Albert, Dose to organs of the head from the X-ray treatment of tinea capitis, Archives of Environmental Health 17 (1968) 935–950.
[14] I. Satokata, K. Iwai, T. Matsuda, Y. Okada, K. Tanaka, Genomic characterization of the human DNA excision repair-controlling gene XPAC, Gene 136 (1993) 345–348.
[15] C. Hensel, R. Xiang, A. Sakaguchi, S. Naylor, Use of the single strand conformation polymorphism technique and PCR to detect *p53* gene mutations in small cell lung cancer, Oncogene 6 (6) (1991) 1067–1071.

Differences of molecular alteration between radiation-induced and N-ethyl-N-nitrosourea-induced thymic lymphomas in B6C3F1 mice

Shizuko Kakinuma [a,*], Mayumi Nishimura [a], Ayumi Kubo [a], Jun-ya Nagai [a], Kazukei Mita [a], Toshiaki Ogiu [a], Hideyuki Majima [a], Yoshimoto Katsura [b], Toshihiko Sado [c], Yoshiya Shimada [a,*]

[a] National Institute of Radiological Sciences, Chiba, 263-8555, Japan
[b] Institute for Frontier Medical Sciences, Kyoto University, Kyoto, 606-8507, Japan
[c] Oita University of Nursing and Health Sciences, Oita, 870-1201, Japan

Abstract

Murine T-cell leukemia (TLs) can be reproducibly induced by carcinogenic agents such as radiation and alkylating agents. We previously found a unique locus with a high frequency of LOH (~ 50%) in the centromeric region of chromosome 11, in X-ray-induced TLs of B6C3F1 mice, and mapped *Ikaros* at this locus. The aim of the present study was to compare the involvement of Ikaros inactivation in TLs of different etiology. Therefore, we concomitantly examined both expression status and genetic changes of *Ikaros*, and determined the contribution of LOH to Ikaros inactivation in X-ray-induced and ENU-induced TLs. The results obtained were as follows: (1) the frequency of LOH at the *Ikaros* locus was much higher in X-ray-induced TLs (54%) than that in the ENU-induced TLs (8%). (2) Ikaros was inactivated by multiple mechanisms (null expression, expression of dominant negative isoforms, point mutation, and insertion). (3) The X-ray-induced TLs exhibited all types of Ikaros inactivation, whereas the ENU-induced TLs showed only point-mutation type. (4) Furthermore, the creation of point mutation was frequently accompanied by the existence of LOH at the *Ikaros* locus in X-ray-induced TLs, but not in ENU-induced TLs. These results suggest that Ikaros inactivation is a preferential pathway for radiation-induced lymphomagenesis, and that its inactivation mechanism is different between X-ray-induced and ENU-induced TLs. © 2002 Elsevier Science B.V. All rights reserved.

Keywords: T-cell leukemia; Radiation; *N*-ethyl-*N*-nitrosourea; LOH; Ikaros

---

* Corresponding authors. Tel.: +81-43-206-3224; fax: +81-43-206-3221.
E-mail addresses: skakinum@nirs.go.jp (S. Kakinuma), y_shimad@nirs.go.jp (Y. Shimada).

0531-5131/02 © 2002 Elsevier Science B.V. All rights reserved.
PII: S0531-5131(01)00754-3

1. Introduction

Murine T-cell leukemia (TLs) can be reproducibly induced by carcinogenic agents such as radiation and alkylating agents, and this murine model is considered to be useful in the search for and characterization of genes involved in the development of human lymphocytic leukemia. The genetic alterations that have been identified include the activation of N-*ras* and K-*ras* oncogenes [1], deficient expression of the *p15* and *p16*, and mutation of *p53* mutations. Extensive LOH studies have recently identified putative tumor suppressor genes on chromosomes 4, 11, 12, 16 and 19. Specifically, in the centromeric region of chromosome 11, we found a unique locus with a high frequency of LOH (~ 50%) in X-ray-induced TLs in B6C3F1 mice [2]. Linkage analysis revealed that the markers of regions with frequent LOH are genetically linked to *Ikaros*, a Krüppel-type zinc-finger transcription factor that plays a critical role in both lineage commitment and differentiation of the lymphocytes [3]. This suggests that *Ikaros* is a potent tumor suppressor gene in lymphomagenesis. Indeed, genomic analysis radiogenic TLs has revealed point mutations in the coding region [4,5]. On the other hand, the LOH on chromosome 11 was hardly observed in ENU-induced or spontaneously developing TLs, suggesting that it was radiation-associated molecular change. Thus, it is interesting to compare the involvement of Ikaros inactivation in ENU-induced with that in X-ray-induced TLs. The aim of the present study was to determine the molecular alteration of the *Ikaros* gene with special reference to LOH in X-ray- and ENU-induced TLs, which develop in B6C3F1 mice.

2. Materials and method

For induction of TLs, the mice were exposed weekly to 1.6 Gy whole-body X-irradiation for 4 consecutive weeks or were given ENU (400 ppm) in drinking water for 6 consecutive weeks. DNA and the total RNA were extracted from the primary TLs. The LOH status was determined at *Ikaros* alleles (*D11Mit62*, *Ikaros*, and *D11Mit2*), and expression status and genetic changes of *Ikaros* were analyzed using RT-PCR, Western blotting and nucleotide sequencing.

3. Results

We first examined the expression status of the *Ikaros* in TLs by semi-quantitative RT-PCR. X-ray-induced TLs exhibited null expression (5/37, 14%), and shorten size Ikaros isoforms (4/37, 11%). Sequence analysis revealed that the shortened size products were dominant-negative *Ikaros* isoforms lacking DNA binding domain; Ik-4, Ik-4(del), IK-8, Ik-8(del) (Fig. 1). Furthermore, all but one of the ENU-induced TLs, however, expressed normal-type *Ikaros*. One TL exhibited new size products concomitantly together with normal-type *Ikaros*. In order to identify the coding region mutations in the *Ikaros* gene of the TLs, the nucleotide sequences of the coding region in genomic DNA and cDNA were next analyzed for the existence of point mutations, deletions or insertions. In X-ray-induced TLs, eight point mutations (8/37, 22%) and one insertion (1/37, 3%) were found

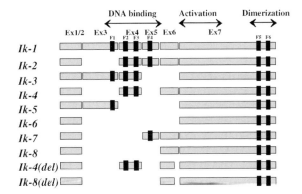

Fig. 1. Schematic representation of Ikaros isoforms. translated exons itilized differently by Ikaros isoforms are indicated. Zinc finger motifs (F1 ? F6) are shown as black boxes. Ikaros contains function domains for DNA-binding (F1 to F4), activation and dimerization (F5 and F6) [3]. Dominant negative function is suggested for Ik-4, Ik-5, Ik-6, Ik-7, Ik-8, Ik-4 (del). Ik-4 (del) and Ik-8 with 30 bp deletion at the end of exon 6, respectively.

in exons 3, 4 and 5. In ENU-induced TLs, five point mutations (5/31; 16%) were found; the four were in exons 4 and 5, and one was in intron 4. All point mutations found in both-type TLs were located in the N-terminal zinc-finger domain that resulted in amino acid substitutions. These mutated Ikaros might lack the DNA-binding ability.

The LOH status at the *Ikaros* locus in X-ray-induced and ENU-induced TLs was determined using the primers for a newly identified polymorphic microsatellite, flanking the upstream region of exon 1 of the *Ikaros* gene and those for micro satellite markers *D11Mit62* and *D11Mit2*. The frequency of LOH at the *Ikaros* locus was much higher in X-ray-induced TLs (54%) than in that of ENU-induced TLs (8%). Furthermore, the point mutation was frequently accompanied by the existence of LOH at the *Ikaros* locus in X-ray-induced TLs (89%) but rarely in ENU-induced TLs (20%).

4. Conclusion

Ikaros was inactivated by multiple mechanisms that fell into the following categories: null expression, expression of dominant negative isoforms, point mutation, and insertion. The X-ray-induced TLs exhibited all types of categories of Ikaros inactivation, whereas the ENU-induced TLs showed only point-mutation type. It is implied that the frequency and the mode of Ikaros inactivation are dependent on the carcinogenic agents.

Acknowledgements

We thank Ms. S. Sasaki, Ms. H. Osada and the staff of the Division of Animal Facility of our institute for their help with the laboratory analysis and maintenance of the animals.

References

[1] Y. Shimada, M. Nishimura, S. Kakinuma, T. Takeuchi, T. Ogiu, G. Suzuki, Y. Nakata, S. Sasanuma, K. Mita, T. Sado, Characteristic association between K-*ras* gene mutation with loss of heterozygosity in X-ray-induced thymic lymphomas of the B6C3F1 mouse, Int. J. Radiat. Biol. 77 (2001) 465–473.

[2] Y. Shimada, M. Nishimura, S. Kakinuma, M. Okumoto, T. Shiroishi, K.H. Clifton, S. Wakana, Radiation-associated loss of heterozygosity at the Znfn1a1 (*Ikaros*) locus on chromosome 11 in murine thymic lymphomas, Radiat. Res. 154 (2000) 293–300.

[3] K. Georgopoulos, S. Winandy, N. Avitahl, The role of the *Ikaros* gene in lymphocyte development and homeostasis, Annu. Rev. Immunol. 15 (1997) 155–176.

[4] H. Okano, Y. Saito, T. Miyazawa, T. Shinbo, D. Chou, S. Kosugi, Y. Takahashi, S. Odani, O. Niwa, R. Kominami, Homozygous deletions and point mutations of the *Ikaros* gene in gamma-ray-induced mouse thymic lymphomas, Oncogene 18 (1999) 6677–6683.

[5] S. Kakinuma, M. Nishimura, S. Sasanuma, K. Mita, G. Suzuki, Y. Katsura, T. Sado, Y. Shimada, Spectrum of Ikaros inactivation and its association with loss of heterozygosity in radiogenic T-cell lymphomas in susceptible B6C3F1 mice, Rad. Res., in press.

Molecular analysis of radiogenic tumors: Human

Transgenic murine models of human cancer: bridging the gap from mouse to man

Jun-Li Luo[a], Boris Zielinski[a], Wei-Min Tong[b], Manfred Hergenhahn[a], Zhao-Qi Wang[b], Monica Hollstein[a,*]

[a]*Department C0700, German Cancer Research Center (Deutsches Krebsforschungszentrum), Im Neuenheimer Feld 280, D-69120 Heidelberg, Germany*
[b]*International Agency for Research on Cancer, Lyons, France*

Abstract

The use of gene targeting technology to develop new mouse models of human cancer has allowed genetic and biochemical data to be put to the test for their biological consequences in vivo. Mice with specifically modified p53 tumor suppressor genes provide an interesting demonstration of this approach. Strains in which p53 is absent are cancer prone, confirming the importance of this protein in inhibiting tumor development [Nature 356 (1992) 215]. The tumorigenicity of murine cells with a mutant p53 transcriptional transactivation domain has clarified the importance of p53-regulated gene expression in the mechanism of cancer suppression by p53 [Nat. Genet. 26 (2000) 37]. Genetically engineered mice also promise to provide new insights into the p53 point mutations found in human tumors. A key question in cancer research has been the relative contributions of various environmental exposures on the one hand, and endogenous biological processes on the other, in the induction of cancer gene mutations in human tumors. Since a mutation spectrum can provide important information on the nature of the mutagenic agents that gave rise to the mutations, internet-accessible mutation databases are being scrutinized for clues to environmental etiological agents, or for confirmation of a role for putative endogenous pro-mutagenic risk factors in contributing to the human cancer burden. To facilitate decoding of human mutation patterns, we have generated a knock-in mouse with human p53 gene sequences. This mouse model (*hupki*, for *h*uman *p*53 *k*nock-*i*n) can be applied to some of the puzzles that the human p53 tumor mutation database presents.
© 2002 Elsevier Science B.V. All rights reserved.

Keywords: Mutation; p53; Gene targeting

---

* Corresponding author. Tel.: +49-6221-42-33-02; fax: +49-6221-42-33-42.
E-mail address: m.hollstein@dkfz-heidelberg.de (M. Hollstein).

1. Introduction

Mutagens produce characteristic patterns of mutations in a defined sequence, which reflect their specific promutagenic interactions with DNA [2]. Mutation profiles in target DNA can thus provide clues to the nature of the DNA damaging agents responsible for base sequence changes.

The p53 gene encodes a transcriptional activator that inhibits cancer development in multiple ways, unless it has acquired a mutation that compromises the function of the protein [3,4]. Point mutations in the p53 tumor suppressor gene leading to dysfunction are found commonly in almost all kinds of human cancers [5]. Unlike tumor mutations in other suppressor genes and oncogenes, p53 mutations in human cancers can be highly diverse in nature and location. These observations, and an internet-accessible database of > 15,000 human tumor p53 mutations (http://www.iarc.fr/p53) present a unique opportunity to examine more closely the long-standing hypothesis that exposure to mutagenic cancer risk factors contributes to cancer by mutating critical genes [6]. The clearest link between a mutagenic risk factor and a tumor mutation profile has been provided by studies on non-melanoma skin cancer ([7]; reviewed in Ref. [5]). Unlike internal cancers of all subtypes, basal and squamous cell carcinomas of the skin, in sun exposed individuals, harbor p53 mutations that are almost exclusively C to T transitions at dipyrimidine sites of the gene. Roughly 10% of these are CC to TT tandem base substitutions, a 'signature' on DNA of exposure to UV light. Additional plausible correlations between a cancer risk factor and specific characteristics of a p53 mutation pattern have been discussed at length in the literature [5]. Striking nevertheless is the fact that despite a decade of research activity there are few such examples. For many cancer types and patient groups, the mutation patterns remain an enigmatic collection of DNA sequence changes. This may be attributable at least in part to the diversity of mutagenic insults that characterize various human exposure situations, and the likelihood that spontaneously arising mutations, for example those due to replication errors [8], increasingly dominate a mutation profile when exposures to exogenous mutagenic agents are low [6].

Some p53 mutation patterns thus probably will remain undistinguished, but others may become more informative if appropriate measures to test hypotheses on the origins of human mutations in tumors were at hand. An experimental test system offering the human p53 gene sequence as a DNA target may be revealing in certain instances where conflicting mutation data have been presented in the literature.

While overall, the large volume of information in the IARC mutation data repository provide internally corroborating and consistent profiles, there are also unexplained discrepancies. PCR artifacts and other methodological pitfalls that have led to several incongruities in the database have been discussed at length in recent articles [9,10]. P53 gene mutation analyses of human tumors from patients exposed to α-particles present a puzzling example of heterogeneous data. Whereas one study [11] showed that a single base substitution at the second position of codon 249 dominates large cell and squamous cell lung tumor p53 mutation profiles in Colorado Plateau uranium miners, neither subsequent analysis of lung adenocarcinomas from the same patient cohort [12], nor screening of lung tumors (various histological subtypes) from two other uranium miner cohorts corroborated the existence of this putative molecular dosimeter of radon gas exposure [13–15]. A similar disparity in data is seen in studies of liver cancer patients administered with the radiocontrast agent thorotrast. Three different laboratories reported the absence of p53 mutations in thorotrast-associated liver tumors [13,16,17], while two other groups found that most tumors, and even normal tissues of thorotrast patients harbored p53 mutations (see other contributions to this volume). Diagnostic procedures with thorotrast have long since been discontinued so it is not likely that this matter will be pursued extensively in the future, whereas the debate over the existence of a mutagenic signature of tobacco smoke in the p53 gene of lung cancers is receiving wide attention because of the broad public health implications. Here the issues focus not only on the validity of the primary information, but also on interpretation and analysis of the data [10].

2. Testing hypotheses on origins of human p53 tumor mutations in experimental systems

A specific hypothesis, for example the proposal that G to T substitutions on the non-transcribed strand at codon 158 and 248 of the p53 gene in tumor specimens from smokers are due largely to benzo[*a*]pyrene (B[a]P) in tobacco smoke, can be explored by inducing tumors with B[a]P experimentally in laboratory mice or rats and examining the B[a]P p53 mutations against those found in spontaneous tumors or tumors induced by a different class of agent. There are limitations to this approach however: (1) although the p53 gene is highly conserved in evolution, the precise DNA sequence of the human p53 gene is not represented in the genome of laboratory strains used for standard cancer tests, and it is clear that the exact base sequence context has a fundamental role in determining the mutation spectrum of a compound. At the murine DNA sequence equivalent to human p53 codons 248–250 (encompassing the most heavily mutated site of the human p53 gene, all cancers considered together), four of the nine bases are different in humans (human sequence: CGG AGG CCC; murine sequence: CG**C** C**GA** CC**T**). To attempt to recapitulate human p53 mutation patterns experimentally, a model is needed in which mutation spectra are induced in vivo in human p53 gene sequences. Although human cell lines can be exposed to mutagens in vitro for this purpose, there is no appropriate selection method for isolating mutant p53 cells that would be equivalent to protocols in standard mutation assays, such as procedures for culturing treated cells in special medium to recover HPRT mutants for example. (2) Secondly, in attempting to generate experimental mutation spectra in animals used for carcinogen bioassays, one is confronted with the general paucity of p53 mutations found in rodent tumors (reviewed in Ref. [18]). In many organs and strains, either the endogenous p53 gene is rarely mutated, or the mutations do not have the same biological impact on tumor development in different organs. Mapping and characterization of cancer susceptibility alleles in different laboratory mice strains (see this volume) may provide some clues to this curious apparent difference between mouse and man in the p53 pathways. What is similar to humans, however, is the observation in mice that the majority of skin tumors induced by UV or chemical treatments do harbor p53 point mutations [18]. The skin can thus be used as a surrogate organ in murine carcinogenesis/mutagenesis experiments.

3. A new transgenic mouse model for investigating human p53 mutations and their consequences

Using gene-targeting technology we have generated a 'humanized' p53 knock-in mouse strain for molecular biology, molecular epidemiology, and cancer therapy studies. In this mouse the murine endogenous sequences encoding the p53 DNA binding domain were replaced by the equivalent human p53 sequences, yielding a knock-in chimeric p53 gene under control of the endogenous murine p53 promoter [19].

Various biochemical and biological parameters indicate that we have been able to retain normal p53 functions in the knock-in mice, as desired. In *hupki* mice homozygous for the knock-in allele, we showed that the p53 message is present at normal levels in various tissues (i.e., equivalent to endogenous, non-recombinant p53 transcript levels), is properly spliced, and encodes a p53 protein that binds to p53 consensus sequences in gel mobility shift assays [19]. These are properties of the wild type, functional protein. P53-regulated genes are induced following DNA damage, and γ-irradiation-induced apoptosis in thymocytes, a p53-dependent process [20], proceeds normally in the *hupki* mice. The fact that this new strain is not prone to spontaneous lymphomas (in contrast to p53 null mice) [1], is a further indication that the mice have retained various essential p53 tumor suppressor functions.

A mouse model for generating mutation patterns along the human p53 gene was our first goal in developing the *hupki* mice. We reasoned that the p53 gene mutation spectra induced in the human p53 sequences of *hupki* mice would parallel human tumor spectra more closely than mutations induced in the murine gene of normal (i.e., non-transgenic) mice. As a first proof-of-principle experiment, we exposed shaved dorsal areas of *hupki* mice to UVB radiaton to see whether *hupki* skin keratinocytes respond to chronic UVB exposure by forming p53 patches (clones of cells with nuclei strongly staining with anti-p53 antibody), as has been observed in sun-exposed areas of human skin [21]. We also analyzed exposed epidermis for a UV-dependent increase in mutation load in locations of the *hupki* p53 gene, where p53 mutations in human skin tumors are found most often. Clones of cells with immuno-reactive nuclei, and UV signature p53 mutations at human hotspots were found in treated skin of *hupki* mice as predicted (unpublished data).

With the p53 knock-in mouse strain for inducing 'humanized' p53 mutation spectra it should be possible not only to test hypotheses on the causes of human tumor mutations, but also to test new pharmaceuticals designed to regenerate tumor suppressor functions of a mutated p53 [22–24]. When a missense p53 mutation arises in *hupki* mice that corresponds, for example, to a hotspot mutation in human cancers, then these mutant *hupki* cells will have the precise core domain in the protein that needs to be remodeled in human tumors to restore DNA binding activity. The *hupki* mouse thus may be an appropriate test system for the development of drugs to rescue p53 function.

Acknowledgements

The gene-targeting studies were supported by grant number R01CA79493 to M.H. from the National Cancer Institute (USA). The contents of this article are solely the responsibility of the authors, and do not necessarily represent the official views of the National Cancer Institute.

References

[1] L.A. Donehower, M. Harvey, B.L. Slagle, M.J. McArthur, C.A. Montgomery, J.S. Butel, A. Bradley, Mice deficient for p53 are developmentally normal but susceptible to spontaneous tumours, Nature 356 (1992) 215–221.
[2] J.H. Miller, Mutational specificity in bacteria, Annu. Rev. Genet. 17 (1983) 215–238.
[3] C. Prives, P.A. Hall, The p53 pathway, J. Pathol. 187 (1999) 112–126.
[4] B. Vogelstein, D. Lane, A. Levine, Surfing the p53 network, Nature 408 (2000) 307–310.
[5] P. Hainaut, M. Hollstein, P53 and human cancer: the first ten thousand mutations, Adv. Cancer Res. 76 (2000) 81–137.
[6] M. Hollstein, G. Moeckel, M. Hergenhahn, B. Spiegelhalder, M. Keil, G. Werle-Schneider, H. Bartsch, J. Brickmann, On the origins of tumor mutations in cancer genes: insights from the p53 gene, Mutat. Res. 405 (1998) 145–154.
[7] D.E. Brash, J.A. Rudolph, J.A. Simon, et al., A role for sunlight in skin cancer: UV-induced p53 mutations in squamous cell carcinoma, Proc. Natl. Acad. Sci. 88 (1991) 10124–10128.
[8] T. Lindahl, Instability and decay of the primary structure of DNA, Nature 362 (1993) 709–715.
[9] T.M. Hernandez-Boussard, R. Montesano, P. Hainaut, Sources of bias in the detection and reporting of p53 mutations in human cancer: analysis of the IARC p53 mutation database, Genet. Anal. 14 (1999) 229–233.
[10] P. Hainaut, G.P. Pfeifer, Patterns of p53 G → T transversions in lung cancers reflect the primary mutagenic signature of DNA-damage by tobacco smoke, Carcinogenesis 22 (2001) 367–374.
[11] J.A. Taylor, M.A. Watson, T.R. Devereux, R.Y. Michels, G. Saccomanno, M. Anderson, p53 mutation hotspot in radon-associated lung cancer, Lancet 343 (1994) 86–87.
[12] J.W. McDonald, J.A. Taylor, M.A. Watson, G. Saccomanno, T.R. Devereux, P53 and K-ras in radon-associated lung adenocarcinoma, Cancer Epidemiol., Biomarkers Prev. 4 (1995) 791–793.
[13] M. Hollstein, H. Bartsch, H. Wesch, E.H. Kure, R. Mustonen, K.-R. Muehlbauer, A. Spiethoff, K. Wegener, T. Wiethege, K.-M. Mueller, p53 gene mutation analysis in tumors of patients exposed to α-particles, Carcinogenesis 18 (1997) 511–516.
[14] K.H. Vähäkangas, J.M. Samet, R.A. Metcalf, J.A. Welsh, W.P. Bennett, D.P. Lane, C.C. Harris, Mutations of p53 and ras genes in radon-associated lung cancer from uranium miners, Lancet 339 (1992) 576–579.
[15] Q. Yang, H. Wesch, K.-M. Mueller, H. Bartsch, K. Wegener, M. Hollstein, Analysis of radon-associated squamous cell carcinomas of the lung for a p53 gene hotspot mutation, Br. J. Cancer 82 (2000) 763–768.
[16] M. Andersson, M. Jonson, L.L. Nielsen, M. Vyberg, J. Visfeldt, H.H. Storm, H. Wallin, Mutations in the tumor suppressor gene p53 in human liver cancer induced by α-particles, Cancer Epidemiol., Biomarkers Prev. 4 (1995) 765–770.
[17] Y. Soini, J.A. Welsh, K.G. Ishak, W.P. Bennett, P53 mutations in primary tumors not associated with vinyl chloride exposure, Carcinogenesis 16 (1995) 2879–2881.
[18] M. Hollstein, M. Hergenhahn, Q. Yang, H. Bartsch, Z.-Q. Wang, P. Hainaut, New approaches to understanding p53 gene tumor mutation spectra, Mutat. Res. 431 (1999) 199–209.
[19] J.-L. Luo, Q. Yang, W.-M. Tong, M. Hergenhahn, Z.-Q. Wang, M. Hollstein, Knock-in mice with a chimeric human/murine p53 gene develop normally and show wild-type p53 responses to DNA damaging agents: a new biomedical research tool, Oncogene 20 (2001) 320–328.
[20] S.W. Lowe, E.M. Schmitt, S.W. Smith, B.A. Osborne, T. Jacks, p53 is required for radiation-induced apoptosis in mouse thymocytes, Nature 362 (1993) 847–849.
[21] A.S. Jonason, S. Kunala, G.J. Price, R.J. Restifo, H.M. Spinelli, J.A. Persing, D.J. Leffell, R.E. Tarone, D.E. Brash, Frequent clones of p53 mutated keratinocytes in normal human skin, Proc. Natl. Acad. Sci. 93 (1996) 14025–14029.

[22] B.A. Foster, H.A. Coffey, M.J. Morin, F. Rastinejad, Pharmacological rescue of mutant p53 conformation and function, Science 286 (1999) 2507–2510.
[23] G. Selivanova, L. Ryabchenkio, L. Jansson, W. Iotsova, K.G. Wiman, Reactivation of mutant p53 through interaction of a C-terminal peptide with the core domain, Mol. Cell Biol. 9 (1999) 3395–3402.
[24] T.R. Hupp, D.P. Lane, K.L. Ball, Strategies for manipulating the p53 pathway in the treatment of human cancer, Biochem. J. 352 (2000) 1–17.

Cancers induced by alpha particles from Thorotrast

Yuichi Ishikawa [a,*], Ikuo Wada [a], Manabu Fukumoto [b]

[a] *Department of Pathology, The JFCR Cancer Institute, 1-37-1 Kami-ikebukuro, Toshima-ku, Tokyo 170-8455, Japan*
[b] *Department of Pathology, Institute for Development, Aging and Cancer, Tohoku University, Sendai 980-8575, Japan*

Abstract

To investigate human cancers induced by alpha-radiation, patients who had an injection of Thorotrast, a radioactive colloidal X-ray contrast medium composed of thorium dioxide (ThO_2) used in Europe, US and Japan during 1930–1955, have been studied. The thorotrast patients died mainly of liver cancer, liver cirrhosis, leukemia and other cancers. Among the three histologies of liver cancer (cholangiocarcinoma, hepatocellular carcinoma and angiosarcoma), angiosarcoma was visible for alpha-radiation. In increased blood neoplasms, erythroleukemia and myelodysplastic syndrome were remarkable. Thorotrast patients exhale an extremely high concentration of radon (Rn-220), typically 20,000 Bq/m^3, but there has been no excess of lung cancer. Analyses of p53 mutations and loss-of-heterozygosity (LOH) at the 17p locus were performed to characterize the Thorotrast-induced liver tumors. Interestingly, LOH supposedly corresponding to large deletions was not frequent and most mutations were transitions, as seen in tumors of the general population. Therefore suggesting that genetic changes of Thorotrast-induced cancers are mainly "delayed mutations", not results of direct effects of radiation. © 2002 Elsevier Science B.V. All rights reserved.

Keywords: Radiation; Alpha particle; Thorotrast; Cancer; Genetic changes

1. Introduction

Thorotrast is the trade name of an X-ray contrast medium that is composed of 25% thorium-dioxide (ThO_2), which was used in Western countries and Japan, during the years

---

* Corresponding author. Fax: +81-3-5394-3923.
 E-mail address: ishikawa@jfcr.or.jp (Y. Ishikawa).

1930–1955. Th-232 is a naturally occurring, alpha-emitting radionuclide with a 14-billion-year half-life. Since Thorotrast is a colloidal solution, it was deposited mainly in the reticuloendothelial tissues, including the liver, spleen and bone marrow, after the injection. Injected Thorotrast was hardly eliminated from the body and therefore, the tissues of the deposited organs were irradiated by alpha-radiation during the remainder of the patients lives. Eighty-eight percent of the emitted energy was from alpha particles. Irradiation of the liver and bone marrow developed cancers and premalignant lesions after a long latent period.

2. Epidemiology

Recent results on epidemiological studies of Thorotrast patients are available for German, Japanese, Danish, Portuguese and Swedish populations. The German population is the largest (2326 patients and 1890 controls) and covers both sexes and a variety of underlying diseases necessitating Thorotrast injection. The Japanese population (412 patients and 1649 controls), on the other hand, consisted mostly of war-wounded ex-servicemen and, therefore, they were largely male. This in turn makes the Japanese cohort particularly important due to the absence of underlying diseases except physical injuries.

The recent results of the Japanese study, which was summarized by Mori et al. [1], demonstrated a remarkable excess of liver cancer ($O/E = 35.9$), as well as significant increases of liver cirrhosis (6.9) and leukemia (12.5). In contrast, the German study conducted by van Kaick et al. [2] demonstrated that liver cancer developed to a tremendous excess ($O/E = 122.9$), which is probably due to a higher incidence of hepatocellular carcinoma by hepatitis viruses in Japan than in Germany.

Thorotrast patients exhale radon (Rn-220) continuously and its concentration is extremely high, typically 20,000 Bq/m^3, similar for uranium miners, as detailed in the next section. However, all the major epidemiological studies consistently showed no increase in lung cancer; $O/E = 0.7$ for German [2], 2.0 for Japanese [1] and 1.6 for Danish [3]. Only increased percentages of small cell lung carcinoma (SCLC) were noted in the Japanese and Danish studies [3,4]. As is well known, lung cancer cases were high in underground miners, particularly SCLC. To solve the lung cancer problem in Thorotrast patients, we should examine the major confounding factor, smoking.

3. Dosimetry

Dosimetry for organs with Thorotrast deposits is very complicated. In order to assess the absorbed energy in an organ, we have to know the amount of parent Th-232 and steady-state activity ratios of its 11 progenies. The German group and we determined the activity ratios independently and they agreed well [5,6]. Estimates of the absorbed dose based on injection volume of Thorotrast need the knowledge of organ partition of the Th-232. The former distribution ratio (liver:spleen:red bone marrow:others = 59:29:9:3) [6] was recently revised to 53:14:25:8, implying leukemia risks have decreased markedly [7].

Risk assessment of alpha particles based on Thorotrast patients should be revised using the new organ partition ratio.

4. Pathology

Thorotrast develops three types of liver cancers: cholangiocarcinoma, angiosarcoma and hepatocellular carcinoma. Although cholangiocarcinoma is most frequent, angiosarcoma is characteristic because of its rarity in the general population.

Incidences of cholangiocarcinoma, angiosarcoma and hepatocellular carcinoma were 44%, 22% and 15%, respectively [8]. Liver cancers with multiple histologies were 19%. Chronological changes of the incidences were evident between 1945–1975 and 1976–1998: e.g., cholangiocarcinoma decreased from 57% to 33% and multiple histologies increased from 6.5% to 28%.

Histology of blood neoplasms arising in Thorotrast patients is characteristically erythroleukemia and myelodysplastic syndrome [9]. This reminds us of the increased number of chronic myeloid leukemia cases in A-bomb survivors. A different quality of radiation, i.e., alpha particles in Thorotrast patients and gamma rays in A-bomb survivors, may cause different histologies of leukemia.

5. Transgenerational carcinogenesis

One of the most striking findings in the dosimetry of Thorotrast patients was that the highest concentration of Th-232 in the whole body, except the major reticuloendothelial organs, was found at the testis [10]. Irradiation to germ cells of the patients might lead to carcinogenesis of the offspring. Andersson et al. [11] examined the cancer incidence among the offspring of Danish Thorotrast patients of both sexes. Although no significant excess of cancers was observed, two interesting findings were noteworthy: one sporadic retinoblastoma was identified in the offspring ($n = 143$) of the female Thorotrast patients, and a positive trend was found between the testicular-dose and the standardized incidence ratio (SIR) of cancers in the offspring ($n = 226$) of male patients. Certainly, further examinations including meta-analysis in countries where major epidemiological studies have been performed are needed.

6. Molecular analysis

To characterize the tumorigenesis process and to identify possible carcinogen-specific signatures of liver cancers induced by Thorotrast, we examined the mutation and the LOH status of the p53 gene in the 19 liver tumors (11 hepatocellular, 5 cholangiocellular carcinomas and 3 angiosarcomas) [12].

Interestingly, LOH was not frequent, only 27% ($=4/15$), in three hepatocellular carcinoma and one angiosarcoma. Eight cases ($8/19 = 42\%$) showed nine mutations in the exons, mostly transitions. These findings were similar to those seen in cancers of the general population, implying that genetic changes of Thorotrast related cancers are mainly "delayed mutations", not the result of direct effects of radiation.

Acknowledgements

We thank Drs. Takesaburo Mori and Yoshio Kato, for leading us to this study, and Prof. Rikuo Machinami, Dr. Hiroshi Tanooka and Prof. Otsura Niwa for their advice and collaboration. We are also indebted to all the doctors in charge of patients whose clinical and pathological records were submitted for research. The studies were supported partly by Grants-in-Aids in Scientific Research from Ministry of Education, Culture, Sports, Science and Technology and by grants from Ministry of Health, Labor and Welfare, Japan, and from the Smoking Research Foundation.

References

[1] T. Mori, C. Kido, K. Fukutomi, Y. Kato, S. Hatakeyama, R. Machinami, Y. Ishikawa, T. Kumatori, F. Sasaki, Y. Hirota, K. Kiyasawa, S. Hayashi, H. Tanooka, T. Sobue, Summary of entire Japanese Thorotrast follow-up study: updated 1998, Radiat. Res. 152 (Suppl. 6) (1999 Dec.) S84–S87 (supplement).
[2] G. van Kaick, A. Dalheimer, S. Hornik, A. Kaul, D. Liebermann, H. Luehrs, A. Spiethoff, K. Wegener, H. Wesch, The German Thorotrast study: recent results and assessment of risks, Radiat. Res. 152 (Suppl. 6) (1999 Dec.) S64–S71 (supplement).
[3] M. Andersson, Long-term effects of internally deposited alpha-particle emitting radionuclides: epidemiological, pathological and molecular biological studies of Danish Thorotrast-administered patients and their offspring, Danish Med. Bull. 44 (1997) 169–190.
[4] Y. Ishikawa, T. Mori, Y. Kato, E. Tsuchiya, R. Machinami, H. Sugano, T. Kitagawa, Lung cancers associated with Thorotrast exposure: high incidence of small cell carcinoma and implications for estimation of radon risk, Int. J. Cancer 52 (1992) 570–574.
[5] Y. Ishikawa, Y. Kato, T. Mori, R. Machinami, T. Kitagawa, Alpha-particle-dose to the liver and spleen tissues of Japanese Thorotrast patients, Health Phys. 65 (1993) 497–506.
[6] A. Kaul, Z. Noffz, Tissue dose in Thorotrast patients, Health Phys. 35 (1978) 113–121.
[7] Y. Ishikawa, J.A.H. Humphreys, C.G. Collier, N.D. Priest, Y. Kato, T. Mori, R. Machinami, Revised organ partition of Th-232 in Thorotrast patients, Radiat. Res. 152 (1999) S102–S106.
[8] T. Mori, K. Fukutomi, Y. Kato, S. Hatakeyama, R. Machinami, H. Tanooka, Y. Ishikawa, T. Kumatori, 1998 results of the first series of follow-up studies on Japanese Thorotrast patients and their relation to the autopsy series, Radiat. Res. 152 (1999) S72–S80.
[9] R. Kamiyama, Y. Ishikawa, S. Hatakeyama, T. Mori, H. Sugiyama, Clinicopathological study of hematological disorders after Thorotrast administration in Japan, Blut 56 (1988) 153–160.
[10] Y. Ishikawa, T. Mori, Y. Kato, R. Machinami, N.D. Priest, T. Kitagawa, Systemic deposits of thorium in Thorotrast patients with particular reference to sites of minor storage, Radiat. Res. 135 (1993) 244–248.
[11] M. Andersson, K. Juel, Y. Ishikawa, H. Storm, Effects of preconceptional irradiation on mortality and cancer incidence in the offspring of patients injected with thorium dioxide, J. Natl. Cancer Inst. 86 (1994) 1866–1870.
[12] I. Wada, H. Horiuchi, M. Mori, Y. Ishikawa, M. Fukumoto, T. Mori, Y. Kato, T. Kitagawa, R. Machinami, High rate of small p53 mutations and infrequent loss of heterozygosity in malignant liver tumors associated with Thorotrast: implications for alpha-particle carcinogenesis, Radiat. Res. 152 (1999) S125–S127.

The *p53* and *M6P/IGF2r* genes of Thorotrast- and atomic bomb-induced liver cancers: a glimpse into the mechanisms of radiation carcinogenesis

Keisuke S. Iwamoto*

Department of Radiobiology, Radiation Effects Research Foundation, 5-2 Hijiyama Park, Minami-ku, Hiroshima 732-0815, Japan

Abstract

Recipients of the radiographic contrast agent, Thorotrast, endured localized doses of high-LET alpha particle irradiation over a period of decades, whereas the A-bomb survivors were exposed to an instantaneous whole-body dose of low-LET radiation. To determine common and divergent mechanisms for chronic high- and acute low-LET radiation-induced carcinogenesis, retrospective molecular analyses of archival liver cancers from these populations were conducted. The tumor suppressor genes *p53* and *M6P/IGF2r* were examined. The loss of *p53* wild-type function is well recognized as an important step in the development of most human cancers. *M6P/IGF2r* is a putative tumor suppressor gene in liver cancer, as well as other cancers. In the Thorotrast cases, 19 out of 20 cases harbored *p53* mutations. The accompanying nontumor tissues from these patients also had *p53* mutations, albeit at lower frequency. Interestingly, in both groups, point mutations were the predominant aberrations found. Approximately half the cases in both groups had *M6P/IGF2r* mutations in the 3′UTR. The A-bomb cases had either *p53* mutations or *M6P/IGF2r* mutations, but rarely both. Moreover, the frequency of cases with *M6P/IGF2r* mutations actually decreased with dose, while those for *p53* increased. In contrast, no such relationship was evident in the tumors of the Thorotrast cases. © 2002 Elsevier Science B.V. All rights reserved.

Keywords: Thorotrast; A-bomb; *p53*; *M6P/IGF2r*; Radiation

1. Introduction

The period of time between radiation exposure and the observation of a statistically significant number of radiation-attributable solid cancers in humans is generally on the

---

* Present/permanent address: Roy E. Coats Research Laboratories, Department of Radiation Oncology, UCLA School of Medicine, 10833 Le Conte Ave., Box 951714 Los Angeles, CA 90095-1714, USA.
  E-mail address: kiwamoto@radonc.ucla.edu (K.S. Iwamoto).

order of half a lifetime. Although the relative risks differ, this phenomenon is independent of whether the victims are A-bomb survivors, who received an acute dose of low-LET radiation, or Thorotrast recipients, who received a chronic dose of high-LET radiation. The events occurring during this time of apparent silence before clinical detection of the tumor are unclear. From the perspective of direct biological damage, a single acute high-dose rate exposure to sparsely ionizing gammas and neutrons is distinct from a chronic low-dose rate exposure to densely ionizing alpha particles. However, from the wider perspective of carcinogenesis, there may be greater similarities between the two radiation types. Accordingly, the underlying mechanisms for both types of radiation-induced cancer processes may be similar and thus comparative analysis of molecular changes in tissues from both exposed populations should provide a more global perspective on radiation carcinogenesis than either one alone.

1.1. Archival liver tissues

The commonest Thorotrast-related tumors are liver cancers, predominantly hepatic angiosarcomas (HAS) and cholangiocarcinomas (ChC) with some hepatocellular carcinomas (HCC) because over 60% of the thorium dioxide deposits in this organ [1]. Unlike tumors in other radiation-exposed populations, all the cancers in Thorotrast recipients can be attributed to the effects of ionizing radiation. Thus, these are valuable tumors to analyze for understanding the mechanisms of radiation carcinogenesis. Likewise, the tissue samples from the A-bomb survivors are valuable because the population is unequaled for its documentation of exposed dose, pathology, medical history and other demographics. The major liver tumors in A-bomb survivors showing increased risk are HCCs. Thus, two tumor suppressor genes reported to be critical in hepatocarcinogenesis were analyzed for radiation-induced damage in these populations.

1.2. Tumor suppressor genes

Mutations in the *p53* tumor suppressor gene are known to be important in most human cancers including liver neoplasms. The p53 protein plays a critical role in various aspects of damage surveillance; a nonfunctional p53 is known to abrogate growth control and apoptosis [2]. The M6P/IGF2r is a member of the IGF family of growth factors, receptors, and binding proteins [3]. The *M6P/IGF2r* gene is lost or mutated in cancers of the liver, breast, colon and other organs [4–6]. *M6P/IGF2r* is a negative regulator with multiple functions that include degradation of IGFII and activation of the pro-apoptotic molecule TGFβ [7,8]. Consequently both *p53* and *M6P/IGF2r* genes were analyzed in paired normal/tumor tissue samples from both populations in hopes of generating two sets of data that could complement each other to reveal the big picture of radiation carcinogenesis.

2. Exposure to Thorotrast

Both chronically irradiated tumor and chronically irradiated nontumor tissues were analyzed for insights into early and late events in tumorigenesis. The most outstanding feature of the molecular analysis of the tissues is the high frequency of point mutations in the *p53* tumor suppressor gene. The detection of multiple point mutations in single tissues is probably the result of an accumulation of mutations via a selective process in system that is under constant exposure to radiation. Moreover, the unexpectedly large number of point mutations found in the nontumor tissues of the cases suggest important clues into the early stages of carcinogenesis.

2.1. Point mutations

Nineteen of the twenty (95%) cases had *p53* point mutations [9] in the tumor sections, whereas 47% had mutations in *M6P/IGF2r*. The *p53* mutations clustered in the conserved domains of the four hotspot exons. These domains have been shown to be important in the formation of structural elements critical in *p53* binding to DNA [10]. The *M6P/IGF2r* mutations were in the 3′UTR, an area known to be critical in the regulation of transcript stability [11]. Sixty-one percent of the cases had *p53* mutations in the nontumor sections indicating that cells possessing such mutations had gained some clonal expansion ability and survival advantage over the wild-type cells. Interestingly, unlike the tumor *p53* mutations, the nontumor *p53* mutations were not clustered in the conserved domains. However, none was extremely rare or unreported in the human cancer *p53* database [12].

2.2. Autoradiographs

Autoradiographs of tissue sections from Thorotrast tumors confirmed the existence of thorium dioxide deposits. An area within a tumor focus was void of any thorium dioxide, indicating that the tumor clonally expanded at a rate faster than the redistribution of the deposits such that it pushed aside the deposits as it grew. In contrast, there were no such void areas in the clonally expanded areas of the nontumor tissues. These observations suggest that the outward growth of the mutated nontumor cells was slower than the rate of thorium dioxide redistribution by phagocytic cells that constantly attempt to remove the deposits before being killed by the high doses.

2.3. Dynamics in radiation carcinogenesis

The molecular analyses of the Thorotrast tissues illustrate a dynamic cycle of clonal expansion, accumulation of critical mutations, death and survival of cells. Cells adjacent to deposits encounter lethal doses of radiation, whereas cells a few cell diameters away most likely experience sublethal damage, and those cells far enough away are essentially unexposed. The loss of many 'point-blank' range cells and the clonal outgrowth of 'mid-range' (sublethally damaged) cells that have gained some survival advantage would redistribute the local cell population. Some of the mid-range clones would therefore become 'point-blank' range clones or 'far-range' clones, reinitiating the cycle. During the process, radioresistant cells may be selected out resulting in a greater number of reproductively viable cells within the 'point-blank' range. Induction of genomic instability following irradiation-induced damage of key components in maintaining genome stability, such as faithful repair or replication, would compound the accumulation of mutations by the chronic irradiation. Eventually, there would exist a large spectrum of cells, from clonally expanded nontumor cells to malignant cells, with varying levels of damage. Those that have gained the necessary mutations for malignant transformation would go on to become a cancer. The chronic nature of the irradiation has allowed observation of the stages of change that occur at various times following a single acute irradiation. In other words, the tumors with multiple mutations represent the survivors of the earliest normal cells that were irradiated and the clonally expanded nontumor cells represent the most recent normal cells that were irradiated.

3. Exposure to the A-bomb

Although the tumors can be considered radiation-induced, one drawback to the Thorotrast study is the lack of a dose response due to a paucity accurate absorbed dose information for each of the cases. In contrast, each of the liver cancers from the A-bomb survivors have well defined absorbed dose information, ranging from 0 to 1569 mSv [13]. Similar to the Thorotrast cases, the HCCs of the A-bomb survivors harbored point mutations. Additionally, there was no dose dependence of *p53* deletions, damage most likely the result of the direct effects of ionizing radiation. The HCC cases had either *p53* mutations or *M6P/IGF2r* mutations, but rarely both. Moreover, the frequency of cases with *M6P/IGF2r* mutations actually decreased with dose, while those for *p53* increased. This implies two independent selection processes leading to liver cancer and that in radiation-induced HCC tumors the spectrum of molecular changes is different from that in "background" tumors.

3.1. Mutator phenotype and posttranscriptional regulation

The dose-dependent enrichment of cells with *p53* mutations in the tumors is probably caused by expansion of cells with *p53* mutations plus mutations in other genes that allow unregulated growth. The direct radiation target is more likely to be a gene that is changed into a mutator by a radiation-induced mutation. The induction of a mutator gene would be expected to increase with dose and would allow a single cell or its progeny to accumulate multiple mutations necessary for conversion of a normal to a cancer cell. Interestingly, *M6P/IGF2r* mutations were detected as increases in the number of repeats in the 3′UTR. These would be expected to affect mRNA stability. The mutations in *p53* were in the ORF and clustered within the conserved domains as was the case in the Thorotrast tumors. Both types of mutations, *p53* ORF and *M6P/IGF2r* 3′UTR, could lead to resistance to apoptosis and tumor growth, but by completely different mechanisms. ORF mutations would destroy wild-type p53 protein resulting in loss of function, whereas 3′UTR mutations would not affect protein function per se but could deregulate protein levels by affecting transcript half-life. The finding that high dose radiation exposure selected against tumors expressing mutations in *M6P/IGF2r* 3′UTR indicates that radiation can modulate posttranscriptional regulatory pathways and compensate for the loss in stability resulting from the mutations.

4. Summary

Molecular analyses of tissues from Thorotrast and A-bomb exposed populations have provided evidence that ionizing radiation can induce a mutator phenotype which can potentially lead to a destabilization of the genome resulting in a propagation of "spontaneous-like" mutations. Examination of nontumor/tumor pairs from the Thorotrast cases has furnished evidence for radiation-induced selection of cells possessing various levels of survival advantage. The very nature of chronic irradiation has allowed observation of such cells at various stages and times following their irradiation. The A-bomb liver cancer cases have provided evidence that ionizing radiation can affect posttranscriptional regulation as well as inflict genetic damage. In addition, the data suggest that all cells, normal and all precancer cells at various stages of development toward malignancy, are potential targets of radiation-induced carcinogenesis. For example, preexisting cells that have gained defects in genes coding for molecules critical in responding to radiation insult could gain further defects following irradiation that would increase its survival advantage and thus raise it closer toward cancer.

Acknowledgements

The A-bomb survivor HCC molecular study was made possible by the cooperative efforts of hospitals throughout the cities of Hiroshima and Nagasaki. Special thanks go to Drs. Masayoshi Tokunaga, Toshiyuki Fukuhara, Masami Yamamoto, Hideo Itakura, Takayoshi Ikeda, Masao Kishikawa, Yasuyuki Fujita, Nori Nakamura, Terumi Mizuno, Shoji Tokuoka, Kiyohiko Mabuchi for the collection and pathological reviews of the tissues and/or for their many helpful suggestions. The Thorotrast recipient liver cancer molecular study was made possible by the cooperative efforts of hospitals throughout Japan. Special thanks go to Drs. Akihiko Kurata, Makoto Suzuki, Tohru Hayashi, Yuji Ohtsuki, Yuhei Okada, Michihiko Narita, Masanori Takahashi, Sadahiro Hosobe, Kenji Doishita, Toshiaki Manabe, Sakae Hata, Ichiro Murakami, Satoru Hata, Shinji Itoyama, Seiya Akatsuka, Nobuya Ohara, Keisuke Iwasaki, Hisamasa Akabane, Megumu Fujihara, Toshio Seyama and Takesaburo Mori for the collection and pathological reviews of the tissues and/or for their many helpful suggestions. These studies could not have been accomplished without the technical support of Shiho Yano, Norie Ishii, Chiyoe Saito, and Tomoko Shinohara, the tissue preparation by Mutsumi Mizuno, Chiyako Ohmoto, and Kazuaki Koyama, and the statistical support of Dr. John B. Cologne and Sachiyo Funamoto.

This publication is based on research performed at the Radiation Effects Research Foundation (RERF), Hiroshima, Japan. RERF is a private nonprofit foundation funded equally by the Japanese Ministry of Health and Welfare and the United States Department of Energy through the National Academy of Sciences.

References

[1] Y. Ishikawa, Y. Kato, T. Mori, R. Machinami, T. Kitagawa, Alpha-particle dose to the liver and spleen tissues of Japanese Thorotrast patients, Health Phys. 65 (5) (1993) 497–506.

[2] L.R. Livingstone, A. White, J. Sprouse, E. Livanos, T. Jacks, T.D. Tlsty, Altered cell cycle arrest and gene amplification potential accompany loss of wild-type p53, Cell 70 (6) (1992) 923–935.
[3] M.J. Ellis, F. Garmroudi, K.J. Cullen, Insulin receptors (IGF1 and IGF2), in: J. Bertino (Ed.), Encyclopedia of Cancer, vol. 2, Academic Press, San Diego, 1997, pp. 929–939.
[4] A.T. DeSouza, G.R. Hankins, M.K. Washington, T.C. Orton, R.L. Jirtle, *M6P/IGF2r* gene is mutated in human hepatocellular carcinomas with loss of heterozygosity, Nat. Genet. 11 (4) (1995) 447–449.
[5] G.R. Hankins, A.T. DeSouza, R.C. Bentley, M.R. Patel, J.R. Marks, J.D. Iglehart, R.L. Jirtle, M6P/IGF2 receptor: a candidate breast tumor suppressor gene, Oncogene 12 (9) (1996) 2003–2009.
[6] R.F. Souza, S. Wang, M. Thakar, K.N. Smolinski, J. Yin, T.T. Zou, D. Kong, J.M. Abraham, J.A. Toretsky, S.J. Meltzer, Expression of the wild-type insulin-like growth factor II receptor gene suppresses growth and causes death in colorectal carcinoma cells, Oncogene 18 (28) (1999) 4063–4068.
[7] M.J. Ellis, B.A. Leav, Z. Yang, A. Rasmussen, A. Pearce, J.A. Zweibel, M.E. Lippman, K.J. Cullen, Affinity for the insulin-like growth factor-II (IGF-II) receptor inhibits autocrine IGF-II activity in MCF-7 breast cancer cells, Mol. Endocrinol. 10 (3) (1996) 286–297.
[8] P.A. Dennis, D.B. Rifkin, Cellular activation of latent transforming growth factor b requires binding to the cation-independent mannose 6-phosphate/insulin like growth factor type II receptor, Proc. Natl. Acad. Sci. U. S. A. 88 (2) (1991) 580–584.
[9] K.S. Iwamoto, S. Fujii, A. Kurata, M. Suzuki, T. Hayashi, Y. Ohtsuki, Y. Okada, M. Narita, M. Takahashi, S. Hosobe, K. Doishita, T. Manabe, S. Hata, I. Murakami, S. Hata, S. Itoyama, S. Akatsuka, N. Ohara, K. Iwasaki, H. Akanabe, M. Fujihara, T. Seyama, T. Mori, p53 mutations in tumor and non-tumor tissues of Thorotrast recipients: a model for cellular selection during radiation carcinogenesis in the liver, Carcinogenesis 20 (7) (1999) 1283–1291.
[10] Y. Cho, S. Gorina, P.D. Jeffrey, N.P. Pavletich, Crystal structure of a p53 tumor suppressor–DNA complex: understanding tumorigenic mutations, Science 265 (5170) (1994) 346–355.
[11] J. Ross, mRNA stability in mammalian cells, Microbiol. Rev. 59 (3) (1995) 423–450.
[12] C. Beroud, T. Soussi, p53 gene mutation: software database, Nucleic Acids Res. 26 (1) (1998) 200–204.
[13] K.S. Iwamoto, T. Mizuno, S. Tokuoka, K. Mabuchi, T. Seyama, Frequency of p53 mutations in hepatocellular carcinomas from atomic bomb survivors, J. Natl. Cancer Inst. 90 (15) (1998) 1167–1168.

Molecular epidemiology of childhood thyroid cancer around Chernobyl

Shunichi Yamashita*, Yoshisada Shibata, Hiroyuki Namba, Noboru Takamura, Vladimir Saenko

Atomic Bomb Disease Institute, Nagasaki University School of Medicine, 1-12-4 Sakamoto, Nagasaki 8528523, Japan

Abstract

Fifteen years after the Chernobyl accident, the accumulative data strongly suggest the direct involvement of radiation fallout exposure on human health, especially thyroid tumorigenesis. Based on the clinical data from Chernobyl, the summary of the second Chernobyl Sasakawa project has been introduced first. The comparative study of thyroid diseases demonstrates the likelihood of short-lived radioactive iodine on thyroid cancer in the children born before the Chernobyl accident. Next, at the standpoint of recent molecular analysis of thyroid carcinogenesis, many reports now indicate evidence of a high incidence of ret/PTC gene rearrangement in childhood thyroid cancer tissues. Besides ret/PTC gene rearrangement, the disturbance of the response of intracellular signal transduction to radiation exposure is also important in thyroid cells, and results demonstrate that radiation exposure could cause abnormal thyroid cell proliferation specifically through constitutive activation of intracellular target molecules via membrane lipid breakdown, and subsequently disturb the apoptosis-prone pathway. It is now urgent and to search for radiation-induced signature genes and/or target molecules using the newly established Chernobyl Thyroid Tissue Bank. Therefore, the late effect of radiation, even in the lower dose on the human thyroid glands, should be monitored carefully for the radiation-sensitive vulnerable group for a longer period, especially around Chernobyl. © 2002 Elsevier Science B.V. All rights reserved.

Keywords: Chernobyl; Screening; Thyroid cancer; Gene rearrangement; Signal transduction

1. Introduction

Epidemiological studies including Atomic Bomb survivors and children around Chernobyl suggest that the thyroid gland, as well as bone marrow, seem to be one of

---

* Corresponding author. Tel.: +81-95-849-7114; fax: +81-95-849-7117.
E-mail address: shun@net.nagasaki-u.ac.jp (S. Yamashita).

the most sensitive organs to the carcinogenic effects of external radiation. The accurate levels of radioactive iodine exposure to the thyroid glands remain to be further clarified among the children around Chernobyl. However, the vulnerable population is highly restricted in the group aged between 0 and 5 years old at the time of the accident. The first joint program of the Chernobyl Sasakawa project was completed in 1996 with the help of three countries, Belarus, Russia and the Ukraine, among which thyroid data are compiled and analyzed [1].

Since 1997, the second Chernobyl Sasakawa project has been conducted mainly in the Gomel region, and international cooperation has been welcomed in the various epidemiological and molecular studies. The proceedings of the international symposium of the Chernobyl Sasakawa project held in Moscow will be published next year (ICS 1234). Here, we will focus on the results of the epidemiological studies and molecular research on Chernobyl-related thyroid cancer.

2. Epidemiological study around Chernobyl

Since the high incidence of childhood thyroid diseases has been demonstrated around Chernobyl, especially in the Gomel region, Belarus, we have continued several medical aid projects focusing on the early diagnosis and careful follow-up for children, who have been detected with thyroid abnormalities after an ultrasound examination. The lack of reliable data on the individual thyroid doses has hindered conclusions about the direct effects of the Chernobyl accident on thyroid diseases. The first joint project with the International Agency of Research on Cancer, Lyon, is the case control study of childhood thyroid cancer in Belarus and in Russia, where data analysis is now underway.

In order to overcome the shortage of accurate dosimetrical data, the comparative study has been carried out in the Gomel region for children who were born before and after the accident. In order to minimize study bias, the medical screening has been done for school children born from January 1, 1983 to December 31, 1989, and the study has been carried out in four districts and Gomel city in the Gomel region. A total of 21,610 children were examined and 32 (0.15%) thyroid cancers were detected. The prevalence of thyroid cancer by sex and age at examination and by study group are shown in Tables 1 (boys) and 2 (girls), respectively. Combined with the number of boys and girls, no thyroid cancer was observed among the 9472 children born from January 1, 1987 to December 31, 1989 (Group I), while one and 31 thyroid cancers were found among the 2409 children born from April 27, 1986 to December 31, 1986 (Group II) and 9720 children born from January 1, 1983 to April 26, 1986 (Group III), respectively. For reference, Group IV shows the results of the first Chernobyl Sasakawa project, which examined the children born from April 26, 1976 to April 26, 1986 in the period from May 1991 to the end of April 1996 in the Gomel region. According to the study design with school-based, the differences in the environmental factors after the accident were deemed small. The major difference in the background was that the children in Group I were not exposed to the short-lived fallout by the Chernobyl accident, while the children in Group II and III were probably exposed to such a fallout in utero or directly, respectively. Furthermore, there is no relationship between the incidence of childhood thyroid cancer and the whole body Cs

Table 1
Prevalence of thyroid cancer by sex, age at examination and study group (Gomel, boys)

Study group		Age at examination													Total
		5	6	7	8	9	10	11	12	13	14	15	16	17	
I	Case				0	0	0	0	0	0					0
	Exam	–	–	–	67	651	1383	1523	897	305	–	–	–	–	4826
II	Case							0	0	0	0				0
	Exam	–	–	–	–	–	–	156	501	437	164	–	–	–	1258
III	Case							0	1	1	3	4	0	1	10
	Exam	–	–	–	–	–	–	28	333	989	1449	1339	579	93	4810
IV	Case	0	0	0	0	0	1	4	2	0	0	1	0	0	8
	Exam	21	138	331	688	939	984	895	868	838	739	310	70	29	6876

I: Children born from 1 January 1987 to 31 December 1989. II: Children born from 27 April 1986 to 31 December 1986. III: Children born from 1 January 1983 to 26 April 1986. IV: Children born from 26 April 1976 to 26 April 1986 and were examined in the period from May 1991 to the end of April 1996 in the Chernobyl Sasakawa Health and Medical Cooperation Project.

137 levels. Another unique cohort study has been conducted in the Kaluga, Oreol and Tula regions in the Russian Federation, in cooperation with the Medical Radiological Research Center, RAMS, Obninsk. Furthermore, there is a relatively highly exposed population in the western region of Bryansk [2], where screening will take place with the aid of our cooperative program.

3. Molecular mechanism of radiation-induced thyroid carcinogenesis

Concerning the operated childhood thyroid cancer tissues, it would be worthy to analyze the signature genes or proteins induced by radiation exposure. Indeed, gene rearrangements

Table 2
Prevalence of thyroid cancer by sex, age at examination and study group (Gomel, girls)

Study group		Age at examination													Total
		5	6	7	8	9	10	11	12	13	14	15	16	17	
I	Case				0	0	0	0	0	0					0
	Exam	–	–	–	75	667	1287	1457	858	302	–	–	–	–	4646
II	Case							0	0	1	0				1
	Exam	–	–	–	–	–	–	149	406	437	159	–	–	–	1151
III	Case							0	1	4	6	8	2	0	21
	Exam	–	–	–	–	–	–	26	345	906	1460	1319	754	100	4910
IV	Case	1	0	1	1	3	3	3	1	2	2	0	0	1	18
	Exam	21	97	317	676	930	971	948	934	868	787	354	80	19	7002

I: Children born from 1 January 1987 to 31 December 1989. II: Children born from 27 April 1986 to 31 December 1986. III: Children born from 1 January 1983 to 26 April 1986. IV: Children born from 26 April 1976 to 26 April 1986 and were examined in the period from May 1991 to the end of April 1996 in the Chernobyl Sasakawa Health and Medical Cooperation Project.

of ret/PTC subtypes have been reported as common genetic damage of radiation-induced thyroid cancer [3]. Unfortunately, there is no specific and selective marker to identify the radiation-induced signature genes. Therefore, the Chernobyl Thyroid Tissue Bank has just started to promote international scientific cooperation [4]. Tissues from the thyroid tumors, which are not needed for diagnostic purposes are valuable research resources, representing a large number of tumors directly related to exposure to the same mutagen at the same time. The post-Chernobyl Thyroid Tissues, Nucleic Acids and Data Bank have just been established in the three countries (Minsk, Kiev and Obninsk) as an internationally supported cooperative research resource (http://www.wrl.cam.ac.uk/nisctb), which is now expected to avoid direct competition for limited tissue resources. Owing to international cooperation on the matter, the infrastructure of scientific research on radiation-associated human thyroid carcinogenesis will be established.

In order to understand the molecular mechanism of radiation-induced thyroid carcinogenesis, we must at first realize that the human thyroid cells are relatively resistant to apoptosis caused by ionizing radiation [5,6], although the latter can cause both DNA damage and breakdown of the cell membrane. Our experimental results have demonstrated that the radioresistant properties of the human thyroid cells is in part due to the dominance of antiapoptotic signals evoked by growth factors and diacylglycerol, which override the apoptotic effect of ceramide released from the human thyroid cell membrane on exposure to ionizing radiation [7]. In addition, the downstream target molecules, such as c-JUN N-terminal kinase (JNK) phosphorylation pattern, are unique in the human thyroid cells, which may be involved in cell survival, not in apoptosis [8,9]. An improved understanding of JNK-mediated apoptotic signaling may provide novel strategies in the prevention and treatment of cancers. The reason why we need to pay special attention to such cell survival and escape mechanism from apoptosis is that low-dose irradiation can cause abnormal cell proliferation through constitutive activation of intracellular target molecules and disturb the apoptosis-prone pathways, besides the direct genetic alteration. In addition, JNK activation by irradiation may contribute to cell survival. The unique response of the thyroid cells, especially the intracellular signal transduction system, may contribute to abnormal cell proliferation after low-dose irradiation.

It is obvious that a high prevalence of ret rearrangements (62.3%) with a significant predominance of ELE1/ret (PTC3) over H4/ret (PTC1) rearrangement was found in childhood papillary thyroid carcinomas of the first Chernobyl decade [10]. Furthermore, we have also confirmed the low frequencies of the other types of gene rearrangements and the absence of point mutations. Therefore, attention has been focused on the direct relationship between the rapid increase of childhood thyroid cancer and ret/PTC gene rearrangements, implicating the characteristics of post-Chernobyl thyroid cancers, biological, phenotypical and clinical from genetic alternation. In contrast, the role and significance of ret/PTC may be far away from our current understanding because of the high frequency of these rearrangements observed in the papillary thyroid carcinoma in New Caledonia and Australia [11]. The age-related ret/PTC rearrangement should be, therefore, analyzed in thyroid papillary cancers. Another important target of radiation-associated thyroid cancer is gene rearrangement, not point mutation of mitochondrial DNA. The results for point mutation of mitochondrial DNA will be presented elsewhere soon.

To further understand the molecular genetic aberrations and microsatellite instabilities, careful and advanced analysis has been undertaken using modern molecular biological techniques.

4. Conclusion

One of the lingering problems of the Chernobyl accident is the concern regarding the various health effects among the several populations directly affected by the accident, especially the emergency workers and people living in radiation-contaminated areas. Among the latter, there has been a dramatic increase in thyroid cancer among those who were children or adolescents at the time of the accident, the risk of which is expected to be long lasting. Based on our own activities and findings of the Chernobyl Sasakawa project, which was carried out in Belarus, Russia and the Ukraine using a uniform protocol for the diagnosis and treatment of the related health effects, and for associated research. The international community is invited to participate in the effort to maintain the long-term follow-up of irradiated victims, to support field oriented radiation research, and to improve the health care of children and others affected by the Chernobyl accident. Finally the direct signature gene(s) and/or target molecules of radiation-induced thyroid cancer should be clarified using the Chernobyl Thyroid Tissue Bank through investigation of advanced molecular biological techniques.

References

[1] J. Farahati, E.P. Demidchik, J. Biko, C. Reiner, Inverse association between age at the time of radiation exposure and extension of disease in cases of radiation-induced thyroid carcinomas in Belarus, Cancer 88 (2000) 1470–14476.
[2] S. Yamashita, S. Shibata, Chernobyl: A Decade, ICS vol. 1156, Excerta Medica, Amsterdam, 1997, 613 pp.
[3] H.M. Rabes, S. Klugbauer, Molecular genetics of childhood papillary thyroid carcinomas after irradiation: high prevalence of RET rearrangement, Recent Results Cancer Res. 154 1998.
[4] G.A. Thomas, E.D. Williams, Members of the Scientific Project Panel, Thyroid tumor bank, Science 289 (2000) 2283.
[5] H. Namba, T. Hara, T. Tsukazaki, et al., Radiation-induced G1 arrest is selectively mediated by the p53-WAF1/Cip1 pathway in human thyroid cells, Cancer Res. 55 (1995) 2075–2080.
[6] T. Yang, H. Namba, T. Hara, et al., p53 induced by ionizing radiation medicates DNA end-joining activity, but not apoptosis of thyroid cells, Oncogene 14 (1997) 1511–1519.
[7] Y. Sautin, N. Takamura, S. Shkylaev, et al., Ceramide-induced apoptosis of human thyroid cancer cells resistant to apoptosis by irradiation, Thyroid 10 (2000) 733–740.
[8] N. Mitsutake, H. Namba, S.S. Shklyaev, et al., PKC delta mediates ionizing radiation-induced activation of c-Jun NH2 terminal kinase through MKK7 in human thyroid cells, Oncogene 20 (2001) 989–996.
[9] S. Shklyaev, H. Namba, N. Mitsutake, et al., Transient activation of c-Jun NH2-terminal kinase by growth factors influences survival but not apoptosis in human thyrocytes, Thyroid, 2000, in press.
[10] H.M. Rabes, E.P. Demidchik, J.D. Sidorov, et al., Pattern of radiation-induced RET and NTRK1 rearrangements in 191 post-Chernobyl papillary thyroid carcinomas: biological, phenotypic and clinical implications, Clin. Cancer Res. 6 (2000) 1103–10930.
[11] E.L. Chua, W.M. Wu, K.T. Tran, et al., Prevalence and distribution of ret/ptc 1, 2 and 3 in papillary thyroid carcinoma in New Caledonia and Australia, J. Clin. Endocrinol. Metab. 85 (2000) 2733–2739.

Molecular analysis of radiation-induced thyroid carcinomas in humans

Hartmut M. Rabes

Institute of Pathology, Ludwig Maximilians University of Munich, Thalkirchner Street 36, D-80337 Munich, Germany

Abstract

Correlations have been found between radiation exposure and thyroid carcinoma development, particularly in children. Recent studies on a large cohort of radiation-induced papillary thyroid carcinomas (PTC) after the Chernobyl reactor accident disclosed a common type of underlying genetic alteration. A high prevalence of rearrangements of the receptor tyrosine kinase (TK) c-RET was observed, besides some rearrangements involving NTRK1. Radiation-induced RET rearrangements in PTC consist most frequently of fusions to the H4 gene (RET/PTC1) or to the ELE1 (ARA70) gene (RET/PTC3). Both fusions are formed by balanced paracentric inversions on chromosome 10. An analysis of the fused genes in ELE1/RET rearrangements revealed DNA double-strand breaks spread over a distance of about 2.3 kb in two introns and the interposed exon of ELE1, exon 11 and intron 11 of RET, without significant clustering in these parts of the genes. Topoisomerase I sites were found exactly at or in close vicinity to all breakpoints, suggesting a role for this enzyme in formation of DNA strand breaks or inversions. The genes fuse at short regions of sequence homology and short direct or inverted repeats (microhomology-mediated DNA end joining). A minority of PTC cases contain novel types of RET rearrangement, with RIα, GOLGA2, HTIF, HTIF homolog, RFG8, ELKS, KTN1 and PCM-1 as the 5′-fused genes. These novel types of gene fusions are formed by interchromosomal translocation. The formation of these rare types of rearrangement seems to be highly related to radiation as they have rarely been found in sporadic PTC. All RET gene fusions seem to act similarly on RET function: The strict physiological control of RET TK activity is suspended through constitutive activation by 5′-fused parts of genes containing coiled-coil domains with dimerization potential. RET expression in thyrocytes, which under normal conditions, lack RET TK activity apparently triggers clonal expansion and early invasion of the affected cells. RET-fused genes, some of which are transcriptional coactivators, are important determinators of the peculiar phenotype of the tumour and for its clinical course. This is most significant in RET/PTC3 rearrangements with ELE1 as the RET-fused gene: this type of rearrangement leads more often to the phenotype of a solid variant of PTC, and to rapid tumour development and early lymph node metastasis. Up to now, no other genetic aberration has more

E-mail address: hm.rabes@lrz.uni-muenchen.de (H.M. Rabes).

0531-5131/02 © 2002 Elsevier Science B.V. All rights reserved.
PII: S0531-5131(01)00750-6

frequently been observed in PTC than RET rearrangement, thus suggesting that RET rearrangement represents a genetic marker lesion of radiation history in the development of a PTC. © 2002 Elsevier Science B.V. All rights reserved.

Keywords: Papillary thyroid carcinoma; Radiation; Chernobyl; Gene rearrangement; RET; NTRK1; Genotype/phenotype correlations

1. Introduction

External or internal irradiation of the thyroid gland is a risk factor for thyroid carcinogenesis. This is concluded from epidemiological studies on patients exposed to accidental or therapeutic irradiation. Age at exposure determines the risk with a lower age associated with a higher risk [1]. For radiation exposure at a young age, an excess relative risk of 7.7 per Gy has been calculated. This is one of the highest risk coefficients observed in any organ with evidence for an increased risk at about 0.1 Gy (see Ref. [2]). The dose–response curve for thyroid neoplasms reached a maximum of 25–29 years after exposure [1]. Though declining afterwards, a long-lasting tumorigenic effect for at least 45 years has to be expected [3].

The underlying molecular mechanisms of radiation-induced thyroid carcinogenesis are far from being understood. Radiation as a DNA damaging factor is expected to induce so-called multiple DNA damage sites [4], and subsequent repair and putatively persisting DNA alterations that may end up in malignant transformation. Several genes have been envisaged to be involved in thyroid tumorigenesis, mainly RAS, GSP, RET, NTRK1 and p53 (see Refs. [5,6]). However, until recently, relations between alterations in structure and function of these genes and irradiation have not been established unequivocally. In most molecular studies, the number of investigated thyroid tumours was small. Thus, it was difficult to draw firm conclusions about molecular changes that might be characteristic for radiation as the inducing carcinogenic factor.

During the last decade, a large number of thyroid cancers were observed which developed as a consequence of a limited and documented exposure to radiation after the Chernobyl reactor accident on April 26, 1986. Large areas were contaminated with radioactive fallout, mainly in southern Belarus where huge amounts of iodine radio-isotopes were deposited and incorporated by ingestion and inhalation [7]. The estimated thyroid doses in children living in the most heavily contaminated regions, the Oblast Gomel in Belarus, were in the range of 1 Gy or greater for about 1% of the children, and higher than 0.1 Gy for more than 20%. A steep increase of childhood papillary thyroid carcinomas (PTC), normally a very rare childhood tumour, was registered in the highly contaminated areas [8,9]. In a prospective study on thyroid carcinoma incidence to be expected during the lifetime of children exposed at the age of 0–4 years in the Oblast Gomel region, a total number of 51,000 thyroid cancers was calculated [10].

These childhood thyroid carcinomas, occurring at such high incidence [11], provide an unprecedented chance to elucidate the molecular genetic mechanisms of thyroid carcinogenesis, in a cohort of patients who had been exposed to radioiodine during a defined period of time. Thus, it was possible to search for typical molecular genetic defects in the tumours, to try to study the underlying mechanisms of genetic alteration and repair and to look for functional consequences and genotype/phenotype correlations.

2. Structural genetic aberrations

Papillary thyroid carcinomas of children and young adults who had been exposed to the radioactive fallout after the Chernobyl reactor accident were studied in various laboratories. Molecular analyses revealed an almost complete lack of point mutations in H-, K- and N-RAS, p53 and GSP in these tumors [12–15]. However, in rare cases a few NTRK1 rearrangements were observed, among them fusions of the NTRK1 tyrosine kinase domain with either the aminoterminal part of the tropomyosin gene (TPM3/NTRK1), or the TPR gene (TPR/NTRK1) [16]. Among the tumours, which had tested negative for other forms of gene rearrangements (RET, see below), the prevalence of NTRK1 rearrangements was low [16], in agreement with the findings from other laboratories [17–19].

The first report on a putative role for the RET proto-oncogene in radiation-induced PTC was published in 1994 by Ito et al. [20], who described in a small series of post-Chernobyl PTC in four of a total of seven cases a rearrangement of the RET gene, without further specifying the type of rearrangement. In a more detailed analysis, we were able to confirm and extend these early findings. Analysing 59 PTC from patients who had been exposed to radioiodine after the reactor accident at ages up to 18 years, we observed in 36 of the 59 PTCs an RET rearrangement (61.0%) [21–23]. Two other reports with smaller cohorts of children or young adults showed almost identical results (4/6 and 29/38 RET rearrangement-positive cases) [24,25].

Three types of RET rearrangement have been described for PTC, irrespective of radiation history: PTC1 in which the aminoterminal of the H4 gene is fused to the carboxyterminal of RET [26,27]; PTC2 with a fusion of RET to the regulatory subunit of the cAMP-dependent protein kinase A [28]; and PTC3 with the aminoterminal part of the ELE1 (ARA70) gene fused to the 3′ end of RET [29–31]. The prevalence of these three types of RET rearrangement is remarkably different in radiation-induced PTC after the Chernobyl reactor accident. Whilst PTC2 has only been found in a single case [24], PTC3 appears to predominate in the series with a short latency period after exposure [25,32].

Table 1
Changes in prevalence and type of RET rearrangements in papillary thyroid carcinomas of 336 children and young adults after the Chernobyl reactor accident as a function of the tumor latency period (interval between radiation exposure in April/May 1986 and diagnosis/thyroidectomy)

Latency period (years)	Total number of PTC	RET rearr. positive	%	RET rearr. negative	%	PTC 1	%	PTC 3	%	PTC 5, 6, 7, 8	%
Total (7–12.9)	337[a]	146[a]	43.3	191	56.7	81[a]	24.0	54[a]	16.0	11	3.3
≤10	61	38	62.3	23	37.7	9	14.8	24	39.3	5	8.2
10.1–11.7	134	57	42.5	77	57.5	40	29.9	14	10.5	3	2.2
>11.7–12.9	142[a]	51[a]	35.9	91	64.1	32[a]	22.5	16[a]	11.3	3	2.1

[a] Analysed tumor sample of one patient contained both PTC1 and PTC3.

Among the RET rearrangement-positive cases, about two-thirds exhibit ELE1/RET rearrangements during the first decade after the reactor accident [33]. At later intervals, the fraction of PTC showing RET rearrangements declines steadily, concomitantly with a shift from PTC3 to PTC1, H4/RET rearrangements as the predominant type. The PTC1 fraction among RET rearrangement-positive PTC increases from about one-fourth during the first decade to about two-thirds later on (Table 1).

3. Mechanisms of gene rearrangement

Transcript analysis in PTC harbouring RET fusions revealed that in most cases not only the orthologous, but also the reciprocal transcript was formed [22,34]. Presence of both transcripts suggested a balanced reciprocal translocation as the underlying mechanism. It has been assumed that fusion by paracentric inversion on chromosome 10 is the most probable mode of generating the functionally active type of RET rearrangement. The participating genes H4 and ELE1 in the predominant types of RET rearrangement are both located near to the RET gene, H4 at 10q21, ELE1 as well as RET at 10q11.2. In order to obtain a closer insight in the topology and mechanism of radiation-induced breakpoint formation and repair, an analysis was performed with genomic DNA of PTC with chimeric ELE1/RET and the reciprocal RET/ELE1 genes. The breakpoint distribution revealed a typical pattern: In the RET gene, all breakpoints were located upstream of exon 12, mainly in intron 11 (84.6%), with a few breaks also in exon 11 (15.4%). They were spread over a distance of about 2.1 kb. In the ELE1 gene, the breakpoints were confined to a stretch of about 2.3 kb over the 522-bp intron K (15.4%), the 144-bp exon L (3.8%) and the 1670-bp intron L (80.8%). There was no indication of any specific breakpoint cluster within these DNA regions. However, breakpoints were found at or in close vicinity to topoisomerase I recognition sites suggesting a role for this enzyme in either breakpoint formation or fusion processes in the chimeric genes [35].

The sequence of the chimeric genes showed only minor deviations as compared to the wild type sequence of the participating genes. Major deletions or insertions were missing. Remarkably, small patches of sequence homologies between both genes were constantly present at or near the fusion points, as well as direct or inverted repeats. We conclude that ELE1/RET inversions are formed by microhomology-mediated strand annealing either by direct annealing of two unrelated single-strands after two double-strand breaks, or after a single double-strand break by invasion of single-strands into an intact unrelated DNA duplex at patches of homology. Details of this study and a complete discussion are published elsewhere [35].

4. Rare novel types of RET rearrangement

Besides the most predominant types of RET rearrangement, PTC1 and PTC3, several other gene fusions with RET at the carboxyterminal portion have been found recently. PTC5 consists of a fusion of the gene for a Golgi integral membrane protein (GOLGA5) [36] to RET [37]; PTC6 and 7 contain HTIF or an HTIF homolog, respectively,

Table 2
PTC type, function and chromosomal location of RET-fused genes (for references see text)

Type of RET rearrangement	RET-fused gene	Function	Chromosomal location of RET-fused gene
PTC1	H4	unknown	10q21
PTC2	RIα	catalytic domain of cAMP-dependent protein kinase A	17q23
PTC3	ELE1 (ARA70)	transcription coactivator of androgen receptor	10q11.2
PTC5	GOLGA5	Golgi integral membrane protein	14q
PTC6	HTIF1	transcriptional coactivator of nuclear receptors	7q32
PTC7	RFG7	HTIF homolog	1p13
PTC8	RFG8	unknown	18q21-22
	ELKS	unknown	12p13
	KTN1	influence on micro-tubular organelle movement	14q22.1
	PCM-1	coding for a centrosomal protein	8p21-22

transcriptional coactivators of nuclear receptors [38], located on chromosome 7q32 and 1p13, respectively [39]; PTC8 with a fusion of RET to a gene with unknown function (RFG8), located at 18q21-22 [40]; furthermore, a fusion of KTN1, which influences the micro tubular organelle movement and is located at 14q22.1, to RET has been described in a single PTC [39]; ELKS, a gene of unknown function, located at 12p13, was found to be fused to RET in another single case of PTC [41]. Most recently, an additional novel fusion with a balanced translocation was demonstrated involving the 5′ portion of PCM-1, a gene that codes for a centrosomal protein with distinct cell cycle distribution, located on chromosome 8p21-22 [42] (Table 2).

5. Functional consequences of RET rearrangement

The different types of RET rearrangements have in common the preservation of the intact RET tyrosine kinase (TK) domain at the carboxyterminal portion of the fusion gene. However, the upstream regulatory part of RET with the transmembrane and the extracellular, ligand-binding domains are replaced by 5′ portions of other genes that contain coiled-coil domains with a putative dimerization potential. They suspend the strict control of RET activity, which under normal conditions is the result of heterotrimeric interaction of the RET ligand, GDNF-receptor and RET receptor domain. A constitutive ligand-independent atypical RET tyrosine kinase activation, for which the exact mechanisms, probably involving dimerization of the fusion protein, are still unknown, is assumed to represent the critical first step in obtaining capacity for clonal expansion and malignant transformation.

However, not only regulation, expression and function of the proto-oncogene RET is perturbed. The concomitant alteration in the RET-fused gene itself may change its expression and possibly function. All RET-fused genes that have been observed so far appear to be expressed in the thyroid gland. Loss of the functional domains due to rearrangement could alter the intracellular level of the coded product. For tumours containing the novel fusion gene RET/PCM-1, a drastic decrease of the PCM-1 protein and an alteration of its subcellular localization have been described [42]. However, persisting heterozygosity indicates that the non-rearranged allele of PCM-1 was not lost, and that the translocation was balanced [42].

6. Genotype/phenotype correlations

Differences in the phenotype and in clinical progression of PTC that exhibit different types of RET rearrangement suggest that RET-fused genes take part in the determination of the biology of developing tumours. Comparative two-dimensional gel electrophoresis (2DE) of the two main tumour types PTC1 and PTC3 revealed differences in the protein pattern. Further proteomic analysis by combined 2DE and mass spectrometry of proteins associated with specific genotypes are in progress. At present, two differences in the biological behaviour between PTC1 and PTC3 became already evident: tumours of the ELE1/RET type of rearrangement reveal a shorter latency period between exposure and clinical manifestation. At the time of diagnosis, tumours of the PTC3 type were in a significantly more progressed stage than PTC1 tumours. This observation might have an impact for treatment concepts.

The genotype is reflected in the morphology of PTC harbouring RET rearrangements, too. Three main histological subgroups have been described in the radiation-induced papillary thyroid carcinomas: the classical papillary variant, the follicular variant, and the solid variant of papillary thyroid carcinoma [33,43]. While PTC1 is connected with either the histology of a classical papillary or follicular variant of papillary thyroid carcinoma, PTC3 carcinomas belong more often to the group of solid variants of PTC [33].

7. Conclusion

Molecular analysis of papillary thyroid carcinomas that developed after exposure of the thyroid gland to iodine radioisotopes, particularly in children, permitted a closer view on the mechanisms of radiation-induced thyroid carcinogenesis. Radioiodine incorporation leads in the juvenile thyroid gland in a dose-dependent mode to multiple DNA damage, which is randomly distributed over the genome. There is no chance to analyse such damage in detail in an exposed thyroid gland. However, thyrocytes that are capable of expanding clonally due to a critical DNA lesion become accessible to further analysis. Thyrocytes, bearing a RET rearrangement, belong to this group. The results argue for the hypothesis, that selection by clonal growth is an attribute of cells that underwent dysregulation of RET (or, in rare cases, NTRK1) receptor tyrosine kinases by fusion to other genes. Their aminoterminal regulatory coiled-coil domains with putative dimerization potential will lead, when fused to RET to replace the aminoterminal portion of RET, to an uncoupling of the stringent physiological control and a ligand-independent RET activation in cells that under normal conditions are devoid of RET tyrosine kinase activity. Ectopic RET activation represents by all probability the first and most important step to malignant transformation in thyrocytes. Its persistence in various parts of a primary tumour and even in lymph node metastases (unpublished observations) proves not only the clonality of PTC arising after irradiation, but documents also the relevance of RET fusion products for further progression, with the RET-fused genes obviously playing a modulating role in this process.

Radiation-induced papillary thyroid carcinomas, after the Chernobyl reactor accident, belong to the rare, but highly informative examples of human tumours that are characterized by a direct relation between inducing carcinogenic factor, here: radioiodine; and a specific molecular alteration, here: RET rearrangement; and based on this persistent gene alteration, the development of a tissue type-specific tumour, the papillary thyroid carcinoma. This unique model will help to further clarify mechanisms of human carcinogenesis.

Acknowledgements

I am most grateful to Professors E.P. Demidchik, E. Lengfelder and D. Hoelzel and to Drs. S. Klugbauer, C. Beimfohr, E. Zeindl-Eberhart, P. Pfeiffer, J. Sidorov and A. Jauch for their excellent cooperation and most helpful discussions. The original work reported in this review was supported by grants from Deutsche Krebshilfe, Dr. Mildred Scheel-Stiftung für Krebsforschung, Wilhelm Sander-Stiftung, Matthias Lackas-Stiftung and Weigand-Stiftung.

References

[1] A.B. Schneider, E. Ron, J. Lubin, M. Stovall, T.C. Gierlowski, Dose–response relationships for radiation-induced thyroid cancer and thyroid nodules: evidence for the prolonged effects of radiation on the thyroid, J. Clin. Endocrinol. Metab. 77 (1993) 362–369.
[2] E. Ron, J.H. Lubin, R.E. Shore, K. Mabuchi, B. Modan, L.M. Pottern, A.B. Schneider, M.A. Tucker, J.D. Boice, Thyroid cancer after exposure to external radiation: a pooled analysis of seven studies, Radiat. Res. 141 (1995) 259–277.
[3] R. Shore, N. Hildreth, P. Dvoretsky, E. Andresen, M. Moseson, B. Pasternak, Thyroid cancer among persons given X-ray treatment in infancy for an enlarged thymus gland, Am. J. Epidemiol. 137 (1993) 1068–1080.
[4] P.E. Bryant, The signal model: a possible explanation for the conversion of DNA double-strand breaks into chromatid breaks, Int. J. Radiat. Biol. 73 (1998) 243–251.
[5] M.A. Pierotti, I. Bongarzone, M.G. Borrello, A. Greco, S. Pilotti, G. Sozzi, Cytogenetics and molecular genetics of carcinomas arising from thyroid epithelial follicular cells, Genes, Chromosomes Cancer 16 (1996) 1–14.
[6] H.M. Rabes, S. Klugbauer, Molecular genetics of childhood papillary thyroid carcinomas after irradiation: high prevalence of RET rearrangement, Recent Results Cancer Res. 154 (1998) 248–264.
[7] S. Nagataki, K. Ashizawa, S. Yamashita, Cause of childhood thyroid cancer after the Chernobyl accident, Thyroid 8 (1998) 115–117.
[8] K. Baverstock, B. Egloff, A. Pinchera, C. Ruchti, D. Williams, Thyroid cancer after Chernobyl, Nature (London) 359 (1992) 21–22.
[9] V.S. Kazakov, E.P. Demidchik, L.N. Astakhova, Thyroid cancer after Chernobyl, Nature (London) 359 (1992) 21.
[10] E. Cardis, E. Amoros, A. Kesminiene, I.V. Malakhova, S.M. Poliakov, N.N. Piliptsevitch, E.P. Demidchik, L.N. Astakhova, V.K. Ivanov, A.P. Konogorov, E.M. Parshkov, A.F. Tsyb, Observed and predicted thyroid

cancer incidence following the Chernobyl accident. Evidence for factors influencing susceptibility to radiation-induced thyroid cancer, in: G. Thomas, A. Karaoglou, E.D. Williams (Eds.), Radiation and Thyroid Cancer, World Scientific Publishing, Singapore, 1999, pp. 395–405.
[11] F. Pacini, T. Vorontsova, E.P. Demidchik, E. Molinaro, L. Agate, C. Romei, E. Shavrova, E.D. Cherstvoy, Y. Ivashkevitch, E. Kuchinskaya, M. Schlumberger, G. Ronga, M. Filesi, A. Pinchera, Post-Chernobyl thyroid carcinoma in Belarus children and adolescents: comparison with naturally occurring thyroid carcinoma in Italy and France, J. Clin. Endocrinol. Metab. 82 (1997) 3563–3569.
[12] Y.E. Nikiforov, M.N. Nikiforova, D.R. Gnepp, J.A. Fagin, Prevalence of mutations of ras and p53 in benign and malignant thyroid tumors from children exposed to radiation after the Chernobyl nuclear accident, Oncogene 13 (1996) 687–693.
[13] J. Smida, H. Zitzelsberger, A.M. Kellerer, L. Lehmann, G. Minkus, T. Negele, F. Spelsberg, L. Hieber, E.P. Demidchik, E. Lengfelder, M. Bauchinger, p53 mutations in childhood thyroid tumours from Belarus and in thyroid tumours without radiation history, Int. J. Cancer 73 (1997) 802–807.
[14] V. Waldmann, H.M. Rabes, Absence of $G_S\alpha$ gene mutations in childhood thyroid tumors after Chernobyl in contrast to sporadic adult thyroid neoplasia, Cancer Res. 57 (1997) 2358–2361.
[15] B. Suchy, V. Waldmann, S. Klugbauer, H.M. Rabes, Absence of *RAS* and *p53* mutations in thyroid carcinomas of children after Chernobyl in contrast to adult thyroid tumours, Br. J. Cancer 77 (1998) 952–955.
[16] C. Beimfohr, S. Klugbauer, E.P. Demidchik, E. Lengfelder, H.M. Rabes, NTRK1 rearrangement in papillary thyroid carcinomas of children after the Chernobyl reactor accident, Int. J. Cancer 80 (1999) 842–847.
[17] W. Wajjwalku, S. Nakamura, Y. Hasegawa, K. Miyazaki, Y. Satoh, H. Funahashi, M. Matsuyama, M. Takahashi, Low frequency of re-arrangements of the ret and trk proto-oncogenes in Japanese thyroid papillary carcinomas, Jpn. J. Cancer Res. 83 (1992) 671–675.
[18] I. Bongarzone, P. Vigneri, L. Mariani, P. Collini, S. Pilotti, M.A. Pierotti, RET/NTRK1 rearrangements in thyroid gland tumors of the papillary carcinoma family: correlation with clinicopathological features, Clin. Cancer Res. 4 (1998) 223–228.
[19] A. Bounacer, M. Schlumberger, R. Wicker, J.A. Du-Villard, B. Caillou, A. Sarasin, H.G. Suarez, Search for NTRK1 proto-oncogene rearrangements in human thyroid tumours originated after therapeutic radiation, Br. J. Cancer 82 (2000) 308–314.
[20] T. Ito, T. Seyama, K.S. Iwamoto, T. Mizuno, N.D. Tronko, I.V. Komissarenko, E.D. Cherstovoy, Y. Satow, N. Takeichi, K. Dohi, M. Akiyama, Activated RET oncogene in thyroid cancers of children from areas contaminated by Chernobyl accident, Lancet 344 (1994) 259.
[21] S. Klugbauer, E. Lengfelder, E.P. Demidchik, H.M. Rabes, High prevalence of *RET* rearrangement in thyroid tumors of children from Belarus after the Chernobyl reactor accident, Oncogene 11 (1995) 2459–2467.
[22] S. Klugbauer, E. Lengfelder, E.P. Demidchik, H.M. Rabes, A new form of RET rearrangement in thyroid carcinomas of children after the Chernobyl reactor pattern, Oncogene 13 (1996) 1099–1102.
[23] H.M. Rabes, S. Klugbauer, Radiation-induced thyroid carcinomas in children: high prevalence of RET rearrangement, Verh. Dtsch. Ges. Pathol. 81 (1997) 139–144.
[24] L. Fugazzola, S. Pilotti, A. Pinchera, T.V. Vorontsova, P. Mondellini, I. Bongarzone, A. Greco, L. Butti, M.G. Butti, E.P. Demidchik, F. Pacini, M.A. Pierotti, Oncogenic rearrangements of the RET proto-oncogene in papillary thyroid carcinomas from children exposed to the Chernobyl nuclear accident, Cancer Res. 55 (1995) 5617–5620.
[25] Y.E. Nikiforov, J.M. Rowland, K.E. Bove, H. Monforte-Munoz, J.A. Fagin, Distinct pattern of ret oncogene rearrangements in morphological variants of radiation-induced and sporadic thyroid papillary carcinomas in children, Cancer Res. 57 (1997) 1690–1694.
[26] M.A. Pierotti, M. Santoro, R.B. Jenkins, G. Sozzi, I. Bongarzone, M. Grieco, N. Monzini, M. Miozzo, M.A. Herrmann, A. Fusco, I.D. Hay, G. Della Porta, G. Vecchio, Characterization of an inversion on the long arm of chromosome 10 juxtaposing D10S170 and RET and creating the oncogenic sequence RET/PTC, Proc. Natl. Acad. Sci. U. S. A. 89 (1992) 1616–1620.
[27] M. Grieco, M. Santoro, M.T. Berlingieri, R.M. Melillo, R. Donghi, I. Bongarzone, M.A. Pierotti, G. Della Porta, A. Fusco, G. Vecchio, PTC is a novel rearranged form of the ret proto-oncogene and is frequently detected in vivo in human thyroid papillary carcinoma, Cell 60 (1990) 557–563.

[28] I. Bongarzone, N. Monzini, M.G. Borrello, C. Carcano, G. Ferraresi, E. Arighi, P. Mondellini, G. Della Porta, M.A. Pierotti, Molecular characterization of a thyroid tumor-specific transforming sequence formed by the fusion of ret tyrosine kinase and the regulatory subunit RIα of cyclic AMP-dependent protein kinase A, Mol. Cell Biol. 13 (1993) 358–366.
[29] I. Bongarzone, M.G. Butti, S. Coronelli, M.G. Borello, M. Santoro, P. Mondellini, S. Pilotti, A. Fusco, G. Della Porta, M.A. Pierotti, Frequent activation of ret protooncogene by fusion with a new activating gene in papillary thyroid carcinomas, Cancer Res. 54 (1994) 2979–2985.
[30] M. Santoro, N.A. Dathan, M.T. Berlingieri, I. Bongarzone, C. Paulin, M. Grieco, M.A. Pierotti, G. Vecchio, A. Fusco, Molecular characterization of RET/PTC3; a novel rearranged version of the RET protooncogene in a human thyroid papillary carcinoma, Oncogene 9 (1994) 509–516.
[31] F. Minoletti, M.G. Butti, S. Coronelli, M. Miozzo, G. Sozzi, S. Pilotti, A. Tunncliffe, M.A. Pierotti, I. Bongarzone, The two genes generating RET/PTC3 are localized in chromosomal band 10q11.2, Genes, Chromosomes Cancer 11 (1994) 51–57.
[32] H.M. Rabes, S. Klugbauer, Molecular genetics of childhood papillary thyroid carcinomas after irradiation: high prevalence of RET rearrangement, Recent Results Cancer Res. 154 (1998) 248–264.
[33] H.M. Rabes, E.P. Demidchik, J.D. Sidorow, E. Lengfelder, C. Beimfohr, D. Hoelzel, S. Klugbauer, Pattern of radiation-induced RET and NTRK1 rearrangements in 191 post-Chernobyl papillary thyroid carcinomas: biological, phenotypic, and clinical implications, Clin. Cancer Res. 6 (2000) 1093–1103.
[34] S. Klugbauer, E.P. Demidchik, E. Lengfelder, H.M. Rabes, Molecular analysis of new subtypes of *ELE/RET* rearrangements, their reciprocal transcripts and breakpoints in papillary thyroid carcinomas of children after Chernobyl, Oncogene 16 (1998) 671–675.
[35] S. Klugbauer, P. Pfeiffer, H. Gassenhuber, C. Beimfohr, H.M. Rabes, RET rearrangements in radiation-induced papillary thyroid carcinomas: high prevalence of topoisomerase I sites at breakpoints and micro-homology-mediated end joining in ELE1 and RET chimeric genes, Genomics 73 (2001) 149–160.
[36] R.A. Bascom, S. Srinivasan, R.L. Nussbaum, Identification and characterization of golgin-84, a novel Golgi integral membrane protein with a cytoplasmic coiled-coil domain, J. Biol. Chem. 274 (1999) 2953–2962.
[37] S. Klugbauer, E.P. Demidchik, E. Lengfelder, H.M. Rabes, Detection of a novel type of RET rearrangement (PTC5) in thyroid carcinomas after Chernobyl and analysis of the involved RET-fused gene RFG5, Cancer Res. 58 (1998) 198–203.
[38] S. Klugbauer, H.M. Rabes, The transcription coactivator *HTIF*1 and a related protein are fused to the RET receptor tyrosine kinase in childhood papillary thyroid carcinomas, Oncogene 18 (1999) 4388–4393.
[39] K. Salassidis, J. Bruch, H. Zitzelsberger, E. Lengfelder, A.M. Kellerer, M. Bauchinger, Translocation t(10;14)(q11.2;q22.1) fusing the Kinectin to the RET gene creates a novel rearranged form (PTC8) of the RET proto-oncogene in radiation-induced childhood papillary thyroid carcinoma, Cancer Res. 60 (2000) 2786–2789.
[40] S. Klugbauer, A. Jauch, E. Lengfelder, E. Demidchik, H.M. Rabes, A novel type of RET rearrangement (PTC8) in childhood papillary thyroid carcinomas and characterization of the involved gene (RFG8), Cancer Res. 60 (2000) 7028–7032.
[41] T. Nakata, Y. Kitamura, K. Shimizu, S. Tanak, M. Fujimori, S. Yokoyama, K. Ito, M. Emi, Fusion of a novel gene, ELKS, to RET due to translocation t(10;12)(q11;13) in a papillary thyroid carcinoma, Genes Chromosomes Cancer 25 (1999) 97–103.
[42] R. Corvi, N. Berger, R. Balczon, G. Romeo, RET/PCM-1: a novel fusion gene in papillary thyroid carcinoma, Oncogene 19 (2000) 4236–4242.
[43] Y.E. Nikiforov, D.R. Gnepp, Pediatric thyroid cancer after the Chernobyl disaster, Cancer (Philadelphia) 74 (1994) 748–766.

Influence of XPD variant alleles on p53 mutations in lung tumors of nonsmokers and smokers

Sai-Mei Hou [a],*, Annamaria Kannio [b], Sabrina Angelini [a,1], Susann Fält [a], Fredrik Nyberg [c], Kirsti Husgafvel-Pursiainen [b]

[a] *Department of Biosciences at Novum, CNT/Novum, Karolinska Institutet, S-14157 Huddinge, Sweden*
[b] *Laboratory of Molecular and Cellular Toxicology, Finnish Institute of Occupational Health, FIN-00250 Helsinki, Finland*
[c] *Division of Environmental Epidemiology, Institute of Environmental Medicine, Karolinska Institutet, Box 210, S-171 77 Stockholm, Sweden*

Abstract

The DNA repair protein XPD is involved in the transcription-coupled nucleotide excision repair of DNA lesions induced by many tobacco and environmental carcinogens. Common polymorphisms in XPD exon 10 (G > A, Asp312Asn) and exon 23 (A > C, Lys751Gln) have been identified, and lung cancer cases homozygous for the variant allele in either exon have been reported to have a reduced repair capacity against benzo[a]pyrene-induced DNA damage. We therefore investigated a possible effect of these variant alleles on the p53 mutation frequency and spectrum among 97 Swedish lung cancer patients. Transversions were found to occur more frequently among patients with at least one variant allele than among wild type homozygotes. The XPD variant alleles may thus be associated with reduced DNA repair proficiency. This finding is not in accord with the previous report that associated the exon 23 wild type allele with increased frequency of X-ray-induced chromatid aberrations. © 2002 Elsevier Science B.V. All rights reserved.

Keywords: Nucleotide excision repair; XPD genotype; p53 mutation; Lung cancer

1. Introduction

Functional and common sequence variations of DNA repair genes may be potential cancer susceptibility factors in the general population exposed to environmental carci-

---

Abbreviations: NER, nucleotide excision repair; SNP, single nucleotide polymorphism
* Corresponding author. Tel.: +46-8-608-9253; fax: +46-8-608-1501.
E-mail address: Saimei.hou@cnt.ki.se (S.-M. Hou).
[1] Permanent address: Dipartimento di Farmacologia, Via Irnerio 48, 40126 Bologna (BO), Italy.

nogens. Several non-synonymous single nucleotide polymorphisms (SNPs) have been identified in the XPD gene, including those in exon 10 (G>A, Asp312Asn) and exon 23 (A>C, Lys751Gln) [1]. These variant alleles exist at frequencies of around 30% in the USA [1,2]. The XPD protein takes part in the transcription-coupled nucleotide excision repair (NER) pathway that rapidly repairs damage on the transcribed strand of active genes.

The XPD exon 23 AA (wild type) genotype has been suggested to be a risk factor for basal cell carcinoma [3] and susceptibility to X-ray-induced chromatid aberrations [2]. Recently, however, the exon 23 CC variant genotype has been associated with an increased risk of head and neck cancer [4], and reduced repair capacity against benzo[*a*]pyrene-induced DNA damage in lung cancer cases homozygous for the variant allele in either exon [5]. We therefore investigated a possible effect of these variant alleles on p53 mutations between never smoking and smoking lung cancer patients. The 53 mutations have previously been analyzed with regard to the effect of exposure to environmental tobacco smoke in a multicenter study on lung cancer [6].

2. Materials and methods

2.1. Study subjects

Cases were recruited at the three major hospitals in Stockholm County responsible for the diagnosis and treatment of lung cancer. Each newly diagnosed never-smoking case was used as an index case for next diagnosed ever-smoking case of the same gender and age in the same hospital.

2.2. p53 mutation analysis

The method for p53 mutation analysis has been described previously [6]. DNA obtained from tumor tissue samples was screened for mutations in p53 exons 4–9 and 11 using PCR and DGGE. The mutations were then identified by direct sequencing.

2.3. XPD genotyping

Analysis of the exon 10 and 23 polymorphisms was performed by restriction analysis (*Taq*I or *Pst*1) of genomic PCR products as described previously [7]. A forward primer with a C>T mismatch was designed in order to create a restriction site for *Taq*I ($\underline{T}^{V}CGA$) in the PCR product of the exon 10 G-allele.

3. Results

A total of 97 Swedish lung cancer patients (56 never-smokers and 41 age-, gender- and hospital-matched ever-smokers) were screened for *p53* mutation and genotyped with regard to XPD. The *p53* gene was mutated in 4 never-smokers and 11 ever-smokers.

Transversions predominated over transitions among smokers (8:3), however, not among never-smokers (1:3). Overall, transversions occurred more frequently among patients with at least one variant allele than among wild type homozygotes (8/11 vs. 1/4 of all mutations for exon 10 SNP; 7/9 vs. 2/6 for exon 23 SNP).

4. Discussion

Our results suggest that the XPD variant alleles may be associated with an increased frequency of transversion mutations in the *p53* gene in the lungs. This appeared to be true for both never-smokers (1/2 vs. 0/2) and ever-smokers. The transversion identified in the never smoker with XPD variant alleles may be attributable to environmental exposure to tobacco smoke or radon.

This finding is consistent with our recent study that suggested an association of the exon 23 C allele with depressed repair of UV-specific cyclobutane pyrimidine dimers in the skin of older volunteers [7]. We conclude that the variant alleles of XPD exon 10 and 23 may be associated with reduced DNA repair proficiency among older individuals.

Acknowledgements

This study was supported by the Swedish Cancer Society and the Swedish Match Medical Research Fund.

References

[1] M.R. Shen, I.M. Jones, H. Mohrenweiser, Nonconservative amino acid substitution variants exist at polymorphic frequency in DNA repair genes in healthy humans, Cancer Res. 58 (4) (1998) 604–608.
[2] R.M. Lunn, K.J. Helzlsouer, R. Parshad, D.M. Umbach, E.L. Harris, K.K. Sanford, D.A. Bell, XPD polymorphisms: effects on DNA repair proficiency, Carcinogenesis 21 (4) (2000) 551–555.
[3] M. Dybdahl, U. Vogel, G. Frentz, H. Wallin, B.A. Nexo, Polymorphisms in the DNA repair gene XPD: correlations with risk and age at onset of basal cell carcinoma, Cancer Epidemiol., Biomarkers Prev. 8 (1) (1999) 77–81.
[4] E.M. Sturgis, R. Zheng, L. Li, E.J. Castillo, S.A. Eicher, M. Chen, S.S. Strom, M.R. Spitz, Q. Wei, XPD/ERCC2 polymorphisms and risk of head and neck cancer: a case-control analysis, Carcinogenesis 21 (12) (2000) 2219–2223.
[5] M.R. Spitz, X. Wu, Y. Wang, L.E. Wang, S. Shete, C.I. Amos, Z. Guo, L. Lei, H. Mohrenweiser, Q. Wei, Modulation of nucleotide excision repair capacity by XPD polymorphisms in lung cancer patients, Cancer Res. 61 (4) (2001) 1354–1357.
[6] K. Husgafvel-Pursiainen, P. Boffetta, A. Kannio, F. Nyberg, G. Pershagen, A. Mukeria, V. Constantinescu, C. Fortes, S. Benhamou, p53 mutations and exposure to environmental tobacco smoke in a multicenter study on lung cancer, Cancer Res. 60 (11) (2000) 2906–2911.
[7] K. Hemminki, G. Xu, S. Angelini, E. Snellman, C.T. Jansen, B. Lambert, S.-M. Hou, XPD exon 10 and 23 polymorphisms and DNA repair in human skin in situ, Carcinogenesis 22 (8) (2001) 1185–1188.

Genetic instability in Thorotrast induced liver cancers

Duo Liu[a], Hirohito Momoi[b], Li Li[a], Yuichi Ishikawa[c], Manabu Fukumoto[a,*]

[a]*Institute of Development, Aging and Cancer (IDAC), Tohoku University, 4-1 Seiryto-machi, Aoba-ku, Sendai 980-8575, Japan*
[b]*Department of Gastroenterological Surgery, Graduate School of Medicine, Kyoto University, Sakyo, Kyoto 606-8507, Japan*
[c]*Department of Pathology, The Cancer Institute, Toshima, Tokyo 170-8455, Japan*

Abstract

Thorotrast, a colloidal suspension of radioactive $^{232}ThO_2$ that emits α-particles, was used as a radiographic contrast during World War II. Thorotrast is known to induce liver cancers, particularly intrahepatic cholangiocarcinoma (ICC), decades after injection. We analyzed microsatellite instability (MSI) in Thorotrast induced ICC by PCR at 10 loci, which are generally determined for the evaluation of MSI. The incidence of MSI-positive (MSI+) cases was 64% in Thorotrast ICC whereas it was only 23% in non-Thorotrast ICC. Promoter regions of DNA mismatch repair (MMR) genes, *hMLH1* and *hMSH2*, were highly methylated in tumor parts compared with adjacent non-tumor parts. The MSI+ phenotype was not associated with the amounts of Thorotrast deposited or the incubation period after the administration of Thorotrast. These suggest that exposure to Thorotrast induces genetic instability by inactivation of mismatch repair genes through methylation of their promoter regions. © 2002 Elsevier Science B.V. All rights reserved.

Keywords: Thorotrast; α-particle; Intrahepatic cholangiocarcinoma; Genetic instability; Micro satellite

1. Background

Identification of the genetic changes in radiation-induced cancer can offer promising contributions to risk assessment and the prevention of cancer. Thorotrast is the trade name for a 25% colloidal suspension of radioactive $^{232}ThO_2$, which naturally emits α-particles

---

* Corresponding author. Tel.: +81-22-717-8507; fax: +81-22-717-8512.
  E-mail address: fukumoto@ldac.tohoku.ac.jp (M. Fukumoto).

(90%), β-particles and γ ray (10%). Thorotrast was used as a radiographic contrast agent in the 1930s–1950s. Intravascularly injected Thorotrast is deposited life long in the reticuloendothelial system. More than half of the total Thorotrast is located in the liver, and this organ is chronically irradiated by α-particles. After several decades of injection, Thorotrast has been known to induce liver cancers with intrahepatic cholangiocellular carcinoma (ICC) being the most prominent [1].

Genetic instability is a hallmark of neoplasia. The subsequent mutations after genetic destabilization have been proposed to account for the multi step nature of carcinogenesis. By analyzing *p53* mutations in Thorotrast induced liver cancers, we noticed that transitional mutations are predominant (Table 1) [2,3], and that monoclonal mutation was found in some cases of Thorotrast burdened livers. The mismatch repair (MMR) system is one of the stabilizing mechanisms of genome through the recognition and repair of mistakes made by the DNA polymerases during replication. Loss of MMR leads to an accumulation of replication errors that results in elevated mutation rates and micro satellite instability (MSI). MMR is composed of a series of molecules involved in the recognition of mismatch and repair. The first step of the MMR pathway is the recognition of mispairs or insertion-deletion loops either by the hMtSa composed of hMSH2 and hMSH3. A second heterodimer hMutLa, composed of hMLH1 and hPMS2, subsequently binds to the protein/DNA complex and is involved in the repair of DNA mismatches. Mutational inactivation of these genes is involved in various human cancers including hereditary nonpolyposis colorectal cancer [4].

Although genetic instability is known to be induced by ionizing radiation, the end point of genetic instability is obscure. In order to elucidate whether long-term irradiation of low-dose α-particles induces genetic instability or not, we, the investigated MSI in Thorotrast, induced ICC (Thorotrast ICC).

2. Methods

Archival tissue sections of ICC from 32 Thorotrast patients were analyzed. As a negative control for Thorotrast, 22 cases of ICC without Thorotrast administration (non-Thorotrast ICC) were used. DNA was extracted from paraffin-embedded tissues and PCR were performed according to the method previously described [5]. Briefly, tissues from tumors and non-tumors were separately collected from several serial sections, using a laser-assisted micro dissection (PALM and Carl Zeiss, Germany), after deparaffinization

Table 1
p53 mutations in thorotrast associated cancers [2,3]

	Base changes	Intrahepatic cholangiocarcinoma	Hepatocellular carcinoma
Transition	G:C → A:T	3 (25%)	4 (36%)
	A:T → G:C	8 (67%)	3 (27%)
Transversion	C → A	1 (8%)	1 (9%)
	A → C		1 (9%)
Deletion			2 (18%)

and staining with hematoxylin and eosin. We confirmed that the tumor cell component exceeded 80% in the tumor portion and that there were no tumor cells in the non-tumor portion. Ten microsatellite markers comprising of two mononucleotide markers (BAT25 and BAT26) and eight dinucleotide markers (D2S123, D3S1029, D3S1611, D5S346, D8S87, D13S175, D16S402 and D20S00) were selected according to previous reports on the detection of MSI [6]. One of each primer pairs was fluorescence dye labeled, and PCR was analyzed by an ABI PRISM 310 genetic analyser (PE Biosystems, USA). MSI was considered positive when additional bands that were not present in DNA from non-tumorous tissues were observed in DNA from the corresponding tumorous portions. If a tumor had more than 2 loci of MSI among cases in which PCR was succeeded in more than 5 loci, we defined it as an MSI-positive (MSI+) tumor. Among MSI+ tumors, cases with more than 40% of loci showing MSI were defined as MSI-high (MSI-H), and others were MSI-low (MSI-L). Methylation specific PCR (MSP) for *hMLH1* and *hMSH2* promoters after sodium bisulfite treatment of DNA was performed according to the method of Herman et al. [7]. In addition, we determined the LOH of the *p53* and *p73* loci to compare Thorotrast ICC and non-Thorotrast ICC [8].

3. Results

The MSI status in ICC is summarized in Table 2. Analysis of MSI was successfully performed in 25 out of the 32 (78%) Thorotrast cases, and MSI+ was observed in 16 cases (64%). On the other hand, MSI+ was found in 5 out of the 22 (22.7%) cases of non-Thorotrast ICC and four cases revealed MSI-H. In non-Thorotrast cases, MSI was observed at the chromosome loci of mismatch repair genes. In Thorotrast cases, the MSI frequency was higher in other loci than in the loci of mismatch repair genes. In contrast to non-

Table 2
MSI Status in ICC

Locus	Map	MSI Frequency	
		Thorotrast (%)	non-Thorotrast (%)
D2S123	hMSH2	4/20 (25.0)	3/22 (13.6)
BAT26	hMSH2	0/21 (0)	3/22 (13.6)
D3S1029	hMLH1	2/14 (14.2)	4/22 (18.2)
D3S1611	hMLH1	3/20 (15.0)	3/22 (13.6)
BAT25	c-kit	1/13 (7.7)	3/22 (13.6)
D5S146	APC	8/20 (40.0)	5/22 (22.7)
D8S87		8/19 (42.0)	2/22 (9.0)
D13S175		9/22 (40.9)	2/22 (9.0)
D16S402		5/12 (41.7)	0/22 (0.0)
D20S100		7/14 (50.0)	2/22 (9.0)
Total	MSI-H	2/25 (8.0)	3/22 (13.6)
	MSI-L	12/25 (48.0)	2/22 (9.0)
	MSS[a]	11/25 (44.0)	17/22 (77.3)

[a] MSS: micro satellite stable.

Thorotrast ICC, MSI in BAT25 and BAT26, which are composed of mononucleotide repeats, was not found in Thorotrast ICC.

Compared with non-tumorous tissues, tumorous parts of Thorotrast ICC revealed a more advanced methylation status of the *hMLH1* promoter in 48% of the cases and that of the *hSMH2* promoter in 24% of the cases. In non-Thorotrast cases, five of each cases showed higher methylation status in the promoter region of *hMLH1* and *hMSH2*. MSI was significantly correlated with methylation of the *hMLH1* promoter region in Thorotrast cases. However, the MSI+ trait was not associated with the duration from the injection of Thorotrast to death from ICC, or from the deposited amount of Thorotrast.

The LOH frequency of the *p73* locus in Thorotrast ICC and in non-Thorotrast ICC was 36.4% and 73.3%, respectively. That of the *p53* locus in Thorotrast ICC and in non-Thorotrast ICC was 17.4% and 13.6%, respectively.

4. Conclusions

DNA mismatch repair genes have been implicated in the pathogenesis and predisposition of certain malignancies through a mutator phenotype [9]. In the present study, the LOH frequency of *p73* was lower whereas the MSI frequency was higher in Thorotrast-induced ICC than non-Thorotrast ICC. These suggest that the carcinogenic process is distinctively different between Thorotrast and non-Thorotrast ICCs. Disruption of the MMR system has been found to play an important role in sporadic human cancers and two of the MMR genes, *hMLH1* and *hMSH2* are well characterized. In cancers of colorectum, stomach, endometrium, and pancreas, disruption of the MMR system is mainly caused by hypermethylation of the *hMLH1* promoter region [10]. However, MSI and *hMLH1* methylation are not frequent in the hepatocellular carcinoma [11]. In this study, we found MSI in Thorotrast-induced ICC is significantly associated with hypermethylation of the *hMLH1* promoter. These suggest that MSI, if it occurs, is attributed to the inactivation of *hMLH1* instead of *hMSH2* during carcinogenesis of Thorotrast ICC.

Generally, MSI is thought to occur at the early steps of carcinogenesis [9,10]. In the present study, the frequency of MSI observed in individual cases was not dependent on the duration of cancer induction. It would be interesting to clarify whether MSI occurs at early stages and is maintained further, or fluctuates during radiation induced carcinogenesis.

Acknowledgements

This work was supported in part by a Grant-in-Aid from the Ministry of Education, Sports and Culture, and Health Labour and Welfare of Japan, and under contracts with the Nuclear Safety Research Association of Japan, and Japan Space Forum.

References

[1] T. Mori, C. Kido, K. Fukutomi, Y. Kato, S. Hatakeyama, R. Machinami, Y. Ishikawa, T. Kumatori, F. Sasaki, Y. Hirota, K. Kiyosawa, S. Hayashi, H. Tanooka, T. Sobue, Related articles. Summary of entire Japanese thorotrast follow-up study: updated, Radiat. Res. 152 (6 Suppl.) (1998) S84–S87.

[2] T. Kamikawa, M. Amenomori, T. Itoh, H. Momoi, H. Hiai, R. Machinami, Y. Ishikawa, T. Mori, Y. Shimahara, Y. Yamaoka, M. Fukumoto, Analysis of genetic changes in intrahepatic cholangiocarcinoma induced by thorotrast, Radiat. Res. 152 (6 Suppl.) (1999) S118–S124.

[3] I. Wada, H. Horiuchi, M. Mori, Y. Ishikawa, M. Fukumoto, T. Mori, Y. Kato, T. Kitagawa, R. Machinami, High rate of small TP53 mutations and infrequent loss of heterozygosity in malignant liver tumors associated with thorotrast: implications for alpha-particle carcinogenesis, Radiat. Res. 152 (6 Suppl.) (1999) 125–127.

[4] A.-L. Lu, Biochemistry of mammalian DNA mismatch repair, in: J.A. Nickoloff, M.F. Hoekstra (Eds.), DNA Damage and Repair vol. II, Humana Press, 1998, pp. 95–145.

[5] H. Momoi, T. Itoh, Y. Nozaki, Y. Arima, H. Okabe, S. Satoh, Y. Toda, E. Sakai, K. Nakagawara, P. Flemming, M. Yamamoto, Y. Shimahara, Y. Yamaoka, M. Fukumoto, Microsatellite instability and alternative genetic pathway in intrahepatic cholangiocarcinoma, J. Hepatol. 35 (2001) 235–244.

[6] C.R. Boland, S.N. Thibodeau, S.R. Hamilton, D. Sidransky, J.R. Eshleman, R.W. Burt, S.J. Meltzer, M.A. Rodriguez-Bigas, R. Fodde, G.N. Ranzani, S. Srivastava, A National Cancer Institute workshop on microsatellite instability for cancer detection and familial predisposition: development of international criteria for the determination of microsatellite instability in colorectal cancer, Cancer Res. 58 (22) (1998) 5248–5257.

[7] J.G. Herman, J.R. Graff, S. Myohanen, B.D. Nelkin, S.B. Baylin, Methylation-specific PCR: a novel PCR assay for methylation status of CpG islands, Proc. Natl. Acad. Sci. U. S. A. 93 (18) (1996) 9821–9826.

[8] H. Momoi, H. Okabe, T. Kamikawa, S. Satoh, I. Ikai, M. Yamamoto, A. Nakagawara, Y. Shimahara, Y. Yamaoka, M. Fukumoto, Comprehensive allelotyping of human intrahepatic cholangiocarcinoma, Clin. Cancer Res. (in press) 2001.

[9] L.A. Loeb, A mutator phenotype in cancer, Cancer Res. 61 (8) (2001) 3230–3239.

[10] M. Esteller, Epigenetic lesions causing genetic lesions in human cancer: promoter hypermethylation of DNA repair genes, Eur. J. Cancer 36 (2000) 2294–2300.

[11] Y. Kondo, Y. Kanai, M. Sakamoto, M. Mizokami, R. Ueda, S. Hirohashi, Genetic instability and aberrant DNA methylation in chronic hepatitis and cirrhosis—a comprehensive study of loss of heterozygosity and microsatellite instability at 39 loci and DNA hypermethylation on 8 CpG islands in microdissected specimens from patients with hepatocellular carcinoma, Hepatology 32 (5) (2000) 970–979.

Molecular and cellular mechanisms of radiation responses

Transmission of damage signals from irradiated to nonirradiated cells

John B. Little *, Hatsumi Nagasawa,
Sonia M. de Toledo [1], Edouard Azzam

*Laboratory of Radiobiology, Department of Cancer Biology, Harvard School of Public Health,
665 Huntington Avenue, Boston, MA 02115 USA*

Abstract

Confluent monolayer cultures of human and rodent cells were exposed to very low fluences of alpha particles, fluences whereby as few as 1% of the cell nuclei were traversed by an alpha track and received any radiation exposure. An elevated frequency of micronuclei, a surrogate measure of DNA damage, and of HPRT mutations were observed in the nonirradiated "bystander" cells. This was associated with an ATM-dependent upregulation of the p53 damage-response pathway. Damage signals were transmitted from irradiated to nonirradiated cells in the population by gap junction mediated intercellular communication. The biological effects in bystander cells appeared to be associated with oxidative stress, consistent with the observation that most of the mutants induced by very low fluences of alpha particles were a result of point mutations rather than deletions. Preliminary evidence suggests that activation of oxidative stress-responsive signal pathways may also occur in bystander cells. These findings, particularly the evidence for the induction of mutations in nonirradiated bystander cells after irradiation with very low fluences of alpha particles, could be of significance in estimations of the risk of residential radon exposure in human populations. © 2002 Elsevier Science B.V. All rights reserved.

Keywords: Bystander effect; Gap junction intracellular communication; Signal transduction; Mutagenesis; Oxidative stress

---

Abbreviations: GJIC, gap junction intercellular communication; CHO, Chinese hamster ovary; SOD, superoxide dismutase; DPI, diphenylionium.
* Corresponding author. Tel.: +1-617-432-1184; fax: +1-617-432-0107.
 E-mail address: jlittle@hsph.harvard.edu (J.B. Little).
[1] Current address: Department of Radiology MSB F-466, UMDNJ-New Jersey Medical School 185 South Orange Avenue, Newark, NJ 07103, USA.

0531-5131/02 © 2002 Elsevier Science B.V. All rights reserved.
PII: S0531-5131(01)00776-2

1. Introduction

The conventional wisdom concerning the effects of radiation on mammalian cells have included the assumptions that the cell nucleus is the target for the biological effects of ionizing radiation, and that these effects arise in irradiated cells as a direct consequence of DNA damage—specifically double strand breaks—that remain unrepaired or misrepaired. Mutations would thus arise at the site of such damage in the irradiated cell. As a consequence, the initiating event in radiation carcinogenesis would presumably arise from a mutational event occurring in a radiation-damaged cell.

We earlier presented evidence to suggest that genetic changes may occur in cells in an irradiated population that in themselves received no radiation exposure [1]. We irradiated densely seeded cultures of Chinese hamster ovary (CHO) cells with very low fluences of Plutonium-238 alpha particles from a specially constructed irradiator, fluences whereby only 0.1–1.0% of the cell nuclei were actually traversed by an alpha particle and thus received any radiation exposure. However, 20–40% of the cells showed increased frequencies of sister chromatid exchanges, indicating that genetic damage was occurring in nonirradiated cells in the population. These findings were subsequently confirmed by Lehnert et al. [2,3], who proposed that the effect was associated with enhanced oxidative stress.

These findings thus suggest that damage signals can be transmitted from irradiated to nonirradiated neighboring cells in a population leading to the occurrence of genetic changes in cells that receive no radiation exposure. This phenomenon has been termed the "bystander effect". In the present paper, we describe some of our more recent findings designed to further characterize the effect and how these signals are transmitted, as well as the possible role of oxidative stress.

2. Materials and methods

All experiments were carried out with confluent or near confluent monolayer cultures of human or rodent cells. These cultures were irradiated with Plutonium-238 alpha particles derived from a specially constructed irradiator [4]. Micronuclei were measured by the cytokinesis-block technique [5]. At least 500 cells were examined; only micronuclei occurring in binucleate cells were scored. Otherwise, the methods employed have been described in detail elsewhere [6–9].

3. Results

3.1. Occurrence of DNA damage and mutations in bystander cells

The induction of micronuclei in bystander cells was employed as a surrogate measure of DNA damage. As can be seen in Fig. 1, a 3–4-fold enhancement in the frequency of cells with micronuclei occurred following exposure to mean doses of 1–2 cGy (approximately 4–8% of cell nuclei traversed by an alpha particle); only a small further increase occurred

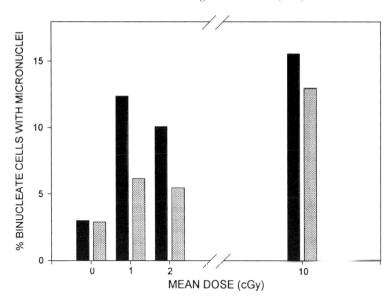

Fig. 1. Induction of micronuclei in human diploid fibroblasts by low fluences of alpha particles. Solid bars are controls and hatched bars are cultures incubated with SOD. The differences between control and SOD at 1 and 2 cGy are significant at the $p<0.01$ and $p<0.03$ levels, respectively.

with 10 cGy when 40–50% of the nuclei were hit. A similar enhancement was observed in the frequency of HPRT mutants in CHO cells exposed to very low fluences of alpha particles, leading to a supralinear dose response curve at mean doses below 3–4 cGy [6]. When the data are presented in terms of the mutation frequency per alpha particle track traversing a cell nucleus, as shown in Fig. 2, this frequency rises markedly at low alpha particle fluences. Thus, at very low fluences, when only a small fraction of the cell nuclei are traversed by an alpha particle track, each track is 4–5-fold more efficient in producing mutations. Presumably, this is the result of mutations occurring in neighboring bystander cells.

In order to further examine this hypothesis, we examined the spectrum of molecular changes associated with mutations arising spontaneously or in cultures exposed to 0.5 or 10 cGy [7]. These results are shown in Table 1. As can be seen, an enhanced frequency of point mutations occurred in cells from cultures exposed to 0.5 cGy. When the contribution of spontaneously arising mutations was subtracted and the results thus normalized for the occurrence of induced mutations only, over 90% of the mutations induced by 0.5 cGy were point mutations. On the other hand, high frequencies of both partial and total deletions were observed among mutations arising in cultures exposed to 10 cGy. Thus, the type of mutations induced in bystander cells differs significantly from those arising in cells receiving direct radiation exposure.

Consistent with the occurrence of DNA damage, upregulation of the p53 damage-response pathway was also observed in bystander cells [8]. Western analysis showed significant upregulation of p53 and $p21^{WAF1}$ in human diploid fibroblasts following mean radiation doses as low as 0.16 cGy, when less than 1% of the nuclei were traversed by an

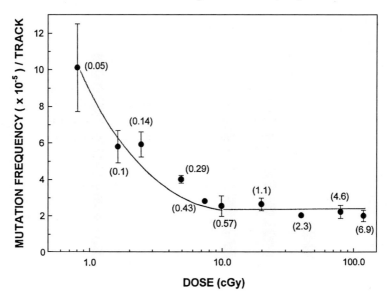

Fig. 2. Induction of HPRT mutations in CHO cells by various mean doses of alpha radiation, expressed as the mutation frequency induced by an alpha particle track traversing the nucleus of a cell. Numbers in parenthesis are mean number of tracks per nucleus. The background mutation frequency was $0.14 \pm 0.02 \times 10^{-5}$. Reproduced from Nagasawa and Little [6].

alpha particle. This result was confirmed by in situ immunofluorescence staining and confocal microscopy with an antibody for $p21^{WAF1}$; upregulation of p21 clearly occurred in clusters of neighboring cells. This effect was associated with an upregulation of serine-15 phosphorylation of p53, indicating that the effect was ATM-dependent and thus the result of the DNA damage occurring in the bystander cells.

3.2. Intercellular transmission of damage signals

These experiments were designed to examine how these damage signals are transmitted from irradiated to nonirradiated cells, specifically the role of gap junction mediated intracellular communication (GJIC). Preliminary experiments were carried out in human diploid fibroblasts by incubation with Lindane, an inhibitor reported to block GJIC by

Table 1
Molecular structural analysis of HPRT mutants arising spontaneously or induced by alpha radiation in CHO cells[a]

Mean dose (cGy)	Percentage of nuclei hit by α particle[b]	Total number of mutant colonies analyzed	Point mutations	Partial deletions	Total deletions
0	0	145	75 (51%)	49 (34%)	21 (15%)
0.5	3	211	157 (74%)	38 (18%)	16 (8%)
10	44	434	176 (41%)	175 (40%)	83 (19%)

[a] Data from Huo et al [7].
[b] Percentage of nuclei traversed by one or more alpha tracks based on Poisson analysis.

degrading the connexin proteins that mediate such communication [8]. Lindane was found to markedly suppress upregulation of the p53 damage-response pathway in bystander cells, as determined both by Western blotting and in situ immunofluorescence. A similar suppression was observed in the induction of micronuclei.

Genetically manipulated cell lines were utilized to examine more directly the role of GJIC in this effect [9]. In the first series of experiments, two isogenic cell lines derived from rat liver epithelial cells were obtained from Dr. James Trosko at Michigan State University. One of these lines was unable to carry out GJIC owing to a mutation in the connexin 43 gene, whereas the other communicated efficiently. This was confirmed by Lucifer yellow transfer experiments. As can be seen in Fig. 3, upregulation of p21 occurred after exposure to low fluences of alpha particles in the GJIC competent line, whereas no effect was observed in the communication deficient line after mean radiation doses below 10 cGy where only a small fraction of the cells was irradiated. This result was confirmed with a connexin 43 mouse embryo fibroblast knockout cell line [9]. These results thus indicate that damage signals are transmitted from irradiated to nonirradiated cells in confluent monolayer cultures by gap junction mediated intercellular communication.

3.3. Role of oxidative stress

The earlier observation that oxidative stress appeared to be involved in the induction of sister chromatid exchanges in bystander cells [3] led us to examine the possible role of oxidative stress in the changes in gene expression observed in these cells. Incubation of human diploid fibroblasts with superoxide dismutase (SOD) (100 µg/ml) markedly suppressed the radiation-induced upregulation of the p53 damage-response pathway in bystander cells. This is consistent with the results shown in Fig. 1 indicating a significant suppression in micronucleus formation in cultures incubated with SOD and exposed to low alpha particle fluences. A lesser suppressive effect was seen with catalase. Almost complete suppression of p21 expression in bystander cells occurred with incubation with diphenylionium (DPI) (0.2 µM), a specific flavoprotein inhibitor that blocks superoxide anion generation by NADP(H) oxidase. These results suggest that upregulation of the p53

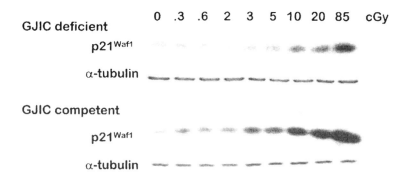

Fig. 3. Expression of p21^{WAF1} in two isogenic cell lines, one proficient and one deficient in gap junction mediated intercellular communication, after various mean doses of alpha radiation. Data from Ref. [9].

damage-response pathway in bystander cells may be associated with enhanced levels of reactive oxygen species. This would be consistent with the results shown in Table 1 indicating that most mutations induced in bystander cells are point mutations rather than deletions, as are usually associated with cells irradiated with higher fluences of alpha particles. These findings focus attention in particular on the superoxide anion, as the effect was suppressed by an agent (DPI) which blocks superoxide formation as well as an agent (SOD) which facilitates its breakdown.

We have carried out preliminary experiments to examine whether other signal transduction pathways responsive to oxidative stress may be activated in bystander cells. These were examined by both Western analysis and electrophoretic mobility shift DNA binding assays. Both the JNK kinase and the P38 MAPK pathways as well as their downstream transcription factors were activated in bystander cells. Evidence for ATF2 activation occurred within 1 min of radiation exposure and was maximal by 15 min. As these pathways are activated by signals arising in the cell membrane, it remains to be determined whether their activation is secondary to increased oxidative stress in the bystander cells or may involve signals transmitted directly to the membrane from the irradiated cells.

4. Discussion/conclusions

These results clearly indicate that damage signals can be transmitted from irradiated to nonirradiated cells in a population leading to DNA damage, the induction of mutations and changes in gene expression in these bystander cells. In the model of confluent cultures of human and rodent cells we have employed, these signals are transmitted by gap junction mediated intercellular communication. These findings are in general consistent with those of the elegant microbeam studies carried out in Columbia University and reported elsewhere in these Proceedings [10], as well as those from the Gray Laboratory [11]. It has also been reported that irradiated cells may release factors into the culture medium that, when transferred to nonirradiated cell populations, can lead to changes in cell survival [12] and gene expression [13]. Bystander effects have also been described in mixing experiments whereby irradiated cells are cultured with nonirradiated cells [14].

These findings challenge the conventional wisdom that mutations and other biological effects arise only in directly irradiated cells as a consequence of DNA damage occurring in those cells. If the existence of such a phenomenon occurs in vivo, our results could be of significance in terms of the risk of exposure to residential radon. Radon levels measured in homes are estimated to yield approximately 1 alpha traversal per 100 bronchial epithelial cells each year [15], a fluence comparable to that utilized in our experiments. Unfortunately, epidemiologic studies of the relationship between lung cancer and residential radon exposure have yielded very wide confidence limits in this low dose range [16]. Although consistent with a linear back extrapolate from the underground miner data, supralinearity of the dose response curve cannot be excluded in this low dose range. The occurrence of a bystander effect for the induction of mutations in vivo would thus imply that one could not predict low dose effects based on the number of cells actually hit by a particle. The assumption of a linear dose response relationship could thus underestimate the potential carcinogenic risk of exposure to very low fluences of alpha particles.

Acknowledgements

This research was supported by Research Grant DE FG02-98ER62685 from the US Department of Energy and Center Grant ES-00002 from the US National Institute of Environmental Health Sciences.

References

[1] H. Nagasawa, J.B. Little, Induction of sister chromatid exchanges by extremely low doses of alpha particles, Cancer Res. 52 (1992) 6394–6396.
[2] A. Deshpande, E.H. Goodwin, S.M. Bailey, B. Marrone, B. Lehnert, Alpha particle-induced sister chromatid exchanges in normal human lung fibroblasts: evidence for an extranuclear target, Radiat. Res. 45 (1996) 260–267.
[3] P. Narayanan, E.H. Goodwin, B. Lehnert, Alpha particles initiate biological production of superoxide anions and hydrogen peroxide in human cells, Cancer Res. 57 (1997) 3963–3971.
[4] N.F. Metting, A.M. Koehler, H. Nagasawa, J.M. Nelson, J.B. Little, Design of a benchtop alpha particle irradiator, Health Phys. 68 (1995) 710–715.
[5] M. Fenech, The in vitro micronucleus technique, Mutat. Res. 455 (2000) 81–95.
[6] H. Nagasawa, J.B. Little, Unexpected sensitivity to the induction of mutations by very low doses of alpha-particle radiation: evidence for a bystander effect, Radiat. Res. 152 (1999) 552–557.
[7] L. Huo, H. Nagasawa, J.B. Little, HPRT mutants induced in bystander cells by very low fluences of alpha particles result primarily from point mutations, Radiat. Res., (2001) 521–525.
[8] E.I. Azzam, S.M. de Toledo, T. Gooding, J.B. Little, Intercellular communication is involved in the bystander regulation of gene expression in human cells exposed to very low fluences of alpha particles, Radiat. Res. 150 (1998) 497–504.
[9] E.I. Azzam, S.M. de Toledo, J.B. Little, Direct evidence for the participation of gap junction-mediated intercellular communication in the transmission of damage signals from alpha-particle irradiated to non-irradiated cells, Proc. Natl. Acad. Sci. U. S. A. 98 (2001) 473–478.
[10] T.K. Hei, H. Zhou, M. Suzuki, G. Randers-Pehrson, C.A. Waldren, E.J. Hall, The yin and yan of bystander versus adaptive response: lessons from the microbeam studies. These Proceedings, 2001.
[11] O.V. Belyakov, A.M. Malcolmson, M. Folkard, K.M. Prise, B.D. Michael, Direct evidence for a bystander effect of ionizing radiation in primary human fibroblasts, Br. J. Cancer 84 (2001) 674–679.
[12] C. Mothersill, C. Seymour, Medium from irradiated human epithelial cells but not human fibroblasts reduces the clonogenic survival of unirradiated cells, Int. J. Radiat. Biol. 71 (1997) 421–427.
[13] R. Iyer, B. Lehnert, Factors underlying the cell growth-related bystander responses to alpha particles, Cancer Res. 60 (2000) 1290–1298.
[14] A. Bishayee, D.V. Rao, R.W. Howell, Evidence for pronounced bystander effects caused by nonuniform distributions of radioactivity using a novel three-dimensional tissue culture model, Radiat. Res. 152 (1999) 88–97.
[15] M. Charles, R. Cox, D. Goodhead, A. Wilson, CEIR forum on the effects of high-LET radiation at low doses/dose rates, Int. J. Radiat Biol. 58 (1990) 859–885.
[16] J.H. Lubin, J.D. Boice Jr., Lung cancer risk from residential radon: meta-analysis of eight epidemiologic studies, JNCI, J. Natl. Cancer Inst. 89 (1997) 49–57.

Specific gene expression by extremely low-dose ionizing radiation which related to enhance proliferation of normal human diploid cells

Masami Watanabe*, Keiji Suzuki, Seiji Kodama

Laboratory of Radiation and Life Science, Department of Health Sciences, School of Pharmaceutical Sciences, Nagasaki University, 1-14 Bunkyo-machi, Nagasaki 852-8521, Japan

Abstract

We demonstrated that X-ray irradiation at low doses of between 2 and 5 cGy stimulated proliferation of a normal human diploid. At low doses of between 2 and 5 cGy, ERK1/2 was phosphorylated as efficiently as at higher doses between 50 and 100 cGy of X-rays, while the p53 protein level was not increased by doses below 50 cGy. On the other hand, the p53 protein was efficiently accumulated at higher doses of X-ray more than 100 cGy. ERK1/2 was phosphorylated by doses over 50 cGy with increasing doses. We found that activated ERK1/2 augmented phosphorylation of the Elk-1 protein. Furthermore, over expression of ERK2 in NCI-H1299, and human lung carcinoma cells, potentiated enhanced proliferation, while down-regulation of ERK2 using the anti-sense ERK2 gene abrogated the stimulative effect of low-dose irradiation. These results indicate that a limited range of low-dose ionizing radiation differentially activate ERK1/2 kinases, which causes enhanced proliferation of cells receiving very low doses of ionizing radiation. © 2002 Elsevier Science B.V. All rights reserved.

Keywords: MAP kinase; ERK; Elk-1; Low dose of X-ray; p53

1. Background

Several reports have indicated that extremely low doses of ionizing radiation cause an unpredicted response in cells [1–3]. For example, a low dose of ionizing radiation such as 2 cGy of X-rays alleviates the lethal and mutagenic effects of subsequent higher doses of radiation [4]. Induction of gene transcriptions or proteins has been found after low-dose

---

* Corresponding author. Tel/fax: +81-958-44-5504.
E-mail address: nabe@net.nagasaki-u.ac.jp (M. Watanabe).

irradiation [5], indicating that the induction of gene transcription through the activation of signal transduction may be involved in the low-dose effects [6]. Because doses of more than 10 cGy do not cause the effects, only a limited range of low doses may induce the differential stimulation of certain signal transduction pathways. To date, most signal transduction studies have used doses of ionizing radiation greater than 100 cGy; therefore, very little is known about the effect of low-dose ionizing radiation on the activation of signal transduction pathways.

Ionizing radiation induces DNA double strand breaks in the nucleus. Therefore, it is very likely that ionizing radiation stimulates multiple signal transduction pathways simultaneously [7]. One such pathway originates in the nucleus and tranduces the signal to the p53 protein. The p53 protein is a tumor suppressor gene product whose function is involved in cell cycle arrest, apoptosis, DNA repair, and senescence. Another signal transduction pathway stimulated by ionizing irradiation is mediated by MAP kinases.

In the present study, we examined the effects of low doses of ionizing radiation on cell proliferation of normal human diploid cells, and determined whether MAP kinases and p53 are activated in a dose-dependent manner at doses of between 1 cGy and 6 Gy [8].

2. Results and conclusions

The effects of low-dose X-rays less than 1 Gy on cell proliferation are investigated. Exponentially growing cells were irradiated and the numbers of cells were counted to determine the initial cell number. The rest of the cells were incubated for 24 h before counting the cell number. Cell proliferation was significantly enhanced by low-dose irradiation at doses of between 2 and 5 cGy.

We found that dose-dependent increase in the phosphorylation of ERK1/2 at doses higher than 2 Gy. The levels decreased with dose down to 1 Gy, but increased again at doses of between 0.02 Gy and 0.1 Gy. The increases were comparable to those observed at 4 Gy or higher. In contrast, p53 accumulation was not detected at doses below 0.5 Gy, and between 1 and 6 Gy, the level of the p53 protein increased in a dose-dependent manner. We demonstrate that low-dose ionizing radiation stimulates proliferation of normal human diploid cells in this study. Since the effect was observed in cells irradiated with very low doses of X-rays, a limited range of low doses was suggested to activate certain signal transduction pathways. Activated ERK1/2 phosphorylated Elk-1, which involves the induction of growth-related genes, suggests that the stimulative effects of low-dose irradiation were mediated by ERK1/2 activation. In contrast to ERK1/2, accumulation of p53 did not occur with X-ray doses lower that 50 cGy. It can be concluded that very low doses of ionizing radiation stimulate only ERK1/2, and enhance cell proliferation, while higher doses activate not only ERK1/2 but also p53, which antagonizes the proliferative effect of ERK1/2 activation and results in cell cycle arrest.

Recent study has shown that ERK1/2 phosphorylate Histone H3, whose phosphorylation is hypothesized to be involved in the transcriptional activation of immediate-early genes through chromatin remodeling. Therefore, it is very likely that activation of ERK1/2 by a limited range of low doses of X-rays prior to subsequent higher dose irradiation may induce gene expression related to DNA damage repair or cell survival, or facilitate DNA repair by remodeling the chromatin structure. Thus, the present results provide the possibility that the activation of ERK1/2 may be one mechanism of low-dose effects in normal human diploid cells.

References

[1] S. Wolff, The adaptive response in radiobiology: evolving insights and implications, Environ. Health Perspect. 106 (1998) 277–283.
[2] L.E. Feinendegen, The role of adaptive responses following exposure to ionizing radiation, Hum. Exp. Toxicol. 18 (1999) 426–432.
[3] O. Rigaud, E. Moustacchi, Radioadaptation for gene mutation and the possible molecular mechanisms of the adaptive response, Mutat. Res. 358 (1996) 127–134.
[4] M. Watanabe, M. Suzuki, K. Suzuki, K. Nakano, K. Watanabe, Effect of multiple irradiation with low doses of gamma-rays on morphological transformation and growth ability of human embryo cells in vitro, Int. J. Radiat. Biol. 62 (1992) 711–718.
[5] K. Suzuki, S. Kodama, M. Watanabe, Suppressive effect of low-dose preirradiation on genetic instability induced by X rays in normal human embryonic cells, Radiat. Res. 150 (1998) 656 662.
[6] K. Suzuki, S. Kodama, M. Watanabe, Effect of low-dose preirradiation on induction of the HSP70B-LacZ fusion gene in human cells treated with heat shock, Radiat. Res. 149 (1998) 195–201.
[7] K. Suzuki, S. Kodama, M. Watanabe, Recruitment of ATM protein to double strand DNA irradiated with ionizing radiation, J. Biol. Chem. 274 (1999) 25571–25575.
[8] K. Suzuki, S. Kodama, M. Watanabe, Extremely low-dose ionizing radiation causes activation of MAP kinase pathway and enhances proliferation of normal human diploid cells, Cancer Res. (2001) 5396–5401.

The Yin and Yan of bystander versus adaptive response: lessons from the microbeam studies

Hongning Zhou[a], An Xu[a], Masao Suzuki[a], Gerhard Randers-Pehrson[a], Charles A. Waldren[b], Eric J. Hall[a], Tom K. Hei[a],*

[a]*Center for Radiological Research, College of Physicians and Surgeons, VC 11-205, Columbia University, 630 West 168th Street, New York, NY 10032, USA*
[b]*Department of Radiological Health Science, Colorado State University, Fort Collins, CO 80523, USA*

Abstract

Two conflicting phenomena, bystander effect and adaptive response, are important in determining the biological responses at low doses of radiation and have the potential to impact on the shape of the dose–response relationship. Using the Columbia University charged-particle microbeam and the highly sensitive A_L cell mutagenic assay, we show here that non-irradiated cells acquire mutagenesis through direct contact with cells whose nuclei had been traversed with two alpha particles each. Pre-treatment of cells with a 0.1-Gy dose X-rays 4 h before alpha particle irradiation significantly decreased this bystander mutagenic response. Results from the present study address some of the fundamental issues regarding both the actual target and radiation dose effect and can impart on our current understanding in radiation risk assessment. © 2002 Elsevier Science B.V. All rights reserved.

Keywords: Radiation; Bystander effect; Adaptive response; Mutagenesis

1. Introduction

The risk of developing radiation-induced cancer has traditionally been estimated from cancer incidence among Japanese A-bomb survivors. These data provide the best estimate of cancer risk over the dose range from 20 to 250 cGy. The cancer risk at doses below 20 cGy, however, remains uncertain and has been the subject of controversy for decades in the absence of definitive data. Two conflicting phenomena appear to be of important at

---

Abbreviations: DMSO, dimethyl sulfoxide; ROS, reactive oxygen species.
* Corresponding author. Tel.: +1-212-305-8462; fax: +1-212-305-3229.
E-mail address: TKH1@Columbia.edu (T.K. Hei).

low doses of radiation and have the potential to impact on the shape of the dose–response relationship. First, there is the bystander effect, the term used to describe the biological effects observed in cells that are not themselves traversed by a charged particle, but are neighbors of cells that are. Second, there is the adaptive response, whereby exposure to a low level of DNA damage renders cells resistant to a subsequent high doses exposure.

Using the Columbia University charged-particle microbeam, we showed recently that in human–hamster hybrid (A_L) cells where only 20% of the cells were irradiated with a near lethal dose of alpha particles, the resultant mutant fraction was threefold higher than expected assuming no interaction between the irradiated and non-irradiated cells. In other words, irradiated cells clearly induced a bystander mutagenic response in neighboring cells not directly traversed by alpha particles [1]. To ascertain if this effect can be demonstrated with lower doses of alpha particles, we show here that irradiating confluent monolayer cultures of A_L cells with two alpha particles each through the nucleus induces a large bystander mutagenic effect in neighboring, non-irradiated cells. Furthermore, treatment with DMSO had little effect on such response. On the other hand, pretreatment of cells with a 0.1-Gy dose X-rays 4 h before alpha particle irradiation significantly decreased the mutagenic yield.

2. Materials and methods

2.1. Cell culture

The human–hamster hybrid A_L cells that contain a standard set of Chinese hamster ovary-K1 chromosomes and a single copy of human chromosome 11 were used in this study [2,3]. Cells were maintained in Ham F-12 medium supplemented with 8% heat-inactivated fetal bovine serum, 25 µg/ml gentamycin, and 2×10^{-4} M glycine at 37 °C in a humidified 5% CO_2 incubator, and passaged as described [4–6].

2.2. Irradiation procedure

Cells were irradiated with alpha particles using the Columbia University charged particle microbeam as described [1,5,7]. Briefly, exponentially growing cells were plated on specially constructed microbeam dishes. Two days after plating, the nuclei of attached cells were stained with a 50-nM solution of Hoechst 33342 dye for 30 min. The image analysis system then located the centroid of each nucleus and irradiated them randomly one at a time with an exact number of alpha particles. After irradiation, cells were maintained in the dishes for 3 days before being removed by trypsinization and replated into culture flasks. After culturing for 4–5 days, the cells were trypsinized and replated to measure both the survival and mutation as described [4–6]. For determining the adaptive response, cells were irradiated with 0.1 Gy X-rays 4 h before the alpha particle irradiation.

2.3. Cytotoxicity and quantification of mutations at the CD59 locus

Irradiated and control cultures were trypsinized immediately after alpha particle exposure and replated into 60-mm-diameter petri dishes for colony formation [4–6]. Additional control and irradiated cultures were further incubated for five more days before the mutagenic assay. 5×10^4 cells were plated into each of six 60-mm dishes in 2 ml of growth medium. Cultures were incubated for 2 h to allow for cell attachment, after which 0.3% *CD59* antiserum and 1.5% (v/v) freshly thawed complement were added to each dish as described [4–6]. The cultures were further incubated for 7–8 days. At this time, the cells were fixed and stained, and the number of *CD59*$^-$ mutant colonies was scored. The cultures derived from each treatment dose together with the appropriate controls were tested for mutant yield for two consecutive weeks to ensure full expression of the mutations.

2.4. Predictions for the yield of mutants

Predictions of the yield of mutants in an experiment where a known fraction of cells were randomly irradiated through the nucleus with an exact number of alpha particles were based on the assumption that there was no bystander effect as described previously [1].

2.5. Treatment with DMSO

To examine the role of reactive oxygen species (ROS) in mediating bystander mutagenesis, cells were treated with the radical scavenger dimethyl sulfoxide (DMSO, 0.5% v/v) for 24 h before irradiation and continued throughout the expression period. DMSO at dose used in these experiments was non-toxic and non-mutagenic and had been shown to be an effective free radical scavenger [8,9]. After treatment, cultures were washed, trypsinized and replated for both survival and mutagenesis as described above.

2.6. Statistical analysis

All numerical data were calculated as means and standard deviations. Comparisons of survival fractions and induced mutation frequencies between treated groups and controls were made by Student's *t*-test. A *p* value of 0.05 or less between groups was considered to be significant.

3. Results and discussion

When the nucleus of individual A_L cells was traversed by two alpha particles, the survival fraction decreased to 0.50 ± 0.08 and the mutation fraction at the *CD59* locus was 211 ± 81 per 10^5 survivors. These data were consistent with our previous studies [1,5]. Using a precision charged particle microbeam and image analysis system, we irradiated 5–20% of randomly selected A_L cells with two alpha particles each. Under the experimental condition, about 70% of the non-irradiated cells were in direct contact with an irradiated one. The induced mutation fractions from these irradiated populations were significantly higher than the expected yields assuming there were no bystander modulation effects ($p < 0.05$, Fig. 1). These results clearly indicated that irradiated cells induced a bystander mutagenic response in neighboring cells not directly traversed by the alpha particles. Furthermore, there was no significant difference in mutant induction between

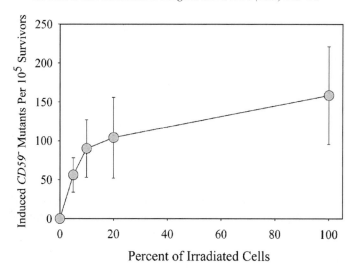

Fig. 1. Induced mutation fraction obtained from population of A_L cells in which a fixed proportion of the cells were traversed by two alpha particles through the nucleus. Induced mutant yield equals total mutant induced minus background incidence, which was 52 ± 10 mutants per 10^5 survivors in A_L cells used in these experiments. Data are pooled from five to seven independent experiments. Bar represents $\pm$ SD.

populations in which all cells were irradiated and those where only 10% or 20% of cells were hit.

ROS such as superoxide anion, hydroxyl radicals, and hydrogen peroxides are the intermediates formed during oxidative metabolism. DMSO has been shown to be an effective free radical scavenger, particularly of hydroxyl radicals, and it protects mammalian cells against the cytotoxic and genotoxic effects of a variety of chemical and physical agents where mechanisms of action are mediated by oxyradicals [8,9]. Fig. 2 shows that in cells pretreated with DMSO (0.5% v/v) 24 h before irradiation and remains in culture throughout the expression period, the resultant mutant yield does not decreased significantly. DMSO treatment by itself was non-toxic and non-mutagenic to A_L cells under the experimental conditions. These data indicate that free radicals generated from irradiation have limited effects on the induction of bystander mutagenic response.

Adaptive response is characterized by a reduction of radiobiological response in cells pretreated with a low dose radiation followed by exposure to a challenging higher dose. Numerous experimental data have shown the existence of such a response with a variety of endpoints [10]. However, there is no data available comparing the bystander effect versus adaptive response. A_L cells were pretreated with a 0.1-Gy dose of X-rays, 4 h later, 10% of randomly selected cells was irradiated with a single alpha particle through the nucleus. Our preliminary data showed that the mutants yield from the population where 10% of randomly selected cells were irradiated with single alpha particle decreased significantly if the cells were pretreated with a low, 0.1-Gy dose of X-ray (Fig. 3). The result implies that in the presence of low dose radiation stress, the bystander mutagenesis is modulated by the adaptive response, though the mechanism(s) is unclear.

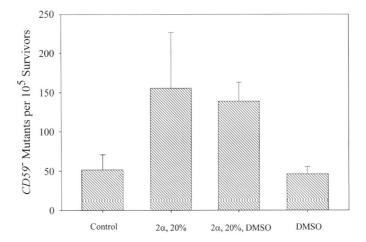

Fig. 2. Effect of the free radical scavenger, DMSO, on mutant yield in A_L cells in which 20% had been irradiated with two alpha particles through the nucleus. Data are pooled from four independent experiments. Bar represents ± SD.

It has long been accepted that the important genetic effects of radiation in mammalian cells are the direct result of DNA damage. Since only a small fraction of the bronchial epithelial cells, the presumed target for lung cancer in domestic radon exposure, are actually hit by alpha particles, the possible contribution to radiation risk due to bystander effect has attracted considerable attention. It is of interest to note that the bystander mutagenic response becomes saturated at level when only 10% of cells is irradiated by two alpha particles. Compared with the mutant yield where all of the cells in the population were traversed with

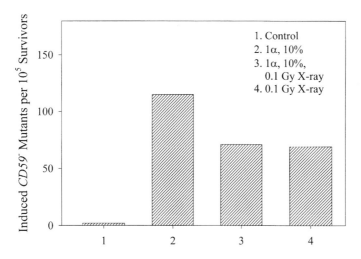

Fig. 3. Induced mutation fraction in A_L cells in which 10% had been irradiated with a single alpha particle through the nucleus with or without X-ray pretreatment. X-ray pretreatment was given 4 h before alpha particle irradiation at a dose of 0.1 Gy. Data are pooled from two independent experiments.

two alpha particles through the nucleus, the mutation fraction induced in population with a 10% hit was not significantly different. These findings suggest the presence of a plateau in the bystander response and that the damage signals from a 10% irradiated population can modulate response in all the non-irradiated neighboring cells. If such a bystander mutagenesis occurs in vivo, our current model used in radiation risk assessment would need to be reexamined.

Although published reports in support of a bystander effect appear to be consistent [1,11–16], the mechanisms of the bystander effects are still not clear. There is evidence that secretion of cytokines or other growth-promoting factors by irradiated cells lead to enhanced production of ROS in bystander cells [13,14]. On the other hand, there is evidence that gap junction mediated cell–cell communication plays a critical role in the bystander responses [1,15,16]. Our studies provide clear evidence that low dose alpha particle irradiation can induce a huge bystander mutagenic response in neighboring cells not directly traversed by alpha particles. Furthermore, in the presence of low dose radiation stress, the bystander effect can be modulated by the adaptive response. Results from the present study imply that the target for radiation-induced genetic damage is larger than an individual cell, and that the radiobiological effect at low dose is a complex interplay between the adaptive response and the bystander effect.

Acknowledgements

This work was supported in part by NIH grants CA 49062, CA 75384. The Columbia Microbeam Facility is funded by NIH Research Resource Center grant RR 11623.

References

[1] H. Zhou, G. Randers-Pehrson, C.A. Waldren, D. Vannais, E.J. Hall, T.K. Hei, Induction of a bystander mutagenic effect of alpha particles in mammalian cells, Proc. Natl. Acad. Sci. U. S. A. 97 (2000) 2099–2104.
[2] C.A. Waldren, C. Jones, T.T. Puck, Measurement of mutagenesis in mammalian cells, Proc. Natl. Acad. Sci. U. S. A. 76 (1979) 1358–1362.
[3] C.A. Waldren, L. Correll, M.A. Sognier, T.T. Puck, Measurement of low levels of X-ray mutagenesis in relation to human disease, Proc. Natl. Acad. Sci. U. S. A. 83 (1986) 4839–4843.
[4] T.K. Hei, C.A. Waldren, E.J. Hall, Mutation induction and relative biological effectiveness of neutrons in mammalian cells, Radiat. Res. 115 (1988) 281–291.
[5] T.K. Hei, L.J. Wu, S.X. Liu, D. Vannais, C.A. Waldren, Mutagenic effects of a single and an exact number of alpha particles in mammalian cells, Proc. Natl. Acad. Sci. U. S. A. 94 (1997) 3765–3770.
[6] T.K. Hei, C.Q. Piao, Z.Y. He, D. Vannais, C.A. Waldren, Chrysotile fiber is a strong mutagen in mammalian cells, Cancer Res. 52 (1992) 6305–6309.
[7] L.J. Wu, G. Randers-Pehrson, A. Xu, C.A. Waldren, C.R. Geard, Z. Yu, T.K. Hei, Targeted cytoplasmic irradiation with alpha particles induces mutations in mammalian cells, Proc. Natl. Acad. Sci. U. S. A. 96 (1999) 4959–4964.
[8] M. Watanabe, M. Suzuki, K. Suzuki, Y. Hayakawa, T. Miyazaki, Radioprotective effects of dimethyl sulfoxide in golden hamster embryo cells exposed to gamma rays at 77 K: II. Protection from lethal, chromosomal, and DNA damage, Radiat. Res. 124 (1990) 73–78.
[9] L.G. Littlefield, E.E. Joiner, S.P. Colyer, A.M. Sayer, E.L. Frome, Modulation of radiation-induced chromosome aberrations by DMSO, an OH radical scavenger: 1. Dose–response studies in human lymphocytes exposed to 220 kV X-rays, Int. J. Radiat. Biol. Relat. Stud. Phys., Chem. Med. 53 (1988) 875–890.

[10] O. Rigaud, E. Moustacchi, Radioadaptation for gene mutation and the possible molecular mechanisms of the adaptive response, Mutat. Res. 358 (1996) 127–134.
[11] H. Nagasawa, J. Little, Induction of sister chromatid exchanges by extremely low doses of α-particles, Cancer Res. 52 (1992) 6394–6396.
[12] A. Deshpande, E.H. Goodwin, S.M. Bailey, B.L. Marrone, B.E. Lehnert, Alpha-particle-induced sister chromatid exchange in normal human lung fibroblasts: evidence for an extranuclear target, Radiat. Res. 145 (1996) 260–267.
[13] C. Mothersill, C.B. Seymour, Cell–cell contact during gamma irradiation is not required to induce a bystander effect in normal human keratinocytes: evidence for release during irradiation of a signal controlling survival into the medium, Radiat. Res. 149 (1998) 256–262.
[14] P.K. Narayanan, E.H. Goodwin, B.E. Lehnert, α Particles initiate biological production of superoxide anions and hydrogen peroxide in human cells, Cancer Res. 57 (1997) 3963–3971.
[15] E.I. Azzam, S.M. de Toledo, T. Gooding, J.B. Little, Intercellular communication is involved in the bystander regulation of gene expression in human cells exposed to very low fluencies of alpha particles, Radiat. Res. 150 (1998) 497–504.
[16] E.I. Azzam, S.M. de Toledo, J.B. Little, Direct evidence for the participation of gap junction-mediated intercellular communication in the transmission of damage signals from alpha-particle irradiated to non-irradiated cells, Proc. Natl. Acad. Sci. U. S. A. 98 (2001) 473–478.

Deficient PCNA expression and radiation sensitivity

Gayle E. Woloschak*, Tatjana Paunesku, Miroslava Protić

Biosciences Division, Argonne National Laboratory, 9700 South Cass Avenue, Argonne, IL, 60439-4833, USA

Abstract

Wasted mice express radiosensitivity in lymphoid and nervous system tissues, displaying increased apoptosis, increased transcriptional activation, and altered gene induction in response to radiation when compared to control littermates. In order to evaluate altered gene induction linked with radiosensitivity in spinal cord tissue of the wasted mice, we isolated total RNA from the spinal cords of radiation-exposed and unexposed wasted and control mice and hybridized it to a gene expression microarray. Previous studies by our group had demonstrated that lymphoid tissues of wasted mice lack proliferating cell nuclear antigen (PCNA) expression. This protein is involved in almost every aspect of DNA replication and repair, and radiosensitivity in lymphoid tissues of wasted mouse can easily be linked to proliferating cell nuclear antigen (PCNA) deficiency. In total spinal cord tissue, however, PCNA mRNA expression was not altered in wasted mice relative to controls. Nevertheless, mRNAs for many proteins that were known to interact with PCNA were modulated in wasted spinal cords compared to the tissues from BCF1 controls. © 2002 Elsevier Science B.V. All rights reserved.

Keywords: Proliferating cell nuclear antigen (PCNA); Gamma irradiation; Gene expression microarrays; Wasted mouse

1. Introduction

Wasted mutant mice show a triad of symptoms: radiosensitivity, immunodeficiency, and motor neuron degeneration [1,2]. We have shown that these animals have no detectable proliferating cell nuclear antigen (PCNA) mRNA or protein in thymus and other lymphoid tissues [3]. PCNA is involved in a plethora of cellular processes from cell cycle regulation to repair, it interacts with many very diverse proteins, and it may be the protein that is executing "decisions" made by p53 [4]. It is therefore possible to link symptoms of

---

* Corresponding author. Tel.: +1-708-252-3312; fax: +1-708-252-3387.
E-mail address: woloschak@anl.gov (G.E. Woloschak).

0531-5131/02 © 2002 Elsevier Science B.V. All rights reserved.
PII: S0531-5131(01)00775-0

radiosensitivity in lymphoid tissues of wasted mouse (such as increased apoptosis, increased transcriptional activation, and altered gene induction) to PCNA deficiency.

In order to evaluate altered gene induction linked with radiosensitivity in the spinal cord tissue of the wasted mice, we isolated total RNA from the spinal cords of wasted and control animals that were either exposed or not exposed to γ-rays; this RNA was hybridized with Affymetrix expression microarrays.

2. Methods

Total RNA was isolated, 1 h post irradiation, from non-irradiated and γ-irradiated (3 Gy gut dose) wasted mice, their healthy littermates, and BCF1 control mice according to the method usually used in our laboratory [2,3]. Mice were between 25 and 30 days old. Biotinilated RNA probes were synthesized according to the Affymetrix protocol and hybridized with the mouse expression microarray chips. Hybridization results were analyzed with Affymetrix software. A "probe mask" listing several dozens of housekeeping and ribosomal protein mRNAs was used for scaling of hybridization signals; "signal-target intensity" was set at 3000.

3. Results

Expression patterns of many mRNAs in response to γ-irradiation are well established in the literature, and we used that information to internally check the quality of our expression microarray hybridization data. Genes that were presented in Table 1 were found, by us or others, to be affected in response to γ-irradiation.

The effect of whole body γ-irradiation on the spinal cord tissue from BCF1 mice and wasted mice was distinctly different, notwithstanding the genes differentially expressed in non-irradiated wasted mice when compared to the non-irradiated controls. Some genes that demonstrated expression patterns altered by irradiation in the control BCF1 mice did not

Table 1
Results from microarray studies: confirmation of genes previously shown to be modulated in response to radiation

Gene name	Affymetrix microarray hybridization	Northern hybridization	Reference
β-Actin	↓	↓	[5]
Histone H3	↓	↓	[6]
Sp100	↓	↓	[7]
H-*ras*	↓	↓	[8]
hsp40	↑	↑	[9]
hsp70	↑	↑	[10]
α-Tubulin	↑	↑	[5]
Growth-arrest specific gene 2 (gas-2)	↑	↑	[11]
α Integrin	↑	↑	[12]

show modified expression in irradiated wasted mice, and vice versa. Other genes that were differentially expressed in wasted non-irradiated animals showed the same expression pattern as in the irradiated control animals.

Some examples (all compared to non-irradiated control mice) are as follows: casein kinase II alpha subunit, c-Jun N-terminal kinase (JNK), DEAD (aspartate-glutamate-alanine-aspatrate) box polypeptide 6, and inner centromere protein transcripts were down-regulated in irradiated control and wasted animals (1.2-, 3.2-, 2.3-, and 3.6-fold, respectively), as well as in non-irradiated wasted animals (2.2-, 3.2-, 2.5-, and 3.9-fold, respectively). Conversely, RAD51-like protein mRNA was induced in irradiated control and non-irradiated wasted animals (2.4- and 2.1-fold, respectively), showing no further change in wasted animals post-irradiation. In wasted irradiated and non-irradiated animals, mRNA for Nitric oxide synthase 1 (NOS1) was reduced to 2.3- and 3.9-fold, respectively, but it was not affected by irradiation of the control animals. In some cases, wasted animals and the controls showed exactly opposite responses to irradiation. For example, irradiation in control animals caused increased expression of heat shock proteins hsp70 and hsp40 (2.3- and 1.5-fold, respectively). In irradiated wasted animals, mRNAs for both heat shock proteins were decreased (2.7- and 1.3-fold, respectively).

PCNA mRNA expression itself was not altered in spinal cords of wasted mice in comparison to controls, although the expression pattern in specific neural subpopulations of the spinal cord is not known. Nevertheless, mRNAs for many proteins interacting with PCNA were modulated in wasted spinal cords compared to the spinal cords from BCF1 controls. For example, cyclin kinase inhibitor p21 (Cip1/Waf1) mRNA in non-irradiated wasted mice was increased 12-fold when compared to the non-irradiated control animal. One hour following γ-irradiation, p21 mRNA levels dropped to the same level as in the irradiated and non-irradiated control animals. In addition to p21, Gadd45 mRNA also increased 2-fold in the non-irradiated, but not irradiated, wasted animals. Other established and hypothetical PCNA interacting proteins [4] showed the same expression pattern in the various treatment groups. For example, flap structure specific endonuclease 1 (FEN1) and Brca1 mRNAs expressed 1.8- and 2.2-fold increases, respectively, compared to the control animals. On the other hand, DNA polymerase α, and Werner syndrome helicase were reduced 2.5- and 1.8-fold, respectively, in wasted non-irradiated animals in comparison to non-irradiated control and irradiated wasted and control animals.

4. Discussion

We conducted a comparative study of mRNA expression patterns in spinal cords of wasted and control mice exposed and unexposed to γ-rays. Comparison of these data with the previously published results (ours and others) led us to the conclusion that gene expression patterns following whole body exposure closely resembles the established gene expression pattern reported from in vitro studies following γ-irradiation of cells in culture.

In this comprehensive study, 15,000 genes were screened at the same time, and a wealth of data was obtained. Gene expression patterns in wasted mice with or without prior γ-irradiation were different. Some genes that are considered to be radioprotective, such as hsp70 and hsp40, were induced in non-irradiated wasted mice (to the extent comparable with the γ-irradiated control animals), but reduced upon γ-irradiation. Similar was the case with the expression pattern of RAD51-like gene, a gene that is linked to repair functions. This mRNA was induced in irradiated control and in the wasted mouse, regardless of irradiation—γ-irradiation of the wasted mouse did not further increase expression of RAD51-like gene. These data corroborate the picture of the wasted mouse as a model of radiosensitivity.

One possible explanation for the radiosensitivity of wasted mouse is the lack of functional PCNA in the cells. This protein is one of the main "executive" molecules in the life of the cell, and it is not expressed at all in the lymphoid tissues of wasted mice. PCNA expression in wasted mouse spinal cord did not appear to be different from the one in the control mice. Many proteins that bind to PCNA, however, had displayed differential expression patterns.

Cell cycle arrest proteins p21 and Gadd45 are over-expressed in the spinal cords of the wasted mouse. It is a matter of dispute whether PCNA is able to fulfill its tasks when linked to these proteins, even when they are present in the cell in "physiologic" concentrations. When over-expressed, these proteins can cause not only cell cycle arrest but also cell death (as reviewed in Ref. [4]). One hour following γ-irradiation concentrations of mRNAs coding these proteins in the wasted mouse spinal cord tissue dropped to the same levels as in the control mice.

Based on the possibility of interaction/recognition of the interdomain-connecting loop of PCNA, we and others created a list of possible protein partners of PCNA. Those included some of the well-known PCNA interacting proteins, and some proteins for which interaction with PCNA is still hypothetical. Among the established PCNA interacting proteins that are found to be differentially expressed in wasted mice, we found increased mRNA levels for FEN1, a protein involved in replication and base excision repair, in non-irradiated wasted mice [13]; and decreased levels of mRNAs for Werner syndrome helicase involved in repair [14]. One of the hypothetical PCNA-interacting proteins, DNA polymerase α [4], one of the main replication proteins, was also reduced in non-irradiated wasted mice compared to controls.

It can be hypothesized that a need for functional repair drives increased expression of FEN1 as an antidote for the overexpression of p21. Down-regulation of replication by down-regulation of DNA polymerase α may contribute to the same goal. Nevertheless, alternative explanations for the expression patterns of PCNA protein partners are also possible, and further analysis and experiments will be necessary to establish firm answers about PCNA involvement in the etiology of the wasted mouse disease.

Acknowledgements

This work was supported by NIH grants CA81375, CA73042, and NS21442.

References

[1] L.D. Shultz, H.O. Sweet, M.T. Davisson, D.R. Coman, "Wasted", a new mutant of the mouse with abnormalities characteristic to ataxia telangiectasia, Nature 297 (5865) (1982) 402–404.

[2] G.E. Woloschak, C.M. Chang-Liu, J. Chung, C.R. Libertin, Expression of enhanced spontaneous and gamma-ray-induced apoptosis by lymphocytes of the wasted mouse, Int. J. Radiat. Biol. 69 (1) (1996) 47–55.

[3] G.E. Woloschak, T. Paunesku, C.R. Libertin, C.M. Chang-Liu, M. Churchill, J. Panozzo, D. Grdina, M.A. Gemmell, C. Giometti, Regulation of thymus PCNA expression is altered in radiation-sensitive wasted mice, Carcinogenesis 17 (11) (1996) 2357–2365.

[4] T. Paunesku, S. Mittal, M. Protić, J. Oryhon, S.V. Korolev, A. Joachimiak, G.E. Woloschak, Proliferating cell nuclear antigen (PCNA): ringmaster of the genome, Int. J. Radiat. Biol., in press.

[5] G.E. Woloschak, Radiation-induced responses in mammalian cells, in: G.E. Koval (Ed.), Stress-Inducible Processes in Higher Eukaryotic Cells, Plenum, New York, 1997, pp. 185–219.

[6] M.F. Lavin, J. Houldsworth, S. Kumar, J.L. Stein, G.S. Stein, Coupling of histone mRNA levels to radioresistant DNA synthesis in ataxia–telangiectasia cells, Mol. Cell. Biochem. 73 (1) (1987) 45–54.

[7] T. Paunesku, C.M. Chang-Liu, P. Shearin-Jones, C. Watson, J. Milton, J. Oryhon, D. Salbego, A. Milosavljevic, G.E. Woloschak, Identification of genes regulated by UV/salicylic acid, Int. J. Radiat. Biol. 76 (2) (2000) 189–198.

[8] A. Anderson, G.E. Woloschak, Cellular proto-oncogene expression following exposure of mice to γ-rays, Radiat. Res. 130 (1992) 340–344.

[9] M.P. Achary, W. Jaggernauth, E. Gross, A. Alfieri, H.P. Klinger, B. Vikram, Cell lines from the same cervical carcinoma but with different radiosensitivities exhibit different cDNA microarray patterns of gene expression, Cytogenet. Cell Genet. 91 (1–4) (2000) 39–43.

[10] M. Nogami, J.T. Huang, S.J. James, J.M. Lubinski, L.T. Nakamura, T. Makinodan, Mice chronically exposed to low dose ionizing radiation possess splenocytes with elevated levels of HSP70 mRNA, HSC70 and HSP72 and with an increased capacity to proliferate, Int. J. Radiat. Biol. 63 (6) (1993) 775–783.

[11] A.J. Fornace, Mammalian genes induced by radiation: activation of genes associated with growth control, Annu. Rev. Genet. 26 (1992) 507–526.

[12] J.M. Onoda, M.P. Piechocki, K.V. Honn, Radiation-induced increase in expression of the alpha IIb beta 3 integrin in melanoma cells: effects on metastatic potential, Radiat. Res. 130 (3) (1992) 281–288.

[13] U. Chen, S. Chen, P. Saha, A. Dutta, p21Cip1/Waf1 disrupts the recruitment of human Fen1 by proliferating-cell nuclear antigen into the DNA replication complex, Proc. Natl. Acad. Sci. U. S. A. 93 (21) (1996) 11597–11602.

[14] M. Lebel, E.A. Spillare, C.C. Harris, P. Leder, The Werner syndrome gene product co-purifies with the DNA replication complex and interacts with PCNA and topoisomerase I, J. Biol. Chem. 274 (53) (1999) 37795–37799.

Mutation induction by continuous low dose rate gamma irradiation in human cells

J. Kiefer*, M. Kohlpoth, M. Kuntze

Strahlenzentrum der Justus-Liebig-Universität, Leihgesterner Weg 217, D-35392 Giessen, Germany

Abstract

Human teratocarcinoma cells P3 were subjected to ^{60}Co-gamma-rays in continuous exponential growth using a "chemostat" set-up. Mutations at the HPRT-locus were measured with dose rates of 5 mGy/h, 50 mGy/h and after acute exposure. There was a clear reduction of mutation induction efficiency with 50 mGy/h compared to the high dose rates, however, with 5 mGy/h the same results were obtained as with acute exposure, in other words mutation induction increased again with very low dose-rates. This phenomenon of an "inverse dose rate effect" is restricted to cycling cells, as it was not found in plateau phase cells exposed under identical conditions. These results confirm previous findings by others and us in rodent cells, which were so far not available for human cells. In order to explore a possible contribution of "genomic instability", cells irradiated acutely with a dose of 3 Gy were grown for up to 40 days in an exponential culture, and the mutant frequencies were determined every second day. A rise in mutant frequency after 20–30 days was consistently found in three independent experiments but time course and mutant numbers were not reproducible and seemed to follow a stochastic behaviour. The results cast doubt on the common assumption that a reduction in the dose rate will always lead to smaller effects, although this somewhat artificial system cannot be generalised to the human situation. © 2002 Elsevier Science B.V. All rights reserved.

Keywords: Dose rate; Gamma rays; Mutation; Genomic instability

1. Introduction

It is generally assumed that with sparsely ionizing radiation the efficiency of damage induction decreases when the dose-rate is reduced. This is satisfyingly documented for cell inactivation (see, e.g., Ref. [1]), however, it is less clear for other endpoints like mutation

---

* Corresponding author. Tel.: +49-641-99-15300; fax: +49-641-99-15009.
E-mail address: juergen.kiefer@strz.uni-giessen.de (J. Kiefer).

induction or neoplastic transformation. There are not many studies with mammalian cells in culture, and the technical problems involved with long-term exposures under well-defined conditions are not easy to overcome, since very low dose-rates require extended irradiation times. There seems to be general agreement that in rodent cells with intermediate dose-rates (above about 30 mGy/h) mutation induction effectiveness is reduced [2–6]. With even lower dose-rates (about 3–5 mGy/h), however, an inversion was frequently seen, i.e., an *increased* effectiveness with further reduced dose-rates, as in V79 Chinese hamster cells [6,7] and mouse L cells [4]. No dose-rate or fractionation effect was found in human TK6 cells [8]. All these studies were performed in growing populations, no inverse dose-rate effect was found in the resting cells but again only in mouse cells [9]. We report here on experiments with human teratocarcinoma cells where the effects of long-term exposures were investigated both in exponentially growing and stationary cells, with dose-rates between 5 and 50 mGy/h. The results show in plateau phase cells a marked reduction in mutagenicity compared to acute exposures, however, again an inverse dose-rate effect in exponentially growing cultures is seen.

Long-term exposure of growing cell cultures poses experimental problems. In order to avoid entering the stationary phase, monolayers have to be trypsinized and reseeded at regular intervals, and in suspensions, the cell density has to adjust by diluting. Both procedures may affect the physiological state of the system. This problem could be overcome by the technique of "continuous culture"—well established with microbial systems—which was introduced by our laboratory into radiobiological research [6,7]. It allows one to keep cells in the exponential growth phase over long periods of time with constant distributions of cell cycle phases, and under well-defined conditions.

Previous investigations were carried out with V79 Chinese hamster [6] and human TK6 cells [8]. Since human and rodent systems may differ in their response and the TK6 line being of lymphoblastoid origin, which is not necessarily representative for other tissues, it was felt worthwhile to extend our investigations to other human cells. Here we report on experiments with the teratocarcinoma line P3 [10,11].

2. Materials and methods

2.1. Cells and culture conditions

The cells used were a subline of the human teratocarcinoma cell line termed SP3. They were adapted in our laboratory for growth in suspension culture from P3 [10,11], originally obtained from Dr. Huberman (Argonne National Laboratory). They can also be maintained attached to tissue culture dishes. Contrary to the original, the new subline forms monocellular layers, which can be kept viable over many weeks with changes of medium from time to time. One advantage of P3 as well as SP3 cells is their fairly constant karyotype: the chromosome number is 46 ± 1. The cell is of female origin; it contains two X-chromosomes, one of which is presumably inactive.

The medium used throughout was MEM spinner medium (GIBCO) supplemented with 10% foetal calf serum (GIBCO). Penicillin and Streptomycin were added at 50 µg/ml and 60 µg/ml, respectively. To maintain a pH of 7.2 HEPES buffer was used at a concentration of 10 mM. The cells were incubated at 37 °C in 5% CO_2 and a relative humidity of 100%. The stock cultures were maintained in normal medium. Because of the usually low background mutant frequencies, it was felt unnecessary to store them in a HAT medium. All cultures were monitored regularly for the presence of mycoplasmal infection using the Mycoplasma Detection Kit (Boehringer) containing antimycoplasmal antibodies.

For long-term low dose-rate exposure of exponential cells, they were grown in continuous culture with constant dilution of the medium (Fig. 1). The method which allows the maintenance of stable conditions over many weeks has been described previously [6,7]. The medium used was "MEM spinner medium" without Mg- and Ca-ions, supplemented with 15% foetal calf serum (FKS). To neutralize Mg- and Ca-ions, 4 ml of 5% Na-Citrate solution was added to 75 ml FKS before mixing.

To obtain "stable" monolayers, cells were plated in culture flasks (Greiner, 75 cm^3) at a concentration of 7×10^4/flask. They formed a closed monolayer after about 1 week. In this time and throughout the experiment no subcultivation took place, and only the medium was replaced. For this, the flasks were turned around, and the medium was sucked off with a Pasteur pipette connected to a vacuum pump. It is important never to touch the confluent monolayer that is very susceptible to damage by which the attachment is easily lost. About 1 week after reaching confluence, there is a density inhibition of growth.

Survival was measured by the standard colony-forming assay after 10 days incubation. For the determination of mutant fraction, 1.5×10^6 cells/dish were plated into two 14-cm Petri dishes to permit phenotypic expression of the induced mutants. After 4 and 8 days incubation, they were then transferred to three 14 cm Petri dishes at a density of 10^6 cells/

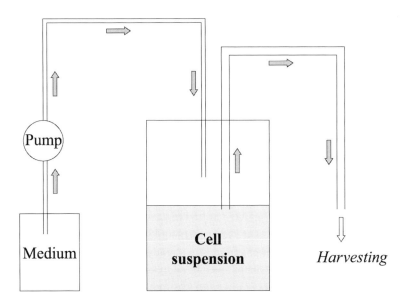

Fig. 1. The principle of continuous culture.

dish in the selection medium containing 10 µg/ml 6-thioguanine (6-TG) (FLUKA), and into four 10-cm Petri dishes at a concentration of 300 cells/dish in the nonselective medium to determine the survival fraction. There was no significant difference between the 4- and 8-day expression times, and the results were pooled. This time is adequate for the phenotypic expression of the plateau-phase as well as exponential SP3 cells. The colonies formed in the selective medium were counted after being incubated for 12 days, and those formed in the 10-cm Petri dishes after 10 days of incubation.

For acute exposures, monolayers were prepared in 10-cm Petri dishes, and a total of about 2×10^7 were irradiated for each dose point. Otherwise, the treatment was as described above. The protocol described is necessarily a compromise. Since the frequency of mutants is quite low, it has to be ascertained that a sufficient number of surviving cells is plated. On the other hand, the density should not be too high in order to avoid "metabolic cooperation" resulting in the underscoring of mutants. This could, in principle, be avoided by using more plates, but there are practical limits. In view of these considerations, the present scheme was standardized to study mutant induction under different experimental conditions. It cannot be excluded that the "true" frequency of mutants is higher than actually measured, however, this was not really the point of interest. Any possible error introduced is not expected to change the conclusions in a systematic way.

Mutant fraction was defined as the number of 6-TG-resistant colonies per 10^6 colony-forming cells. The background mutation frequency typically lay around 5×10^{-6} per surviving cell.

To check for genomic instability, cells were acutely irradiated with 3 Gy of X-rays and then grown in continuous culture and then the mutant frequencies were determined.

2.2. Radiation and dosimetry

All low-dose rate irradiations were given by a ^{60}Co-gamma-source. The culture flasks containing the stable monolayers were irradiated within an incubator (at 37 °C in 5% CO_2 and a relative humidity of 100%). Because of the different dose-rates and occasional infections, the total irradiation time was different for each dose-rate (5 mGy/h 29 days, 10 mGy/h 30 days, 20 mGy/h 15 days). The maximum accumulated dose was about 7 Gy with a dose-rate of 20 mGy/h. The number of culture flasks set up corresponded to half the number of days expected for the duration of the experiment for each dose-rate. Additional to these, some more flasks were kept as replacement in case of an infection. Every other day, the medium was changed in all culture flasks and one flask was held back for the determination of mutant fractions. The other flasks were left in the incubator.

Due to technical limitations, it was not possible to perform the acute irradiation with ^{60}Co-gamma-rays. In this case, 300 kV X-rays (Müller MOD 300 X-Ray-Machine) with a dose rate of 1.5 Gy/min were used. This small difference in radiation quality is not expected to cause any significant changes. The slightly higher mean LET could at most lead to *increased* mutant fractions which would not invalidate the conclusions drawn from the low dose-rate experiments.

Doses were measured by a thimble ionization chamber and checked by ferrous sulphate dosimetry at reference positions.

3. Results

3.1. Exponential cultures

SP3 cells grow very well in the continuous culture with a fairly constant plating efficiency around 60%. Mutation induction was also checked in unexposed cultures over several weeks, and no significant deviation from the original value was seen.

With 50 mGy/h continuous exposure mutation induction effectiveness is significantly reduced in exponentially growing cells. With 5 mGy/h, however, this effect is reversed, and no significant difference to the acute data can be found. This is in line with results in TK6 [8] and mouse L cells [9]. The large increase reported for V79 Chinese hamster cells [6] can, on the other hand, not be confirmed. Nevertheless, the present finding demonstrates once more that with regard to mutation induction, it cannot generally be assumed that lowering the dose-rate leads also to smaller effectiveness.

3.2. Plateau phase cultures

Plating efficiencies and background mutant fractions as a function of incubation time in unirradiated control cultures, maintained for 37 days, were checked and did not show any significant systematic changes. Survival after acute exposure did not significantly differ

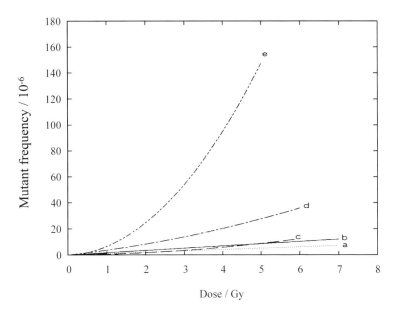

Fig. 2. Comparison of mutation induction experiments with different dose-rates in plateau-phase and exponentially growing cultures: (a) stable monolayers continuously irradiated, dose-rate 10 mGy/h; (b) as (a), but with a dose-rate of 20 mGy/h; (c) exponential cells in continuous culture, dose-rate 50 mGy/h; (d) exponential cells in continuous culture, dose-rate 5 mGy/h (this is identical within error limits to that found after acute exposure in exponential cells); (e) mutation induction after acute exposure in plateau-phase cells, dose-rate 1.5 Gy/min.

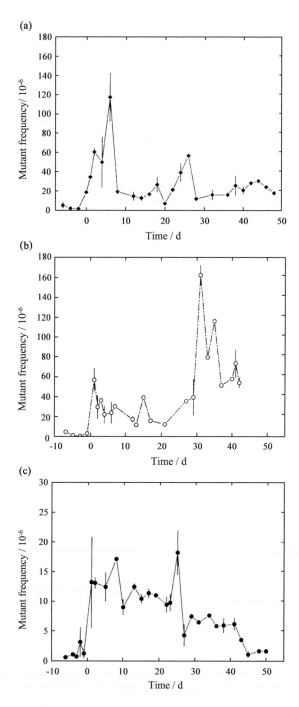

Fig. 3. Mutant frequencies as function of incubation time in continuous culture after an acute dose of 3 Gy at time zero to investigate genomic instability. Panels a–c display results of different experiments.

from that of the exponential cultures. This is not the case, however, for mutation induction: plateau-phase cultures display considerably higher mutant fractions than exponentially growing ones. One reason for this may be the higher proportion of G1-cells in plateau-phase cultures, which are more sensitive in terms of mutant induction by sparsely ionizing radiations [12].

Contrary to exponential cultures, mutation induction efficiency is always reduced when the dose-rate is lowered. Fig. 2 summarizes the results of mutation experiments with very low dose-rates in plateau-phase and exponentially growing cultures in a schematic way. The differences are obvious: While in non-growing cells lowering the dose-rate leads to drastically reduced mutation induction efficiencies, the situation is different in continuous culture populations. At 50 mGy/h, one finds less mutants compared to acute exposures, but not at 5 mGy/h, i.e., there is an inversion of the effect at very low dose-rates.

3.3. Genomic instability

There is a possibility that the inverse dose-rate effect in continuous cultures may be related to the induction of genomic instability. Furthermore, the occurrence of mutants after acute irradiation was followed over several weeks. The results are sown in Fig. 3(a–c): In all cases, the initial peak in mutant fraction reflecting delayed expression is followed by a drop to nearly background values. At later times, however, new mutants arise but in an irregular and not reproducible manner as seen from a comparison of parallel experiments, which were all conducted in the same way. It is noteworthy that the rise in mutant fractions is not represented by single points (which could indicate an experimental artifact), but suggest a real trend. The fact that the number of mutants decreases again suggests that they are at a growth disadvantage and diluted out on further continuous cultivation. It has to be pointed out that these "delayed" mutations arise after comparative later times and can thus not account for the inverse dose rate effect which is seen also with lower doses and with smaller incubation times.

4. Discussion

The results of our investigations demonstrate the existence of an "inverse dose rate effect" for mutation induction also in human cells. It is, however, restricted to exponentially growing cultures and is not found in plateau phase cells. The same conclusion was reached by Furuno-Fukushi [5] for mouse cells. A detailed discussion of possible underlying mechanisms is out of the scope of the present short report and will be given elsewhere. "Genomic instability" cannot be held responsible since it occurs too late and the total number of mutants created this way is too small. We have shown, however, that it occurs also in terms of mutation induction, thus confirming work by others [13,14] who used different techniques.

It has been suggested that radiation induces a repair system which is not "switched on" at very low dose-rates to explain the unusually high mutation rates in continuously exposed Chinese hamster cells [6]. This at the time very speculative ideas received some support by experiments which showed that very low doses reduce survival to a much larger degree than usually assumed [15]. This does, of course, not prove the existence of an inducible repair system, but is in line with our present and previous results [6,7].

The differences in the response to changes in dose-rate with regard to mutation induction by gamma-rays, which are found between plateau-phase [9,16] and actively dividing cultures [2–7,17], have to be attributed to the proliferative status. It is possible that under stationary conditions, there is more time for repair so that initial lesions do not lead to mutations.

Whatever the explanation may be, the results presented here appear to be also of practical importance. If the so-called "dose-rate effect" (which forms the basis of the "dose and dose-rate effectiveness factor DDREF [18]) depends on the growth fraction of a tissue (assuming bravely that an extrapolation to the in vivo situation is allowed), it should be different in different organs of the human body which would have implications for risk estimates for radiation workers.

References

[1] E.J. Hall, The dose-rate factor in radiation biology, Int. J. Radiat. Biol. 59 (1991) 595–610.
[2] N. Nakamura, S. Okada, Dose-rate effects of gamma-ray-induced mutations in cultured mammalian cells, Mutat. Res. 83 (1981) 127.
[3] A.M. Ueno, I. Furuno-Fukushi, H. Matsudaira, Induction of cell killing, micronuclei, and mutation to 6-thioguanine resistance after exposure to low-dose-rate gamma rays and tritiated water in cultured mammalian cells (L5178Y), Radiat. Res. 91 (1982) 447.
[4] I. Furuno-Fukushi, A.M. Ueno, H. Matsudaira, Mutation induction by very low dose rate gamma rays in cultured mouse leukemia cells L5178Y, Radiat. Res. 115 (1988) 273.
[5] I. Furuno-Fukushi, K. Aoki, H. Matsudaira, Mutation induction by different dose rates of gamma rays in near-diploid mouse cells in plateau- and log-phase cultures, Radiat. Res. 136 (1993) 97–102.
[6] N.E.A. Crompton, B. Barth, J. Kiefer, Inverse dose-rate effect for the induction of 6-thioguanine-resistant mutants in chinese hamster V79-S cells by 60-Co-gamma-rays, Radiat. Res. 124 (1990) 300–308.
[7] N.E.A. Crompton, F. Zoelzer, E. Schneider, J. Kiefer, Increased mutant induction by very low dose-rate gamma-irradiation, Naturwissenschaften 72 (8) (1985) 439.
[8] F. König, J. Kiefer, Lack of dose-rate effect for mutation induction by gamma-rays in human TK6 cells, Int. J. Radiat. Biol. 54 (1989) 891.
[9] I. Furuno-Fukushi, K. Aoki, H. Matsudaira, Mutation induction by different dose rates of gamma rays in near-diploid mouse cells in plateau- and log-phase cultures, Radiat. Res. 136 (1993) 97–102.
[10] E. Huberman, C.K. McKeown, C.A. Jones, D.R. Hoffman, S. Murao, Induction of mutations at the hypoxanthine-guanine-phosphoribosyl-transferase-locus in human epithelial teratoma cells, Mutat. Res. 130 (1984) 127–137.
[11] J. Zeuthen, J.O.R. Norgaard, P. Avner, M. Fellous, J. Wartiovaara, A. Vaheri, A. Rosen, B.C. Giovanella, Characterization of a human ovarian teratocarcinoma-derived cell line, Int. J. Cancer 25 (1980) 19–32.
[12] H.J. Burki, Ionizing radiation-induced 6-thioguanine-resistant clones in synchronous CHO cells, Radiat. Res. 81 (1980) 76–84.
[13] J.B. Little, H. Nagasawa, T. Pfenning, H. Vetrova, Radiation-induced genomic instability: delayed mutagenic and cytogenetic effects of X-rays and alpha particles, Radiat. Res. 148 (1997) 299–307.
[14] J.J.B. Boesen, S. Stuivenberg, C.H.M. Thyssens, H. Panneman, F. Darroudi, P.H.M. Lohman, J.W.I.M. Simons, Stress response induced by DNA damage leads to specific, delayed and untargeted mutations, Mol. Gen. Genet. 234 (2) (1992) 217–227.
[15] V. Marples, M.C. Joiner, The response of Chinese hamster V79 cells to low radiation doses: evidence of enhaced sensitivity of the whole cell population, Radiat. Res. 133 (1993) 41–50.
[16] J. Thacker, A. Stretch, Recovery from lethal and mutagenic damage during post-irradiation holding and low-dose-rate irradiations of cultured hamster cells, Radiat. Res. 96 (1983) 380.

[17] H.H. Evans, M. Nielsen, J. Mencl, M.-F. Horng, M. Ricanati, The effect of dose rate on X-radiation-induced mutant frequency and the nature of DNA lesions in, mouse lymphoma L5178Y cells, Radiat. Res. 122 (1990) 316–325.
[18] ICRP, Recommendation of the International Commission on Radiological Protection, ICRP Publication, vol. 60, Pergamon, Oxford, 1991.

DNA-PK activity plays a role on radioadaptation for radiation-induced apoptosis

Takeo Ohnishi *, Akihisa Takahashi, Ken Ohnishi

Department of Biology, Nara Medical University, 840 Shijo-cho, Kashihara, Nara 634-8521, Japan

Abstract

Concerning human health, studies of radioadaptive responses are quite important in radiation biology. Recently, we reported the effects of chronic pre-irradiation on the response of p53 after acute challenge irradiation in human cultured cells, and in murine whole body. When mice were previously exposed to chronic irradiation with a low-dose rate, p53-dependent apoptosis induced by acute irradiation was significantly suppressed. In contrast, the incident of radiation-induced apoptosis was not changed after chronic irradiation in *Scid* mice (DNA-PKcs-deficient mice), although we detected the apoptosis in the spleen after carrying out acute irradiation alone. These data suggested that DNA-PK activity might play a major role on the radioadaptive response by pre-irradiation with a low-dose rate. Here, we reviewed our recent results with radioadaptive response. © 2002 Elsevier Science B.V. All rights reserved.

Keywords: Low-dose rate; DNA-PK; Apoptosis; Whole body; p53

1. Background

Human beings are constantly exposed to low levels of natural background radiations. In addition, exposure to low-dose radiation from medical sources and the possibility of exposure to radiation from materials associated with nuclear power plants are becoming worldwide problems. Although epidemiological data regarding the cancer incidence in areas with high background radiation levels seem to indicate a beneficial effect of chronic low-dose rate radiation, the molecular mechanisms remain unclear. Prediction of future risk estimates should be made with acuity and an open mind. Therefore, the biological effects of low-dose rate radiation on cellular responses have been increasingly recognized as an important influence on human health.

---

* Corresponding author. Tel.: +81-744-22-3051x2264; fax: +81-744-25-3345.
E-mail address: tohnishi@naramed-u.ac.jp (T. Ohnishi).

0531-5131/02 © 2002 Elsevier Science B.V. All rights reserved.
PII: S0531-5131(01)00746-4

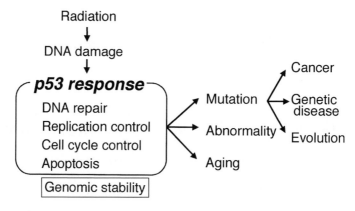

Fig. 1. Radiation-induced cellular response.

To date, it has been shown in many papers that a series of continuous chemical reactions is induced by irradiation. Both stabilization and functional activation of proteins are regulated by post-transcriptional modifications such as phosphorylation, acetylation and/or poly(ADP-ribosyl)ation. Subsequently, cells synthesized de novo proteins of genomic stability-related down-stream genes, influencing activities such as DNA repair, replication, cell cycle and cell death. It is thought that the cells thus escaped from cancer, abnormality and aging through genomic stability. In general, mutations are well understood to induce cancer events, genetic disease and evolution (Fig. 1). The p53 molecule plays important roles in maintaining genomic stability during the cell cycle checkpoint in G1 and G2/M transition, and as an effecter of DNA repair and apoptosis [1]. However, it remains unknown whether chronic irradiation at a low-dose rate is harmful to health, for example, inducing cancer via interference with p53.

2. In vitro study

We reported the effects of chronic pre-irradiation at a low-dose rate on the response of p53, after acute challenge irradiation in human cultured cancer cells [2,3]. We clearly detected a dose-dependent increase in the p53 cellular content (Fig. 2A). Immediately after chronic irradiation of 1.44 Gy for 24 h, challenge of acute irradiation at 0.1–3.0 Gy did not have any further efficacy in inducing accumulation of p53, even if accumulation was observed during chronic irradiation (Fig. 2B). Similar results are obtained for WAF1 induction of the downstream gene product of p53 [2,3]. Therefore, we aim to clear the effects of pre-irradiation with a low-dose rate on the cellular response, by analysis of apoptosis and the related proteins. Acute challenge irradiation (5.0 Gy) immediately after chronic pre-irradiation (1.5 Gy, 0.001 Gy/min) resulted in lower levels of p53, Bax, active-caspase-3 and p85 kDa PARP fragment than those observed after exposure to the acute challenge irradiation without pre-irradiation in human cultured cancer cells (Fig. 3A) [4]. Radiation-induced DNA ladders were also diminished by chronic pre-irradiation at 1.5 Gy for 25 h (Fig. 3B) [4]. There are two possible mechanisms for the suppression of radiation-induced

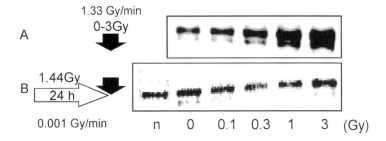

Fig. 2. p53 inducibility in human cultured cells. (A) p53 accumulation following various doses of the acute irradiation (1.33 Gy/min). (B) p53 accumulation following various doses of the acute irradiation immediately after chronic irradiation (0.001 Gy/min, 1.44 Gy).

apoptosis by chronic pre-irradiation. One is the lack of successful emergency signals to p53 function due to complete repair of DNA damage during chronic pre-irradiation. Another is the perturbation of signaling in response to acute radiation. The absence of successful emergency signals may be induced by the reduction in damage due to radical detoxification and/or repair systems. Not only DNA damage but also protein or membrane degeneration might be involved in this possibility. The perturbation after chronic pre-irradiation may affect post-translational modifications of p53. In unstressed cells, p53 is present in a latent state and is maintained at low levels through a feedback system by Mdm2. Radiation-activated signaling pathways transiently stabilize the p53 protein that causes the accumulation of p53 in the nucleus, then activate p53 as a transcription factor. The molecular mechanisms of activation and stabilization are still not understood, but accumulating evidence points to roles for multiple post-translational modifications of p53 through several chemical reactions. In addition, p53 molecules are phosphorylated by certain kinds of

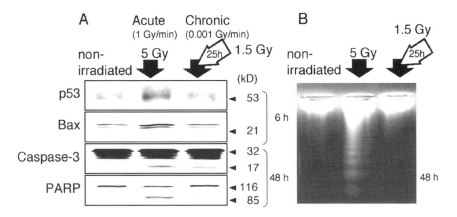

Fig. 3. Radiation-induced apoptosis-related proteins and apoptosis. (A) Western blot analysis of cellular contents of p53, Bax, Caspase-3 and PARP. The cells were challenged acute irradiation after chronic irradiation, then cultured for 6 h (p53 and Bax) or 48 h (Caspase-3 and PARP) before extraction of total cell proteins. (B) DNA fragmentation analysis by agarose gel electrophoresis. ⬇, challenging irradiation (1 Gy/min for 5 min); ⇨, chronic irradiation (0.001 Gy/min for 25 h).

protein kinases, including ATM, ATR, cdk and Chk1/Chk2. Thus, the responsiveness of radiation-induced signal transduction may depend on the conditions of these modulators.

3. In vivo study

Recently, we also reported that chronic pre-irradiation with a low-dose rate suppressed Bax-mediated apoptosis in the C57BL/6N mouse spleen [5]. These findings suggested that the radioadaptive response in the mouse spleen might be due to the suppression of p53-mediated apoptosis. Here, this study investigated the role of DNA-PK activity in the radioadaptive response. *Scid* mice display increased sensitivity to ionizing radiation compared with the parental CB-17 mice (*12*). The 460-kDa catalytic subunit of DNA-PK (DNA-PKcs) has previously been identified as a strong candidate for the *Scid* gene (*13*). It has become clear that the DNA-PK holoenzyme composed of DNA-PKcs and two subunits of the Ku auto antigen, 70 and 86 kDa (Ku70 and Ku80, respectively), is required for phosphorylation of target molecules (*15*). As an index of the radioadaptive response, we measured the frequency of radiation-induced Bax- and apoptosis-positive cells in the mouse spleen. Based on immunohistochemical analyses in the present study, we detected dose- and time course-dependent induction of Bax and apoptosis after acute irradiation, especially in splenic white pulps of both *Scid* and parental mice. Both Bax induction and apoptosis were significantly suppressed immediately after the low-dose rate (0.001 Gy/min for 25 h, 1.5 Gy) irradiation in the parental mice, but not in the *Scid* mice (Fig. 4).

These data suggested that DNA-PKcs (expressed in the parental mice, but not in *Scid* mice) might play a major role on the suppression of the acute radiation-induced apoptosis in

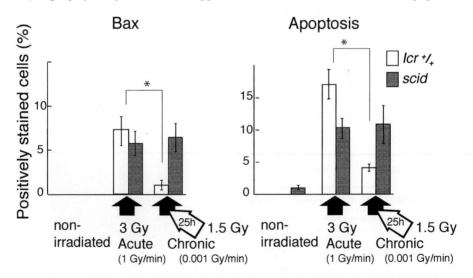

Fig. 4. The rate of cells positively stained for Bax (A) and apoptosis (B) 12 h after acute or chronic pre-irradiation in the splenic white pulp. Open column, normal parental mice; closed column, *Scid* mice. The error bars indicate standard deviations. *, a highly significant difference ($P < 0.01$) by Student's *t*-test. ▲, challenging irradiation (1 Gy/min for 3 min); ◁, chronic irradiation (0.001 Gy/min for 25 h).

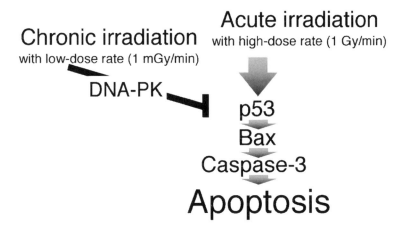

Fig. 5. The suppression of the radiation-induced apoptosis by pre-chronic irradiation.

the spleen by low-dose rate irradiation. It was also suggested that post-translational modifications of target molecule(s) by DNA-PK indirectly depressed the Bax-mediated apoptosis (Fig. 5). Shen et al. [6] reported that over-expression of DNA-PK is a novel cellular adaptation, mechanistically contributing to the resistance of cancer cells to the DNA-damaging agents. However, some conflicting reports have shown that DNA-PK does not play a major role in adaptive response in mammalian cultured cells [7]. Furthermore, Sasaki [11] has also postulated that the radiation-induced chromosomal aberrations were not depressed by pre-irradiation at a low dose in cultured *Scid* cells, and the normal parental cells (personal communication). Although these contradictory studies used culture cell systems, we examined the radioadaptive response in a whole-body system. Therefore, it was suggested that immune responses might participate in the radioadaptive response to chronic irradiation. Recently, it was reported that conditioning of the radioadaptive response in survival (2 weeks after 0.5 Gy or 8 weeks after 0.075 Gy) induced short-term thymocyte apoptosis [8], and long-term augmentation of T-cell-dependent immune responses [9]. In fact, since it is known that the *Scid* mouse has immature T-cells [10], the radioadaptive response through such T-cells may not yet be defined. In addition, we assume that the adaptive response to low doses requires a certain minimal dose before it becomes active, and that it occurs only within a relatively small-dose window, is dose-rate dependent, and depends on the genetic background of the organisms [11,12].

4. Conclusions

Pre-irradiation with low-dose rate, depressed radiation-induced Bax and apoptosis in the spleens of the parental mice, but not *Scid* mice were found. These findings suggest that the experience of pre-exposure with a low-dose rate alters the radiation sensitivity through *p53*-dependent signal transduction. Since chronic pre-irradiation interfered with the p53 response, it is possible that the frequencies of mutation and chromosomal aberration may be changed under different conditions of irradiation. If *p53*-dependent radiation-induced apoptosis can prevent cancer events, chronic pre-irradiation with a low-dose rate may increase the frequency of cancer events. Therefore, it is possible that there are different maximal permissible dose equivalents for radiation workers and the general population. We expect that the present findings will provide useful information for the clarification of radioadaptive responses.

Acknowledgements

This work was partly supported by grants from the Ministry of Education, Culture, Sports, Science and Technology of Japan and the Central Research Institute of Electric Power Industry in Japan. This study was also funded in part by "Ground Research for Space Utilization" promoted by NASDA and the Japan Space Forum.

References

[1] D. Lane, p53, guardian of the genome, Nature 358 (1992) 15–16.
[2] T. Ohnishi, X. Wang, A. Takahashi, K. Ohnishi, Y. Ejima, Low-dose-rate radiation attenuates the response of the tumor suppressor TP53, Radiat. Res. 151 (1999) 368–372.
[3] A. Takahashi, K. Ohnishi, T. Ohnishi, Radiation response of p53 and WAF1 accumulation after chronic or acute irradiation, in: T. Yamada, C. Mothersill, B.D. Michael, C.S. Potten (Eds.), International Meeting on Biological Effects of Low Dose Radiation, Cork, Ireland, 25–26 July 1999. Elsevier, Amsterdam, 2000, pp. 59–66.
[4] A. Takahashi, Different inducibility of radiation- or heat-induced *p53*-dependent apoptosis after acute or chronic irradiation in human cultured squamous cell carcinoma cells, Int. J. Radiat. Biol. 77 (2001) 215–224.
[5] A. Takahashi, K. Ohnishi, I. Asakawa, N. Kondo, H. Nakagawa, M. Yonezawa, A. Tachibana, H. Matsumoto, T. Ohnishi, Radiation response of apoptosis in C57BL/6N mouse spleen after whole-body irradiation, Int. J. Radiat. Biol., in press.
[6] H. Shen, M. Schultz, G.D. Kruh, K.D. Tew, Increased expression of DNA-dependent protein kinase confers resistance to adriamycin, Biochem. Biophys. Acta 1381 (1998) 131–138.
[7] E. Odegaard, C.R. Yang, D.A. Boothman, DNA-dependent protein kinase does not play a role in adaptive survival responses to ionizing radiation, Environ. Health Perspect. 106 (1998) 301–305.
[8] M. Yonezawa, J. Misonoh, Y. Hosokawa, Two types of X-ray-induced radio-resistance in mice—Presence of 4 dose ranges with distinct biological effects, Mutat. Res. 358 (1996) 237–243.
[9] J. Matsubara, V. Turcanu, P. Poindron, Y. Ina, Immune effects of low-dose radiation: short-term induction of thymocyte apoptosis and long-term augmentation of T-cell-dependent immune responses, Radiat. Res. 153 (2000) 332–338.
[10] J. Reimann, A. Rudolphi, M.H. Claesson, Selective reconstitution of T lymphocyte subsets in *Scid* mice, Immunol. Rev. 124 (1991) 75–95.
[11] M.S. Sasaki, On the reaction kinetics of the radioadaptive response in cultured mouse cells, Int. J. Radiat. Biol. 68 (1995) 281–291.
[12] S. Wolff, The adaptive response in radiobiology: evolving insights and implications, Environ. Health Perspect. 106 (1998) 277–283.

The responses of the haemopoietic system to ionizing radiation

E.G. Wright[*]

Department of Molecular and Cellular Pathology, University of Dundee,
Ninewells Hospital and Medical School, Dundee DD1 9SY, UK

Abstract

All blood cells are ultimately derived from a common stem cell population and it is at this cellular level that all haemopoietic function is ultimately regulated. It is also at this level that dysfunctions involved in the pathogenesis of leukaemias frequently arise. The induction of leukaemia by ionising radiation is conventionally attributed to the induction of irreversible mutational changes in stem cells during the processing of the DNA damage or during DNA replication immediately after irradiation. However, there is now considerable evidence that cells that are not themselves irradiated but are the progeny of stem cells exposed to ionizing radiation many cell divisions previously may express delayed gene mutations and chromosomal aberrations. These delayed effects challenge the conventional models for radiation-induced genetic lesions and are collectively regarded as manifestations of radiation-induced genomic instability. The expression of instability is strongly influenced by genetic factors and although the mechanism is unclear, excessive production of reactive oxygen species has been implicated. Recently, instability in the descendants of non-irradiated stem cells has been demonstrated as a consequence of their interactions with irradiated cells or the progeny of irradiated cells. The findings highlight the potential importance of cellular interactions in radiation leukaemogenesis. © 2002 Elsevier Science B.V. All rights reserved.

Keywords: Ionizing radiation; Haemopoietic stem cells; Genomic instability; Leukaemogenesis; Macrophages

1. Stem cells in the haemopoietic system

The haemopoietic system may be regarded as a three-tiered hierarchy in which the maintenance of constant numbers of mature functional cells in the blood is achieved by the proliferation and differentiation of lineage-committed precursor cells and these precursors are, in turn, derived from a small self-maintaining population of multipotential stem cells

---

[*] Tel.: +44-1383-632169; fax: +44-1382-633952.
E-mail address: e.g.wright@dundee.ac.uk (E.G. Wright).

[1,2]. The orderly production of blood cells is regulated by feedback mechanisms that involve complex interactions of the cells and cellular products of both the haemopoietic system and their stromal microenvironment. At the level of the stem cell, locally acting microenvironmental signals that regulate stem cell proliferation result in the majority of cells being quiescent, residing in the G_0 phase of the cell cycle. The reason for this extensive dormancy is unclear but may relate to the 'genetic housekeeping' required for the maintenance of the genomic integrity of the functionally immortal stem cell pool. Following cytotoxic insult to the haemopoietic system, normally quiescent stem cells very rapidly enter cell cycle and these proliferating cells serve to repopulate the depleted haemopoietic system after which they return to a proliferatively quiescent state. Both positive and negative proliferative control of stem cells are suggested by these findings; a great deal is now known about inhibitors and stimulators of haemopoietic stem cell proliferation on the basis of the restoration of steady-state haemopoiesis after a variety of perturbations [2]. While these regulatory factors are expressed in vivo, their precise role in physiological steady-state regulation is less well characterized.

The haemopoietic stem cell compartment itself is hierarchically organized as an age-structured continuum in which the most primitive members have the greatest capacity for self-replication and long-term repopulating ability. These developmentally early long-term repopulating stem cells are relatively resistant to proliferation and differentiation stimuli and when they are recruited into proliferation, generate stem cells that in subsequent divisions generate short-term repopulating stem cells with decreasing self-renewal capacity and increasing sensitivity to the factors involved in commitment to specific lineage differentiation. This developmentally structured stem cell compartment has important functional significance in vivo as transplantation experiments have demonstrated two phases of engraftment; an initial but transient engraftment, essential for survival following the conditioning regime, followed by a delayed but long-term reconstitution of the haemopoietic system. These two phases can be attributed respectively to the later and earlier members of the continuum [3,4].

The haemopoietic stem cell compartment occupies a pivotal position within the haemopoietic hierarchy and it is at this cellular level that all haemopoietic functions are ultimately regulated. It is also at this level that dysfunctions involved in the pathogenesis of leukaemias and myeloproliferative disorders frequently arise and stem cells are important targets for the well-established association between radiation exposure and leukaemic transformation [5].

2. The concept of radiation-induced genomic instability

The biological consequences of exposure to ionizing radiation include cell death, gene mutation and chromosome aberrations. These effects are all conventionally attributed to irreversible changes fixed during the processing of the DNA damage by enzymatic repair processes or during DNA replication. Accordingly, it has been generally accepted that most of these changes take place during the cell cycles immediately following exposure and as malignant transformation is generally regarded as being initiated by a mutational change, the initiating lesion for leukaemogenesis has similarly been attributed to direct

DNA damage at the time of radiation exposure. However, there is now considerable evidence that the cellular consequences of radiation exposure may be manifested not in the irradiated cells but in their descendant progeny. For example, there may be an enhanced death rate in the progeny of irradiated cells that persists for many generations. This delayed death is genetically distinct from cellular senescence since it is demonstrable in established cell lines that normally have unlimited replicative potential. In addition to delayed death, there is a substantial body of evidence that the progeny of cells exposed to ionizing radiation exhibit delayed mutational responses including specific gene mutations and chromosome aberrations. These delayed responses, can be effectively demonstrated as non-clonal mutations/aberrations in the clonal descendants of irradiated haemopoietic stem cells and are generally regarded as the consequences of a destabilisation of the genome collectively termed radiation-induced genomic instability. A key feature of the evolution of malignancy is genetic instability and this raises important questions about the contribution of inducible instability to radiogenic malignancies.

3. Radiation-induced chromosomal instability in haemopoietic cells in vitro

Initial experiments demonstrating inducible instability in haemopoietic cells were designed to study the effects of low doses of α-particle irradiation (~ one track per cell) on haemopoietic stem cells [6]. An in vitro clonogenic assay was used to obtain clonal cell populations derived from stem cells present in α-irradiated suspensions of mouse bone marrow and it was found that up to half the colonies had karyotypic abnormalities but in an individual colony, not all cells exhibited abnormalities. Typically, up to 20% of metaphases had single or multiple, non-identical aberrations; that is, the aberrations were non-clonal and a high frequency of chromatid-type aberrations was consistent with ongoing chromosome breakage during colony development. Using an assay for comparable human bone marrow clonogenic cells, similar findings were demonstrated with some, but not all, samples of human bone marrow from haematologically normal donors [7]. These findings suggested that a transmissible chromosomal instability could be induced in a stem cell, resulting in a diversity of aberrations in its clonal progeny many cell divisions later.

4. Genetic factors influencing the expression of chromosomal instability

In the studies of human haemopoietic cells [8], chromosomal instability was induced in some, but not all, bone marrow samples investigated (Table 1) and it was suggested that this might be attributed to genetic differences between individuals. This interpretation was supported by studies of the induction of chromosomal instability by α-irradiation using different inbred mouse strains [8]. CBA/H and DBA/2 strains may be regarded as susceptible and the C57BL/6 strain as relatively resistant to the induction of chromosomal instability in bone marrow cells (Table 1). Low levels of expression in stem cell-derived clones from F_1 hybrid bone marrow (similar to the C57BL/6 cells) suggest that susceptibility is genetically recessive in this system. Similar findings have been reported for mammary epithelial cells [9]. These laboratory studies clearly demonstrate the importance

Table 1
The expression of the chromosomal instability phenotype in vitro in colonies derived from haemopoietic stem cells in human or mouse bone marrow cell suspensions exposed to α-particle irradiation

Source of bone marrow cells	Percentage metaphases exhibiting chromosomal instability
Human bone marrow[a]	
1	<1
2	12
3	<1
4	15
Mouse bone marrow[b]	
CBA/H	12
DBA/2	11
C57BL/6	4
(C57BL/6 × CBA/H] F1	4
(C57BL/6 × DBA/2] F1	3

[a] Data summarized from Kadhim et al. [7,11].
[b] Data summarized from Watson et al. [8].

of "genetic predisposition" to inducible instability and raise the important issue of the need to consider the possibility of predisposition genes in the human population.

5. Other delayed effects of ionizing radiation in haemopoietic cells

In addition to chromosomal instability, a 5- to 10-fold increase in arising non-clonal mutations at hypoxanthine-guanine-phosphoribosyl-transferase locus was demonstrated in the clonal descendants of murine haemopoietic stem cells in vitro after α-, neutron and X-irradiation [10]. Furthermore, in the clonal descendants of irradiated haemopoietic stem cells, there was a significant increase in the frequency of cells dying by apoptosis [11]. It is possible that the apoptotic cells are those in which instability-derived lesions in DNA significantly exceed the repair capacity of the cell and result in an apoptotic response. Alternatively, delayed apoptosis may result from signal antonymy, a cellular state where a cell simultaneously engages in incompatible pathways of proliferation and cell cycle arrest and in response to these conflicting signals, initiates an apoptotic response [12]. Whether this is the basis of delayed reproductive death/lethal mutations is a matter for speculation. However, it is of interest that there may be a genotype-dependent inverse correlation between the incidence of chromosomal aberrations and death by apoptosis in haemopoietic cells. The genetic background that produces a more effective apoptotic response may well be more resistant to the development of malignancy due to a more effective elimination of unstable and potentially malignant cells.

6. Radiation-induced chromosomal instability in vivo

Using a bone marrow transplantation protocol, it was shown that the instability initiated by in vitro α-irradiation persists in cells of donor origin for up to a year after transplantation

Table 2
The expression of chromosomal instability persists in vivo in mice transplanted with long-term repopulating stem cells exposed in vitro to 3 Gy X-rays, 0.5 Gy α particles or 0.5 Gy Californium neutrons

Months post-transplantation	Percentage metaphases expressing instability			
	Control	X-rays	α Particles	Neutrons
3–4.5	<1	7	7	5
6–7.5	<1	2	17	5
9–12	1	3	6	6

Data summarized from Watson et al. [13,18].

[8]. As shown in Table 2, similar instability has also been demonstrated for another high-LET irradiation, that is, neutron irradiation and also for low-LET X-irradiation [13]. More importantly, total body irradiation experiments (Table 3) demonstrate that the phenotype can be initiated and perpetuated in vivo after both high- and low-LET irradiations [13]. The induced instability is ongoing for at least 24 months post-irradiation, consistent with it persisting for the rest of the animal's life. Overall, there is good agreement between the frequency of cells expressing instability many months after both whole-body irradiation and after transplantation of irradiated bone marrow. This correlation supports the concept that the same populations of long-term repopulating stem cells are being studied in the two protocols.

An important feature of α-emitting radionuclides is that, no matter how low the total dose to a tissue, a substantial dose of radiation (~ 0.5 Gy) is delivered to an individual cell if it happens to be traversed by a single α particle [14] and it is conventionally assumed that cells that are not traversed are unaffected by the radiation. Because our laboratory conditions of dose, particle fluence and linear energy transfer were precisely defined [14], we were able to calculate the mean number of α particles per target cell and obtain data corresponding to a mean of approximately one α particle per clonogenic stem cell. Inevitably at these low doses some cells, by chance, would not have been irradiated and the data were consistent with instability being expressed in the progeny of more clonogenic cells than were the survivors of an α-particle traversal. Subsequent experimental investigations of this apparent discrepancy confirmed that not only was there an absence of a conventional dose response but it was indeed the case that instability was expressed in the descendants of unirradiated stem cells [15]. These studies took advantage of the Poisson distribution of α particles to design experiments in which the absence or presence of a shielding grid between the source of α particles and the target cells produced

Table 3
The expression of chromosome instability in bone marrow derived from long-term repopulating stem cells after 3 Gy X-rays or 0.5 Gy Californium neutrons whole-body irradiation

Months post-irradiation	Percentage metaphases expressing instability		
	Control	X-rays	Neutrons
3–6	<1	3	4
9–12	<1	4	5
18–24	<1	4	4

Data summarized from Watson et al. [13].

a significant difference in the number of viable clonogenic cells. Survival data demonstrated that the area of shielding resulted in the expected reduction in the number of clonogenic cells traversed and killed by α particles but there was not the expected reduction in the number of descendant clones exhibiting chromosomal instability. Thus, instability was demonstrated in the descendants of non-irradiated stem cells and must be attributed to an indirect mechanism. There are now a number of reports of effects in cells that are not themselves irradiated but in the vicinity of irradiated cells, a consequence of irradiation generally known as a radiation-induced bystander effect [16].

To investigate the potential for an indirect mechanism in vivo, we transplanted mixtures of non-irradiated cells with cells exposed to neutrons (a densely ionizing radiation like α particles) to model the mixture of irradiated and non-irradiated cells in the α-irradiation experiments [17]. A sex mismatch transplantation protocol using CBA/H mice and congenic CBA/H mice that have a stable reciprocal chromosomal translocation provided a three-way marker system that allowed us to distinguish not only host-derived cells from donor-derived cells but also cells derived from the irradiated or non-irradiated donor stem cells. After transplantation of a mixture of irradiated and non-irradiated bone marrow, chromosomal instability was demonstrated in approximately 4% of the 40XY cells; that is, in cells derived from transplanted, irradiated stem cells (Table 4). However, at all times, post-transplantation, chromosomal instability was also demonstrated in cells carrying the translocation marker; that is, in cells derived from the non-irradiated, transplanted 40XYT6T6 stem cells. The overall frequency of 40XYT6T6 cells expressing instability (~ 2%), although lower than that in the 40XY cells (~ 4%), was significantly greater than controls (0.4%) transplanted with non-irradiated 40XYT6T6 marrow. Using this congenic system, it is evident that chromosomal instability in 40XYT6T6 cells cannot be explained by a direct transmission of instability from an irradiated stem cell to its descendants. The findings support an indirect mechanism in which the descendants of 40XY-irradiated stem cells are able to induce instability in the 40XYT6T6 descendants of non-irradiated stem cells over a period of many months after irradiation, a situation rather more complex than the radiation-induced bystander effects referred to above where effects are observed in non-

Table 4
Expression of chromosomal instability in bone marrow cells obtained from mice transplanted with a mixture of 40% 0.5 Gy neutron-irradiated (40XY) and 60% non-irradiated (40XYT6T6) CBA/H, long-term repopulating stem cells

Months post-transplantation	Percentage metaphases expressing instability	
	Cells derived from irradiated 40XY stem cells	Cells derived from non-irradiated 40XYT6T6 stem cells
3	0	2
4.5	6	2
7.5	0	2
9	4	2
12	4	2
Overall	4	2

In controls, the expression of instability was consistently <1% of cells.
Data summarized from Watson et al. [13,18].

irradiated neighbouring cells at the time of, or shortly after, the irradiation of the exposed cells.

Overall, therefore, the data are consistent with two distinct mechanisms for the induction of chromosomal instability, one being a direct transmission of instability from an irradiated stem cell to its progeny and the other a complex indirect interaction mechanism. The demonstration of instability in cells derived from irradiated donor stem cells is, of course, not necessarily inconsistent with an indirect mechanism downstream of the irradiated stem cells where a cell derived from an irradiated stem cell might induce instability in a bystander cell derived from a different irradiated stem cell.

7. Differences between the in vitro and in vivo expression of inducible chromosomal instability

When instability in haemopoietic cells has been demonstrated using the short-term in vitro [6,8,17] and in vivo [13] assays, the frequencies of cells expressing chromosomal instability and the frequencies of cells with more than one unstable aberration are greater than is found 3 or more months after transplantation [8,17] or whole-body irradiation [13]. The more effective physiological recognition and removal of abnormal cells in steady-state haemopoiesis at these later times may explain these important differences. It is also important to note that summarising the pooled data masks the important observation of significant inter-individual variation in the number of cells expressing instability [18]. Examples of this variation are shown in Table 5. Many cells carrying unstable aberrations will be blood cell precursors destined to become post-mitotic and physiologically programmed for death. Chromosomal instability in such a cell would be of little consequence and there is a

Table 5
Inter-individual variation in the expression of stable and unstable aberrations in the progeny of irradiated haemopoietic stem cells

Replicate animals	Percentage metaphases with aberrations	Percentage metaphases with unstable aberrations	Percentage metaphases with stable aberrations
17.5 months post-whole-body X-irradiation			
1	21	3	18
2	6	3	3
3	23	8	15
24 months post-whole-body X-irradiation			
1	15	4	11
2	96	20	91
3	51	1	50
24 months post-transplantation of X-irradiated bone marrow			
1	19	2	17
2	1	0	2
3	30	4	26

Data summarized from Watson et al. [18].

high proportion of such mature and maturing cells in haemopoietic tissues compared with the long-term repopulating stem cells that are approximately 0.001% of the total haemopoietic cells [2].

The data from steady-state in vivo studies reflect the complex structure and regulation of the haemopoietic system, the physiological recognition and removal of aberrant cells and the relatively small number of stem cells maintaining haemopoiesis at any one time [13,17]. It can be argued that, even in situations where there is a high level of induced instability, these factors all contribute to reducing the probability of pathological consequences.

8. Potential mechanisms underlying radiation-induced genomic instability in haemopoietic cells

At present, the mechanisms underlying the phenomenon of radiation-induced genomic instability are not understood. In all studies, the various delayed effects of radiation have been demonstrated at frequencies considerably greater than conventional mutation frequencies and this finding argues against instability being due to the mutation of a 'genome stability gene'; rather it tends to favour the possibility of epigenetic mechanisms. In cultures where induced genomic instability in haemopoietic cells has been demonstrated, there are increases in intracellular oxidants and oxidative DNA base damage [19]. As it is known that reactive oxygen species can be clastogenic and produce high frequencies of chromatid-type aberrations [20], the findings are consistent with free radicals contributing, at least in part, to the underlying mechanism(s). This suggestion is supported by the observations that bone marrow cells obtained from a mouse strain sensitive to the expression of radiation-induced chromosomal instability exhibited high rates of superoxide production following biochemical stimulation than cells from a resistant strain and cytogenetic analysis confirmed the clastogenic activity of reactive oxygen species in such cultures [8].

9. Implications of inducible instability for radiation leukaemogenesis

The ability to maintain genome integrity in the face of DNA damage is critical for healthy survival and the means by which organisms achieve this are complex. There are human diseases in which homeostatic processes have broken down, resulting in complex and often multi-system effects including malignancy, immunodeficiency, neurological disorders and growth and development abnormalities. Collectively, they are identified as chromosome instability syndromes [21] and include disorders such as Fanconi anaemia, Bloom's syndrome and ataxia telangiectasia, all of which are predisposing conditions for haemopoietic malignancies [22]. This predisposition prompts the hypothesis that chromosomal instability whether genetically determined or induced by ionizing radiation produces lesions in the haemopoietic stem cells of certain individuals that may contribute to the subsequent development of leukaemia. Radiogenic leukaemias may be studied using mouse models as the haemopoietic systems of mouse and man are remarkably similar with respect to their organization and regulation [2]. Although there is evidence for genetic instability in radiation-induced murine myeloid leukaemias, it is not yet possible to distinguish between the instability associated with the leukaemic process and instability that might reflect the delayed effects of exposure to ionizing radiation [23].

As discussed in Sections 3 and 6, mutational changes in haemopoietic cells arising as a consequence of induced instability may be a consequence of a directly transmitted instability or a more complex, cell interaction-mediated instability. It is possible that the enhanced oxidative processes in the myelo-monocytic progeny of irradiated stem cells associated with instability [19] may be related to the changes in the irradiated haemopoietic microenvironment that result in production of mutagenic reactive oxygen species, altered expression of adhesion molecules and growth factors and the ability to alter the overall growth and phenotypic characteristics of co-cultured non-irradiated stem cells [24]. Thus, after irradiation, macrophages or other accessory cells that contribute to the microenvironmental regulation of stem cells may have many of the characteristics of the activated macrophages found in inflammatory conditions. Activated macrophages are known to produce clastogenic factors via the intermediacy of superoxide [25] and are able to produce gene mutations [26], DNA base modifications [27], DNA strand breaks [28,29] and cytogenetic damage [30] in neighbouring cells. Furthermore, activated phagocytic cells have been demonstrated to transform co-cultured non-haemopoietic cells [31]. Thus, it is possible that activated accessory cells, generated as a consequence of induced instability, may contribute to genetic changes in other haemopoietic cells, including stem cells. Potentially, such cells may produce genetic damage in non-haemopoietic cells. Direct and indirect mechanisms would not be mutually exclusive and the suggestion that activated accessory cells may produce genetic lesions in neighbouring cells is similar to the mechanisms proposed to explain the relationship between inflammation and carcinogenesis [32]. The reports of patients treated for haemopoietic disorders by bone marrow transplantation following preparative whole body irradiation relapsing with disease in the donor-derived cells [33,34] and the leukaemic transformation of non-irradiated stem cells transplanted into syngeneic irradiated mice [35] are also consistent with indirect mechanisms and highlight the potential importance of cellular interactions in radiation leukaemogenesis.

Acknowledgements

The Medical Research Council, The United Kingdom Coordinating Committee on Cancer Research, The Leukaemia Research Fund, The Kay Kendall Leukaemia Fund and the Department of Health have supported the author's research. The author acknowledges the members of his laboratory, past and present, who have made important contributions to the emerging story of radiation-induced genomic instability.

References

[1] D. Metcalf, Molecular Control of Blood Cells, Harvard Univ. Press, Cambridge, MA, 1988.
[2] G.J. Graham, E.G. Wright, Haemopoietic stem cells: their heterogeneity and regulation, Int. J. Exp. Pathol. 78 (4) (1997) 197–218.
[3] R.J. Jones, P. Celano, S.J. Sharkis, L.L. Sensenbrenner, Two phases of engraftment established by serial bone marrow transplantation in mice, Blood 73 (2) (1989) 397–401.

[4] R.J. Jones, J.E. Wagner, P. Celano, M.S. Zicha, S.J. Sharkis, Separation of pluripotent haematopoietic stem cells from spleen colony-forming cells, Nature 347 (6289) (1990) 188–189.
[5] UNSCEAR, Sources and Effects of Ionizing Radiation, United Nations, New York, 1994.
[6] M.A. Kadhim, D.A. Macdonald, D.T. Goodhead, S.A. Lorimore, S.J. Marsden, E.G. Wright, Transmission of chromosomal instability after plutonium alpha-particle irradiation [see comments], Nature 355 (6362) (1992) 738–740.
[7] M.A. Kadhim, S.A. Lorimore, M.D. Hepburn, D.T. Goodhead, V.J. Buckle, E.G. Wright, Alpha-particle-induced chromosomal instability in human bone marrow cells, Lancet 344 (8928) (1994) 987–988.
[8] G.E. Watson, S.A. Lorimore, S.M. Clutton, M.A. Kadhim, E.G. Wright, Genetic factors influencing alpha-particle-induced chromosomal instability, Int. J. Radiat. Biol. 71 (5) (1997) 497–503.
[9] B. Ponnaiya, M.N. Cornforth, R.L. Ullrich, Radiation-induced chromosomal instability in BALB/c and C57BL/6 mice: the difference is as clear as black and white, Radiat. Res. 147 (2) (1997) 121–125.
[10] K. Harper, S.A. Lorimore, E.G. Wright, Delayed appearance of radiation-induced mutations at the Hprt locus in murine hemopoietic cells, Exp. Hematol. 25 (3) (1997) 263–269.
[11] M.A. Kadhim, S.A. Lorimore, K.M. Townsend, D.T. Goodhead, V.J. Buckle, E.G. Wright, Radiation-induced genomic instability: delayed cytogenetic aberrations and apoptosis in primary human bone marrow cells, Int. J. Radiat. Biol. 67 (3) (1995) 287–293.
[12] U. Hibner, A. Coutinho, Signal antonymy: a mechanism for apoptosis induction, Cell Death Differ. 1 (1994) 33–37.
[13] G.E. Watson, D.A. Pocock, D. Papworth, S.A. Lorimore, E.G. Wright, In vivo chromosomal instability and transmissible aberrations in the progeny of haemopoietic stem cells induced by high- and low-LET radiations, Int. J. Radiat. Biol. 77 (4) (2001) 409–417.
[14] S.A. Lorimore, D.T. Goodhead, E.G. Wright, Inactivation of haemopoietic stem cells by slow alpha-particles, Int. J. Radiat. Biol. 63 (5) (1993) 655–660.
[15] S.A. Lorimore, M.A. Kadhim, D.A. Pocock, D. Papworth, D.L. Stevens, D.T. Goodhead, et al., Chromosomal instability in the descendants of unirradiated surviving cells after alpha-particle irradiation, Proc. Natl. Acad. Sci. U. S. A. 95 (10) (1998) 5730–5733.
[16] C. Mothersill, C. Seymour, Radiation-induced bystander effects: past history and future directions, Radiat. Res. 155 (6) (2001) 759–767.
[17] G.E. Watson, S.A. Lorimore, D.A. Macdonald, E.G. Wright, Chromosomal instability in unirradiated cells induced in vivo by a bystander effect of ionizing radiation [In Process Citation], Cancer Res. 60 (20) (2000) 5608–5611.
[18] G.E. Watson, S.A. Lorimore, E.G. Wright, Long-term in vivo transmission of alpha-particle-induced chromosomal instability in murine haemopoietic cells, Int. J. Radiat. Biol. 69 (2) (1996) 175–182.
[19] S.M. Clutton, K.M. Townsend, C. Walker, J.D. Ansell, E.G. Wright, Radiation-induced genomic instability and persisting oxidative stress in primary bone marrow cultures, Carcinogenesis 17 (8) (1996) 1633–1639.
[20] T. Duell, E. Lengfelder, R. Fink, R. Giesen, M. Bauchinger, Effect of activated oxygen species in human lymphocytes, Mutat. Res. 336 (1) (1995) 29–38.
[21] M.S. Meyn, Chromosome instability syndromes: lessons for carcinogenesis, Curr. Top. Microbiol. Immunol. 221 (1997) 71–148.
[22] E.G. Wright, Inherited and inducible chromosomal instability: a fragile bridge between genome integrity mechanisms and tumourigenesis, J. Pathol. 187 (1) (1999) 19–27.
[23] M. Plumb, H. Cleary, E. Wright, Genetic instability in radiation-induced leukaemias: mouse models, Int. J. Radiat. Biol. 74 (6) (1998) 711–720.
[24] J.S. Greenberger, M.W. Epperly, A. Zeevi, K.W. Brunson, K.L. Goltry, K.L. Pogue-Geile, et al., Stromal cell involvement in leukemogenesis and carcinogenesis, In Vivo 10 (1) (1996) 1–17.
[25] I. Emerit, Reactive oxygen species, chromosome mutation, and cancer: possible role of clastogenic factors in carcinogenesis, Free Radical Biol. Med. 16 (1) (1994) 99–109.
[26] S.A. Weitzman, T.P. Stossel, Mutation caused by human phagocytes, Science 212 (4494) (1981) 546–547.
[27] M. Dizdaroglu, R. Olinski, J.H. Doroshow, S.A. Akman, Modification of DNA bases in chromatin of intact target human cells by activated human polymorphonuclear leukocytes, Cancer Res. 53 (6) (1993) 1269–1272.
[28] H.C. Birnboim, DNA strand breakage in human leukocytes exposed to a tumor promoter, phorbol myristate acetate, Science 215 (4537) (1982) 1247–1249.

[29] E. Shacter, E.J. Beecham, J.M. Covey, K.W. Kohn, M. Potter, Activated neutrophils induce prolonged DNA damage in neighboring cells, Carcinogenesis 9 (12) (1988) 2297–2304.
[30] A.B. Weitberg, S.A. Weitzman, M. Destrempes, S.A. Latt, T.P. Stossel, Stimulated human phagocytes produce cytogenetic changes in cultured mammalian cells, N. Engl. J. Med. 308 (1) (1983) 26–30.
[31] S.A. Weitzman, A.B. Weitberg, E.P. Clark, T.P. Stossel, Phagocytes as carcinogens: malignant transformation produced by human neutrophils, Science 227 (4691) (1985) 1231–1233.
[32] S.A. Weitzman, L.I. Gordon, Inflammation and cancer: role of phagocyte-generated oxidants in carcinogenesis, Blood 76 (4) (1990) 655–663.
[33] M. Lawler, Leukaemogenesis, gene interplay, and the role of the haemopoietic environment, Radiat. Oncol. Invest. 5 (3) (1997) 154–157.
[34] S.A. Giralt, R.E. Champlin, Leukemia relapse after allogeneic bone marrow transplantation: a review, Blood 84 (11) (1994) 3603–3612.
[35] U. Duhrsen, D. Metcalf, A model system for leukemic transformation of immortalized hemopoietic cells in irradiated recipient mice, Leukemia 2 (6) (1988) 329–333.

Does radiation enhance promotion of already-initiated cells in protracted high-LET carcinogenesis via a bystander effect?

S.B. Curtis*, E.G. Luebeck, W.D. Hazelton, S.H. Moolgavkar

Fred Hutchinson Cancer Research Center, 1100 Fairview Avenue North, MP 665, P.O. Box 19024, Seattle, WA 98109-1024, USA

Abstract

The application of the two-stage clonal expansion (TSCE) model to lung cancer mortality of two populations of miners (the Colorado Plateau miners and the Chinese tin miners) after protracted high-LET radiation exposure (alpha particles of radon-progeny decay) results in the conclusion that the carcinogenesis process in *protracted exposures* is dominated by the promotion of already-initiated cells. This means that the increase in tumor induction is caused mainly by the modification (i.e., increase) of the net cell proliferation of already-initiated cells during the time of exposure. We hypothesize that this phenomenon is caused by a "bystander effect" whereby normal cells being hit by alpha particles send mitogenic signals to surrounding cells. If the signals are received by (unhit but) already-initiated cells, which are less controlled by homeostasis than the normal cell population, the cells respond by increasing their net cell proliferation, thus increasing the probability of the emergence of a fully malignant cell. Thus, tissue homeostasis (or the lack thereof) plays an important role. Possible modulators at the cellular level include reactive oxygen species (ROS), transforming growth factor-beta (TGF-β) within the extracellular matrix and the subsequent up and/or down regulation of genes controlling mitogenic factors within the cell. This process of proliferative stimulation causes an "inverse dose-rate effect" which is more appropriately termed as a *protraction effect* to distinguish it from the dose-rate effects usually attributed to cellular repair and other mechanisms (e.g., a radiosensitive window in the cell cycle). © 2002 Elsevier Science B.V. All rights reserved.

Keywords: Multistage models; Bystander effects; High-LET radiation; Promotion; Carcinogenesis

---

* Corresponding author. Tel.: +1-206-667-2685; fax: +1-206-667-7004.
E-mail address: sbcurtis@fhcrc.org (S.B. Curtis).

1. Introduction

The two-stage clonal expansion (TSCE) model has been used to analyze several experimental and epidemiological data sets which include acute and protracted radiation exposures [1–4]. A pictorial representation of the model is shown in Fig. 1. For protracted exposure to alpha particles from radon-progeny decay, the analyses consistently result in the emergence of *promotion* as the dominant process in radiation-induced lung carcinogenesis. In fact, radiation-induced initiation is found to be negligible compared to radiation-induced promotion for protracted exposure to high-LET radiation for both the Colorado Plateau miners [3] and the Chinese tin miners [4]. This is shown in Fig. 2 where we compare the relative contributions of initiation and promotion to the excess cases of lung cancer observed in the two miner populations. In each population, promotion is seen to dominate while radiation-induced initiation is found to have little influence. Therefore, for protracted exposure, the major impact of promotion (i.e., the modification of net proliferation of intermediate cells) is on *already-initiated cells* in the tissue.

2. Discussion

Further analysis has shown that the model predicts that for very short (i.e., acute) exposures, initiation does become dominant over promotion. This is seen in Fig. 3 where the prediction of lifetime excess absolute risk (at age 70) is shown as a function of protraction interval for two total doses, 2 and 200 WLM (working level months), corresponding roughly to 1 cGy and 1 Gy, respectively, for the Colorado Plateau miner population [5]. Here, the dashed lines show the contribution of promotion.

The dominance of promotion during the exposure period results in an "inverse dose-rate effect", but by a mechanism distinct from other hypothesized mechanisms, such as the "sensitive-window-in-the-cell-cycle" hypothesis. Thus, we call this a *protraction effect*.

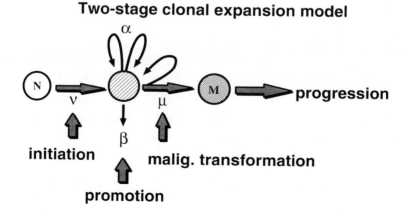

Fig. 1. Pictorial representation of the two-stage clonal expansion (TCE) model.

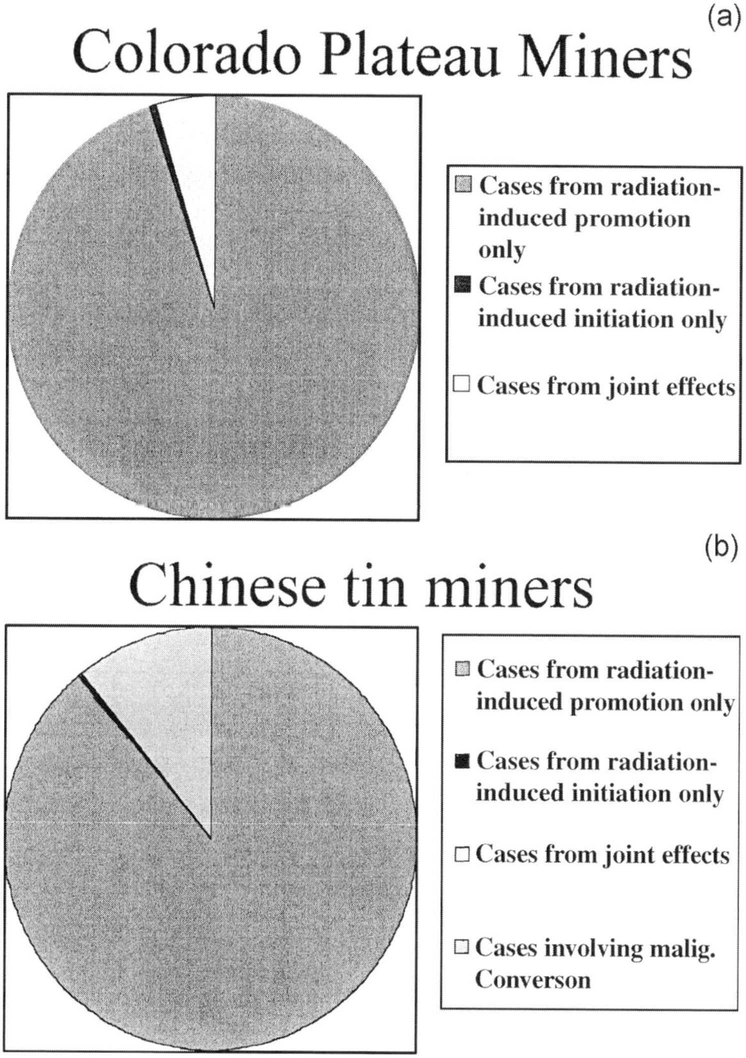

Colorado Plateau Miners (a)

- Cases from radiation-induced promotion only
- Cases from radiation-induced initiation only
- Cases from joint effects

Chinese tin miners (b)

- Cases from radiation-induced promotion only
- Cases from radiation-induced initiation only
- Cases from joint effects
- Cases involving malig. Converson

Fig. 2. (a) Percentage of cases from radiation-induced promotion alone, initiation alone and joint effects for the Colorado Plateau miners [3]. (b) Percentage of cases from radiation-induced promotion, initiation and joint effects plus malignant conversion for the Chinese tin miners [4].

For a given total dose, the risk increases as the exposure is protracted due to the increased net proliferation of the intermediate cells during the protraction interval.

Promotion is assumed to terminate at the end of exposure, and so, could be caused by processes external to the cells in question. This suggests that "bystander effects" could play a role in promoting the initiated cells. Hit cells send out signals to surrounding cells indicating that they have been damaged. If there are already-initiated (but unhit) cells in the neighborhood and if they lack the homeostatic control of their normal cell counterparts,

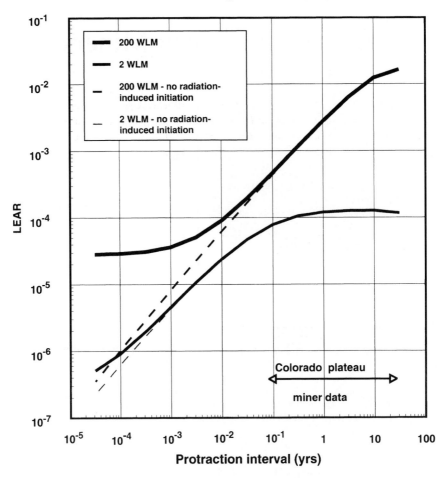

Fig. 3. Lifetime excess absolute risk (LEAR) at age 70 as a function of protraction interval for exposure of 2 (thin solid line) and 200 WLM (thick solid line) corresponding to doses of roughly 1 cGy and 1 Gy, respectively. In each case, the exposure was centered at 42 years of age. The dashed lines indicate the contribution of the promotion term for each exposure.

they could respond preferentially to mitogenic stimuli which, in turn, could result in their gaining proliferative advantage, the hallmark of the promotion process.

Recently, there has been a growing awareness of the importance of effects irradiated cells and the extracellular matrix have on nonirradiated cells as reviewed, for example, by Little [6], and as studied, for instance, by Barcellos-Hoff and Ravani [7] and Iyer and Lehnert [8]. In the latter paper (with normal human lung fibroblasts, in vitro), the following were shown.

(1) Supernatant irradiated with low doses of alpha particles (1 cGy) can induce proliferation in unirradiated cells and can induce intracellular reactive oxygen species (ROS) bystander responses.

(2) Low levels of alpha-particle exposure can increase activation of transforming growth factor-beta (TGF-β) in the extra-cellular matrix (ECM), which leads to an increase in (a) intracellular ROS in bystander cells and (b) proliferating cell nuclear antigen (PCNA) as well as a decrease in p53 and p21, thus leading to enhancement of proliferation in bystander cells.

An implication of the above analyses for radiation protection is that quantities such as dose and dose-rate effectiveness factor (DDREF) and r_W (radiation weighting factor) may become functions of the protraction interval since entirely different processes (initiation and promotion) are dominant for acute as opposed to protracted exposure intervals.

Acknowledgements

This research was supported by grants from the National Institutes of Health: NIH RO1 ES09683 and NCI5 PO1 CA76466.

References

[1] M. Kai, E.G. Luebeck, S.H. Moolgavkar, Analysis of the incidence of solid cancer among atomic bomb survivors using a two-stage model of carcinogenesis, Radiat. Res. 148 (1997) 348–358.
[2] E.G. Luebeck, S.B. Curtis, F.T. Cross, S.H. Moolgavkar, Two-stage model of radon-induced malignant lung tumors in rats: effects of cell killing, Radiat. Res. 145 (1996) 163–173.
[3] E.G. Luebeck, W.F. Heidenreich, W.D. Hazelton, H.G. Paretzke, S.H. Moolgavkar, Biologically-based analysis of the Colorado uranium miners cohort data: age, dose, dose-rate effects, Radiat. Res. 152 (1999) 339–351.
[4] W.D. Hazelton, E.G. Luebeck, W.F. Heidenreich, S.H. Moolgavkar, Analysis of a historical cohort of Chinese tin miners with arsenic, radon, cigarette smoke and pipe smoke exposures using the biologically based two-stage clonal expansion model, Radiat. Res. 156 (2001) 78–94.
[5] S.B. Curtis, E.G. Luebeck, W.D. Hazelton, S.H. Moolgavkar, The role of promotion in carcinogenesis from protracted high-LET exposure, Phys. Med. 17 (Suppl. 1) (2001) 157–160.
[6] J.B. Little, Radiation carcinogenesis, Carcinogenesis 21 (2000) 397–404.
[7] M.-H. Barcellos-Hoff, S.A. Ravani, Irradiated mammary gland stroma promotes the expression of tumorigenic potential by unirradiated epithelial cells, Cancer Res. 60 (2000) 1254–1260.
[8] R. Iyer, B.E. Lehnert, Factors underlying the cell growth-related bystander responses to alpha particles, Cancer Res. 60 (2000) 1290–1298.

A pulsed laser generated soft X-ray source for the study of gap junction communication and 'bystander' effects in irradiated cells

R.A. Meldrum [a,*], G.O. Edwards [a], J.K. Chipman [a], C.W. Wharton [a], S.W. Botchway [b], G.J. Hirst [b], W. Shaikh [b]

[a] *School of Biosciences, University of Birmingham, Edgbaston, Birmingham, B15 2TT, UK*
[b] *Central Laser Facility, CCLRC, Rutherford Appleton Laboratory, Chilton, Didcot, Oxfordshire, OX11 0QX, UK*

Abstract

The killing of individual cells exposed to ionizing radiation has, in historic terms, been considered to be a consequence of direct hits by the radiation, but recent research has shown that cells in populations do not have to be directly exposed to radiation in order to suffer damage. In irradiated cell populations, this 'bystander' effect can be mediated by long-range transmission of toxic molecules or by direct transfer of molecules between cells. An important dynamic cellular process that controls the spread of the toxic effects of damage through a cell population is gap junction intercellular communication (GJIC). A critical limitation in the study of this phenomenon in irradiated cell populations is the lack of experimental systems that enable examination of the incidence of intercellular transmission of toxic effects, without clouding the definition of the mechanism by which it occurs. For instance, if high-energy radiation is used, the dispersal of the tracks means that interaction of the radiation cannot be confined to a single cell, which is in contact with other cells. This means that gap junction intercellular communication cannot be defined or studied under these circumstances. We describe the development and application of a laser plasma generated soft X-ray source, along with micro-masks manufactured by laser machining techniques, for the study of gap junction intercellular communication between adjacent, irradiated and nonirradiated cells. © 2002 Elsevier Science B.V. All rights reserved.

Keywords: Laser plasma; Pulse train; Soft X-rays; Bystander effects; Gap junction intercellular communication

---

Abbreviations: GJIC, Gap junction intercellular communication.
* Corresponding author. Tel.: +44-121-414-43392; fax: +44-121-414-5925.
E-mail address: r.meldrum@bham.ac.uk (R.A. Meldrum).

1. Introduction

An increasing number of experimental observations reveal that toxic effects can be spread between cells in multicellular organisms even when neighboring cells have not come in contact with the toxin. It is easily appreciated how important it is to understand this phenomenon since the effects of very low doses of toxins might be underestimated if a direct dose relationship is assumed. Radiation is no exception in this case.

The phenomenon where toxic effects are passed from exposed cells to nonexposed cells in close proximity is facilitated by several different cellular responses which are collectively known as the 'bystander' effect.

It is also clear that this 'bystander' effect can occur via several mechanisms following radiation exposure. The toxic factors can, for example, be spread or transferred by diffusion through aqueous media in organisms or cell cultures (reviewed in Ref. [1]).

In gene therapy experiments, where gene products have been observed to transfer from DNA transfected cells to neighbouring cells in contact [2], gap junction intercellular communication (GJIC) appears to be the prime mediator of the 'bystander' effect. GJIC has been implicated in experiments concerning radiation effects [3]. What makes it difficult to distinguish clearly between GJIC mechanisms in cells exposed to radiation and other mechanisms of transfer is the lack of precise control over the exposure of cells.

In experiments with a low energy plasma X-ray source, which was developed at the Lasers for Science Facility at the Rutherford Appleton Laboratory [4,5], we observed a situation where toxic effects were passed from irradiated to nonirradiated cells [unpublished work]. This was possible because the energy of the X-rays (1 keV) was so low that the interaction of the radiation with cellular components could only take place in cells on which the radiation directly impinged. The X-ray spectrum from the laser plasma source is sufficiently soft for long-range scattering to be very weak and the penetration of the 1.1 keV copper X-rays into cells is less than 5 μm.

A monolayer of V79 Chinese hamster ovary cultured cells was partially shielded from the X-rays, but death of the cells in the shielded areas also took place. Amplification of cell death took place only in confluent irradiated cell cultures, suggesting that the mechanism required cell to cell contact. Most cells in these cultures were nonproliferating, therefore, only cells at the edges, which may have been in cycle, could be affected by longer-range transmission of molecules [1].

2. The laser plasma X-ray source

The development of laser plasma generated soft X-ray source began in the late 1980s and it is described in detail in Refs. [4,5]. In its present state, the source uses a system of eximer lasers, which produce trains of picosecond pulses. These are focused down to a 10-μm spot onto a moving tape target material. The ablation of the material, which is heated typically to 10^7 K, results in the thermal generation of X-rays. The target material can easily be changed and thus, the system is capable of generating X-rays of different wavelengths. The material used for exposure of biological cells was copper, which gives rise to 1.2-nm

X-rays (CuLα). Cells were grown on Hostaphan ($\sim$ mylar) of 1-μm thickness and the penetration of the 1.1-keV X-rays into the cells would be less than 5 μm. The spectrum of these X-rays is sufficiently soft that the possibility of the X-rays impinging on shielded neighbouring cells is negligible. A study of the spectra of X-rays produced by this system from target materials of copper, steel ($\sim$ iron), aluminium, titanium and Mylar (carbon) is described in Ref. [6].

3. Sharp edged focused radiation from the laser plasma soft X-rays

A 25 μm × 3 mm slit was manufactured in a disc of stainless steel of 13-μm thickness. This disc was fitted into the wall of the chamber directly above the point of the laser plasma (Fig. 1). The stainless steel is completely opaque to the 1.1-keV X-rays. An aluminium filter (3 μm thick) placed between the X-ray source and the X-ray outlet excludes all visible and UV light. The 1-μm Hostaphan base of the cell culture dish is also opaque to UV light. A film with a layer of radiochromic chemical, a few microns thick on one surface (Gafchromic Film, ISP Technologies) was placed face down on Hostaphan over the micro-slit in the stainless steel. This was then exposed to the X-ray plasma for several minutes. A sharp edged image exactly 25 μm wide and 3 mm long was obtained on the film (Fig. 2).

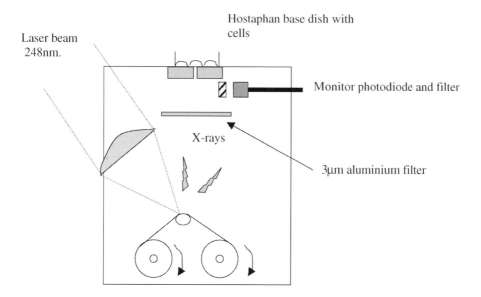

Fig. 1. A 248-nm laser beam is focused onto a moving copper tape in a sealed target chamber. The chamber is filled with helium at atmospheric pressure or lower. A 3-μm aluminium filter excludes all visible and UV light from reaching the outlet slit. A monitor photodiode, also shielded by an aluminium filter, measures the X-ray dose at the slit.

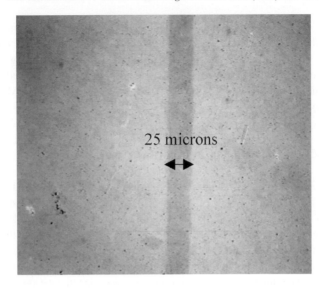

Fig. 2. Radiochromic film was exposed to the 1.1-keV Cu soft X-rays through a 25 μm × 3 mm slit and a layer of 1-μm Hostaphan. The image shows the sharp, well-defined edges obtained with these X-rays.

This supports the idea that these X-rays will not spread to cells that are not directly above the slit in the shielding material, as the average diameter of the cells used for the experiments is 30 μm.

4. Gap junction intercellular communication

GJIC has been studied in primary rat hepatocytes [7] and an immortalised cell line, MH_1C_1. Gap junctions act like valves between cells, which are in contact and control the passage of stimulatory or inhibitory molecules between them. They are not highly selective, as to the molecules they allow to pass, but they can be 'opened' or 'closed'. In this way, they control the cooperation between cells of the multicellular organism and play a vital role in tissue development and function. They are comprised of a hexameric arrangement of connexin molecules that allows the intercell transfer of low molecular weight materials ($< \sim 1$ kDa). Their role is thought to be important in maintaining tissue homeostasis and coordination and they are particularly important in the control of cell proliferation [8]. Gap junctions may also be able to mediate the transfer of potentially damaging molecules such as reactive oxygen species (ROS) and Ca^{2+}. Under conditions of extreme toxicity, gap junctions close, presumably to protect the tissue against the spread of toxic species, which promote necrosis.

Tumour cells show little or no GJIC but when transfected with connexin genes, can exhibit functional gap junctions which restores cell proliferation control.

The MH_1C_1 cell line originates from the epithelium of rat liver hepatoma, but despite its tumorogenic phenotype, it was shown to express connexin 32 (Cx32) [9]. This led to the expectation that these cells may have functioning gap junctions, but when tested by a scrape loading assay and compared to another cell line, WB F344 (nontransformed rat liver epithelial cells), there was virtually no migration of the lucifer yellow dye through the monolayer of contacting cells. WB F344 cells have been shown to express connexin 43, whereas MH_1C_1 cells do not [9].

In preliminary experiments, MH_1C_1 cells in a confluent monolayer grown on Hostaphan were irradiated through a slit, 3 mm in length and 200 µm in width. The incident dose of radiation was 100 Gy. The irradiated cells over the slit was stained and fluoresced with 488-nm excitation in 10 µM dihyroethidium, whereas contacting cells shielded from the radiation, showed no staining. Dihyroethidium is oxidised by the radiation induced superoxide anion to produce ethidium bromide, which intercalates with DNA. No spread of the dye to nonirradiated cells was observed. Cells were also irradiated over a 25 µm × 3 mm slit with an incident dose of 100 Gy. This time, the helium in the interior of the target chamber had been replaced by a vacuum, which ensured close contact of the Hostphan base of the dish containing the cells, which were placed over the micro-machined slit in the stainless steel support. This time, an area of cells greater than 25 µm in width stained with the dihyroethidium dye. Since the MH_1C_1 cells had been shown to lack GJIC, it was concluded that the downward pull created by the vacuum on the Hostaphan membrane over the slit imposed a stress on adjoining cells causing them to produce reactive oxygen species.

Caution must therefore be exercised when using markers such as dyes that identify reactive oxygen species or dyes that respond to calcium ions. Changes in the cellular levels of these species can be produced by stress induced by the technical procedures themselves. The spread of these changes amongst contacting cells may also reflect membrane potential effects as well as gap junction transport.

5. Conclusions

The laser plasma X-ray source is a highly appropriate and useful tool to use for the study of 'bystander' effects in radiation. It can produce X-rays, which are spatially very well-defined. This means that single cells, lines of cells of only one cell width, groups of several cells or a number of single cells spatially separated, may be exposed to radiation with no spread of radiation to neighbouring unexposed cells. The number and pattern of the cells irradiated can be dictated by the design of the mask through which the X-rays reach the cell monolayer. It is essential that the mask be machined to produce very well-defined sharp edges when the X-rays travel through the micro-spaces. This technology provides a means whereby effects transmitted from irradiated cells to nonirradiated cells, may be quantified as the number of cells and the geometry of an area of cells exposed can be determined with some reliability. Because of this spatial control of the X-rays, it is an especially suitable source to use to determine the role that GJIC plays in radiation 'bystander' effects. A further advantage over most soft X-ray sources is that because of the pulsed nature of the source, the dose of X-rays can be controlled and as the system is capable of delivering 3 Gy s^{-1}, a substantial dose may be given in a short time.

Reactive oxygen species might be expected to form a good marker of early effects of radiation in cells and define which cells have been irradiated and those which have not. In preliminary experiments, however, we found that technical manipulations can cause cells to stain in the presence of a dye, which identifies ROS. Techniques must therefore be very well controlled to avoid this or a more distinctive marker of radiation damage should be used.

Acknowledgements

This work is funded by the Engineering and Physical Sciences Research Council, UK.

References

[1] C. Mothersill, C. Seymour, Radiation-induced bystander effects: past history and future directions, Radiation Research 155 (2001) 759–767.
[2] F.L. Moolten, Tumour chemosensitivity conferred by inserted herpes thymidine kinase genes. Paradigm for a prospective cancer control strategy, Cancer Research 46 (1986) 5276–5280.
[3] E.I. Azzam, S.M. de Toledo, T. Gooding, J.B. Little, Intercelluar communication is involved in the bystander regulation of gene expression in human cells exposed to very low fluences of alpha particles, Radiation Research 150 (1998) 497–504.
[4] I.C.E. Turcu, C.W. Wharton, R.A. Meldrum, et al., Picosecond excimer laser–plasma source for microscopy, biochemistry and lithography, SPIE 2015 (1994) 243–260.
[5] I.C.E. Turcu, C.W. Wharton, R.A. Meldrum, et al., Journal of the Society of Photo-Optical Instrumentation Engineers U.S.A. 2015 (1994) 243–287.
[6] A. Nottola, G.J. Tallents, D.T. Goodhead, M.A. Hill, D.L. Stevens, I.C.E. Turcu, G. Hirst, W. Shaikh, J. Westhall, Spectra in the range 5 to 50A^0 from the X-ray source, Annual Report, Central Laser Facility, Rutherford Appleton Laboratory (1999) 156–157.
[7] F.J. Elcock, E. Deag, R.A. Roberts, J.K. Chipman, Nafenopen causes protein kinase c-mediated serine phosphorylation and loss of connexin 32 protein in rat hepatocytes without aberrant expression or localization, Toxicological Sciences 56 (2000) 86–94.
[8] J.K. Chipman, Commentary, Human Experimental Toxicology 15 (1996) 995.
[9] P. Ren, A.W. de Feijter, D.L. Paul, R.J. Ruch, Enhancement of liver cell gap junction protein expression by glucocorticoids, Carcinogenesis 15 (1994) 1807–1813.

Induction of radioresistance by a nitric oxide-mediated bystander effect

Hideki Matsumoto [a,*], Sachiko Hayashi [a], Zhao-Hui Jin [a], Masanori Hatashita [a], Hiroki Shioura [b], Toshio Ohtsubo [c], Ryuhei Kitai [d], Yoshiya Furusawa [e], Osami Yukawa [f], Eiichi Kano [a]

[a]*Department of Exp Radiol and Health Phys, Fukui Med Univ, Fukui, 910-1193, Japan*
[b]*Department of Radiol, Fukui Med Univ, Fukui, 910-1193, Japan*
[c]*Department of Otorhinolaryngol, Fukui Med Univ, Fukui, 910-1193, Japan*
[d]*Department of Neurosurg, Fukui Med Univ, Fukui, 910-1193, Japan*
[e]*Heavy-Ion Radiobiol Res Gr, Natl Inst Radiol Sci, Chiba, 263-8555, Japan*
[f]*Division of Biol and Oncol, Natl Inst Radiol Sci, Chiba, 263-8555, Japan*

Abstract

To elucidate whether nitric oxide excreted from irradiated cells affects cellular radiosensitivity, we examined (a) the accumulation of inducible nitric oxide synthase, p53 and Hsp70, (b) the concentration of nitrite in the medium of cells after irradiation with X-rays, and (c) cellular radiosensitivity using two human glioblastoma cell lines, A-172, having the wild type *p53* gene, and transfectant of A-172 cells, A-172/m*p53*, bearing a mutated *p53* gene. Accumulation of inducible nitric oxide synthase was caused by X-ray irradiation in the mutant *p53* cells but not in the wild type *p53* cells. Accumulation of p53 and Hsp70 was observed in the wild type *p53* cells after exposure to the conditioned medium from X-irradiated mutant *p53* cells, and the accumulation was abolished by the addition of a specific nitric oxide scavenger to the medium. The radiosensitivity of wild type *p53* cells was reduced when the cells were cultured in the conditioned medium from X-irradiated mutant *p53* cells, as compared with the conventional fresh growth medium. These findings indicate that one of the possible mechanisms of the radiation-induced bystander effect is an intercellular signal transduction initiated by nitric oxide radicals. © 2002 Elsevier Science B.V. All rights reserved.

Keywords: Radioresistance; Nitric oxide; Irradiated cells

1. Background

Recently, there have been several reports that unirradiated cells exhibit biological responses when they are cocultivated with irradiated cells [1,2], or exposed to the

---

* Corresponding author.

conditioned medium (CM) harvested from irradiated cells [3–5]. This phenomenon has been termed the *bystander effect*. However, the mechanism of the bystander effect induced by radiation is still unclear. Although it has reported a relationship between nitric oxide (NO) and cellular responses after irradiation in vitro and in vivo [6], the role of NO that is excreted from irradiated cells as part of the stress response caused by exposure to ionizing radiation is still also unclear. We have previously found that the accumulation of stress-induced proteins and thermoresistance in NO-recipient w$tp53$ cells cocultivated with heat-shocked NO-donor m$p53$ cells, is induced through an intercellular signal transduction pathway initiated by NO [7,8].

In the present study, to elucidate whether NO excreted from irradiated cells affects cellular radiosensitivity, we examined (a) the kinetics of accumulation of inducible nitric oxide synthase (iNOS) after X-irradiation, (b) the kinetics of accumulation of Hsp70 and p53 in non-irradiated NO-recipient cells cocultivated with X-irradiated NO-donor cells, and (c) the modification of radiosensitivity of NO-recipient cells in the CM from X-irradiated NO-donor cells using the human glioblastoma cell lines, A-172 and A-172/m$p53$.

2. Methods

2.1. Cells

Human glioblastoma cell line, A-172, was purchased from the JCRB Cell Bank (Setagaya, Tokyo, Japan). Two human glioblastoma cell lines A-172 and A-172/m$p53$, which is transfectant of A-172 with the mutant $p53^{Trp248}$ gene, were cultured in Dulbecco's modified Eagle's medium containing 10% fetal bovine serum (DMEM-10).

2.2. Irradiation with X-rays

Exponentially growing cells were seeded in dishes or in flasks containing DMEM-10 without irradiated feeder cells. Cells were irradiated with X-rays (1.0–10.0 Gy at 1.0 Gy/min) in DMEM-10 using High Technical System X-ray (Model HW-150, Hitex, Tokyo, Japan).

2.3. Cocultivation of cells on slide glasses with cells in dishes

Exponentially growing cells were seeded in dishes containing DMEM-10 without irradiated feeder cells. At the same time, cells were seeded on slide glasses. The cells in the dishes were irradiated with X-ray. The slide glasses were then transferred into the dishes, and the cocultures were incubated at 37 °C for up to 10 h.

2.4. Western blot analysis

Cells were suspended in the RIPA buffer, and then frozen and thawed three times. Aliquots of the supernatants obtained after centrifugation were subjected to Western blotting analysis for iNOS, Hsp70 and p53.

2.5. Measurement of nitrite concentration in medium

The nitrite concentration in the medium was measured according to the method of Saltzman [9].

2.6. Survival curves

The surviving cell fraction after irradiation was determined as colony-forming units, and corrected by the plating efficiency of non-treated cells as a control.

3. Results

3.1. Accumulation of iNOS by X-rays

After irradiation with X-rays at 2.5 Gy, the accumulation of iNOS was observed in A-172/m*p53* cells, but not in A-172. In A-172/m*p53* cells, the level of iNOS increased gradually after X-irradiation and reached a level about three-fold greater than the control level at 24 h, whereas the level of iNOS in A-172 cells did not increase after X-irradiation.

3.2. Measurement of nitrite concentration in medium

The nitrite concentration increased gradually to over 2.0 μM 24 h after X-irradiation in A-172/m*p53* cells. The elevation of nitrite concentration in the medium of A-172/m*p53* cells was completely suppressed by the addition of 0.1 mM aminoguanidine (AG) as an iNOS inhibitor. In contrast, in the medium of A-172 cells, the nitrite concentration did not change after X-irradiation.

3.3. Accumulation of Hsp70 and p53 in the NO-recipient cells co-cultivated with the NO-donor cells

Stress-responsive proteins Hsp70 and p53 were accumulated in non-irradiated A-172 cells cocultivated with X-irradiated A-172/m*p53* cells at 5 Gy, and the levels of both proteins reached about two-fold greater than the control level at 10 h. These accumulations were completely abolished by the addition of 0.1 mM AG to the medium.

3.4. Accumulation of Hsp70 and p53 in the NO-recipient cells exposed to CM from the NO-donor cells

The CM from m*p53* cells was prepared by culturing them for 10 h after irradiation with X-rays at 5.0 Gy with or without c-PTIO. The levels of Hsp70 and p53 in A-172 cells increased to about two-fold greater than the control level at 10 h after exposure to the CM from X-irradiated A-172/m*p53* cells without c-PTIO. The accumulation of these proteins was inhibited by the addition of c-PTIO to the CM from X-irradiated A-172/m*p53* cells in a dose-dependent manner.

3.5. Radiosensitivity of the NO-recipient cells exposed to the CM of the NO-donor cells

A-172 cells in the CM from A-172/m$p53$ cells cultured for 10 h after X-irradiation at 5.0 Gy (CM-X) were more radioresistant than those in the fresh growth medium (GM), or in the CM from A-172/m$p53$ cells cultured for 10 h without irradiation (CM-0). In GM, the D_0 of the A-172 cells was 1.5 Gy. In CM-0, the radiosensitivity of the A-172 cells was scarcely different: the D_0 of the A-172 cells was 1.6 Gy. However, in the CM-X, the D_0 of the A-172 cells was 1.9 Gy. The dose modifying factors in D_0 were 1.1 and 1.3 for CM-0 and CM-X, respectively. In addition, the values of D_q of A-172 in GM, CM-0 and CM-X were 0.7, 1.0 and 1.4 Gy, respectively. The dose modifying factors in D_q were 1.4 and 2.0 for CM and CM-X, respectively.

4. Conclusions

Nitric oxide (NO) excreted from the X-irradiated donor m$p53$ cells could induce radioresistance in the recipient non-irradiated wt$p53$ cells. These findings indicate that one of the possible mechanisms of the radiation-induced bystander effect is an intercellular signal transduction initiated by NO radicals [10,11].

References

[1] H. Nagasawa, J.B. Little, Cancer Res. 52 (1992) 6394–6396.
[2] E.I. Azzam, S.M. de Toledo, T. Gooding, J.B. Little, Radiat. Res. 150 (1998) 497–504.
[3] C. Mothersill, C. Seymour, Int. J. Radiat. Biol. 71 (1997) 421–427.
[4] C. Mothersill, C. Seymour, Radiat. Res. 149 (1998) 256–262.
[5] R. Iyer, B.E. Lehnert, Cancer Res. 60 (2000) 1290–1298.
[6] W.K. MacNaughton, A.R. Aurora, J. Bhamra, K.A. Sharkey, M.J. Miller, Int. J. Radiat. Biol. 74 (1998) 255–264.
[7] H. Matsumoto, S. Hayashi, M. Hatashita, H. Shioura, T. Ohtsubo, R. Kitai, T. Ohnishi, E. Kano, Nitric Oxide 3 (1999) 180–189.
[8] H. Matsumoto, S. Hayashi, M. Hatashita, K. Ohnishi, T. Ohtsubo, R. Kitai, H. Shioura, T. Ohnishi, E. Kano, Cancer Res. 59 (1999) 3239–3244.
[9] B.E. Saltzman, Anal. Chem. 26 (1954) 1949–1955.
[10] H. Matsumoto, S. Hayashi, M. Hatashita, K. Ohnishi, T. Ohtsubo, R. Kitai, H. Shioura, T. Ohnishi, Y. Furusawa, O. Yukawa, E. Kano, Int. J. Radiat. Biol. 76 (2000) 1649–1657.
[11] H. Matsumoto, S. Hayashi, M. Hatashita, T. Ohtsubo, R. Kitai, H. Shioura, T. Ohnishi, E. Kano, Radiat. Res. 155 (2001) 387–396.

Roles of protein kinase C in radiation-induced apoptosis signaling pathways in murine thymic lymphoma cells (3SBH5 cells)

Tetsuo Nakajima*, Osami Yukawa, Harumi Ohyama, Bing Wang, Isamu Hayata, Hiroko Hama-Inaba

Research Center for Radiation Safety, Natl. Inst. Radiol. Sci., Anagawa 4-9-1, Inage-ku, Chiba 263-8555, Japan

Abstract

Murine thymic lymphoma cells and 3SBH5 cells are quite sensitive to X-rays and undergo apoptosis shortly after X-irradiation. Phorbol 12-myristate 13-acetate (PMA), an activator of protein kinase C (PKC), blocked the radiation-induced apoptosis in the 3SBH5 cells. On the other hand, chelerythrine, a PKC inhibitor, enhanced the radiation-induced apoptosis. These results suggest that PKC plays a key role in the regulation of radiation-induced apoptosis in 3SBH5 cells. Irradiation alone had no effect on the distribution of PKC subtypes (α, βI, βII and δ) in the 3SBH5 cells. The amounts of PKC βI in the cytosol of the 3SBH5 cells decreased in the cells pretreated with PMA. Irradiation did not change the decrease of PKC βI. In contrast, treatment with PMA had no effect on the distribution of the other PKC subtypes. PMA appears to influence processes of radiation-induced apoptosis through the change in the distribution of PKC βI. In addition, it was demonstrated that immunoprecipitates by anti-PKC α antibody included Raf-1, one of stress response proteins, in 3SBH5 cells after irradiation. These results suggest that PKC α might participate as a regulator in radiation-induced apoptosis, but PKC βI might influence the apoptosis indirectly through the PMA-induced change of the distribution. © 2002 Elsevier Science B.V. All rights reserved.

Keywords: Thymic murine lymphoma; Protein kinase C; Raf-1; Apoptosis; Radiation sensitivity

1. Background

Protein kinase C (PKC) is known to be an important participant in radiation-induced signaling cascades [1,2]. Particularly, from the viewpoint of apoptosis, many studies on PKC function have been performed in various types of apoptosis involving radiation-in-

---

* Corresponding author. Tel.: +81-43-206-3088; fax: +81-43-255-6497.
E-mail address: otetsu@nirs.go.jp (T. Nakajima).

0531-5131/02 © 2002 Elsevier Science B.V. All rights reserved.
PII: S0531-5131(01)00881-0

duced apoptosis. However, its roles still remain unknown and controversial, as it has been reported that PKC activation induces apoptosis in some cases, and promotes cell survival in other cases. This study was performed to demonstrate PKC function in radiation-induced apoptosis regulation, in radiation-sensitive cells.

2. Methods

Cell culture, X-irradiation and apoptosis assay were performed as previously described [3]. PKC localization was evaluated by Western blot, and interaction between PKC and Raf-1 was analyzed by immunoprecipitation and Western blot. PKC antisense inhibition was performed by PKC antisense oligonucleotides (Biognostik, Göttingen, Germany).

3. Results and discussion

3.1. PKC activator and inhibitor influence radiation sensitivity in 3SBH5 cells

Murine thymic lymphoma cells and 3SBH5 are quite sensitive to X-rays and undergo apoptosis shortly (4 h) after irradiation. Phorbol 12-myristate 13-acetate (PMA), an activator of PKC, blocked the radiation-induced apoptosis in 3SBH5. On the other hand, chelerythrine, a potent selective inhibitor of PKC, enhanced the radiation-induced apoptosis.

3.2. PKC localization in 3SBH5 cells

PKC α, βI, βII and δ were detected in 3SBH5 cells. Irradiation alone had no effect on the distribution of the PKCs in the 3SBH5 cells. The amounts of PKC βI in the cytosol of the 3SBH5 cells decreased in the cells pretreated with PMA. Irradiation did not change the decrease of PKC βI. In contrast, treatment with PMA had no effect on the amounts of the other PKCs. Therefore, PMA appears to influence processes of radiation-induced apoptosis through the change in the distribution of PKC βI.

3.3. Interaction of PKC with Raf-1

Raf-1 as well as PKC is known to mediate several conflicting cellular responses such as proliferation, and apoptosis. The kinase activity of Raf is regulated by phosphorylation of a highly conserved serine (Ser259) and the kinases responsible for phosphorylating Ser259 are unknown but might include PKC. Therefore, it was analyzed whether radiation influences interaction of PKC with Raf-1 in 3SBH5 cells. Raf-1 was detected only in the cytosolic fractions in controlled, irradiated cells and even in the cells treated with PMA and chelerythrine. Phosphorylation of Ser259 in Raf-1 did not change in any conditions after irradiation. To determine whether PKC forms a complex with Raf-1, we obtained immunoprecipitates by using anti-PKC α, βI and βII antibodies in the cytosolic fractions from controlled and irradiated cells, and those immunoprecipitates were subjected to immunoblotting with an anti-Raf-1 antibody. Raf-1 was detected in both controlled and irradiated cells in the case of the immunoprecipitates by anti-PKC βI and βII antibodies, whereas the immunoprecipitates by the anti-PKC α antibody included Raf-1 only in irradiated cells. These results suggested that radiation could induce PKC signaling cascades via the interaction of PKC α, with Raf-1 in 3SBH5 cells. In addition, anti-sense PKC α induced more apoptosis in 3SBH5 cells, 4 h after irradiation with 0.5 Gy.

4. Conclusions

This study indicates that PKC α might function directly in radiation-induced apoptosis regulation, presumably through the interaction with Raf-1; however, PKC βI might influence the apoptosis indirectly through the PMA-induced change of the distribution.

Acknowledgements

This study was partly supported by grants from the Radiation Effects Association.

References

[1] T. Nakajima, O. Yukawa, Radiation-induced translocation of protein kinase C through membrane lipid peroxidation in primary cultured rat hepatocytes, Int. J. Radiat. Biol. 70 (1996) 473–480.
[2] T. Nakajima, O. Yukawa, Mechanism of radiation-induced diacylglycerol production in primary cultured rat hepatocytes, J. Radiat. Res. 40 (1999) 135–144.
[3] H. Hama-Inaba, et al., Radio-sensitive murine thymoma cell line 3SB: characterization of its apoptosis-resistant variants induced by repeated X-irradiation, Mutat. Res. 403 (1998) 85–94.

Cellular mechanisms of radiation adaptive response in cultured glial cells

Yuri Miura*, Kazuhiko Abe, Shozo Suzuki

Department of Biochemistry and Isotopes, Tokyo Metropolitan Institute of Gerontology, Sakaecho 35-2, Itabashi, Tokyo 173-0015, Japan

Abstract

We have examined the radiation adaptive response in cultured glial cells conditioned with a low dose and subsequently challenged with a high dose of X-rays. The radiation adaptive response was evident for glial cells cultured from young rats (1 month old), however, not from aged rats (24 months old). In order to investigate the subcellular signaling pathway of the radiation adaptive response, we examined the various effects of inhibitors and activators of potential factors involved in signaling responses to radiation. The inhibitors of protein kinase C (PKC), DNA-dependent protein kinase (DNA-PK) or PI3K suppressed the radiation adaptive response in young glial cells, and the activators of PKC instead of low-dose pre-irradiation demonstrated protective effects against the growth inhibition caused by 2-Gy irradiation. The glial cells cultured from severe combined-immunodeficiency (scid) mice, which lack DNA-PK activity, and from ataxia–telangiectasia mutated (*Atm*)-knockout mice, did not show radiation adaptive response at all. These results suggest that PKC, the ATM protein, DNA-PK or PI3K are involved in the radiation adaptive response in cultured glial cells, and that the response decreases with age. © 2002 Elsevier Science B.V. All rights reserved.

Keywords: Radiation; Adaptive response; Oxidative stress; Aging; Glial cells

1. Background

Oxidative stress causes various types of damage followed by cell death or the activation of repair systems. While low stress, which does not damage cells seriously, also gives rise

---

Abbreviations: ROS, reactive oxygen species; PKC, protein kinase C; DNA-PK, DNA-dependent protein kinase; PI3K, phosphatidylinositol 3-kinase; DOG, 1,2-dioctanoyl-sn-glycerol; FTT, farnesyl thiotrizole; scid, severe combined immunodeficiency; AT, ataxia–telangiectasia.

* Corresponding author. Tel.: +81-3-3964-3241x3062; fax: +81-3-3579-4776.

E-mail address: miura@center.tmig.or.jp (Y. Miura).

to some responses of cultured cells such as induction of preventive systems. Since ROS not only trigger various types of tissue damage, but are also inducers and stimulants of antioxidative and preventive systems, the biological effects of low stress cannot be estimated by extrapolation of those for high stress, and as a result, they remain unclear. Therefore, it is important to study cell responses to low oxidative stress for the disease prevention and/or elucidation of the role of ROS as biological messengers. Ionizing radiation is widely studied for low-dose effects, and it has sometimes been reported to induce hormesis and adaptive response as a stimulant [1]. However, its molecular mechanism or signaling pathway is not yet known and needs to be clarified.

Therefore, we examined the adaptive response of cultured glial cells, which are known to be more resistant to oxidative stress [2,3], and to support the survival of neurons to low-dose radiation [4–6] as a slight oxidative stress. We also studied the molecular mechanism of these phenomena.

2. Methods

2.1. Animals and cell cultures

Wistar rats (female, 1 or 24 months old) were obtained from the Laboratory Animal Facilities, Tokyo Metropolitan Institute of Gerontology. CB-17 scid/scid and their counterpart controls, CB-17+/+ mice (male, 5 weeks) were purchased from CLEA Japan. Ataxia–telangiectasia mutated (*Atm*)-knockout mice and their wild type controls, 129 sv mice were gifts from Dr. Furuse and Dr. Tatsumi. Primary culture of the glial cells was performed following the conventional procedure. The cultured cells were shown to be immunopositive after immunocytochemical staining of the glial fibrillary acidic protein (GFAP), a marker for astrocytes.

TIG-118 and AT-2KY cells were fibroblasts derived from the skin of a normal human and an ataxia–telangiectasia patient, respectively.

2.2. X-irradiation

Irradiation of the cells was performed with X-rays (150 kV, 2 or 5 mA). Low-dose and high-dose irradiation was performed at a dose rate of 0.06 and 0.34 Gy/min, respectively. Cells treated with sham irradiation were taken out of a CO_2 incubator and left under air, which was kept at room temperature during the process of irradiation.

2.3. Radiation adaptive response

Cells (2.5×10^5) were plated in flasks 1 day before irradiation. Four cell groups were set up for these experiments: 0 Gy alone (0–0), a low dose of 0.1 Gy alone (0.1–0), a high dose of 2 Gy alone (0–2) and a low dose of 0.1 Gy followed by a high dose of 2 Gy (0.1–2). The interval between pre- and additional-irradiation was 3 h. Each cell group was incubated at 37 °C under a humidified atmosphere containing 5% CO_2/95% air, and cell growth was measured by counting the number of cells 2 days after the irradiation process.

3. Results

3.1. Radiation adaptive response of glial cells

Fig. 1 shows the growth ratio of the 2-Gy-irradiated cell group to the sham-irradiated one, with or without 0.1-Gy pre-irradiation. The growth of the glial cells was inhibited by 2-Gy irradiation in both groups, with and without pre-irradiation. However, in glial cells cultured from young rats, the 0.1-Gy pre-irradiation of cells suppressed the growth inhibition due to 2 Gy, indicating that pre-irradiation causes radiation adaptive response. Furthermore, 2 Gy less inhibited the growth of cells taken from aged rats than that from young rats. However, 0.1 Gy-pre-irradiation did not significantly influence growth inhibition, thus suggesting that the radiation adaptive response due to 0.1-Gy pre-irradiation was not observed in aged rat cells. These results suggested that the radiation adaptive response was evident for glial cells taken from young rats, and decreased with aging.

3.2. Effects of various inhibitors and activators of PKC or DNA-PK on radiation adaptive response

In order to investigate the molecular mechanisms of radiation adaptive response, we examined the effects of the inhibitors of protein kinase C (PKC) and DNA-dependent protein kinase (DNA-PK), which were reported to be activated by radiation [7–9], and on the radiation adaptive response with regard to cell growth. Calphostin C and chererythrine chloride were used as inhibitors of PKC, and they both significantly suppressed the radiation adaptive response of the cultured glial cells. On the other hand, the treatments

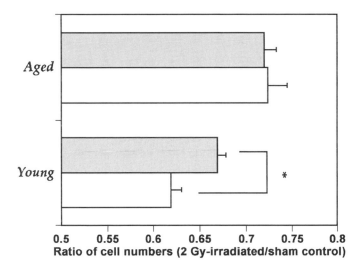

Fig. 1. The radiation adaptive response in glial cells cultured from young and aged rats. The open and shadow columns represent the ratio of cell numbers (2 Gy/0 Gy) with and without 0.1-Gy pre-irradiation, respectively. *$p<10^{-5}$ ($n=25$) statistically significant difference from samples without pre-irradiation.

with PKC activators, FTT or DOG instead of 0.1-Gy pre-irradiation, protected the growth inhibition, which was similar to that found with low-dose pre-irradiation. These results suggest that PKC could be involved in the radiation adaptive response in glial cells from young rats. The inhibitors of DNA-PK or PI3K, wortmannin and LY294002, also suppressed the radiation adaptive response under the present conditions.

3.3. Radiation adaptive response of scid and AT cells

It is necessary to confirm the involvement of DNA-PK in the radiation adaptive response in another way, because wortmannin and LY294002 at high doses (25 and 1 µM, respectively) may possibly inhibit other kinases of the PI3K super family, which includes PI3K and ATM [10,11]. Therefore, we used glial cells of a severe combined-immunodeficiency (scid) mouse (CB-17 scid), which lack any detectable DNA-PK activity [12], and glial cells of an *Atm*-knockout mouse (129 sv AT-KO) or fibroblasts of an AT patient, in which ATM is deficient or mutated. Normal cells (CB-17, 129 sv, and TIG-118 cells) showed radiation adaptive response, while deficient or mutated cells such as scid cells, AT-KO and AT-2KY cells did not. These results indicate that DNA-PK and ATM could be involved in the radiation adaptive response.

4. Conclusions

The glial cells cultured from young rats demonstrated radiation adaptive response regarding cell growth, while those from aged rats did not. The experiments using various inhibitors and deficient cells demonstrated that PKC, DNA-PK and ATM were involved in the radiation adaptive response. In addition, the present study suggested that aging suppressed the cell response to low-dose irradiation, such as activation of PKC, DNA-PK and ATM.

Acknowledgements

We wish to thank Drs. T. Furuse and K. Tatsumi (National Institute of Radiological Sciences) for the 129 sv and *Atm*-knockout mice, Dr. H. Kondo (Tokyo Metropolitan Institute of Gerontology) for the TIG118 cells, and Dr. M. S. Sasaki (Kyoto University) for the AT-2KY cells. This study was supported in part by a Grant-in-Aid for Encouragement of Young Scientists (No. 11771477) from the Ministry of Education, Science, Sports and Culture of Japan, and by a Grant from the Ground Research Announcement for Space Utilization promoted by the Japan Space Forum.

References

[1] S. Wolff, Is radiation all bad? The search for adaptation, Radiat. Res. 131 (1992) 117–123.
[2] J.P. Bolanos, S.J.R. Heales, J.M. Land, J.B. Clark, Effect of peroxynitrite on the mitochondrial respiratory chain: differential susceptibility of neurons and astrocytes in primary culture, J. Neurochem. 64 (1995) 1965–1972.

[3] S.B. Hollensworth, C.-C. Shen, J.E. Sim, D.R. Spitz, G.L. Wilson, S.P. LeDoux, Glial cell type-specific responses to menadione-induced oxidative stress, Free Radical Biol. Med. 28 (2000) 1161–1174.
[4] F. Noel, P.J. Tofilon, Astrocytes protect against X-ray-induced neuronal toxicity in vitro, NeuroReport 9 (1998) 1133–1137.
[5] S. Desagher, J. Glowinski, J. Premont, Astrocytes protect neurons from hydrogen peroxide toxicity, J. Neurosci. 16 (1996) 2553–2562.
[6] J. Tanaka, K. Toku, B. Zhang, K. Ishihara, M. Sakanaka, N. Maeda, Astrocytes prevent neuronal death induced by reactive oxygen and nitrogen species, Glia 28 (1999) 85–96.
[7] R.K. Schmidt-Ullrich, P. Dent, S. Grant, R.B. Mikkelsen, K. Valerie, Signal transduction and cellular radiation responses, Radiat. Res. 153 (2000) 245–257.
[8] I. Szumiel, Monitoring and signaling of radiation-induced damage in mammalian cells, Radiat. Res. 150 (1998) s92–s101.
[9] G.E. Woloschak, C.M. Chang-Liu, P. Shearin-Jones, Regulation of protein kinase C by ionizing radiation, Cancer Res. 50 (1990) 3963–3967.
[10] A. Arcaro, M.P. Wymann, Wortmannin is a potent phosphatidylinositol 3-kinase inhibitor: the role of phophatidylinositol 3,4,5-trisphosphate in neutrophil responses, Biochem. J. 296 (1993) 297–301.
[11] J.N. Sarkaria, R.S. Tibbetts, E.C. Busby, A.P. Kennedy, D.E. Hill, R.T. Abraham, Inhibition of phosphoinositide 3-kinase related kinases by the radiosensitizing agent wortmannin, Cancer Res. 58 (1998) 4375–4382.
[12] T. Blunt, N.J. Finnie, G.E. Taccioli, G.C.M. Smith, J. Demengeot, T.M. Gottlieb, R. Mizuta, A.J. Varghese, F.W. Alt, P.A. Jeggo, S.P. Jackson, Defective DNA-dependent protein kinase activity is linked to V(D)J recombination and DNA repair defects associated with the murine scid mutation, Cell 80 (1995) 813–823.

Radiation-induced genomic instability and delayed activation of p53

Keiji Suzuki*, Seiji Kodama, Masami Watanabe

Laboratory of Radiation and Life Science, School of Pharmaceutical Sciences, Nagasaki University, 1-14 Bunkyo-machi, Nagasaki 852-8521, Japan

Abstract

Recent studies have described that radiation induces genomic instability, which manifests in the progeny of surviving cells as a persistent decrease in plating efficiency (delayed reproductive death), increased chromosome instability, and increased mutation rate. Despite extensive studies into the process of delayed phenotypes, very little is known about the mechanism of genomic instability. Delayed chromosome instability is characterized by dicentric chromosomes, chromatid gaps and breaks, indicating that delayed DNA breakage is induced in the surviving cells several generations after irradiation. To determine the possibility, we have constructed a reporter plasmid containing the p53-responsible promoter and the β-galactosidase (β-gal) gene, and transfected the plasmid into human fibrosarcoma cells containing the wild-type *p53* gene. The cells were irradiated with various doses of X-rays and plated to form the primary colonies. They were collected to perform the secondary colony formation, and we found that the expression of β-gal was frequently detected in the secondary colonies, derived from the primary colonies surviving X-rays. These observations indicate that p53 is frequently activated in the progeny of surviving cells, indicating that delayed DNA damage is responsible for the expression of delayed phenotypes. © 2002 Elsevier Science B.V. All rights reserved.

Keywords: Radiation instability; p53; DNA damage; delayed effects

1. Introduction

Recent studies have described that radiation induces genomic instability, which manifests in the progeny of surviving cells as a persistent decrease in plating efficiency (delayed reproductive death), increased formation of giant cells, micronuclei, and

---

* Corresponding author. Tel.: +81-95-844-5504; fax: +81-95-844-5504.
E-mail address: kzsuzuki@net.nagasaki-u.ac.jp (K. Suzuki).

chromosome bridge in anaphase, increased chromosome instability, and increased gene amplification and mutation rate [1]. It has been implicated that radiation-induced genomic instability could be the driving force underlying the multi-step process of radiation-induced cancer development.

Despite extensive studies into the process of delayed phenotypes, very little is known about the mechanism of genomic instability. Delayed chromosome instability is characterized by dicentric chromosomes, chromatid gaps and breaks, indicating that delayed DNA breakage is induced in the surviving cells several generations after irradiation. It is well known that DNA breakage causes accumulation and activation of p53, a tumor suppressor protein [2]. The activated p53 protein functions as a transcription factor, and it regulates the expression of the downstream effectors, such as the $p21^{WAF1/CIP1}$ gene. In response to DNA damage, ATM kinase is activated and it phosphorylates the p53 protein in its amino-terminal region, which is the modification essential for its accumulation and activation [3]. In the present study, we examined the induction of delayed DNA damage using the reporter plasmid harboring the p53 responsible promoter.

2. Materials and methods

Human fibrosarcoma cells were cultured in Eagle's minimum essential medium supplemented with 10% fetal bovine serum. The reporter plasmid containing the p53-responsible promoter and the β-galactosidase gene was transfected into human fibrosarcoma cells, which have the wild-type *p53* gene. Exponentially growing cells were irradiated with X-rays from an X-ray generator at 150 kVp and 5 mA with a 0.1-mm Cu filter. The β-gal expression was detected by incubating the cells in a substrate buffer containing 1 mg/ml X-gal.

3. Results

The level of β-gal is increased in response to radiation exposure, and the induction kinetics is closely related to that of $p21^{WAF1/CIP1}$, the downstream effector of the p53 protein. This confirms that the reporter plasmid is physiologically functional in human fibrosarcoma cells. We irradiated the cells with various doses of X-rays, and plated cells to form the primary colonies. Then, they were collected to perform the secondary colony formation. We found that the expression of β-gal was frequently detected in the secondary colonies derived from the primary colonies surviving X-rays, however, not in the control colonies. These observations indicate that p53 is frequently activated in the progeny of surviving cells.

4. Discussion

The present study shows that radiation-induced genomic instability induces delayed DNA damage, which has the potential to accumulate genetic alterations in the growth-related genes. However, it is also suggested that delayed p53 reactivation, which causes senescence-like growth arrest or delayed apoptosis, may eliminate cells causing genome destabilization. Thus, if the checkpoint fails, radiation-induced genomic instability accelerates the development of multiple genetic changes.

References

[1] K. Suzuki, R. Takahara, S. Kodama, M. Watanabe, In situ detection of chromosome bridge formation and delayed reproductive death in normal human embryonic cells surviving X irradiation, Radiat. Res. 150 (4) (1998) 375–381.
[2] J.C. Ghosh, Y. Izumida, K. Suzuki, S. Kodama, M. Watanabe, Dose-dependent biphasic accumulation of TP53 protein in normal human embryo cells after X irradiation, Radiat. Res. 153 (3) (2000) 305–311.
[3] K. Suzuki, S. Kodama, M. Watanabe, Recruitment of ATM protein to double strand DNA irradiated with ionizing radiation, J. Biol. Chem. 274 (36) (1999) 25571–25575.

Unstable nature of the X-irradiated human chromosome in unirradiated mouse m5S cells

Seiji Kodama [a,*], Kiyo Yamauchi [a], Taeko Tamaki [a],
Ayumi Urushibara [a], Satoko Nakatomi [a], Keiji Suzuki [a],
Mitsuo Oshimura [b], Masami Watanabe [a]

[a]*Laboratory of Radiation and Life Science, School of Pharmaceutical Sciences, Nagasaki University, 1-14 Bunkyo-machi, Nagasaki 852-8521, Japan*
[b]*Department of Molecular and Cell Genetics, Faculty of Medicine, School of Life Science, Faculty of Medicine, Tottori University, 86 Nishimachi, Yonago 638-8503, Japan*

Abstract

To find out the mechanism of induction for delayed chromosome aberrations, mouse A9 cells containing a human chromosome 11 were irradiated with 6 or 15 Gy of X-rays, and then the human chromosome was transferred into unirradiated mouse m5S cells using a microcell-mediated chromosome transfer. Chromosome aberrations were analyzed by whole chromosome painting using a probe specific for human chromosome 11. In a microcell hybrid transferred with 15 Gy-irradiated chromosomes 11, all chromosomes 11 were fragmented and 45% of the chromosomes 11 were rearranged after chromosome transfer. Similarly, in a microcell hybrid transferred with 6 Gy-irradiated chromosome 11, 25% and 46% of the chromosomes 11 were rearranged to form rings and telomeric fusion chromosomes, respectively. These results suggest that the irradiated chromosome per se is of an unstable nature, and that telonomic instability induced by radiation is possibly involved in the induction of delayed chromosome aberrations. © 2002 Elsevier Science B.V. All rights reserved.

Keywords: Radiation; Chromosome instability; Chromosome transfer; Telomeric fusion; Telonomic instability

1. Introduction

Ionizing radiation (IR) induces persistent delayed chromosomal instability in the surviving progeny of irradiated cells [1]. This instability is characterized by a high frequency

---

Abbreviations: IR, ionizing radiation; DSBs, double-strand breaks; FBB cycle, fusion–bridge–breakage cycle.
* Corresponding author. Tel./fax: +81-95-844-5504.
E-mail address: s-kodama@net.nagasaki-u.ac.jp (S. Kodama).

and transmissible nature over many cell divisions post-irradiation, suggesting that a single gene mutation might not be responsible for this phenomenon [2]. However, because of the complexity and variety of DNA lesions induced by IR, it is difficult to elucidate the primary target and initiating lesion responsible for inducing delayed chromosomal instability. We formerly demonstrated that the initial DNA damage and the efficiency of the subsequent repair process combined with environmental oxidative stress influenced the occurrence of delayed chromosome aberrations [3,4]. It is also shown that the complexity or quality of DNA double-strand breaks (DSBs), but not the DSBs per se, is important in initiating this phenotype [5]. On the basis of these results, we assume that the delayed chromosomal instability induced by IR is initiated with some DNA lesions produced by impaired repair of the DSBs, and that the unstable nature is transmitted via irradiated chromosomes. This hypothesis can be examined by monitoring the stability of a chromosome exposed to IR under an unirradiated environmental condition. It is very likely that the chromosome exposed to IR is rearranged over many cell divisions post-irradiation when the irradiated chromosome per se is unstable. In the present study, we examined this hypothesis using a microcell-mediated chromosome transfer technique.

2. Materials and methods

Mouse A9 cells containing a human chromosome 11 were used as chromosome donor cells and mouse m5S cells, which retained near-diploid karyotype, were used as chromosome recipient cells. The donor cells were irradiated with 6 or 15 Gy of X-rays, and then a human chromosome 11 was introduced into the unirradiated m5S cells by the process of microcell-mediated chromosome transfer as previously described [6]. The cells were cultured in α-MEM medium (Life Technologies, Gaithersburg, MD) with 10% FCS, penicillin (100 units/ml), streptomycin (100 µg/ml) and the microcell hybrids were cultured in the α-MEM medium supplemented with 3 µg/ml blasticidin-S-hydrochloride (Funakoshi, Tokyo). After the microcell hybrids containing a human chromosome 11 were isolated, the stability of the human chromosome in the microcell hybrids over 20 population doublings post-irradiation was examined by FISH using a probe specific for human chromosome 11 as previously described [6]. The existence of telomere sequences at the chromosome termini was examined by telomere FISH using a PNA probe, $(CCCTAA)_3$, as previously reported [7].

3. Results

The morphology of microcell hybrids transferred with unirradiated or irradiated human chromosomes 11 was similar to that of parental m5S cells (data not shown). We analyzed delayed chromosome aberrations in three microcell hybrids transferred with unirradiated chromosome 11, and in two and one microcell hybrids transferred with 6 and 15 Gy-irradiated chromosome 11, respectively. The result is shown in Table 1. Five out of the six microcell hybrids represented near-diploid karyotypes, although a microcell hybrid containing 15 Gy-irradiated chromosomes 11 was near tetraploid. An intact chromosome

Table 1
Stability of the human chromosome 11 transferred into mouse m5S cells by microcell fusion

No. of cells	X-ray dose (Gy)	No. of cells scored	Karyotype of the introduced chromosome 11	No. of cells with the rearranged chromosome 11 after CT[a] (%)	Ploidy of the microcell hybrid (%)
2011-4	0	126	Intact	5 (4%)	Near diploid
2011-13	0	200	Intact	2 (1%)	Near diploid
2011-14	0	128	Intact	1 (0.8%)	Near diploid
6X11-5	6	50	Rearranged	8 (16%)	Near diploid
6X11-11	6	196	Intact	141 (72%)	Near diploid
2X11-1	15	150	Rearranged	67 (45%)	Near tetraploid

[a] Chromosome transfer.

11 existed in all three microcell hybrids transferred with unirradiated chromosome 11 and in a microcell hybrid, 6X11-11, transferred with 6 Gy-irradiated chromosome 11. In contrast, chromosome 11 existed in rearranged forms in microcell hybrids 6X11-5 and 2X11-1. The frequencies of the rearrangements of chromosome 11 after the chromosome transfer showed 16–72% in microcell hybrids containing irradiated chromosomes 11, whereas those frequencies were less than 4% in those containing unirradiated chromosomes 11 (Table 1). This result indicates that the irradiated chromosome per se is of an unstable nature. Of particular interest is the fact that 25% and 46% of the chromosomes 11 were rearranged to form rings and telomeric fusion chromosomes, respectively, in 6X11-11 cells. Similarly, in the 6X11-5 cells and the 2X11-1 cells, several types of deletions and translocations between the human chromosome and mouse chromosomes formed the majority of the chromosome rearrangements, possibly by telomeric fusion. These results suggest that telonomic instability induced by radiation is involved in the induction of delayed chromosome aberrations.

4. Discussion

Recent reports clearly demonstrate that the progeny of irradiated cells show a variety of delayed effects such as reproductive death, giant cell formation, chromosome aberration and gene mutation [2–5]. In the present study, we focused on the mechanism of formation of delayed chromosome aberrations and adopted the chromosome transfer technique to find out the critical target for inducing this phenomenon. It is quite reasonable to conclude that the chromosome transfer technique does not induce chromosome instability, because the chromosome analysis shows that chromosome instability is very low in the microcell hybrids transferred with the unirradiated chromosome. Therefore, the fact that the irradiated chromosome is more unstable than the unirradiated chromosome in the unirradiated recipient m5S cells, indicates that the major cause of delayed chromosomal instability occurred in the descendants of the surviving cells and exists within the irradiated chromosome per se.

The nature of initiation responsible for inducing chromosome instability remains unknown. Limoli et al. [5] examined the induction of delayed chromosomal instability by using several DNA strand-breaking agents, and found that DNA damage in the form of complex DSBs constitutes the signal that initiates the onset of chromosomal instability. However, little is known on how those lesions are transmitted through the chromosome. In the present study, we demonstrated that telomeric fusion is one of the major rearrangements observed in the microcell hybrids, suggesting that dysfunction of the telomeres contributes to the instability of the irradiated chromosome.

On the basis of the present results, we propose the model for the induction of delayed chromosome aberrations, mediated by telonomic instability as follows. Radiation induces dysfunction of telomere (telonomic instability) and this promotes intra- and inter-chromosomal telomeric fusions. Some fused chromosomes are broken at anaphase to produce new chromosomes, which have sticky ends. This cycle was originally proposed by McClintock [8] based on the study using *Zea mays* as the fusion–bridge–breakage cycle (FBB cycle). Recent reports [9,10] also suggest the telomeric fusions as a source of chromosomal instability. Thus, telonomic instability induced by IR is a crucial trigger to promote subsequent genomic rearrangements including delayed chromosome aberrations, and this chromosome-mediated instability can be transmissible over many cell divisions post-irradiation.

Acknowledgements

The authors thank Prof. Takeki Tsutsui, The Nippon Dental University, School of Dentistry, for kindly providing a PNA telomere probe. This work was partly supported by a Grant for Scientific Research from the Ministry of Education, Culture, Sports, Science and Technology of Japan and by a Grant from the Health Research Foundation, Kyoto, Japan.

References

[1] M.A. Kadhim, D.A. Macdonald, D.T. Goodhead, S.A. Lorimore, S.J. Marsden, E.G. Wright, Transmission of chromosomal instability after plutonium α-particle irradiation, Nature 355 (1992) 738–740.
[2] J.B. Little, Radiation-induced genomic instability, Int. J. Radiat. Biol. 74 (6) (1998) 663–671.
[3] K. Roy, S. Kodama, K. Suzuki, M. Watanabe, Delayed cell death, giant cell formation and chromosome instability induced by X-irradiation in human embryo cells, J. Radiat. Res. 40 (1999) 311–322.
[4] K. Roy, S. Kodama, K. Suzuki, K. Fukase, M. Watanabe, Hypoxia relieves X-ray-induced delayed effects in normal human embryo cells, Radiat. Res. 154 (2000) 659–666.
[5] C.L. Limoli, M.I. Kaplan, J.W. Phillips, G.M. Adair, W.F. Morgan, Differential induction of chromosomal instability by DNA strand-breaking agents, Cancer Res. 57 (1997) 4048–4056.
[6] S. Kodama, G. Kashino, K. Suzuki, T. Takatsuji, Y. Okumura, M. Oshimura, M. Watanabe, J.C. Barrett, Failure to complement abnormal phenotypes of simian virus 40-transformed Werner syndrome cells by introduction of a normal human chromosome 8, Cancer Res. 58 (1998) 5188–5195.
[7] P.M. Lansdorp, N.P. Verwoerd, F.M. van de Rijke, V. Dragowska, M.-T. Little, R.W. Dirks, A.K. Raap, H.J. Tanke, Heterogeneity in telomere length of human chromosomes, Hum. Mol. Genet. 5 (5) (1996) 685–691.
[8] B. McClintock, The stability of broken ends of chromosomes in *Zea mays*, Genetics 26 (1941) 234–282.
[9] R. Riboni, A. Casati, T. Nardo, E. Zaccaro, L. Ferretti, F. Nuzzo, C. Mondello, Telomeric fusions in cultured human fibroblasts as a source of genomic instability, Cancer Genet. Cytogenet. 95 (1997) 130–136.
[10] M.I. Kaplan, C.L. Limoli, W.F. Morgan, Perpetuating radiation-induced chromosomal instability, Radiat. Oncol. Invest. 5 (1997) 124–128.

Delayed cell-cycle arrest following heavy-ion exposure

S. Goto [a,b,*], S. Morimoto [a], T. Kurobe [b,c], M. Izumi [a], N. Fukunishi [d], M. Watanabe [b], F. Yatagai [a]

[a] *Division of Radioisotope Technology, The Institute of Physical and Chemical Research, Japan*
[b] *Laboratory of Radiation and Life Science, Department of Health Science, School of Pharmacological Science, Nagasaki University, Japan*
[c] *Fac. Science. Eng., Waseda University, Japan*
[d] *Cyclotron Center, The Institute of Physical and Chemical Research, Japan*

Ionizing radiation produces various kinds of damage, for example, DNA double- or single-strand break, and base damage. Cellular responses after ionizing radiation exposure can be considered to occur for the protection of cells from radiation damage. The purpose of this study is to assess the question, "Are there any different cellular responses between heavy-ions and X-rays?"

The human lymphoblastoid cell, TK6, was used in this study. The Riken Ring Cyclotron at the Institute of Physical and Chemical Research, Japan accelerated the heavy-ion beams. For heavy-ion irradiation in the present study, a LET of C- and Fe-ions was adjusted to 22 and 1000 keV/µm, respectively, at the cell sample site by the experimental procedures as schematically shown in the illustration.

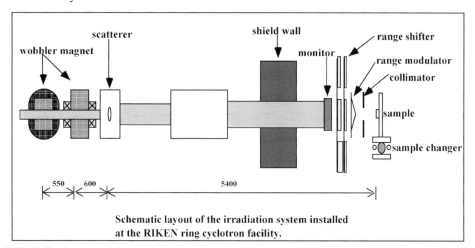

Schematic layout of the irradiation system installed at the RIKEN ring cyclotron facility.

* Corresponding author. Division of Radioisotope Technology, The Institute of Physical and Chemical Research, 2-1 Hirosawa, Saitama 351 0198, Japan.

The cells were irradiated with X-rays and heavy-ions to reduce the similar survival level (3–8%). The cell-cycle progression after irradiation was determined by using Laser Scanning Cytometry (LSC Olympus). The results showed an accumulation of cells at G2/M after 3 Gy X-ray irradiation. Compared to the X-ray case, the cell exposed to heavy-ion beams resulted in the delayed accumulation of G2/M cells.

In order to determine whether this delay is due to cell freezing in the S-phase, cells undergoing DNA replication in the S-phase were prelabeled with bromodeoxyuridine (BrdU) before irradiation to follow their cell-cycle progression. The migration pattern of the cells with incorporated BrdU demonstrated that the heavy-ion exposed cells remained in the S-phase longer. In terms of cell-cycle progression, the cellular responses are clearly different between X-ray and heavy-ion exposure.

These results suggest that heavy-ion-produced DNA damage is difficult to be repaired in the S-phase because the damages are different from those induced by X-ray irradiation. Taken together, the similar level of survival between the two types of radiation, the damages caused by heavy-ions are expected to distribute, not uniformly, but restrictedly through the cells.

Analysis of radiation-inducible hSNK gene in human thyroid cells

Yuki Shimizu-Yoshida*, Keiichi Sugiyama, Tatiana Rogounovitch, Hiroyuki Namba, Vladimir Saenko, Shunichi Yamashita

Department of Nature Medicine, Atomic Bomb Disease Institute, Nagasaki University School of Medicine, 1-12-4 Sakamoto, Nagasaki 852-8523, Japan

Abstract

The late effect of radiation on thyroid tumorigenesis is closely associated with the disturbance of cell cycle regulation. The polo-like kinase (PLKs) family plays an important role in several stages of mitosis. Using the cDNA subtraction method, we recently identified in irradiated cultured human thyroid cells a gene belonging to the PLKs family, the human homologue of mouse serum-inducible kinase (hSNK). Study of the gene structure revealed the presence of 14 exons spanning over 6 kb of genomic DNA, that encoded a 2.9 kb mRNA product ubiquitously expressed to a various extent in different human tissues. The 5′-flanking region of hSNK was examined to study the basic transcriptional activity of the gene. The core promoter with activation elements including the TATA-like motif and GC box resided within the first 120 nucleotides, upstream of an analogous transcription start site. Functional mapping of the regulatory region demonstrated the presence of both enhancer and repressor fragments. Analysis of 0.7 kb of the 5′-upstream region of the gene revealed a number of motifs comprising of putative binding sites for several transcription factors and, notably, the p53 binding homology sequence. In conclusion, the hSNK may be a unique target molecule to influence cell fate following irradiation. © 2002 Elsevier Science B.V. All rights reserved.

Keywords: Radiation; Thyroid; Cancer; hSNK; Polo-like kinase

1. Introduction

The thyroid gland is particularly vulnerable to radiation that manifests in the increased tumorigenesis in irradiated individuals. In order to determine the molecular events evoked by radiation in thyrocytes, we focused on searching for gene(s) that responds to radiation, and thus might be involved in thyroid tumorigenesis.

---

* Corresponding author. Tel.: +81-95-849-7114; fax: +81-95-849-7117.
E-mail address: shun@net.nagasaki-u.ac.jp (Y. Shimizu-Yoshida).

Following on from our recent finding, the human homologue of mouse serum-inducible kinase (hSNK) was rapidly up regulated after irradiation in cultured human thyroid cells. In our present work, we studied the structural organization of this gene and addressed questions of its transcriptional regulation.

2. Materials and methods

2.1. Cell culture

Quiescent primary thyroid cells maintained in monolayer were acutely irradiated with a single dose (0.939 Gy/min) of X-rays.

2.2. RNA and DNA analysis

The total RNA was extracted with Isogen 2 h after the irradiation. The mRNA was processed with a PCR-Select kit (Clontech), and the resultant fragments were cloned into pGEM-T Easy and screened by Northern blotting against 1 μg of the total RNA.

The entire genomic DNA of the hSNK gene was derived from the human DNA PAC library after PCR screening with primers specific for sequence-identified insert.

2.3. Plasmid construction and Luciferase assay

A 0.7 kb of the hSNK 5′-upstream region was sub cloned into pGL3-basic in the sense orientation, and serial deletion fragments were generated. Plasmids were co-tranfected into

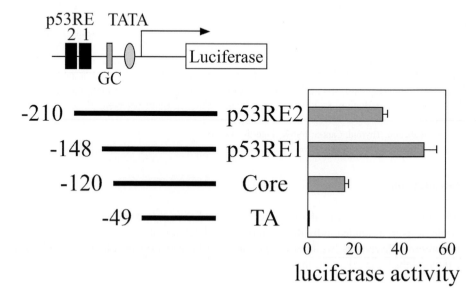

Fig. 1.

HeLa cells together with the reference vector, and 48 h later cell lysates were subjected to Dual-Luciferase assay.

3. Results

As demonstrated by Northern blot analysis, there was an enhanced detection of 2.9 kb of hSNK mRNA in cultured thyroid cells 2 h after irradiation. In addition, its level was found to be changing in a dose- and time-dependent manner.

Study of the gene structure showed the presence of 14 exons spanning over 6 kb of the genomic DNA. Analysis of 0.7 kb of the 5'-upstream region of the gene revealed a number of motifs comprising putative binding sites for several transcription factors (GATA-1/2, c-Ets, SP-1, CREB) and the p53 binding homology sequence (p53RE). In HeLa cells, when pS-TA was used as a control, the pS-Core had an order higher luciferase activity. Furthermore, the pS-p53RE1 displayed the highest basal promoter activity. The pS-p53RE2 construct, however, had repressed enzyme activity; suggesting the presence of basal suppression factor(s) binding site(s), see Fig. 1.

4. Discussion

The subtractive cDNA hybridization/PCR approach allowed us to identify the radiation inducible gene, hSNK, whose mRNA expression was up regulated in irradiated thyroid cells in a time- and dose-dependent manner.

Study of the genomic organization demonstrated the distribution of the coding region to 14 exons spanning over 6 kb of genomic DNA, that encodes a 2.9 kb mRNA product ubiquitously expressed to a various extent in different human tissues.

Sequence analysis of the 0.7 kb 5'-upstream regulatory region and functional mapping revealed the presence of the core promoter and suppressor fragment, as well as the occurrence of binding motifs for a variety of transcription factors including those of p53. Thus, the hSNK gene may be the subject of complex regulation and further experiments are underway to determine key factors responsible for its radiation and/or stimulation induced transcriptional activation.

Cellular response in normal human cells exposed to chronically low-dose radiation in heavy-ion radiation field

Masao Suzuki [a,*], Hiroshi Yasuda [a], Ryonfa Lee [b], Chisa Ohira [b], Hideyuki Majima [a], Yoko Yamaguchi [c], Chizuru Yamaguchi [a], Kazunobu Fujitaka [a]

[a] *International Space Radiation Laboratory, National Institute of Radiological Sciences, 4-9-1 Anagawa, Inage, Chiba 263-8555, Japan*
[b] *Frontier Research Center, National Institute of Radiological Sciences, 4-9-1 Anagawa, Inage, Chiba 263-8555, Japan*
[c] *School of Dentistry, Nihon University, 1-8-13 Kanda-Surugadai, Tokyo 101-8310, Japan*

Abstract

We have been studying the biological effects in normal human fibroblasts exposed to chronically low-dose radiation in a heavy-ion radiation field. The cells were cultured in a CO_2 incubator, which was placed in the irradiation room for biological study of the heavy ions in the Heavy Ion Medical Accelerator in Chiba (HIMAC), at the National Institute of Radiological Sciences (NIRS). We measured the absorbed dose, which was determined by a thermoluminescence dosimeter (TLD) and a Si-semiconductor detector, to be 1.4 mGy per day when operating the HIMAC machine. The total population doubling number (tPDN) of the exposed cells reduced to 77–94% of the non-exposed control cells. Furthermore, the shortening of the speed of the telomere length in the exposed cells was much higher than that in the non-exposed cells at the 26th passage, during which period the exposed cells were accumulated at 0.22 Gy. These findings show that the acceleration of senescence in normal human cells occurs by the chronically low-dose irradiation in a heavy-ion radiation field.
© 2002 Elsevier Science B.V. All rights reserved.

Keywords: Low-dose radiation; Heavy ion; Normal human fibroblasts; Life span; Telomere

1. Introduction

At a time when manned space exploration is more a reality with the planned International Space Station (ISS) well underway, the biological effects of chronically low-dose

---

* Corresponding author. Tel.: +81-43-206-3238; fax: +81-43-251-4531.
  E-mail address: m_suzuki@nirs.go.jp (M. Suzuki).

(rate) irradiation in the field of low-flux galactic cosmic rays (GCR) have become one of the major concerns of space science. It is therefore important to estimate the radiation risk on the human body of high-energy and charged (HZE) particles, especially chronically low-dose irradiation in the HZE field. One useful approach in the risk assessment of the HZE particles is to investigate the cellular responses using normal human cells. We have been studying the cellular response in normal human fibroblasts exposed to chronically low-dose radiation in a heavy-ion radiation field.

2. Materials and methods

2.1. Cell culture

Normal human skin fibroblasts (NB1RGB) distributed by the RIKEN Cell Bank (Cell No. RCB0222) were used in this study. The cells were stored in liquid nitrogen at passage 6 (total population doubling number=16.0) and we used these cells for each experiment at the same passage thawed from the stock in a liquid nitrogen tank. The cells were cultured in Eagle's minimum essential medium (MEM) containing kanamycin (60 mg/l), supplemented with 10% fetal bovine serum in a 5% CO_2 incubator at 37 °C.

2.2. Exposure of chronically low-dose radiation

The cells were inoculated in a plastic flask (75 cm^2) with a density of 1×10^6 cells per flask. The flasks were placed in a CO_2 incubator, which was set in the irradiation room for biological study of the heavy ions in the Heavy Ion Medical Accelerator in Chiba (HIMAC) at the National Institute of Radiological Sciences (NIRS) in Japan. Dosimetry in the incubator was carried out using a thermoluminescence dosimeter (TLD) and a Si-semiconductor detector. We measured the absorbed dose to be 1.4 mGy per day when operating the HIMAC machine [1].

2.3. Life-span determination

The cells were sub-cultured at a density of 1×10^6 cells per flask every 7 days. For every subcultivation, the number of cells in the flask (N) was counted with a hemocytometer and a Coulter Counter. The mean population doubling number (PDN) and the total population doubling number (tPDN) were calculated using the following formulas:

$$PDN = \log(N/10^6)/\log 2$$
$$tPDN = \sum (PDN)_n$$

3. Results

Fig. 1 shows the representative growth curves (exp. 1) of both chronically low-dose exposed and non-exposed control cells. Both cell cultures stopped growing after about 446

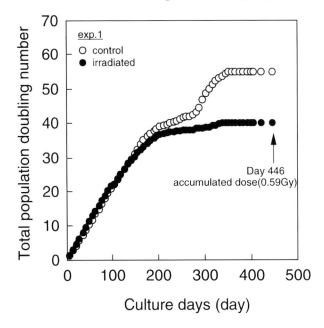

Fig. 1. Representative growth curves (exp. 1) of both chronically low-dose exposed (closed circle) and non-exposed control cells (open circle).

days. The accumulated dose of the exposed cell culture was 0.59 Gy during this period. The total population doubling number (tPDN) of the exposed cell population was significantly smaller than that of the control cell population. The results of the four independent experiments are summarized in Table 1. The tPDN of the exposed cells was reduced to 77–94% when compared to the control cells.

4. Conclusion

Our results clearly show the reduction in the life span of the chronically low-dose exposed cells. This result was the opposite of the data previously recorded using a gamma-ray field [2,3]. The results suggest that the different cellular response for the chronically low-dose irradiation occurs between the gamma-ray and heavy-ion field.

Table 1
Reduced life span of chronically low-dose exposed cells relative to non-exposed control cells

Exp. #	tPDN (exposed)/tPDN (control)
Exp. 1	0.77
Exp. 2	0.94
Exp. 3	0.85
Exp. 4	0.88

References

[1] H. Yasuda, M. Suzuki, K. Ando, K. Fujitaka, Simulation of the low-Earth-orbit dose rates using secondary radiations from the HZE particles at NIRS-HIMAC, Phys. Med. XVII (suppl. 1) (2001) 133–136.
[2] M. Watanabe, M. Suzuki, K. Suzuki, K. Nakano, K. Watanabe, Effect of multiple radiation with low doses of gamma rays on growth ability of human embryo cells in vitro, Int. J. Radiat. Biol. 62 (6) (1992) 711–718.
[3] M. Suzuki, Z. Yang, K. Nakano, F. Yatagai, K. Suzuki, S. Kodama, M. Watanabe, Extension of in vitro life-span of gamma-irradiated human embryo cells accompanied by chromosome instability, J. Radiat. Res. 39 (3) (1998) 203–213.

Inhibition of radiation-induced DNA-double strand break repair by various metal/metalloid compounds

Sentaro Takahashi [a,*], Ryuichi Okayasu [b], Hirosi Sato [a], Yoshihisa Kubota [a], Joel S. Bedford [b]

[a] Environmental and Toxicological Research Group, National Institute of Radiological Sciences, 4-9-1 Anagawa, Inage, Chiba 263-8555, Japan
[b] Department of Radiological Health Science, Colorado St. University, Fort Collins, CO 80523, USA

Abstract

In order to clarify the mechanism underlying the combined effects of radiation and metals, we have studied the cytotoxicity and inhibition of radiation-induced DNA double-strand-break (DNA-DSB) repair by various metals and metalloids. CHO cells were treated with $NiCl_2$, $NaAsO_2$, $ZnCl_2$, $CdCl_2$, $CuCl_2$, and potassium antimonyl tartrate for 2 h, and irradiated with γ-rays at a dose of 40 Gy to induce DNA-DSB. Then, the degree of DNA-DSB was determined by an electrophoresis technique immediately after irradiation and following a 30-min repair period. The DNA-DSB repair was significantly inhibited by exposure to Ni, Cu, Zn, As, Sb, and Cd compounds at concentrations of 200, 2.0, 0.4, 0.08, 0.55, and 1.0 mM, respectively. At these concentrations, the cell viability determined by trypan blue dye exclusion was over 50% for all the chemicals, suggesting that all of these compounds inhibited the repair of radiation-induced DNA-DSB at the concentrations where the acute cytotoxicity was relatively low. More significantly, the plating efficiencies for As, Sb, and Zn compounds at these concentrations were higher than 10%, suggesting that these chemicals inhibited DNA-DSB repair at relatively low concentrations where some of the cells sustained the ability of proliferation. © 2002 Elsevier Science B.V. All rights reserved.

Keywords: DNA-DSB; Repair; Metal; Arsenite; Antimony; Radiation

1. Background

Recent advances in science and technology involve the potential release of various toxic agents, which might increase the human health risk. Since simultaneous exposure to multiple toxicants frequently occur in the actual living environment, studies on the combined and/or

---

* Corresponding author. Tel.: +81-43-206-3158; fax: +81-43-251-4853.
E-mail address: sentaro@nirs.go.jp (S. Takahashi).

competitive effects of toxicants are of great importance. Some heavy metals and arsenic demonstrated supra- or sub-additive effects when combined with ionizing or nonionizing radiation [1,2], and the inhibitory effect of heavy metals on the repair of radiation-induced DNA damage is thought to be responsible for the combined effect. Recently, we found that arsenite inhibited the repair of not only DNA single-strand-breaks (SSB) but also DNA double-strand-break (DSB) induced by ionizing radiation [3]. There is little information available on the influence of metals on DNA-DSB repair. On the other hand, the effects of metals on the repair of DNA single-strand-breaks (SSBs) and base damages have been well studied. DNA-DSBs are thought to be the key lesion in radiation-induced cell death as well as mutation/transformation. Here, we report the data on the cytotoxicity and inhibition of radiation-induced DNA-DSB repair by various metals including arsenite and antimony.

2. Methods

Chinese hamster ovary cells (CHO-K1) were cultured in Ham's F-12 medium and labeled with [^{14}C]-thymidine. The cells were exposed to various concentrations of chemicals for 2 h. The chemicals used were $NiCl_2$, $NaAsO_2$, $ZnCl_2$, $CdCl_2$, $CuCl_2$, and potassium antimonyl tartrate, which have been dissolved in Hank's solution with an adjusted pH. Then the cells were irradiated using a ^{137}Cs irradiator at a dose of 40 Gy to induce DNA-DSB, and processed for further DNA-DSB analyses immediately after irradiation (nonrepair group), or following a 30-min repair incubation period (repair group). The DNA-DSB assay was carried out with pulsed field gel electrophoresis according to methods described elsewhere [4]. The amount of DNA-DSB was expressed as the fraction of activity released (FAR), that is, the disintegration per minute (DPM) of a lane divided by the total DPM lane (proportional to the migrated DNA) + plug (proportional to the control size DNA) per sample. In addition, the repair fraction (fraction of repaired DNA-DSB) was calculated by the following equation:

Repair fraction = $(FAR(0) - FAR(30))/FAR(0)$,

where FAR(0) and FAR(30) denote the FAR values at 0 and 30 min after irradiation, respectively.

The proliferative capability of the cells was determined by a colony-forming assay. The cells were prepared and exposed to each chemical for 2.5 h, which corresponded to the sum of the preincubation time with chemicals (2 h) and repair incubation time (0.5 h). Then they were trypsinized, counted, and replated onto tissue culture dishes so that 100–500 colonies would yield, per dish, 8–9 days later. Acute cytotoxicity of the chemicals was determined by trypan blue dye exclusion. After 2.5 h of exposure to the chemicals, the number and viability of the nonadherent cells collected by centrifugation and the adherent cells on the dish were determined by a dye exclusion test with 0.5% trypan blue.

3. Results

In the control group (no chemical treatment), the FAR values were 21.9% immediately after the 40 Gy irradiation and 8.7% after the 30-min repair culture. This suggests that

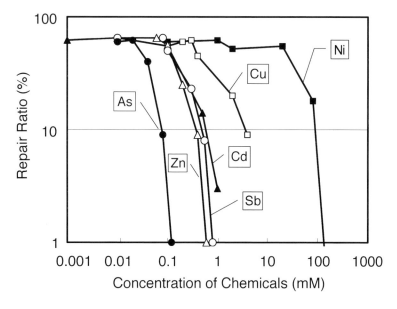

Fig. 1. The effect of Ni, Cu, Cd, Sb, Zn, and As on the repair of radiation-induced DNA-DSB.

approximately 60% of the DNA-DSB had been rejoined during the 30-min repair incubation period, since it has been shown that the FAR was proportional to the number of DNA-DSB in a linear fashion in this dose range (0–40 Gy).

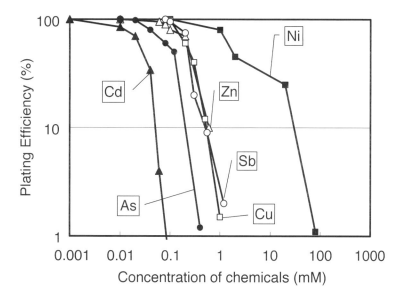

Fig. 2. Plating efficiencies of CHO cells after exposure to Ni, Cu, Cd, Sb, Zn, and As for 2.5 h.

The DNA-DSB repair was inhibited by exposure to all the chemicals used here. The repair fractions were decreased to less than 10% for Ni, Cu, Zn, As, Sb, and Cd at concentrations of 200, 4.0, 0.4, 0.08, 0.55, and 1.0 mM, respectively (Fig. 1). At these concentrations, the viabilities determined by trypan blue dye exclusion were over 50% for all the chemicals, suggesting that all of these metal compounds inhibited the repair of radiation-induced DNA-DSB at the concentrations where the acute cytotoxicity was relatively low. The plating efficiency after the chemical exposure is shown in Fig. 2. The most toxic chemical was Cd and the least toxic was Ni in the present experimental condition. Comparison of the repair ratio (Fig. 1) with the plating efficiency (Fig. 2) indicates that As, Sb, and Zn inhibited DNA-DSB repair at relatively low concentrations where some of the cells sustained the ability of proliferation. In contrast, Ni, Cu, and Cd inhibited the repair only at cytotoxic concentrations where cells lost their proliferative capability.

4. Conclusion

The present studies demonstrate that some metal/metalloid compounds inhibit the repair of DNA-DSB at relatively low biological concentrations. Since DNA-DSBs are thought to be the key lesion in radiation-induced cell death as well as mutation/transformation, these results provide a new insight into the mechanism of the combined effects of radiation and other environmental toxicants.

References

[1] A. Hartwig, Current aspects in metal genotoxicity, BioMetals 8 (1995) 3–11.
[2] A. Hartwig, Carcinogenicity of metal compounds: possible role of DNA repair inhibition, Toxicol. Lett. 102 (1998) 235–239.
[3] S. Takahashi, E. Takeda, Y. Kubota, R. Okayasu, Inhibition of radiation induced DNA-double strand break repair by nickel and arsenite, Radiat. Res. 154 (2000) 686–691.
[4] R. Okayasu, K. Suetomi, R.L. Ullrich, Wortmannin inhibits repair of DNA-double strand breaks in irradiated normal human cells, Radiat. Res. 149 (1998) 440–445.

Induction of a large deletion in mitochondrial genome of mouse cells by X-ray irradiation

T. Ikushima [a,*], T. Andoh [a], T. Kaikawa [a], K. Hashiguchi [b]

[a] *Biology Division, Kyoto University of Education, Kyoto 612-8522, Japan*
[b] *Laboratory of Radiation Biology, Graduate School of Science, Kyoto University, Kyoto 605-8502, Japan*

Abstract

Large deletions and point mutations of mitochondrial DNA (mtDNA) is causally associated with mitochondrial diseases, and accumulates with age in human tissues. In order to find out the role of oxidative stress in the generation of large mtDNA deletions, using normal Balb and severe combined immunodeficiency (SCID) mouse cells, we assessed whether X-ray irradiation induces large mtDNA deletions. Cultured cells were irradiated with X-rays and assayed for a large mtDNA deletion using a PCR technique with a specific pair of primers. X-ray doses as low as 1 Gy were effective for the induction of the large mtDNA deletion (5823 bp long), which corresponds to the human 4977 bp common deletion. The breakpoints were flanked by the 5-bp long tandem repeats, 5′-TACCC-3′. A dose-dependent induction of the large mtDNA deletion was observed, the yield being higher in normal cells than in SCID cells. The fraction of large mtDNA deletion increased in the normal cells but decreased in the SCID cells within 7 days after X-ray irradiation. As the SCID cells carry a mutation in the gene encoding DNA–PKcs, the key enzyme in DNA double-strand break repair, it can be concluded that repair of DNA strand breaks may be involved in the formation of X-ray induced large mtDNA deletions. © 2002 Elsevier Science B.V. All rights reserved.

Keywords: X-rays; Mitochondrial DNA; Deletion; SCID; Mouse

1. Background

Many mitochondrial DNA (mtDNA) point mutations and DNA rearrangement mutations have been found responsible for specific human diseases [1]. Large deletions, such as a deletion of 4977 bp, are associated with myopathies such as Kearns–Sayer syndrome and progressive external ophthalmoplegia [2]. Furthermore, these deletions accumulate during normal aging in different human, mammal and rodent tissues [3]. Damage to the

---

* Corresponding author. Tel.: +81-75-644-8266; fax: +81-75-645-1734.
E-mail address: ikushima@kyokyo-u.ac.jp (T. Ikushima).

mtDNA is caused by an attack of the reactive oxygen species (ROS), and the mtDNA appears to be more prone to oxidative damage than nuclear DNA [4]. ROS are endogenously generated as byproducts of normal cellular respiration in mitochondria as well as externally generated by exposure to ionizing radiation. Though oxidative stress may contribute to the generation of large mtDNA deletions, any direct evidence for this contribution has not yet been provided. We therefore, using normal Balb and severe combined immunodeficiency (SCID) mouse cells, assessed whether large mtDNA deletions are induced by X-ray irradiation. A SCID mutation impairs the recombination activity involved in the joining of V, D, and J gene segments encoding immunoglobulin. As SCID cells carry a mutation in the gene encoding DNA–PKcs, they can repair neither the DNA double-strand breaks nor the single-strand breaks [5].

2. Methods

2.1. Cell culture

The experiments were performed with the mouse lung fibroblast cell line from a SCID mouse, SC3VA2, and its parental normal cell line, Balb YSV [6]. The cells were cultured in α-MEM (Sigma) supplemented with 10% fetal calf serum and antibiotics at 37 °C, in a humidified atmosphere containing 5% CO_2.

2.2. X-ray irradiation and cell survival assay

After actively growing cells were exposed at room temperature to 140 kVp X-rays at a dose rate of 1.1 Gy/min, cell survival was measured using a clonogenic assay.

2.3. DNA isolation, polymerase chain reaction (PCR) and DNA sequencing

A large mtDNA deletion was detected by PCR-based assay. The total DNAs were isolated by standard phenol/chloroform extractions 1, 3 and 7 days after X-irradiation. A specific pair of primers, TF1 (5′-AACAGTAACATCAAACCGACCAGG-3′, 7558–7581) and TR1 (5′-TATAGTTGGAAGGAGGGATTGGGTA-3′, 13,666–13,642) was used. The PCR was initiated with a 3 min incubation at 94 °C, followed by 30 cycles of denaturation at 94 °C for 1 min; annealing at 50 °C for 1 min; extension at 72 °C for 1 min, with a final extension at 72 °C for 1 min. The PCR-amplified fragments were separated by 8% polyacrylamide gel electrophoresis and detected by SYBR Gold staining. The relative yield of each amplified fragment was determined by image analysis using NIH Image software (version 1.61). The target DNA fragments were extracted from the agarose gel slices and sequenced directly by thermal cycle sequencing as previously described [7].

3. Results

In this study, we used two murine cell lines, SC3VA2 (SCID) and Balb YSV (normal). The D_{10} values of SC3VA2 and Balb YSV cells were 4.5 and 6.5 Gy, respectively. A large

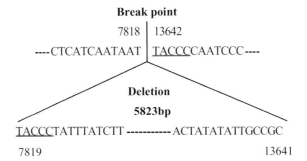

Fig. 1. Nucleotide sequence around the break points of a large mtDNA deletion. Underlines indicate 5 bp long direct repeats.

mtDNA deletion was detected in 3- and 7-day cultures of X-rayed cells, but not in the 1-day culture. The deleted region was 5823 bp long (7819–13,641), about one third of the whole mitochondrial genome of 16,295 bp, and included the same gene array as the human 4977 bp common deletion. The breakpoints were flanked by the 5 bp long direct repeats, 5′-TACCC-3′ (Fig. 1). A dose-dependent induction of the large mtDNA deletion was observed, the yield being higher in normal Balb cells than SCID cells (Fig. 2). These results indicate that X-ray doses as low as 1 Gy are effective for the induction of large mtDNA deletions. Furthermore, the fraction of large mtDNA deletions increased in normal cells, but decreased in SCID cells within 7 days after X-ray irradiation. As SCID cells carry a mutation in the gene encoding DNA–PKcs, the key enzyme in DNA double-strand break repair [5], the results suggest that the repair of the DNA strand breaks might be involved in the formation of X-ray induced large mtDNA deletions probably through the slip-replication mechanism [2].

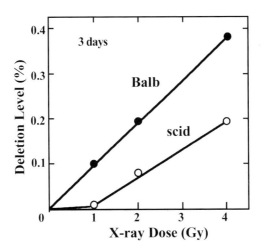

Fig. 2. The yield of X-ray induced large mtDNA deletions in normal Balb and SCID murine cells.

4. Conclusion

Relatively low doses of X-rays are effective for the generation of large mtDNA deletions, possibly through a slip-replication mechanism.

Acknowledgements

We thank Dr. K. Komatsu (Hiroshima University, Hiroshima, Japan) for supplying the murine SCID cell line. This study was partially supported by a Grant-in-Aid for Scientific Research from the Ministry of Education, Science, Sports and Culture of Japan.

References

[1] A.M. Kogenlnik, M.T. Lott, M.D. Brown, S.B. Navathe, D.C. Wallace, MITMAP: a human mitochondrial genome database—1998 update, Nucleic Acid Res. 26 (1998) 112–115.
[2] J.M. Shoffner, M.T. Lott, A.S. Vojavec, S.A. Souedan, D.A. Costigan, D.C. Wallace, Spontaneous Kearns–Sayre/chronic external ophthalmoplegia plus syndrome associated with a mitochondrial DNA deletion: a slip-replication model and metabolic therapy, Proc. Natl. Acad. Sci. U. S. A. 86 (1989) 7952–7956.
[3] S. Simonetti, X. Chen, S. DiMauro, E.A. Schon, Accumulation of deletions in human mitochondrial DNA during normal aging: analysis by quantitative PCR, Biochim. Biophys. Acta 1180 (1992) 113–122.
[4] C. Richter, J.-W. Park, B.N. Ames, Normal oxidative damage to mitochondrial and nuclear DNA is extensive, Proc. Natl. Acad. Sci. U. S. A. 85 (1988) 6465–6467.
[5] R.H. Alrefai, E.J. Beechamt, V.A. Bohr, P.J. Gearhart, Less repair of pyrimidine dimers and single-strand-breaks in genes by scid cells, Biochem. Biophys. Res. Commun. 264 (1999) 878–882.
[6] K. Komatsu, M. Yoshida, Y. Okumura, Murine scid cells complement ataxiatelangiectasia cells and show a normal post-irradiation response of DNA synthesis, Int. J. Radiat. Biol. 63 (1993) 725–730.
[7] K. Hashiguchi, T. Ikushima, Nucleotide changes in mitochondrial 16S rRNA gene from different mammalian cell lines, Genes Genet. Syst. 73 (1998) 317–321.

Effects of increased telomerase activity on radiosensitive human SCID cells

Yoshiko Arase [a,*], Katsuo Sugita [a], Takaki Hiwasa [b], Hiroshi Shirasawa [c], Kazunaga Agematsu [d], Hisao Ito [e], Nobuo Suzuki [a,1]

[a] *Department of Environmental Biochemistry, Graduate School of Medicine, Chiba University, Chiba 260-8670, Japan*
[b] *Department of Biochemistry and Genetics, Graduate School of Medicine, Chiba University, Chiba 260-8670, Japan*
[c] *Department of Molecular Virology, Graduate School of Medicine, Chiba University, Chiba 260-8670, Japan*
[d] *Department of Infectious Immunology and Pediatrics, Graduate School of Medicine, Shinshu University, Matsumoto 390-8621, Japan*
[e] *Department of Radiology, Graduate School of Medicine, Chiba University, Chiba 260-8670, Japan*

Abstract

Human severe combined immune deficiency (SCID) syndrome is a heterogeneous disease, and some SCID mutations are characterized by a high sensitivity to ionizing radiation (RS-SCID). This phenomenon is suspected to be due to impaired repair systems for damaged DNA [J. Exp. Med. 188 (1998) 627]. However, the details are not fully understood. In this work, two RS-SCID cell lines were conferred radioresistance by infection with a recombinant adenovirus encoding *E6/E7* (AxE67), and by transfection with an expression vector bearing human telomerase reverse transcriptase (*hTERT*). Our findings suggest that overexpressed *E6* and *E7* or elevated telomerase activity have some influence on the acquired radioresistance in human RS-SCID cells. © 2002 Elsevier Science B.V. All rights reserved.

Keywords: Telomerase activity; Sensitivity to ionizing radiation; Human SCID cells; *E6*; Transfection

1. Introduction

Human severe combined immune deficiency (SCID) mutations are heterogeneous, and are categorized into two groups in terms of sensitivity to ionizing radiation. A subset of

---

* Corresponding author.
E-mail address: nobuo@med.m.chiba-u.ac.jp (N. Suzuki).
[1] Tel.: +81-43-226-2039; fax: +81-43-226-2041.

0531-5131/02 © 2002 Elsevier Science B.V. All rights reserved.
PII: S 0 5 3 1 - 5 1 3 1 (0 2) 0 0 2 9 3 - 5

SCID has a mutation in either the RAG1 or RAG2 gene [2], which are required for the initiation of the V(D)J recombination [3], and this subset is normosensitive to ionizing radiation. On the other hand, RS-SCID is accompanied by an increased sensitivity to ionizing radiation, and these cells appear to have impaired DSB repair after irradiation [1]. However, the details are not fully understood.

In this work, we used two types of human RS-SCID cells in which the causes are unknown thus far. To begin understanding the radiosensitivity of RS-SCID, we induced radioresistance deliberate by infection with a recombinant adenovirus encoding papillomavirus *E6/E7*, according to previous studies [4,5]. They had extended life spans and increased telomerase activity. A previous report demonstrated that expression of *E6* activates telomerase [6]. Thus, we hypothesized that telomerase activity might be associated with acquired radioresistance. Next, we observed whether the increased telomerase activity by overexpressed human telomerase reverse transcriptase (*hTERT*) resulted in reduced radiosensitivity. Our findings suggest an association between increased telomerase activity and acquisition of radioresistance in human RS-SCID cells.

2. Materials and methods

SCID1 and SCID4 fibroblast cells were established from patients with $T^-B^-NK^+SCID$ syndrome. Viral infection was performed with a recombinant adenovirus AxCACSHPV16-E67 (AxE67), which encodes *E6/E7* genes of papillomavirus. Telomerase activity levels were assayed with PCR-based TRAP assay using a TRAPEZE Telomerase Detection Kit

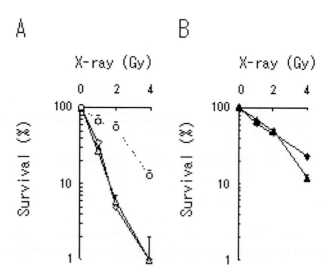

Fig. 1. Colony survival curves. (A) Primary cultured SCID1 (◇), SCID4 (△), and control fibroblast cells (○). (B) Infected SCID1+av (◆) and SCID4+av cells (▲).

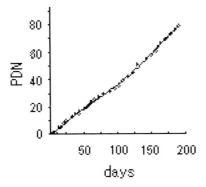

Fig. 2. Proliferating profiles of infected SCID1 cells with AxE6/E7. PDN: population doubling number.

(Oncol, Gaithersburg, DM, USA). Transfection was performed with an expression vector, pBABE-puro-*hTERT*, which was a kind gift from Dr. Robert A. Weinberg.

3. Results

3.1. Infection with AxE67 conferred radioresistance in RS-SCID cells

SCID1 and SCID4 cells showed higher radiosensitivity than control NC1 cells (Fig. 1A). We infected RS-SCID cells with AxE67. Infected SCID cells were radioresistant compared to the primary cultured SCID cells (Fig. 1B).

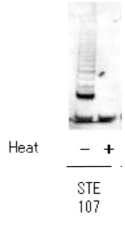

Fig. 3. Increased telomerase activity in *hTERT*-transfected STE107 cells. To rule out false-positive signals from PCR artifacts, cell extracts were incubated at 85 °C for 10 min before PCR-amplification.

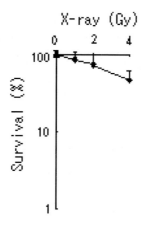

Fig. 4. Colony survival curves of hTERT-transfected STE107 cells (♦).

3.2. Infected cells had prolonged life spans and increased telomerase activity

In addition to the difference in radiosensitivity, infected cells had prolonged life spans compared with primary cultured cells (Fig. 2). To confirm the cause of the prolonged life spans, we examined telomerase activity. Infected cells showed increased telomerase activity.

3.3. Clones overexpressing hTERT were radioresistant

In an effort to investigate the relationship between increased telomerase activity and acquired radioresistance, we transfected RS-SCID cells with a vector bearing hTERT. Transfected cells were cloned and the expression of the introduced hTERT was examined by TRAP assay (Fig. 3). Next, sensitivity to ionizing radiation was examined in a hTERT-transfected clones, STE107 cells. STE107 cells were radioresistant compared with primary cultured cells (Fig. 4).

4. Conclusion

We generated radioresistant SCID cells for comparison. In addition to the acquired radioresistance, the infected cells also had prolonged life spans and increased telomerase activity. We examined the effects of overexpressed telomerase activity. hTERT-transfected cells were radioresistant compared with primary cultured cells.

Our findings suggest that overexpressed *E6* and *E7*, or overexpressed telomerase activity have some role in the reduction in sensitivity to ionizing radiation.

Acknowledgements

Expression vector, pBABE-puro-*hTERT*, was a kind gift from Dr. Weinberg. This work was supported in part by a grant-in-aid from the Smoking Research Foundation, "Ground

Research for Space Utilization" promoted by NASDA and Japan Space Forum, the Nissan Science Foundation, the Japan Atomic Energy Research Institute, by the contract on the Nuclear Safety Research Association, and grant-in aid from the Ministry of Education, Science, Sports and Culture, Japan.

References

[1] N. Nicolas, D. Moshous, M. Cavazzana-Calvo, D. Papadopoulo, R. de Chasseval, F. Le Deist, A. Fischer, J.P. de Villartay, A human severe combined immunodeficiency (SCID) condition with increased sensitivity to ionizing radiations and impaired V(D)J rearrangements defines a new DNA recombination/repair deficiency, J. Exp. Med. 188 (17) (1998) 627–634.

[2] K. Schwarz, G.H. Gauss, L. Ludwig, U. Pannicke, Z. Li, D. Lindner, W. Friedrich, R.A. Seger, T.E. Hansen-Hagge, S. Desiderio, M.R. Lieber, C.R. Bartram, RAG mutations in human B cell-negative SCID, Science 274 (5284) (1996) 97–99.

[3] P. Mombaerts, J. Iacomini, R.S. Johnson, K. Herrup, S. Tonegawa, V.E. Papaioannou, RAG-1-deficient mice have no mature B and T lymphocytes, Cell 68 (5) (1992) 869–877.

[4] N.M. Tsang, H. Nagasawa, C. Li, J.B. Little, Abrogation of p53 function by transfection of HPV16 E6 gene enhances the resistance of human diploid fibroblasts to ionizing radiation, Oncogene 10 (12) (1995) 2403–2408.

[5] M. Shimakage, Y. Miyata, H. Inoue, M. Yutsudo, A. Hakura, Increased sensitivity of EBNA2-transformed rat fibroblasts to ionizing radiation, Int. J. Cancer 68 (5) (1996) 612–615.

[6] A.J. Klingelhutz, S.A. Foster, J.K. McDougall, Telomerase activation by the E6 gene product of human papillomavirus type 16, Nature 380 (6569) (1996) 79–82.

Protein synthesis, cellular defence and *hprt*-mutations induced by low-dose neutron irradiation

Annamaria Dám [a,*], Noemi E. Bogdándi [a], István Polonyi [a], Márta M. Sárdy [a], Imre Balásházy [b], József Pálfalvy [b]

[a]*National Research Institute for Radiobiology and Radiohygiene (NRIRR), National Centre of Public Health, H-1775 Budapest, Hungary*
[b]*KFKI Atomic Energy Research Institute, P.O. Box 49, H-1525 Budapest, Hungary*

Abstract

The mutagenic potential of low-dose neutrons was studied to demonstrate RAR, and was aimed to recognise the correlation of the effectiveness of antioxidant enzymes and the synthesis of certain proteins. The CHO cells were adapted by priming exposure with neutron doses of 0.5–50 mGy and with dose rates of 1.59–10 mGy min^{-1}. A total of 2–4 Gy of gamma radiation after 1–48 h then challenged the cells. The number of induced *hprt*-mutants scored was to be reduced by 33–57% when priming was administered, compared to a single dose of challenge, depending on the adapting dose and dose rate. For adaptation, the most appropriate priming dose was 2 mGy and 5 h was the optimal time for its development. Six main, induced proteins were detected (109, 70, 65, 60, 57, 45 kDa). The protein 70 kDa was identified after the priming and challenge dose, but it disappeared after their combination. The capacity of the endogenous antioxidant (superoxide dismutase, SOD) to remove reactive oxygen species had been shown to be enhanced. It could be concluded that low-dose neutrons also induced RAR and the contribution of induced proteins; furthermore, the antioxidant potential would be in part responsible for the RAR. © 2002 Elsevier Science B.V. All rights reserved.

Keywords: Radio-adaptive response; Low-dose neutrons; Mutation; Superoxide dismutase; Proteins

1. Introduction

The discovery of radio-adaptation processes in cells, i.e., increased resistance to the effects of a "challenge dose" administered after a lower "adapting dose", has fuelled the

Abbreviations: CHO, Chinese hamster ovary cells; SOD, superoxide dismutase; kDa, kilodalton.
* Corresponding author. Tel./fax: +36-1-482-2011.
E-mail addresses: dam@hp.osski.hu (A. Dám), nbogdandi@hotmail.com (N.E. Bogdándi), istvan1@hotmail.com (I. Polonyi), sardy@hp.osski.hu (M.M. Sárdy), ibalas@sunserv.kfki.hu (I. Balásházy), palfalvy@sunserv.kfki.hu (J. Pálfalvy).

debate on possible cellular processes relevant for low-dose exposures and their consequence on the dose risk assessment.

The adaptive response induced by a low dose may result from the induction of a novel, efficient DNA repair mechanism, which leads to less residual damage [1]. Moreover, an adapting dose of X-rays triggers the protection against oxidant stress. This cross-adaptive response suggests that lesions mediated by reactive oxygen species could act as a signal for resistance, and that an increase in the endogenous antioxidant potential could contribute to the protective effect [2].

Radio-adaptation is extensively studied for low LET radiation. Nevertheless, there is not much data available for high LET radiation at very low doses and dose rates. Our study was aimed (1) to demonstrate radio-adaptation induced by low-dose neutrons by the detection of genotoxic damage, i.e., mutation induction on the *hprt* locus; (2) to investigate whether the response triggered by low-dose pre-exposure could be due to the activation of the antioxidant defence system; (3) to verify that the adaptation is related to altered protein synthesis and to special proteins induced de novo.

2. Materials and methods

2.1. Cell culture and irradiation protocols

Monolayers of the Chinese hamster ovary (CHO) cells were irradiated by neutrons produced in the biological irradiation channel of the research reactor, of the Budapest Neutron Centre. The neutron doses ranged between 0.5 and 50 mGy with dose rates of 1.59 and 10 mGy min^{-1}. The challenging gamma dose (2–4 Gy) was delivered 1–3–5–24–48 h later (dose rate: 200 mGy min^{-1}).

2.2. Mutation assay

After irradiation, the cells were trypsinized and the appropriate number of cells were plated in Petri dishes and assayed for the survival and expression of mutations, and by adding 6-thioguanine the induced *hprt*-deficient mutants were selected. The mutation frequency (MF) was expressed as the number of resistant colonies divided by the total number of viable cells, as determined by the cloning efficiency at the time of selection.

2.3. Enzymatic assays and gel electrophoresis

After irradiation, the enzymatic assay and protein gel electrophoresis were carried out in parallel with mutation and survival assays. Enzymatic activity was determined spectrophotometrically by using a Ransod kit (Randox Laboratories), on the basis of the degree of inhibition of the reaction between the superoxide radical and the red formazan dye, 2-(4-iodophenyl)-3-(4-4-nitrophenol)-5-phenyltetrazolium chloride (INT). In case of protein determination after irradiation(s), cells were lysed and subjected to SDS-PAGE electrophoreses. For identification of the proteins, a logarithmic calibration curve was constructed relative to the migration of a protein standard (Low Molecular Weight, LMW), whereon the newly synthesised protein's molecular masses were calculated in conjunction with their migration distance.

3. Results

3.1. Mutation frequency in adapted and non-adapted cells

In the first set of experiments, the CHO cells were pre-treated with 10, 25, and 50 mGy neutron doses (with a dose rate of 10 mGy min^{-1}), and after 1 h were challenged by a 4-Gy gamma dose. The number of induced mutants in the pre-treated cells did not differ considerably from those, which were treated only with the challenge dose. It was assumed that neither of these adapting doses nor the incubation time was adequate for the RAR to fully develop (data not showed).

By varying the irradiation protocol, i.e., by reducing the dose and dose rate and increasing the time between the two irradiations, we found that the most effective dose to achieve the protection of the cells was 2 mGy with a dose rate of 1.59 mGy min^{-1}, as is shown in Fig. 1. The number of induced mutants decreased in the adapted cells by 26%, 45–60%, 30–42%, and 17–28% after 3, 5, 24, and 48 h, respectively. The most appropriate time between the

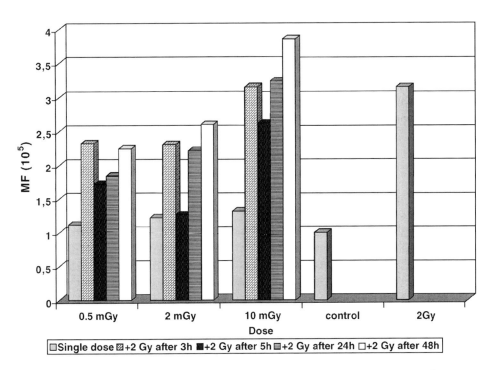

Fig. 1. Mutation frequency in the CHO cells induced by 0.5, 2, 10 mGy (does rate 1.59 mGy min^{-1}) and the 2 Gy gamma challenge dose (dose rate 200 mGy min^{-1}) administered after 3–5–24–48 h of the low dose.

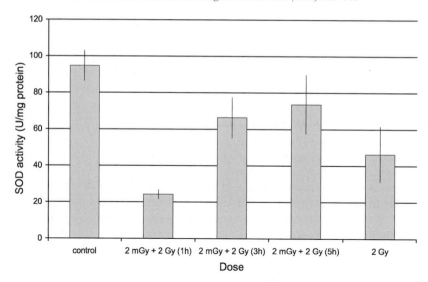

Fig. 2. The activity of SOD measured in irradiated cells. The priming dose was 2 mGy and the challenge dose of 2 Gy was applied after 1–3–5 h.

pre-exposure and the challenge dose was found to be 5 h, during which the adaptation was fully developed. The RAR also persisted after 48 h but it was reduced in size.

3.2. Contribution of superoxide dismutase (SOD) to the adaptive response

The role of SOD is to accelerate the dismutation of the toxic superoxide radical formed in cells exposed to ionising radiation. The activity of SOD was reduced to 53% 1 h after a single dose of 2 Gy, compared to the non-irradiated samples. The data presented in Fig. 2 show that the exposure of pre-treated cells to the 2 Gy challenge dose triggered enhanced SOD activity, as the time between the priming and challenge dose increased. After 5 h, the activity obtained from cells that had been irradiated twice was less then the control level only by 24.6%.

3.3. Detection of newly synthesised proteins

Protein synthesis in the irradiated cells was to some extent suppressed, and new proteins were obtained simultaneously. Their molecular mass was 109, 70, 65, 60, 57 and 45 kDa. All of them appeared in cells exposed to a single dose of 4 Gy. The 60 kDa protein could be detected in all combinations of the priming and challenge doses, while the 70 kDa protein was found in none of the combinations.

4. Discussion

In this study, the cellular effect of low dose and low-dose rate neutron priming treatment mixed with gamma challenge was investigated. In the past decade, many controversies have arisen about the biological effects of the low-doses of neutrons and the role the dose rate plays. In these experiments, we demonstrated that the mutagenic damage (number of *hprt* mutants) produced by the challenge dose (2 Gy) could be reduced when the cells were primed with low-neutron doses. This adaptation experienced in a narrow dose range (0.5–2 mGy) depended on the dose rate (1.59 mGy min^{-1}), and needed a certain time to develop.

It is evident that ionising radiation is capable of inducing the expression of various proteins in cells. Different proteins appear to respond to different radiation qualities and doses [3], however, the role of the radiation-induced proteins in the adaptive response has, as yet, not been characterised. In the course of our study, six induced proteins were detected. Further experimentation is now in progress to look at whether these proteins are involved in the adaptation or in the protection of cells.

It is generally postulated that RAR could result from the activation of damage repair and/or the antioxidant defense system [4]. Our results showed enhanced SOD activity in the adapted cells. This moderate increase could be partly responsible for the protective mechanism triggered by the low-dose neutron irradiation.

Although there is strong evidence that radio-adaptation relies on de novo protein synthesis involving a number of different proteins, it is still an open question whether cells react to radiological impact with a more or less unique programme of gene expression once a certain triggering signal has occurred, or whether there is a broad spectrum of possible responses according to the frequency and type of lesions within certain time periods.

Acknowledgements

This work was supported by the Hungarian Research Foundation OTKA T030353 and T034564 projects.

References

[1] T. Ikushima, H. Aritomi, J. Morisita, Radioadaptive response: efficient repair of radiation-induced DNA damage in adapted cells, Mutat. Res. 358 (1996) 193–198.
[2] F. Cortès, I. Dominguez, M.J. Flores, J. Pinero, J.C. Mateos, Differences in the adaptive response to radiation damage in G_0 human lymphocytes conditioned with hydrogen peroxide or low-dose X-rays, Mutat. Res. 311 (1994) 157–163.
[3] G.E. Woloschak, C.M. Chang-Liu, Differential modulation of specific gene expression following high- and low LET radiations, Radiat. Res. 124 (1990) 183–187.
[4] A. Bavard, C. Luccioni, E. Moustacchi, O. Rigaud, Contribution of antioxidant enzymes to the adaptive response to ionizing radiation of human lymphoblasts, Int. J. Radiat. Biol. 75 (5) (1999) 639–645.

Enhanced induction of mutation by X-irradiation in Werner syndrome cells

Genro Kashino[a], Seiji Kodama[a], Keiji Suzuki[a], Akira Tachibana[b], Mitsuo Oshimura[c], Masami Watanabe[a],*

[a]*Laboratory of Radiation and Life-Science, Department of Health Sciences, School of Pharmaceutical Sciences, Nagasaki University, Nagasaki 852-8521, Japan*
[b]*Radiation Biology Center, Kyoto University, Kyoto 606-0041, Japan*
[c]*Department of Molecular and Cell Genetics, School of Life Science, Faculty of Medicine, Tottori University, Yonago 638-8503, Japan*

Abstract

Werner syndrome (WS) is a rare premature aging syndrome. Because the WS gene, *WRN*, is a member of the RecQ helicase family, it is assumed to be a caretaker of genomic integrity. In the present study, we examined the X-ray-induced mutation frequency at the *hypoxiantin guanine phosphoribosyl transferase (HPRT)* locus in a WS cell line (WS780), and analyzed the different types of mutations by using multiplex PCR. The results indicated that the mutation frequency induced by X-irradiation was higher in the WS cells than in the control cells. It is notable that the majority of mutations observed in the WS cells consist of deletion mutations. We then examined in vitro assays for the end-joining ability of the DNA double-strand breaks (DBSs) in the nuclear extracts prepared from the WS cells, and the control cells. Sequencing analysis revealed that the deletions possibly caused by illegitimate recombination between two short homologies, occurred more frequently in the WS cells than in the control cells. These results suggest that a defect in the WRN gene function promotes the illegitimate recombination, and that this recombination might lead to the induction of large deletion mutations in WS cells. © 2002 Elsevier Science B.V. All rights reserved.

Keywords: Werner syndrome; *WRN*; DNA double-strand breaks repair; Mutation; Illegitimate recombination

---

Abbreviations: WS, Werner syndrome; HPRT, hypoxiantin guanine phosphoribosyl transferase; HAT, 100 μM sodium hypoxantine, 0.4 μM aminopterin, 16 μM thymidine.
* Corresponding author. Tel.: +81-95-844-5504; fax: +81-95-844-5504.
E-mail address: nabe@net.nagasaki-u.ac.jp (M. Watanabe).

0531-5131/02 © 2002 Elsevier Science B.V. All rights reserved.
PII: S 0 5 3 1 - 5 1 3 1 (0 1) 0 0 8 7 5 - 5

1. Introduction

Werner syndrome (WS) is a rare premature aging syndrome [1]. It is evident that chromosome aberrations and large deletion mutations are frequently observed in WS cells [2,3]. These facts suggest that the WRN protein plays a role in the DNA repair pathway, to be a caretaker of genomic integrity. Therefore, in the present study, we examined the fidelity of repair for DNA double-strand breaks (DBSs) in WS cells.

2. Methods

SV40-transformed WS780 cells were used as a WS cell line and SV40-transformed GM638 cells were used as a control cell line. They were cultured in Dulbecco's modified Eagle's medium (DMEM; Nissui Pharmaceutical, Tokyo) supplemented with 10% FCS, penicillin (100 units/ml), and streptomycin (100 μg/ml), and maintained at 37 °C in a humidified atmosphere with 5% CO_2. In the mutation assay, the cells were harvested from about 50 colonies cultured in the HAT medium (100 μM sodium hypoxantine, 0.4 μM aminopterin, 16 μM thymidine), expanded to $1-2 \times 10^6$ cells in a T25 flask and irradiated with X-rays. After X-irradiation, the 10^6 cells were grown in a non-selective medium for 7–14 days (mutation expression period) by successive subcultures and then plated into a medium containing 60 μM of 6-thioguanine (Tokyo Kasei Organic Chemicals, Tokyo), at a density of 10^5 cells per 100-mm dish. The cells were incubated in a CO_2 incubator for 14 days for mutation selection. The types of mutations were determined by multiplex PCR analysis described previously [4]. The in vitro rejoining assay was performed as previously reported [5].

3. Results and discussion

The WS780 cells were susceptible to mutation induction by radiation especially in the low-dose range under 3 Gy, showing 3.1- to 7.5-fold higher mutation frequencies as compared with the control cells. In contrast, the susceptibility to mutation induction in the WS780 cells as compared with the control cells declined to 1.4-fold at a dose of 6 Gy. Multiplex PCR analysis revealed that the majority of X-ray-induced mutations observed in the WS780 cells consisted of deletion mutations. In the assay for the in vitro rejoining ability, sequencing analysis revealed that the deletions possibly caused by illegitimate recombination between two short homologies, occurred more frequently in the WS780 cells than in the control cells. These results suggest that a defect in the WRN gene function promotes the illegitimate recombination mediated by a short homology, and that this recombination might lead to the induction of large deletion mutations in WS cells.

Acknowledgements

This work was supported by a Grant for Scientific Research from the Ministry of Education, Culture, Sports, Science and Technology of Japan.

References

[1] C.J. Epstein, G.M. Martin, A.L. Schultz, A.G. Motulsky, Werner's syndrome. A review of its symptomatology, natural history, pathologic features, genetics, and relationship to the natural aging process, Medicine 45 (1966) 177–221.
[2] H. Hoehn, E.M. Bryant, K. Au, T.H. Norwood, H. Boman, G.M. Martin, Variegated translocation mosaicism in human skin fibroblast cultures, Cytogenet. Cell Genet. 15 (1975) 282–298.
[3] K. Fukuchi, G.M. Martin, R.J. Monnat Jr., Mutator phenotype of Werner syndrome is characterized by extensive deletions, Proc. Natl. Acad. Sci. U. S. A. 86 (1989) 5893–5897.
[4] S. Kodama, G. Kashino, K. Suzuki, T. Takatsuji, Y. Okumura, M. Oshimura, M. Watanabe, J.C. Barrett, Failure to complement abnormal phenotypes of simian virus 40-transformed Werner syndrome cells by introduction of a normal human chromosome 8, Cancer Res. 58 (1998) 5188–5195.
[5] P. North, A. Ganesh, J. Thacker, The rejoining of double-strand breaks in DNA by human cell extracts, Nucleic Acids Res. 18 (1990) 6205–6210.

An ESR and ESEEM study of long-lived radicals which cause mutation in irradiated mammalian cells

Jun Kumagai [a,*], Tetsuo Miyazaki [a], Takayuki Kumada [b], Seiji Kodama [c], Masami Watanabe [c]

[a]*Department of Applied Chemistry, Graduate School of Engineering, Nagoya University, Froh-cho Chikusa-ku, Nagoya 464-8603, Japan*
[b]*Advanced Science Research Center, Japan Atomic Energy Research Institute, Tokai-mura 319-1195, Japan*
[c]*School of Pharmaceutical Sciences, Nagasaki University, Nagasaki 852-8521, Japan*

Abstract

Long-lived radicals (LLR) that exist in γ-ray-irradiated mammalian cells and have a very long lifetime such as 20 h induce mutation and transformation. The long-lived radicals in the irradiated mammalian cells were scavenged using AsA(L-ascorbic acid) as well as by EGCG((−)-epigallocatechin-3-O-gallate) after irradiation, accompanying the decrease in mutation in the irradiated cells. Determining the location of the long-lived radicals in the cells is therefore very important in order to elucidate the mechanism of the induction of mutation by the radicals. The electron spin echo envelope modulation (ESEEM) study indicates that the long-lived radicals are produced in the interior of the biopolymer coils where water molecules scarcely exist. Since DNA and its sugar chains are quite hydrophilic, the long-lived radicals are not produced in them. However, they are produced in proteins as a major content in cells that have several hydrophobic sites in their higher order structure. The long-lived radicals that cause mutations are not generated in DNA and/or in its sugar chains. However, they are probably produced in proteins. © 2002 Elsevier Science B.V. All rights reserved.

Keywords: Long-lived radicals; Mutation; ESR; ESEEM; Protein radical

1. Background

Many researchers studying in the area of radiation biology believe that reactive oxygen species (ROS), such as OH radicals produced by the radiolysis of water, play an important role in expressing the biological effects of radiation in cells such as cell killing and mutation induction [1]. In contrast to the ROS as short-lived radicals, we have found long-lived

---

* Corresponding author. Tel./fax: +81-52-789-2591.
E-mail address: kumagai@apchem.nagoya-u.ac.jp (J. Kumagai).

radicals (LLR) in γ-ray irradiated mammalian cells, which have a very long lifetime, for example 20 h [2]. Recently, we have investigated the biological effects of LLR and found that mutation and transformation are induced by LLR. The LLR in the irradiated mammalian cells were scavenged using AsA(L-ascorbic acid) as well as EGCG(($-$)-epigallocatechin-3-O-gallate) after the irradiation, accompanying the decrease in mutation in the irradiated cells [3–5]. Since AsA and/or EGCG were added to the cells after the irradiation at which the ROS disappeared completely, they did not scavenge the ROS but instead the LLR. Determining the location of the LLR in the cells is, therefore, very important in order to elucidate the mechanism of the induction of mutation by the radicals. In this study, the LLR, which are generated in γ- or X-ray irradiated mammalian cells, are investigated with electron spin resonance (ESR) and electron spin echo envelope modulation (ESEEM) spectroscopy.

2. Materials and methods

2.1. Sample preparation for ESR measurement

Golden hamster embryo cells (GHE) obtained by trypsinization of 13- or 14-day-old embryos were used in this experiment. For the addition of AsA and/or EGCG to the GHE cells, these aqueous solutions were added to the γ-ray-irradiated GHE cells 2 h after the irradiation at 24 °C. The mixture was homogenized rapidly, put into high purity Suprasil quartz tubes and sealed in air.

2.2. Sample preparation for ESEEM measurement

A block of GHE cells was immersed in D_2O at room temperature for 5 days in a sealed glass ample. Almost all the H_2O molecules inside the immersed cells were converted to HDO or D_2O molecules. The cells denoted as "D-GHE" were put into the Suprasil quartz tubes and sealed in air. The tubes were slowly cooled down to 77 K and irradiated with γ-rays (5 kGy). The irradiated tubes were annealed to 111 K for removing OH or OD radicals, and then the tubes were annealed to 210 and 250 K in order to decrease the amount of radicals.

2.3. ESR and ESEEM measurement

The ESR spectra were measured at 70 K by carefully choosing the microwave power and modulation condition to avoid saturation of the signals. The electron spin echo envelope modulation (ESEEM) spectra were measured at 5 K using the two-pulse method. Simulations of the ESEEM spectra were done according to the second order perturbation theory described elsewhere [6].

2.4. Mutation assay

The mutation frequency at the hypoxanthine-guanine phosphoribosyl transferase (HPRT) locus was investigated as described before [3]. Briefly, human embryo cells (HE17) were cultured in Eagle's minimum essential medium (MEM) at 37 °C. The cells were irradiated with 3 Gy of X-rays and were treated with 5 and 0.2 mM of AsA and EGCG, respectively. Then, after the end of irradiation, the cells were subcultured for 2 weeks and then inoculated into the medium containing 6-thioguanine (6TG). After 2 weeks of incubation at 37 °C, the resulting 6TG-resistant colonies that consist of more than 50 cells were scored as mutants. The mutation frequency was calculated by adjusting the plating efficiency.

3. Results and discussion

The decay of the relative rate of mutation and the amounts of LLR by the addition of AsA and EGCG to the irradiated mammalian cells are summarized in Table 1. When AsA and/or EGCG were added to the irradiated mammalian cells, the rates of mutation decreased with a decrease in the amounts of LLR. Since AsA and/or EGCG were added to the cells after the irradiation at which the ROS disappeared completely, they did not scavenge the ROS but instead the LLR. Therefore, the LLR in the cells are very important for inducing mutation in cells.

It is very interesting to know the location where the LLR are produced in the cells by the γ-irradiation. Since the absorbance of radiation energy is proportional to the weight contents in cells and since the weight of the proteins is much higher than that of DNA in the cells, the LLR are not produced in DNA but probably in proteins. The ESEEM study indicates that the LLR are produced in the interior of the biopolymer coils in the irradiated cells, where water molecules scarcely exist by the decomposition of the polymer, which directly absorbed the energy from radiation [7]. Since DNA and its sugar chains are quite hydrophilic, the LLR are not produced in DNA and/or in its sugar chains. On the other hand, proteins as a major content in cells have several hydrophobic sites in their higher order structures. These assignments lead to a very important conclusion that the LLR are produced in proteins and that they cause mutation in irradiated cells. This is a new kind of extra-DNA bystander effect on mutation.

Table 1
Effects of AsA and EGCG addition to irradiated mammalian cells

Scavengers	Relative rate of mutation[a]	Relative yield of LLR[b]
Control	1	1
AsA	0.03[c]	0.02[d]
EGCG	0.40[e]	0.43[f]

[a] HE: human 17 embryo. A dose of X-irradiation is 3 Gy.

[b] GHE: golden hamster embryo. A dose of γ-irradiation is 5 kGy.

[c] The AsA aqueous solution (5 mM) was added to the cells 20 min after the irradiation, and the cells were exposed to the solution for 2 h. See Ref. [3].

[d] AsA (45 mM) was added to the cells 2 h after the irradiation. See Ref. [5].

[e] The EGCG aqueous solution (0.2 mM) was added to the cells 6 h after the irradiation, and the cells were exposed to the solution for 2 h. See Ref. [4].

[f] EGCG (74 mM) was added to the cells 2 h after the irradiation. See Ref. [4].

Acknowledgements

This study was supported in part by the grants from the Ministry of Education, Science, Sports and Culture, Japan and is carried out partly as a part of the "Ground Research Announcement for Space Utilization" promoted by the Japan Space Forum.

References

[1] Ce. Thomas, B. Kalyanaraman, Oxygen Radicals and the Disease Process, Harwood Academic Publishers, Amsterdam, 1997.
[2] T. Yoshimura, K. Matsumoto, T. Miyazaki, K. Suzuki, M. Watanabe, Electron spin resonance studies of free radicals in gamma-irradiated golden hamster embryo cells: radical formation at 77 K and 295 K, and radio-protective effects of vitamin C at 295 K, Radiat. Res. 136 (1993) 361–365.
[3] S. Koyama, M. Watanabe, S. Kodama, K. Suzuki, T. Matsumoto, T. Miyazaki, Radiation-induced long-lived radicals which cause mutation and transformation, Mutat. Res. 421 (1998) 45–54.
[4] J. Kumagai, M. Nakama, T. Miyazaki, T. Ise, S. Kodama, M. Watanabe, Scavenging of long-lived radicals by (−)-epigallocatechin-3-O-gallate and simultaneous suppression of mutation in irradiated mammalian cells, Radiat. Phys. Chem., 2001, in press.
[5] T. Matsumoto, T. Miyazaki, Y. Kosugi, T. Kumada, S. Koyama, S. Kodama, M. Watanabe, Reaction of long-lived radicals and vitamin C in γ-irradiated mammalian cells and their model system at 295 K: tunneling reaction in biological system, Radiat. Phys. Chem. 49 (1997) 547–551.
[6] L. Kevan, Modulation of electron spin-echo decay in solids, in: L. Kevan, R.N. Schwartz (Eds.), Time Domain Electron Spin Resonance, Wiley, New York, 1979, pp. 279–341.
[7] J. Kumagai, T. Kumada, M. Watanabe, T. Miyazaki, Electron spin echo study of long-lived radicals which cause mutation in γ-ray irradiated mammalian cells, Spectrochim. Acta, Part A 56 (2000) 2509–2516.

A role of long-lived radicals in radiation mutagenesis and its suppression by epigallocatechin gallate

Tamaki Ise[a], Seiji Kodama[a], Keiji Suzuki[a], Takashi Tanaka[b], Jun Kumagai[c], Tetsuo Miyazaki[c], Masami Watanabe[a],*

[a]*Laboratory of Radiation and Life Science, School of Pharmaceutical Sciences, Nagasaki University, Nagasaki 852-8521, Japan*
[b]*Laboratory of Pharmacognosy Science, School of Pharmaceutical Sciences, Nagasaki University, Nagasaki 852-8521, Japan*
[c]*Department of Applied Chemistry, Graduate School of Engineering, Nagoya University, Nagoya 464-8603, Japan*

Abstract

Ionizing radiation induces cell death, chromosome aberration and gene mutation. It is widely accepted that most of these biological effects is caused by free radicals such as hydroxyl and hydrogen radicals. However, we formerly demonstrated that ionizing radiation produces organic radicals, which have a long lifespan and whose half-life are several hours. In the present study, we studied the effect of (−)-epigallocatechin-3-O-gallate (EGCG), which scavenges the long-lived radicals on X-ray-induced gene mutation. In order to find out the role that the long-lived radicals play in radiation mutagenesis, the cells were synchronized, irradiated with 3 Gy of X-rays with a combined treatment of 200 μM EGCG for 2 h before and after DNA synthesis, and the cells mutation frequency was examined at the hypoxanthin guanine phosphoribosyl transferase (*HPRT*) locus. The result indicated that the treatment with EGCG before but not after DNA synthesis post-irradiation reduced the mutation frequency. This result implies that the presence of the long-lived radicals during DNA synthesis plays a role in the induction of gene mutation by ionizing radiation. © 2002 Elsevier Science B.V. All rights reserved.

Keywords: Radiation; Human cells; EGCG; Mutation; Long-lived radicals

Abbreviations: EGCG, (−)-epigallocatechin-3-O-gallate; HAT, 100 μM sodium hypoxanthin, 0.4 μM aminopterin, and 16 μM thymidine; HPRT, hypoxanthin guanine phosphoribosyl transferase; 6-TG, 6-thioguanine; ESR, electron spin resonance.
* Corresponding author. Tel.: +81-95-844-5504; fax: +81-95-844-5504.
E-mail address: nabe@net.nagasaki-u.ac.jp (M. Watanabe).

1. Introduction

We formerly demonstrated that ionizing radiation produces organic radicals, that differ from hydroxyl and hydrogen radicals of water origin, which, have a long lifespan in the irradiated cells [1]. We also demonstrated that ascorbic acid efficiently scavenges the long-lived radicals and suppresses radiation-induced mutation frequency, without affecting cell killing and chromosome aberrations [2]. In addition, there is evidence to suggest that green tea possesses anti-carcinogenic effects [3]. Therefore, in the present study, we examined (−)-epigallocatechin-3-O-gallate (EGCG), a major constituent of green tea, for its scavenging ability of the long-lived radicals and the effect on X-ray-induced cell death, micronuclei formation and gene mutation.

2. Materials and methods

Normal human embryo fibroblast cells (HE17) were grown in Eagle's minimum essential medium (MEM) supplemented with 10% fetal bovine serum (Trace Bioscience, Australia), penicillin (100 units/ml) and streptomycin (100 μg/ml). The cells were cultured at 37 °C in a humidified atmosphere with 5% CO_2.

The cell killing effect induced by X-irradiation was determined by colony formation assay. The cells were irradiated with X-rays, seeded into 100-mm dishes, and incubated for 2 weeks. Colonies consisting of more than 50 cells were scored as survivors.

In order to investigate the chromosome aberrations induced by X-irradiation, the induction of micronuclei was examined. The X-irradiated cells were plated into 100-mm dishes, treated with 200 μM EGCG for 2 h followed by the addition of 3 μg/ml cytochalasin B, and cultured for 24 h. After harvesting, the cells were washed with PBS(−) and fixed with Carnoy's (MeOH:Acetic acid, 3:1) solution. At least 1000 binucleated cells were scored for each treatment.

Before the mutation assay, the cells were cultured in the HAT medium (Life Technologies) for 10 days in order to eliminate pre-existent 6-thioguanine (6-TG) resistant cells. After the HAT selection, the cells were arrested at the G_0/G_1 phase by keeping the culture at confluence. Under this condition, the fraction of cells at the S phase was less than 4% as measured by flow-cytometric analysis. The cells were irradiated with X-rays at confluence and then plated into 100-mm dishes at a density of 6×10^5 cells per 100-mm dish, immediately after irradiation. At 6 and 48 h post-irradiation, the cells were treated with 200 μM EGCG for 2 h. After 50 h post-irradiation, the cells were harvested, cultured for 7 days (mutation expression period) with successive subcultures, and then plated into the medium containing 40 μM 6-TG (Tokyo Kasei Organic Chemicals, Tokyo) at a density of 5×10^4 cells per 100-mm dish. The 6-TG resistant colonies were scored as mutants after incubation for 14 days.

3. Results

We found that by adding EGCG after the irradiation the relative yield of long-lived radicals in γ-irradiated Golden hamster embryo (GHE) cells and albumin solution decayed

Table 1
Effect of EGCG treatment on spontaneous and X-ray-induced mutation frequencies at the *HPRT* locus in human embryo cells

Time of treatment post-irradiation	Treatment	Mutation frequency ($\times 10^{-6}$) (SD[i])
6–8 h	Control[a]	4.5 (1.8)
	EGCG[b]	1.0 (1.0)
	3Gy[c]	62.1 (26.7)
	3 Gy/EGCG[d]	28.7 (16.1)
48–50 h	Control[e]	3.7 (4.7)
	EGCG[f]	4.8 (2.6)
	3Gy[g]	43.9 (17.8)
	3 Gy/EGCG[h]	46.9 (33.3)

The mutation frequencies were calculated based on at least three independent experiments.
[a,e] The cells were treated with MEM for 2 h.
[b,f] The cells were treated with 200 μM EGCG for 2 h.
[c,g] The cells were treated with MEM for 2 h at indicated time post-irradiation.
[d,h] The cells were treated with 200 μM EGCG for 2 h at indicated time post-irradiation.
[i] Standard deviation.

much faster than the controls without EGCG. The result obtained using electron spin resonance (ESR) indicated that EGCG scavenges 40% of the long-lived radicals at 3 h post-irradiation, as compared with the controls (data not shown).

Next, we examined the effect of EGCG treatment on cell death and micronuclei induction by X-irradiation. Neither the pre- nor post-treatment with EGCG affected the cell killing by X-irradiation. Similarly, neither the pre- nor post-treatment with EGCG altered the frequency of micronuclei produced by X-irradiation. The result of mutation analysis is shown in Table 1. Furthermore, treatment with 200 μM EGCG for 2 h did not change the spontaneous mutation frequency. In contrast, the treatment of EGCG at 6 h post-irradiation, where the synchronized cell population did not enter the S phase, decreased the X-ray-induced mutation frequency to 50% of that induced with 3 Gy of X-rays. However, the mutation frequency was not decreased by the treatment of EGCG at 48 h post-irradiation, where 70% of the cell population did pass through the S phase.

4. Discussion

The present study demonstrated that EGCG scavenges long-lived radicals produced by radiation. The half-life of these radicals is more than several hours at room temperature and much longer than that of the hydroxyl radicals (70–200 ns). Since more than 80% of the cells are composed of water, the hydroxyl or hydrogen radicals that are produced after irradiation are generally regarded as a cause of the radiation-induced biological effects. We formerly demonstrated that 72% of the initial radicals in γ-irradiated cells are hydroxyl radicals, 12% are hydrogen radicals, and 16% are radicals from organic substances such as DNA, proteins, and lipids. Moreover, it was suggested that the hydroxyl radicals did not contribute to the creation of the organic radicals [4]. However, the mechanisms involved in the formation and scavenge of these organic radicals remains unclear.

In present study, pre-treatment with EGCG before DNA replication reduced the X-ray-induced mutation frequency in the human embryo cells without affecting cell killing and chromosome aberrations, although the treatment with EGCG after DNA replication was no more effective. These results imply that the presence of the long-lived organic radicals during DNA replication is critical for inducing gene mutation in the human embryo cells. This suggests that the long-lived radicals may not directly interact with the DNA molecules; however, they may induce gene mutation by reducing the fidelity of DNA replication possibly by affecting the DNA replication machinery. Further studies need to be carried out in order to find out the molecular structure of the gene mutations induced by the long-lived radicals.

Acknowledgements

This work was supported partly by a Grant for Scientific Research from the Ministry of Education, Culture, Sports, Science, and Technology of Japan and also carried out as a part of the "Ground Research Announcement for Space Utilization" promoted by the Japan Space Forum.

References

[1] T. Yoshimura, K. Matsuno, T. Miyazaki, K. Suzuki, M. Watanabe, Electron spin resonance studies of free radicals in gamma-irradiated golden hamster embryo cells radical formation at 77 and 295 K, and radio-protective effects of Vitamin C at 295 K, Radiat. Res. 136 (1993) 361–365.
[2] S. Koyama, S. Kodama, K. Suzuki, T. Matsumoto, T. Miyazaki, M. Watanabe, Radiaion-induced long-lived radicals which cause mutation and transformation, Mutat. Res. 421 (1998) 45–54.
[3] S.K. Katiyar, C.A. Elmets, Green tea polyphenolic antioxidants and skin photoprotection (Review), Int. J. Oncol. 18 (6) (2001) 1307–1313.
[4] T. Miyazaki, T. Yoshimura, K. Mita, K. Suzuki, M. Watanabe, Rate constant for reaction of Vitamin C with protein radicals in gamma-irradiated aqueous albumin solution at 295 K, Radiat. Phys. Chem. 45 (1994) 199–202.

Low dose of wortmannin reduces radiosensitivity of cells

Kumio Okaichi [a,*], Keiji Suzuki [b], Naoko Morita [a], Megumi Ikeda [a], Naoki Matsuda [c], Haruki Takahashi [d], Masami Watanabe [b], Yutaka Okumura [a]

[a] *Atomic Bomb Disease Institute, Nagasaki University, 12-4 Sakamoto 1-chome, Nagasaki 852-8523, Japan*
[b] *School of Pharmaceutical Sciences, Nagasaki University, 12-4 Sakamoto 1-chome, Nagasaki 852-8523, Japan*
[c] *Radioisotope Center, Nagasaki University, 12-4 Sakamoto 1-chome, Nagasaki 852-8523, Japan*
[d] *Neurosurgery, School of Medicine, Nagasaki University, 12-4 Sakamoto 1-chome, Nagasaki 852-8523, Japan*

Abstract

Wortmannin is an inhibitor of PI3-kinase and acts on cultured cells at doses below 1 µM. Wortmannin also inhibits the PI3-kinase family such as ATM or DNA-PK at doses above 10 µM. There are many reports on the enhancement of cell radiosensitivity by using a high dose of wortmannin inhibiting the proteins of the PI3-kinase family. However, there have been no reports of the effects on the radiosensitivity of low doses of wortmannin inhibiting PI3-kinase. We found that low doses of wortmannin reduced the radiosensitivity of human glioblastoma cells, which had wild-type *p53*. A low dose of wortmannin did not affect the accumulation of *p53* and the phosphorylation of *p53* at ser-15; however, a low dose reduced the induction of Waf-1 and enhanced the induction of GADD45. As the fraction of G2/M cells was reduced, however, the fraction of G1 cells was increased by a low dose of wortmannin after X-ray irradiation. © 2002 Elsevier Science B.V. All rights reserved.

Keywords: Ionizing radiation; PI3-kinase; *p53*; Radiosensitivity; Wortmannin

1. Background

Wortmannin is a microbial metabolite that specifically inhibits p110 PI3-kinase [1]. PI3-kinase is an enzyme that acts as a direct biochemical link between the phosphatidylinositol pathway and the homeostasis of cells. Recently, the ATM or DNA-PK, which are involved in DNA repair, was revealed to be members of a family of PI3-kinase [2]. Wortmannin inhibits

---

* Corresponding author. Department of Radiation Biophysics, Radiation Effect Research Unit, Atomic Bomb Disease Institute, Nagasaki University School of Medicine, Sakamoto, Nagasaki 852-8523, Japan. Tel.: +81-95-849-7102; fax: +81-95-849-7104.
 E-mail address: okaichi@net.nagasaki-u.ac.jp (K. Okaichi).

the activity of ATM or DNA-PK at high concentrations (over 10 μM), in contrast to PI3-kinase, which is inhibited at concentrations below 1 μM in the cells. Subsequently, many reports appeared on the enhancement of the radiosensitivity of cells by high doses of wortmannin [3,4]. We speculated that PI3-kinases would also be involved in DNA repair and affect the radiosensitivity of the cells. We examined the effects of low doses of wortmannin on the radiosensitivity of cells to clarify the role of genuine PI3-kinases in radiosensitivity.

2. Materials and methods

2.1. Cell cultures and irradiation

Human glioblastoma A-172 (wild-type *p53*) and T98G cells (mutant-type *p53*, at codon 237 from Met to Ile) were cultured at 37 °C in Dulbecco's modified Eagle's medium containing 10% (v/v) fetal bovine serum. Exponentially growing cells were irradiated with X-rays from an X-ray generator at 150 kV and 5 mA with a 0.1-mm copper filter, or at 200 kV and 15 mA with a 0.5-mm aluminum and 0.5-mm copper filter.

2.2. Plasmids

The plasmid pC53-SCX3 (provided by Dr. B. Vogelstein, Johns Hopkins Oncology Center, Baltimore, MD, USA) was used for transfection into the A-172 cells [5]. In addition, pC53-SCX3 contains the mutant *p53* (point mutation at codon 248 from Arg to Trp).

2.3. Western blotting analysis

The cell lysate was electrophoresed on SDS-polyacrylamide gels as previously described [6]. The proteins were electrophoretically transferred to a membrane and the membrane was incubated with each antibody for 2 h. The membrane was then incubated with a biotinylated secondary antibody and the band was visualized by streptavidin-alkaline phosphatase. Otherwise, the protein level was analyzed using the enhanced chemiluminesence (ECL) system of Amersham.

2.4. Laser scanning cytometry (LSC)

We used a laser scanning cytometer model LSC 101 (Olympus, Tokyo, Japan) to measure the nuclear DNA.

3. Results

3.1. Radiosensitivity of cells with low doses of wortmannin

The cells were preincubated with a low dose of wortmannin (0.25, 0.5 and 0.75 μM) for 90 min, then exposed to ionizing radiation and incubated for 3 weeks with wortmannin.

The A-172 cells incubated with wortmannin showed decreased radiosensitivity. However, the T98G cells did not show any change in radiosensitivity after the wortmannin treatment. The A-172/248W cells carrying a dominant point-mutated *p53* gene also showed no change in radiosensitivity with wortmannin.

3.2. Effect of wortmannin on p53 signal transduction

Wortmannin at 0.5 μM did not affect the influence of ionizing radiation on the phosphorylation of *p53* at ser-15 or accumulation in the A-172 cells. Wortmannin suppressed the induction of Waf-1 by ionizing radiation and increased the induction of GADD45. The wortmannin dose did not affect Bax, Bcl-2, MDM2, poly(ADP-ribose) polymerase or the phosphorylation of AKT by ionizing radiation.

3.3. Effect of wortmannin on cell cycle

Using LSC, we examined the cell cycle after 3 Gy irradiation in the A-172 cells with 1 μM wortmannin. The percentage of G1 increased in the presence of wortmannin; on the other hand, the percentage of G2/M decreased with wortmannin. Furthermore, the percentage of mitotic cells increased with wortmannin.

4. Conclusion

The results demonstrate that a low dose of wortmannin (less than 1 μM) reduced the radiosensitivity of the A-172 cells, which have wild-type *p53*. This result is very striking because a high dose of wortmannin (more than 10 μM) enhances the radiosensitivity of the cells [3,4]. A low dose of wortmannin reduced the induction of Waf-1 and increased the induction of GADD45. Nevertheless, we were unable to detect a direct effect on *p53*, such as the accumulation of *p53* or the phosphorylation of *p53* at ser-15. A low dose of wortmannin increased the fraction of G1 cells and reduced the number of G2/M cells after irradiation. The percentage of mitotic cells increased with a low dose of wortmannin. This increase corresponded with a decrease in the G2 cells and an increase in the G1 cells. PI3-kinase or another wortmannin-sensitive enzyme, may affect the cell cycle through the signal transduction of *p53* after X-ray irradiation.

References

[1] G. Powis, R. Bonjouklian, M.M. Berggern, A. Gallegos, R. Abraham, C. Ashendel, L. Zalkow, W.F. Matter, J. Dodge, G. Grindley, C.J. Vlahos, Wortmannin, a potent and selective inhibitor of phosphatidylinositol-3-kinase, Cancer Res. 54 (1994) 2419–2423.
[2] K.O. Hartley, D. Gell, G.C.M. Smith, H. Zhang, N. Divecha, M.A. Connelly, A. Admon, S.P. Lees-Miller, C.W. Anderson, S.P. Jackson, DNA-dependent protein kinase catalytic subunit: a relative of phosphatidylinositol 3-kinase and the ataxia telangiectasia gene product, Cell 82 (1995) 849–856.
[3] B.D. Price, M.B. Youmell, The phosphatidylinositol 3-kinase inhibitor wortmannin sensitizes murine fibroblasts and human tumor cells to radiation and blocks induction of *p53* following DNA damage, Cancer Res. 56 (1996) 246–250.

[4] R. Okayasu, K. Suetomi, R.L. Ullrich, Wortmannin inhibits repair of DNA double-strand breaks in irradiated normal human cells, Radiat. Res. 149 (1998) 440–445.
[5] T. Ohnishi, K. Ohnishi, X. Wang, A. Takahashi, K. Okaichi, Restoration of mutant *p53* to normal *p53* function by glycerol as a chemical chaperone, Radiat. Res. 151 (1999) 498–500.
[6] K. Suzuki, S. Kodama, M. Watanabe, Recruitment of ATM protein to double strand DNA irradiated with ionizing radiation, J. Biol. Chem. 274 (1999) 25571–25575.

Defective accumulation of p53 protein in X-irradiated human tumor cells with low proteasome activity

Motohiro Yamauchi, Keiji Suzuki, Seiji Kodama, Masami Watanabe*

Laboratory of Radiation and Life Science, School of Pharmaceutical Sciences, Nagasaki University, 1-14 Bunkyo-machi, Nagasaki 852-8521, Japan

Abstract

Because the loss of p53 function is the most common event in human cancers, *p53* gene therapy is now under clinical trial. Here, we examined whether X-irradiation potentiated the function of the exogenous p53 protein induced in H1299 cells and human non-small cell lung carcinoma cells. We found that the induced p53 protein was not accumulated after X-irradiation, although both phosphorylation of the p53 protein at Ser15 and Ser20, and phosphorylation of MDM2 were observed normally. Next, we examined the kinetics of degradation of the p53 protein in the presence of cycloheximide, a translation inhibitor. The level of the p53 protein in HE49 cells decreased rapidly, but there was no change in the H1299 cells. Furthermore, significant accumulation of the p53 protein was observed only in the HE49 cells after being treated for 2 h with ALLN, a proteasome inhibitor. These results indicate that low proteasome activity in H1299 cells cause defective accumulation of the p53 protein. Furthermore, it is possible that proteasome activity in cancer cells may determine the prognosis of the *p53* gene therapy. © 2002 Elsevier Science B.V. All rights reserved.

Keywords: p53; Accumulation; Gene therapy; Radiotherapy; Proteasome

1. Introduction

Mutations and inactivation of tumor suppressor p53 are the most common events in human cancers and they are deeply involved in genome instability related to tumor progression [1]. Therefore, the introduction of the *p53* gene into such tumor cells has been

---

* Corresponding author. Tel.: +81-95-844-5504; fax: +81-95-844-5504.
E-mail address: nabe@net.nagasaki-u.ac.jp (M. Watanabe).

tried for clinical application of cancer therapy [2]. Furthermore, it has been shown that radiosensitivity is increased by introducing the *p53* gene into some types of tumors [3,4], providing the possibility that a combination of gene therapy with radiotherapy will increase the efficacy of the treatment.

Because the p53 protein is very unstable in most normal cells, it needs to accumulate to its proper function. In this study, we examined the level of p53 protein ectopically expressed in H1299 cells, which lack the endogenous p53 expression after X-irradiation. We also compared phosphorylation of p53 and MDM2 proteins, and the stability of the p53 protein between the cell line and normal human diploid cells.

2. Materials and methods

Normal human embryonic (HE49) cells were cultured in Eagle's MEM supplemented with 10% FBS. NCl-H1299, non-small cell lung carcinoma cells, were purchased from ATCC. The plasmids, pVgRXR and pIND/V5-HisB were purchased from Invitrogen (San Diego, CA); pp53-EGFP was from CLONTECH (Palo Alto, CA). The pp53-EGFP plasmid was digested with *Kpn*I and *Bam*HI, and the 1.2-kb fragment was ligated with the *Kpn*I–*Bam*HI-digested pIND/V5-HisB plasmid to make pINDp53-His. The H1299 cells were transfected with the pVgRXR plasmid to establish 99V9 cells, expressing VgECR and RXR proteins. The 99V9 cells were then transfected with the pINDp53-His plasmid to isolate the clone named 99-p53 His 34. Expression of the p53-His protein was induced by Ponasterone A (Invitrogen), a synthetic analog of insect hormone, Ecdyson. Exponentially growing cells were irradiated with X-rays from an X-ray generator at 150 kVp and 5 mA with a 0.1-mm copper filter. The dose rate was 0.425 Gy/min. The level of p53, phosphorylated p53, and MDM2 were detected by western blot analysis using specific antibodies.

3. Results

After 4 Gy of X-rays, ectopically expressed p53 protein in 99-p53 His cells did not accumulate. We compared phosphorylation of p53 at Ser15 and Ser20 and phosphorylation of MDM2 between 99-p53 His cells and HE49 cells, in which the p53 protein accumulated after irradiation. However, we could not find any difference in the phosphorylation of these proteins. Therefore, we examined the kinetics of p53 degradation in the presence of cycloheximide, a translation inhibitor. The level of p53 protein in HE49 cells decreased rapidly, however, it did not change in the 99-p53 His cells. Furthermore, significant accumulation of p53 was observed in the HE49 cells after being treated with ALLN, a proteasome inhibitor, for 2 h; however, there was little or no accumulation of the p53 protein in 99-p53 His cells. These results indicate that the proteasome activity, which degrades the p53 protein, is extremely low in H1299 cells. Although phosphorylation of p53 inhibits p53 degradation and causes p53 accumulation after irradiation, the low degrading activity of p53 in H1299 may result in defective accumulation of the p53 protein.

4. Conclusion

The present study indicates that the level of p53 protein is affected by its phosphorylation and the proteasome activity of the cells. It is suggested that radiotherapy together with gene therapy is more effective for tumors in which proteasome activity is high, whereas it may not enhance the effect for tumors with low proteasome activity.

Acknowledgements

This work was supported by a grant of Scientific Research from the Ministry of Education, Culture, Sports, Science and Technology of Japan.

References

[1] A.J. Levine, p53, the cellular gatekeeper for growth and division, Cell 88 (3) (1997) 323–331.
[2] T. Fujiwara, E.A. Grimm, T. Mukhopadhyay, D.W. Cai, L.B. Owen-Schaub, J.A. Roth, A retroviral wild-type p53 expression vector penetrates human lung cancer spheroids and inhibits growth by inducing apoptosis, Cancer Res. 53 (18) (1993) 4129–4133.
[3] D. Cowen, N. Salem, F. Ashoori, R. Meyn, M.L. Meistrich, J.A. Roth, A. Pollack, Prostate cancer radiosensitization in vivo with adenovirus-mediated p53 gene therapy, Clin. Cancer Res. 6 (11) (2000) 4402–4408.
[4] D. Gallardo, K.E. Drazan, W.H. McBride, Adenovirus-based transfer of wild type p53 gene increases ovarian tumor radiosensitivity, Cancer Res. 56 (21) (1996) 4891–4893.

High susceptibility and possible involvement of telonomic instability in the induction of delayed chromosome aberrations by X-irradiation in *scid* mouse cells

Ayumi Urushibara[a], Seiji Kodama[a], Keiji Suzuki[a], Fumio Suzuki[b], Masami Watanabe[a,*]

[a]*Laboratory of Radiation and Life Science, School of Pharmaceutical Sciences, Nagasaki University, 1-14, Bunkyo-machi, Nagasaki 852-8521, Japan*
[b]*Department of Regulatory Radiobiology, Research Institute for Radiation Biology and Medicine, Hiroshima University, Hiroshima 734-8553, Japan*

Abstract

Ionizing radiation induces genetic instability in the progeny of irradiated cells. There is evidence to suggest that DNA double-strand breaks (DSBs) and subsequent repair processes are important in the induction of genetic instability. In the present study, we hypothesize that the impaired repair for DSBs increases the susceptibility to the induction of genetic instability. To examine this hypothesis, we studied delayed chromosome aberrations induced by X-irradiation in *scid* mouse cells. The chromosome analysis revealed that the *scid* cells were 2.7-fold more susceptible to the induction of delayed dicentric chromosomes than the wild-type cells, even though they received the equivalent survival dose. In order to find the mechanism for the formation of delayed chromosome aberrations, we examined telomere sequences that remained at the junctional position of two chromosomes in the delayed dicentric chromosomes using the telomere-FISH technique. The *scid* cells showed a higher percentage of telomeric fusions than the wild-type cells. These results suggest that the DNA-dependent protein kinase catalytic subunit may be involved in the maintenance of telomere stability. From the results presented here, we propose that the induction of delayed chromosome aberrations is

Abbreviations: DSBs, double-strand breaks; *scid* mouse, severe combined immunodeficiency mouse; DNA-PKcs, DNA-dependent protein kinase catalytic subunit; t-FISH, telomere-fluorescence in situ hybridization; FBB, fusion–bridge–breakage; NHEJ, nonhomologous end-joining; HR, homologous recombination.
* Corresponding author. Tel.: +81-95-844-5504; fax: +81-95-844-5504.
E-mail address: nabe@net.nagasaki-u.ac.jp (M. Watanabe).

mediated by the fusion–bridge–breakage cycle possibly initiated with telonomic instability, induced by ionizing radiation. © 2002 Elsevier Science B.V. All rights reserved.

Keywords: Telonomic instability; DNA-PKcs; Radiation; Delayed chromosome aberrations; *scid* mouse

1. Introduction

Mammalian cells have at least two repair pathways for DSBs, nonhomologous end-joining (NHEJ) and homologous recombination (HR) [1]. The DNA-dependent protein kinase (DNA-PK) is an essential protein for both NHEJ and V(D)J recombination [2]. In mammals, the DNA-PK is composed of the catalytic subunit DNA-PKcs, and a regulatory factor, Ku, which is a heterodimer of two proteins, Ku70 and Ku80. There is accumulated evidence to show that *scid* mice, which have a defect in the DNA-PKcs function, are hypersensitive to radiation and are susceptible to radiation-induced carcinogenesis [3].

Former reports demonstrated that both *scid* mouse cells and DNA-PKcs knock out mouse cells show an increased frequency of telomeric fusions at the spontaneous condition [4,5]. Because telomeric repeats, together with telomere binding proteins stabilize chromosome ends and protect from end-to-end fusions [6], the evidence to show high telomeric fusions in *scid* and DNA-PKcs KO mouse cells suggests that DNA-PKcs might play a role in maintaining the telomere function.

In the present study, we investigated delayed chromosome aberrations in *scid* cells to find out the effect of a defect in NHEJ on the induction of delayed chromosome aberrations. We focused on the instability of the telomeres and its role in the induction of genomic instability by radiation.

2. Materials and methods

Scid mouse and isogenic wild-type mouse fibroblast cell lines were established from embryos of C.B-17 *scid/scid* and C.B-17+/+ mice. They were cultured in α-MEM medium supplemented with 10% fetal bovine serum (FBS), penicillin (100 units/ml) and streptomycin (100 μg/ml) at 37 °C in a humidified atmosphere with 5% CO_2.

The killing effect of the X-rays was determined by colony formation assay. The cells were irradiated with X-rays, seeded into 100-mm dishes and incubated for 2 weeks. Colonies containing more than 50 cells were scored as survivors.

For studying delayed chromosome aberrations, primary surviving colonies irradiated with 2 or 6 Gy of X-rays were harvested, again incubated for another 2 weeks for colony formation as secondary colonies, and used for chromosome analysis. Chromosome samples were prepared as previously described [7], and the slides were stained using the C-banding method.

In telomere-FISH (t-FISH), the chromosome samples were hybridized with a FITC-labeled telomeric PNA probe (300 ng/ml) for 5 h, washed sequentially and visualized as previously reported [8].

3. Results

The *scid* cells were highly sensitive to X-rays compared with the wild-type cells (data not shown). The D0 values were 0.5 and 1.8 Gy for the *scid* cells and wild-type cells, respectively. Ten percent of the survival doses of the *scid* cells and the wild-type cells were 2 and 6 Gy, respectively.

The spontaneous frequency of the dicentrics was 2.3-fold higher in the *scid* cells than that in the wild-type cells as shown in Fig. 1. The frequencies of the delayed dicentrics in the *scid* cells were 2.7-fold higher than those in the wild-type cells when they were exposed to the equivalent 10% survival dose of X-rays (Fig. 1).

Because most of the delayed dicentrics were not accompanied with fragments, we examined telomere sequences, which remained at the junctional position of two chromosomes using the t-FISH technique. The frequency of the dicentrics by telomeric fusions increased to 10.9% after exposure to 6 Gy of X-rays from the spontaneous level of 5.0% in the wild-type cells. In contrast, the *scid* cells showed a higher frequency of telomeric fusions, 12.0%, spontaneously, however, the frequency did not increase much, showing 15.0% by X-irradiation.

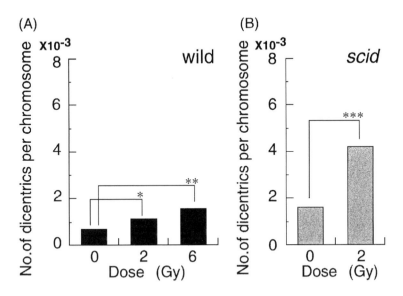

Fig. 1. Frequencies of delayed chromosome aberrations in wild type and *scid* mouse cells. (A) Frequencies of delayed dicentrics in wild-type cells. One hundred metaphases were scored in each group. Induction is not significant ($*p=0.29$, $**p=0.11$) by a statistical test using Poisson distribution. (B) Frequency of delayed dicentrics in the *scid* cells. One hundred metaphases were scored in both groups. Induction is significant ($***p=0.007$) by a statistical test using Poisson distribution.

4. Discussion

The frequencies of the spontaneous and delayed chromosome aberrations in the *scid* cells are higher than those of the wild-type cells. This result suggests that NHEJ repair plays a role in suppressing both spontaneous and radiation-induced delayed chromosome aberrations. Because the frequency of telomeric fusions in the *scid* cells is higher than those in the wild-type cells, the *scid* cells are more susceptible to the induction of telonomic instability by radiation, thus, suggesting that this instability might contribute to the higher induction of delayed chromosome aberrations in the *scid* cells. From the results presented here, we propose a mechanism of delayed chromosome aberrations by radiation as follows: Radiation induces destabilization of telomere function (telonomic instability) and this promotes the cells to enter the fusion–bridge–breakage (FBB) cycle originally proposed by McClintock [9]. This cycle may result in further promotion of chromosomal instability over many cell divisions.

Acknowledgements

The authors would like to thank Prof. Takei Tsutsui, The Nippon Dental University, School of Dentistry, for kindly providing a PNA telomere probe. This work was partly supported by a grant for Scientific Research from the Ministry of Education, Culture, Sports, Science and Technology of Japan and by a grant from the Health Research Foundation, Kyoto, Japan.

References

[1] P. Pfeiffer, W. Goedecke, G. Obe, Mechanisms of DNA double-strand break repair and their potential to induce chromosomal aberrations, Mutagenesis 15 (4) (2000) 289–302.
[2] J.H. Wilson, P.B. Berget, J.M. Pipas, Somatic cells efficiently join unrelated DNA segments end-to-end, Mol. Cell. Biol. 2 (1982) 1258–1269.
[3] R. Araki, A. Fujimori, K. Hamatani, K. Mita, T. Saito, M. Mori, R. Fukumura, M. Morimyo, M. Muto, M. Itoh, K. Tatsumi, M. Abe, Nonsense mutation at Tyr-4046 in the DNA-dependent protein kinase catalytic subunit of severe combined immune deficiency mice, Proc. Natl. Acad. Sci. U. S. A. 94 (1997) 2438–2443.
[4] F.A. Goytisolo, E. Samper, S. Edmonson, G.E. Taccioli, M.A. Blasco, The absence of DNA-Dependent protein kinase catalytic subunit in mice results in anaphase bridges and increased telomeric fusions with normal telomere length and g-strand overhang, Mol. Cell. Biol. 21 (11) (2001) 3642–3651.
[5] S.M. Bailey, J. Meyne, D.J. Chen, A. Kurimasa, G.C. Li, B.E. Lehnert, E.H. Goodwin, DNA double-strand break repair proteins are required to cap the ends of mammalian chromosomes, Proc. Natl. Acad. Sci. U. S. A. 96 (26) (1999) 14899–14904.
[6] J.D. Griffith, L. Comeau, S. Rosenfield, R.M. Stansel, A. Bianchi, H. Moss, T. de Lange, Mammalian telomere end in a large duplex loop, Cell 97 (1999) 503–514.
[7] K. Roy, S. Kodama, K. Suzuki, M. Watanabe, Delayed cell death, giant cell formation and chromosome instability induced by X-irradiation in human embryo cells, J. Radiat. Res. 40 (1999) 311–322.
[8] P.M. Lansdorp, N.P. Verwoerd, F.M. von de Rijke, V. Dragowska, M.-T. Lile, R.W. Dirks, A.K. Raap, H.J. Tanke, Heterogeneity in telomere length of human chromosomes, Hum. Mol. Genet. 5 (5) (1996) 685–691.
[9] B. McClintock, The stability of broken ends of chromosomes in Zea mays, Genetics 26 (1941) 234–282.

Possible role of ATM-dependent pathway in phosphorylation of p53 in senescent normal human diploid cells

Masatoshi Suzuki, Keiji Suzuki, Seiji Kodama, Masami Watanabe*

Laboratory of Radiation and Life Science, Department of Health Sciences, School of Pharmaceutical Sciences, Nagasaki University, 1-14 Bunkyou-machi, Nagasaki 852-8521, Japan

Abstract

The p53 protein is accumulated and activated as a transcription factor in response to various types of stress. Posttranslational modification, such as phosphorylation and acetylation, are involved in this process. After X-irradiation, for example, the p53 protein at Ser15 is phosphorylated by the ATM kinase, which recognizes DNA double-strand breaks. Recently, we found accumulation of the p53 protein in senescent cells. In the present study, we investigated the involvement of phosphorylation by ATM in p53 accumulation in senescent cells. By using antibody recognizing phosphorylated p53 at Ser15, we confirmed that the accumulated p53 protein was phosphorylated at Ser15. Because the level of p53 and its phosphorylation were comparable to those in X-irradiated young cells, we looked into whether ATM is involved in p53 modification after X-irradiation using the ATM kinase inhibitor, wortmannin. We found that the phosphorylation level decreased when the senescent cells were treated with 40, 60 and 80 μM of wortmannin. Because these concentrations of wortmannin also suppressed p53 phosphorylation in response to X-rays, it is highly possible that ATM is stimulated in senescent cells. © 2002 Elsevier Science B.V. All rights reserved.

Keywords: p53; Telomere shortening; ATM; Phosphorylation; Senescence

1. Introduction

The p53 protein accumulates and activates as a transcription factor in response to ionizing radiation. The activated p53 protein promotes the expression of its target genes, whose products are involved in cell cycle arrest and apoptosis. In addition, the p53 protein is phosphorylated at Ser15. This phosphorylation is carried out by ATM, which recognizes

---

* Corresponding author. Tel.: +81-95-844-5504; fax: +81-95-844-5504.
 E-mail address: nabe@net.Nagasaki-u.ac.jp (M. Watanabe).

DNA double-strand breaks. On the other hand, the phosphorylation of the p53 protein at Ser15 is also observed in senescent cells [1]. Senescent cells have shortened telomeres, which play an important role in the protection of DNA termini and the avoidance of the checkpoint mechanism by formation of the t-loop [2]. In this study, we investigated whether ATM-dependent phosphorylation is involved in p53 accumulation in the senescent cells.

2. Materials and methods

Normal human diploid cells, HE49, were cultured in minimum Eagle's medium supplemented with 10% fetal bovine serum. The cells were subcultured every 3–5 days to maintain exponential cell growth. Exponentially growing cells were irradiated with X-rays from an X-ray generator. The cells were lysed in RIPA containing 1 mM 4-(2-aminoethyl) benzensulfonyl fluoride hydrochloride. The cell lysate was cleared by centrifugation at 15,000 rpm for 10 min at 4 °C, and the supernatant was used as the total cellular proteins. Furthermore, the protein concentration was determined by BCA protein assay (Pierce). These proteins were electrophoretically transferred to the polyvinyl difluoride (PVDF) membrane in transfer buffer, and the membrane was incubated with blocking solution overnight. The membrane was incubated with the anti-p53 monoclonal antibody and the anti-phospho p53 (Ser15) antibody for 2 h, and then incubated with the biotinylated secondary antibody, followed by streptavidin-alkalinephophatase. The band was then visualized after the addition of nitro blue tetrazolium/5-bromo-4-chloro-3-indolyl phosphate (NBT/BCIP) as a substrate. Senescence-associated β-galactosidase staining was carried out as previously described [3].

3. Results

To investigate the p53 level, we used young, presenescent and senescent HE49 cells. The senescent cells were subcultured for more than 20 passages. They showed morphological alteration, which is characteristic for senescent cells. More than 50% of the cells were senescence-associated β-galactosidase-positive. Furthermore, we confirmed that the senescent cells have shortened telomeres, whose mean length was approximately 6.0 kb. This is in contrast to approximately 9.0 kb observed in the young cells. In the presenescent and senescent cells, the level of p53 was increased and the accumulated p53 protein was phosphorylated at Ser15. The level of phosphorylation of p53 at Ser15 in the senescent cells was comparable to that in the young cells irradiated with X-rays. We found that treatment of the senescent cells with wortmannin, which is a specific ATM kinase inhibitor, decreased phosphorylation of the p53 protein, and the same result was observed when irradiated cells were treated with wortmannin.

4. Conclusion

The present results indicate that the ATM pathway is involved in the p53 phosphorylation pathway in both senescent and irradiated cells. It has been previously reported that the ATM–p53-dependent pathway is activated by the disruption of the t-loop [4]. It is possible that the t-loop in the senescent cells may be disrupted by its shortening, and ATM kinase may recognize the abnormal t-loop structure as DNA damage. Moreover, activated ATM may phosphorylate p53 at Ser15, which causes accumulation and activation of p53 in the senescent cells.

Acknowledgements

A Grant of Scientific Research from the Ministry of Education, Culture, Sports, Science and Technology of Japan supported this work.

References

[1] K. Webley, J.A. Bond, C.J. Jones, J.P. Blaydes, A. Craig, T. Hupp, D. Wynford-Thomas, Posttranslational modifications of p53 in replicative senescence overlapping but distinct from those induced by DNA damage, Moll. Cell Biol. 20 (8) (2000) 2803–2808.
[2] J.D. Griffith, L. Comeau, S. Rosenfield, R.M. Stansel, A. Bianchi, H. Moss, T. de Lange, Mammalian telomeres end in a large duplex loop, Cell 97 (4) (1999) 503–514.
[3] G.P. Dimri, X. Lee, G. Basile, M. Acosta, G. Scott, C. Roskelley, E.E. Medrano, M. Linskens, I. Rubelj, O. Pereira-Smith, M. Peacocke, J. Campisi, A biomarker that identifies senescent human cells in culture and in aging skin in vivo, Proc. Natl. Acad. Sci. U. S. A. 92 (20) (1995) 9363–9367.
[4] J. Karlseder, D. Broccoli, Y. Dai, S. Hardy, T. de Lange, p53- and ATM-dependent apoptosis induced by telomeres lacking TRF2, Science 283 (5406) (1999) 1321–1325.

Suppressive effects of p53 protein on heat-induced centrosomal abnormality

Mana Miyakoda, Keiji Suzuki, Seiji Kodama, Masami Watanabe*

Laboratory of Radiation and Life Science, School of Pharmaceutical Sciences, Nagasaki University, 1-14 Bunkyo-machi, Nagasaki 852-8521, Japan

Abstract

Accumulation and activation of the p53 protein is necessary for maintenance of genome stability in response to DNA strand breaks. Recently, it has been found that heat shock, which does not directly damage DNA, induces accumulation of p53. However, it is not clear whether activated p53 suppresses heat-induced genome destabilization. Here, we examined centrosome destabilization by heat. We found that centrosomal abnormality, including denatured, dispersed and multiple centrosomes, increased after heat shock at 43 °C for 2 h. Although the frequency of centrosomal abnormality was not significantly deferent in p53 functional and dysfunctional cells, the cells with multiple centrosomes were found to grow in p53-dysfunctional cells. These results suggest that p53 has no effect on the induction of heat-induced centrosome damage; however, it suppresses the growth of cells with abnormal centrosomes. This study also showed that heat-induced centrosome dysfunction leads to abnormal separation of the chromosomes, and it may destabilize the chromosome number. © 2002 Elsevier Science B.V. All rights reserved.

Keywords: Heat shock; p53; Centrosome; Chromosome; Genome destabilization

1. Background

DNA-damaging agents, such as ionising radiation, accumulate and activate the p53 protein as a transcriptional factor. Activated p53 induces cell cycle arrest or apoptosis, leading to inhibition of genome destabilization [1]. Recently, we have found that heat shock, which does not damage DNA, also accumulates and activates p53 [2]. However, the biological significance of p53 activation after heat shock is not clear. In the present study, we examined the protective effects of p53 on centrosomes, which maintain stability of the genome, after heat shock.

2. Materials and methods

Normal human diploid fibroblasts (HE49), p53-functional tumour (HT1080 and MCF7) and p53 non-functional tumour (T24, RD, A431 and H1299) were heat shocked at 43 °C for 2 h. Immunofluorescence staining using the anti-gamma tubulin antibody was performed to visualize the centrosomes. The nucleus and chromosomes were stained with Giemsa's solution.

3. Results

To determine whether p53 plays a role in heat-shocked cells, we first examined the nuclear morphology of the p53-null NCI-H1299 cells. We found that the cells had multi or giant nuclei, resulting from abnormal chromosome separation. Therefore, we next examined centrosome abnormality, which may cause abnormal chromosome separation after heat shock. While one or two centrosomes are detected in cells at the G1 phase or S and G2 phase in non-treated cells, respectively, three types of centrosome abnormality were observed by immunostaining. One was the decreased staining of centrosomes, one was dispersed centrosomes and the last one was the multiple centrosomes. We compared the frequency of these centrosome abnormalities in p53-functional and dysfunctional cells. Twelve hours after heat shock was carried out, more than 40% of the cells showed decreased staining intensity of centrosomes, in all types of cells. The most probable cause is centrosome denaturation or degradation. In fact, the centrosomes became visible but the percentage of cells containing multiple centrosomes was increased thereafter. In addition, we could not find any significant differences between the p53 function-deficient cells (T24, A431, RD, and H1299) and cells with functional p53 (HE49, MCF7, HT1080). We also counted the percentage of colonies in which more than 50% of the cells contained multiple centrosomes, 120 h after heat shock. The percentage was more than 40% in the p53-dysfunctional cells; however, it was less in the p53 functional cells. These results indicate that cells with multiple centrosomes may proliferate in the absence of functional p53.

4. Conclusion

The present study indicates that heat shock denatures or degrades centrosomes immediately after treatment. Although the damaged centrosomes become visible, the number is increased. The proliferation of cells with multiple centrosomes is suppressed by functional p53; otherwise, cells with an abnormal number of chromosomes should increase. It is suggested that activation of p53 by heat shock may result in inhibition of genome destabilization.

Acknowledgements

This work was supported by a grant of Scientific Research from the Ministry of Education, Culture, Sports, Science and Technology of Japan.

References

[1] D.B. Woods, K.H. Vousden, Regulation of p53 function, Exp. Cell Res. 264 (1) (2001) 56–66.
[2] M. Miyakoda, K. Suzuki, S. Kodama, M. Watanabe, Heat induced G1 arrest is dependent on p53 function but not RB dephosphorylation, Biochem. Biophys. Res. Commun. 266 (1999) 377–381.

Susceptibility of calcium-deficient hydroxyapatite-collagen composite to irradiation

Masatoshi Ohta [a,*], Morihiro Yasuda [a], Hisakazu Okamura [b]

[a]*Department of Material Science and Technology, Faculty of Engineering, Niigata University, 8050 Ikarashi 2-no-cho, Niigata 950-2181, Japan*
[b]*Graduate School of Science and Technology, Niigata University, 8050 Ikarashi 2-no-cho, Niigata 950-2181, Japan*

Abstract

We investigated the susceptibility of a human tooth and synthetic calcium-deficient hydroxyapatite compounds to irradiation. The ESR signal intensities at nearly $g=2$ in a human tooth (enamel, dentine) and synthetic calcium-deficient hydroxyapatite compounds, i.e., calcium-deficient hydroxyapatite (DAp), DAp-collagen composites (cp-Dap, cu-DAp), and stoichiometric hydroxyapatite (HAp) after X-ray irradiation were proportional to the absorbed dose in the range from 6 to 39 Gy. These ESR signal intensities at nearly $g=2$ decreased when the X-ray irradiated samples were soaked in the simulated body fluid. This fact suggests that the surface layer contained a high density of ESR active species in all samples dissolved in the simulated body fluid. Furthermore, the fading of these ESR signal intensities was observed in all of the samples. Regarding the dosimeter utilizing such biomaterials as a tooth, care must be sufficiently given to the storage conditions of the irradiated samples, especially for exact evaluation in the low-dose region. © 2002 Elsevier Science B.V. All rights reserved.

Keywords: ESR dosimetry; Enamel; Dentine; Calcium-deficient hydroxyapatite; Simulated body fluid

1. Introduction

The ESR measurement method has only recently been applied to a radiation dosimetry. A detection of the radiation dose in a tooth is used in the evaluation of the accident radiation dose [1]. However, the components of a tooth are not definite. Then, the addition

---

* Corresponding author. Tel./fax: +81-25-262-6774.
E-mail address: mohta@eng.niigata-u.ac.jp (M. Ohta).

dose method is usually applied to the dosimetry-using tooth. The addition dose can only be made on the biomaterial excised from the body so that the irradiation of an addition dose cannot be carried out under a similar environment to the original irradiation.

In this study, the susceptibility to X-ray irradiation was investigated using ESR spectroscopy under various environmental conditions on the enamel and dentine of a human tooth, and on the calcium-deficient hydroxyapatite-collagen composite synthesized like a tooth by the electrolytic deposition method.

2. Experimental

The calcium-deficient hydroxyapatite $(Ca_{10-x}(HPO_4)_x(PO_4)_{6-x}(OH)_{2-x} \cdot nH_2O$, ($0 < x < 1.5$); DAp) and DAp-collagen composite with collagen-p or-u (composite cp-DAp or cu-DAp) were prepared using the electrolytic deposition method [2]. A stoichiometric hydroxyapatite $(Ca_{10}(PO_4)_6(OH)_2$; HAp) was supplied by the Wako. The enamel and dentine of a human were supplied by Ohu University.

The simulated body fluid consists of the components previously reported [3].

The ESR and XRD were measured by the same method previously reported [2].

3. Results and discussion

The susceptibility to X-ray irradiation was investigated by ESR spectroscopy on the enamel and dentine of a human tooth and on the calcium-deficient hydroxyapatite-collagen composite synthesized like a tooth by the electrolytic deposition method. All ESR spectra had the peak at nearly $g = 2$. These ESR signal intensities were proportional to an absorbed dose in the range from 6 to 39 Gy. The initial ESR signal intensity of dentine was about one-tenth of that of the enamel and DAp, and the composites were about one-tenth of that of the HAp. The fading of these ESR signals suggests that the radical is captured at the trap level of various depths and is released by first order reaction, resulting in disappearance.

The crystallinity evaluated by the XRD method decreased in the order: HAp > enamel > DAp > cp-DAp, cu-DAp > dentine, similar to the order of the ESR signal intensity. Therefore, the susceptibility to X-ray irradiation is found to be dependent on the crystallinity in the enamel and dentine of a human tooth and in the synthesized calcium-deficient hydroxyapatite-collagen composite.

The ESR signal intensities in all of the samples were decreased after soaking in the simulated body fluid. The ESR signal intensities of the enamel and HAp, higher crystallinity, were reduced to nearly 82% of their initial intensities, and that of dentine, DAp and the composites, lower crystallinity, were reduced to nearly 78%. These facts suggest that the reduction in the ESR signal intensity is caused by the dissolution of the surface, including the high density of radicals in the simulated body fluid in all of the samples. Therefore, the body fluid influence on the stability of the radicals produced by X-ray irradiation must be sufficiently considered, when biomaterials such as a tooth are applied to the dosimetry, especially in the case of a low dose.

4. Conclusion

The susceptibility to X-ray irradiation and the stability of the radicals in the simulated body fluid depends on the crystallinity of the sample. This means that, these properties of dentine are similar to that of the DAp and the composites, and that of the enamel is similar to that of the HAp. Therefore, the DAp, the composites, and the HAp can be used as substitutes for a human tooth in the radiation dosimetry.

Acknowledgements

The authors are grateful to the Kawaken Fine Chemical for providing the collagen-p and -u with triple helix structure, and are grateful to Prof. C. Miyazawa of the Ohu University for providing the enamel and dentine of a human tooth.

References

[1] M. Ikeya, New application of electron spin resonance, Dating, Dosimetry and Microscopy, Word Scientific, Singapore, 1993.
[2] H. Okamura, M. Yasuda, M. Ohta, Synthesis of calcium-deficient hydroxyapatite-collagen composite, Electrochemistry 68 (6) (2000) 486–488.
[3] M. Ohta, M. Yasuda, H. Okamura, Susceptibility of calcium-deficient hydroxyapatite-collagen composite to irradiation, Radiation Protection Dosimetry 94 (4) (2001) 385–388.

RBE values and dose rate effects on the ratio of translocation to dicentrics yields in neutrons with low-energy spectrum

K. Tanaka [a,*], N. Gajendiran [b], M. Mohankumar [b]

[a]*Department of Radiobiology, Institute for Environmental Sciences, Rokkukasho, Kamaikita, Aomori, 039-3212 Japan*
[b]*Indira Gandhi Centre for Atomic Research, 603-102 Tamil Nadu, India*

Abstract

The biological effects of low-energy neutrons were dependent on the energy spectrum. The FISH method in the exposure of neutrons and γ-rays with 100 different dose rates (0.02–2.0 cGy/min) revealed that the ratio of translocation to the dicentric yield was influenced by LET and the dose rates. No dose-rate effects were found for the induction of dicentric chromosome aberration. However, an increase in the incidence of dicentric chromosomes was observed in the 2.3, 0.79, 0.57, and 0.37 MeV neutrons. © 2002 Elsevier Science B.V. All rights reserved.

Keywords: Neutrons; Monoenergetic neutrons; Chromosome aberration; FISH; RBE

1. Background

The atomic bomb radiation in Hiroshima contained low-energy neutrons. The recent nuclear incidence at Tokai Mura has re-emphasized the need for a greater understanding of neutron exposure. Study of RBE of the neutrons at various dose rates is important for the risk assessment of occupationally exposed nuclear plant workers and for the estimation of the appropriate radiation quantities for neutron therapy. Also, the variations in RBE at different neutron energies and dose rates are of interest in understanding the basic mechanisms of radiation biology. The RBE values of the neutrons with a low-energy spectrum were assessed in human lymphocytes by comet, chromosome and FISH analyses. The biological effects of the low-energy neutrons were dependent on the energy

---

* Corresponding author. Tel.: +81-0175-71-1308; fax: +81-0175-71-1270.
E-mail address: kmtanaka@ies.or.jp (K. Tanaka).

spectrum. The chromosome-aberration rates and DNA damage detected by the comet assay became higher at lower energy levels (2.3–0.56 MeV), and the RBE value reached a maximum of 16 at the energy value of 0.22 MeV [1,2].

2. Methods

A ^{252}Cf source in the linear accelerator and monoenergetic neutrons (1.0, 0.79, 0.57, 0.37, 0.22 and 0.186 MeV) at Hiroshima University was used for the present study. Lymphocytes from five donors were irradiated to several radiation doses by neutrons with different energy spectrums. Conventional chromosome analysis and the metaphase FISH method of chromosome subsets with chromosomes 1, 2, 4, 6, 7 and 9 were used for scoring dicentric and ring chromosomes and translocations, respectively. ^{60}Co- and ^{137}Cs-γ-rays were used for the standard radiation source.

3. Results

The yields of dicentric and ring chromosomes in the lymphocytes exposed to three dose rates (2.0, 0.2, and 0.02 cGy/min) of the ^{252}Cf neutrons at various radiation doses were observed. Furthermore, the frequency of the translocations was obtained by FISH. The ratios of translocation to the dicentric chromosome yield after exposure to the ^{60}Co- and ^{137}Cs-γ-rays, and ^{252}Cf neutrons were obtained in two ways. The first was the ratio of observed translocations and observed dicentric chromosomes within six painted pairs of chromosomes obtained by FISH. The second was the ratio of calculated translocation frequencies based on the FISH results and the observed dicentric chromosomes obtained using the conventional staining method. We obtained the expected translocation rates from the observed translocation rates using Lucas's formula. Enhancement of the translocation frequency was observed at the lower dose rate of 0.02 cGy/min. The mean values of the ratio of translocation to the dicentric chromosomes in ^{252}Cf neutrons and γ-rays were 2.19–1.99, and 5.03 and 1.3 in two different dose rates, respectively. This means that the ^{252}Cf neutrons induced dicentric chromosomes 1.3–2.0 times higher than the γ-rays. The ratio of the yield of translocation to dicentric chromosomes was influenced by LET, dose rates and the exposure dose. High LET radiation, high dose rate and a high dose induced a higher percentage of dicentric chromosomes. The ratios of translocation to the dicentric chromosomes in various monoenergetic neutrons were observed. The ratio in each monoenergetic neutron was 1.08, 0.78, 0.57, and 1.06 compared with 1.10 of the ^{252}Cf neutrons, which has 2.3 MeV in the mean. The ratio of translocation to the dicentric chromosome changed, depending on neutron energy levels and LET.

4. Conclusions

Before developing the FISH method, several reports indicated that the ratio of translocation and dicentric chromosome yields was 1:1, since the opportunity for these two types of chromosomal aberrations to occur during mitosis were approximately equal. Recent FISH results on the ratio of translocation to dicentric chromosome formation after X-rays or ^{60}Co-γ-rays have been controversial. In the present study, the ratios of translocation to dicentric chromosome yields were 1.3–7.2 and 2.7–8.3 in 2 cGy/min of ^{60}Co-γ-rays and 0.02 cGy/min of ^{137}Cs-γ-rays. The present study revealed that the dicentric chromosomes were induced by ^{252}Cf neutrons at a rate 1.3–2.0 times higher than by the ^{60}Co-γ-rays, indicating that the increase of LET was related to the increase in dicentric chromosomes. High LET radiation causes a higher number of chromosome breaks than low LET radiation. It remains to be resolved by means of molecular biological techniques, why high LET radiation induced a higher number of dicentric chromosomes and also why exposure to a low dose rate with high LET radiation induced a higher percentage of translocation.

References

[1] K. Tanaka, N. Gajendiran, S. Endo, K. Komatsu, M. Hoshi, N. Kamada, Neutron energy-dependent initial DNA damage and chromosomal exchange, J. Radiat. Res. 40 (1999) 36–44, Suppl.
[2] N. Gajendiran, K. Tanaka, N. Kamada, Comet assay to sense neutron 'fingerprint', Mutat. Res. 452 (2000) 179–187.

Antioxidants as radioprotecting agents for low-level irradiation

E.B. Burlakova*, A.N. Goloshchapov, A.A. Konradov,
E.M. Molochkina, Yu.A. Treshchenkova, L.N. Shishkina

Emanuel Institute of Biochemical Physics, Russian Academy of Sciences, ul. Kosygina 4, 119991 Moscow, Russia

Abstract

In our work, we discovered the radioprotective effect of a synthetic antioxidant, potassium salt of -β(4-hydroxi-3.5 ditert.butylphenyl)-propionic acid [phenozan], introduced to SHK mice before and after X-ray irradiation (dose 15 cGy, dose rate 0.01 cGy/min). We used natural antioxidants, multivitamins with mineral additives, as a 1-month radioprotective therapy for the participants of the Chernobyl accident liquidation (PAL) exposed to low-level irradiation (15 cGy). For almost all the indices, a tendency towards normalization was observed. The most drastic changes were observed for the structural parameters of the membranes. © 2002 Elsevier Science B.V. All rights reserved.

Keywords: Low-level irradiation; Protection and therapy with antioxidants

1. Introduction

The actions of the low doses of low-level irradiation show a number of specific features when compared to the average (>25 cGy) and high doses, namely:

- a nonlinear, nonmonotonic dose–effect dependence;
- an inverse dependence of the effect on a dose rate in certain dose ranges;
- changes in the sensitivity of an object irradiated with low doses to the action of other endogenic and exogenic damaging factors [1].

The nonlinear and nonmonotonic dose–effect dependence determined in our experiments can be explained in terms of the concept of changes in the relationship between the

---

* Corresponding author. Tel.: +95-137-64-20; fax: +95-137-41-01.
E-mail address: seren@sky.chph.ras.ru (E.B. Burlakova).

damage caused, on the one hand, and repair of the damage on the other, as a result of the action taken by the low doses of low-level irradiation. We suppose that the reparative systems under the action of the irradiation are not induced, are functioning with a low efficiency, or are initiated with a great delay after the induction of radiation related damages in an irradiated object.

By studying many indices, low and high doses of irradiation cause identical changes in the genetic and membrane apparatus of the cell, e.g., cytogenetic effects in the cell, and structural changes in the membranes. However, the mechanisms and pathways of the effects of low and high doses on cellular metabolism are different. With a low-level of irradiation, the role membrane damage plays in the total damage caused to the cells increases [2].

Thus, it is anticipated that membranoprotectors will produce a radioprotective effect both with a prophylactic and a therapeutic introduction for low-level irradiation of animals.

The fact is that antioxidants are versatile modifiers of the composition, structure, and functional activity of the membranes. Therefore, we selected synthetic and natural antioxidants and vitamins as potential protectors against low-level irradiation.

2. Materials/methods/results

We studied the biochemical and biophysical indices of the membrane apparatus of cells, the structural properties (investigated with a spin-probe technique), the composition and antioxidizing activity of the lipids, lipid peroxides and the activity of membrane and cytosol enzymes in organs of irradiated mice (dose 15 cGy, dose rates 0.01, 0.25 and 9.0 cGy/min) 24 h and 7 and 30 days after irradiation. Changes in all the parameters under observation were detected. A degree of deviation from the normal was dependent on the dose rate and time passed since irradiation.

We detected the radioprotecting effect of a synthetic antioxidant, potassium salt of -β(4-hydroxi-3.5 ditert.butylphenyl)-propionic acid [phenosan], introduced to SHK mice before and after X-ray irradiation (dose 15 cGy, dose rate 0.01 cGy/min). The antioxidant was used in two doses: 3×10^{-14} and 6×10^{-6} M/kg. In both cases of introduction, as prophylaxis (20–35 min before irradiation) and therapy (1 h after irradiation), the effects were protecting and normalizing in regard to the parameters of the antioxidant status of the animals, the activity of the membrane and cytosol enzymes, the composition of lipids, and the structural properties of the membranes. The strongest effect was observed for a super low dose 10^{-14} M/kg.

As said above, we are of the opinion that a cellular response to the action of low-level irradiation develops by other mechanisms than the pathways of the effects of high-level ionizing irradiation on cells and organisms. Therefore, we used the radioprotector not only in standard, but also in super low doses (on the order of 10^{-14}–10^{-15} M/kg), since the trends and perhaps the main pathways of the effects of low-level physical factors (irradiation) and chemical agents on organisms are similar.

We also studied the parameters of the antioxidant and immunological status of participants in the Chernobyl NPP accident liquidation (PAL). Many of these parameters differed significantly from the control (Tables 1 and 2). We used natural antioxidants and

Table 1
Indices of the antioxidant status of an organism for the control and PAL groups and the PAL group after the vitamin therapy

Index	Control	PAL	PAL after vitamin therapy
DBp	0.32	0.27	0.39
Dbe	0.303	0.15	0.47
Vit E	20.9	15.59	21.59
Vit A	2.99	2.67	2.82
R Gl	19.53	20.34	15.32
SOD	125.41	137.53	105.30
GP	7.2	9.28	4.93
GR	5.12	5.75	5.97
H1	7.23	5.79	8.65
H2	7.62	11.32	11.04
MDA1	1.93	3.90	1.89
MDA2	1.95	2.58	1.89
TRClip	1.08	2.06	1.04
TRCpr	1.94	2.22	1.63
ZP	1.23	0.80	0.86
TF	0.78	1.03	0.77
Fr Rad	0.69	2.03	1.14

DBp: plasma lipids double bonds (DB number/mg of lipids $\times 10^{18}$). DBe: erythrocyte plasma lipids double bonds (DB number/mg of lipids $\times 10^{18}$). Vit E: vitamin E. Vit A: vitamin A. R Gl: GSH. SOD: superoxidedismutase. GP: glutathionperoxidase. GR: glutathionreductase. H1: erythrocyte hemolysis. H2: erythrocyte hemolysis after LPO. MDA 1: malonic dialdehyde in erythrocytes. MDA 2: malonic dialdehyde in erythrocytes after LPO initiation. TRClip: rotation correlation time of "lipid" spin probe in erythrocyte membranes. TRCpr: rotation correlation time of "protein" spin probe in erythrocyte membranes. ZP: ceruloplasmin. TF: transferrin. Fr Rad: free radicals with 2.0 g-factor.

Table 2
Indices of the immunological status of an organism for the control and PAL groups and the PAL group after the vitamin therapy

Index	Control	PAL	PAL after vitamin therapy
CD-2	76.66	55.77	87.69
CD-3	60.36	38.50	60.93
NK	16.3	24.60	27.19
CD-20	13.85	16.08	10.47
CD-4	36.53	26.12	33.19
CD-8	28.87	27.67	
CD-4/CD-8	1.31	0.97	1.17
IGG	14.21	14.29	14.57
IGA	3.57	3.69	3.26
IGM	1.35	0.94	1.33
IGE	48.50	86.25	161.86

CD-2: T-lymphocytes (marker CD-2) (%). CD-3: T-lymphocytes (marker CD-3) (%). NK: natural killers (markers CD-16, CD-56) (%). CD-20. CD-4: T-helpers/inductors (marker CD-4) (%). CD-8: T-supressors/killers (marker CD-8) (%). CD-4/CD-8: ratio T-helpers/T-suppressors. IGG: immunoglobulin class G level (mg/ml). IGA: immunoglobulin class A level (mg/ml). IGM: immunoglobulin class M level (mg/ml). IGE: immunoglobulin class E level (mg/ml).

multivitamins with mineral additives, which PAL received for one month as a therapy. For almost all the indices, a tendency towards normalization was observed. The most drastic changes were observed for the structural parameters of the membranes.

References

[1] E.B. Burlakova, A.N. Goloshchapov, N.V. Gorbunova, et al., Radiat. Biol. Radioecol. 36 (4) (1996) 610–631 (in Russian).
[2] E.B. Burlakova, A.N. Goloshchapov, G.P. Zhizhina, A.A. Konradov, in: E.B. Burlakova (Ed.), Low Doses of Radiation: Are They Dangerous? Nova Science Publishers, Huntington, NY, 2000, pp. 1–15.

Tissue responses

Intestinal metaplasia induced by X-irradiation: its biological characteristics

Hiromitsu Watanabe*

Research Institute for Radiation Biology and Medicine, Hiroshima University, Kasumi 1-2-3, Minami, Hiroshima 734-8553, Japan

Abstract

The gastric regions of mice, Mongolian gerbils or different strains of rats were irradiated with a total dose of 20 Gy of X-rays given in two fractions. Intestinal metaplasia was induced in rats, however, not in mice or gerbils, and it was greatly influenced by the rat strains and sexes. Alkaline phosphatase (ALP)-positive metaplastic foci were increased by the administration of ranitidine, crude stomach antigens or the subtotal resection of the fundus and decreased by cysteamine, histamine or the removal of the submandibular gland. We suggest that an elevation in the pH of the gastric juice was due to the disappearance of the parietal cells and is one of the principal factors for the development of intestinal metaplasia, which may be reversible. When the gastric tissue was implanted into the duodenum, pepsinogen-positive chimeric glands with goblet cells appeared in the graft. Esophagus grafts were transplanted into the stomach or duodenum and newly differentiated into gastric or duodenal mucosa, respectively. In the gut, the microenvironment might thus be conducive to the development of metaplasia or to new differentiation. Intestinal metaplasia was found to possibly increase the sensitivity to the induction of tumors by carcinogens of the 1,2-dimethylhydrazine (DMH) or azoxymethane (AOM) type. This route, however, is relatively minor compared to the main route of the N-methyl-N'-nitro-N-nitrosoguanidine (MNNG) or N-methylnitrosourea (MNU) which was not affected. © 2002 Elsevier Science B.V. All rights reserved.

Keywords: X-irradiation; Rat; Glandular stomach; Intestinal metaplasia; Gastric cancer

---

Abbreviations: ALP, alkaline phosphatase; DMH, 1,2-dimethylhydrazine; AOM, azoxymethane; MNNG, N-methyl-N'-nitro-N-nitrosoguanidine; MNU, N-methylnitrosourea; DES, dimethyl estradiol; PhIP, 2-amino-1-methyl-6-phenylimidazo[4,5-b]pyridine.

* Tel.: +81-82-257-5814; fax: +81-82-256-7104.
E-mail address: tonko@hiroshima-u.ac.jp (H. Watanabe).

1. Introduction

Intestinal metaplasia in the stomach is characterized by a highly differentiated epithelium, which resembles that of the intestine [1]. It is more prevalent in men than in women and an increase with age has been observed [2]. The frequency of intestinal metaplasia varies widely in different countries, areas and races [3]. Intestinal metaplasia in the stomach has been considered to be a possible pre-cancerous state on the basis of epidemiological surveys carried out on human gastric carcinomas. Several authors have suggested that intestinal metaplasia could play a role in the development of gastric carcinomas or it is not likely to be termed as pre-cancer because it is common in the benign condition [1–4]. However, its pathogenesis remains unclear. The present paper describes findings on the induction of intestinal metaplasia for our eventual establishment of a model of advanced metaplasia for the analysis of its relation to neoplasia.

2. Dose [5–7]

No intestinal metaplasia was induced by four X-ray doses of 1 Gy. A few intestinal metaplasias were induced by six X-ray doses of 5 Gy to a total dose of 30 Gy. An increase in intestinal metaplasia was induced by two X-ray doses of 10 Gy each at a 3-day interval to give a total dose of 20 Gy, however, no gastric tumors appeared after 12 months. However, gastric tumors were induced after a single X-irradiation dose of 20 Gy and the incidence was increased with two 20-Gy doses given at an interval of 1 week. In contrast, the incidence of intestinal metaplasia was decreased. Thus, these results provide evidence that the best induction of intestinal metaplasia was two X-ray doses of 10 Gy each at a 3-day interval to give a total dose of 20 Gy.

3. Sequential development of intestinal metaplasia [5]

Goblet cells in the gastric mucosa (Type A) appeared 1 week after irradiation, then the intestinal type crypt without Paneth cells (Type B) after 2 weeks, and finally with Paneth cells (Type C) which positively had alkaline phosphatase (ALP) 8 weeks after two X-ray doses of 10 Gy each at a 3-day interval to give a total dose of 20 Gy.

4. Strain and spices differences

The strain differences in the susceptibility of rats to the induction of intestinal metaplasia by X-irradiation were examined. The gastric regions of 5-week-old male rats were irradiated with a total dose of 20 Gy of X-rays given in two equal fractions separated by 3 days. Upon sacrifice at 6 months after the last irradiation, the number of intestinal metaplastic crypts positive for ALP was highest in the Donryu case and lowest in the Copenhagen rats [8]. Morphologically, the numbers of crypts with intestinal metaplasia in the glandular stomachs of Donryu, Wistar, SD or F344 rats were higher than in ACI

(MNNG-sensitive strain), Buffalo or Copenhagen rats [9]. Intestinal metaplasia was more frequently observed in the pyloric glands than in the fundic glands. The results demonstrate that the induction of intestinal metaplasia by X-irradiation is greatly influenced by the strains of the rat. However, intestinal metaplasias were not induced in mice [5] and Mongolian gerbils [10] using the same irradiation.

5. Sex differences [11,12]

The influence of the sex hormones on the induction of intestinal metaplasia was examined in 5-week-old Crj/CD rats of both sexes. At the age of 4 weeks, the animals were gonadectomized and given testosterone or dimethyl estradiol (DES). At the termination of the experiment at 6 months after the X-irradiation, the incidence of intestinal metaplasia with ALP-positive foci in males was significantly higher than in females, in the orchidectomized males or orchidectomized plus DES-treated rats. On the other hand, the incidence of intestinal metaplasia with ALP-positive foci in normal females appeared lower than in ovariectomized females and was increased in rats by the treatment with testosterone or decreased by DES. These results suggested a promoting role for testosterone in the development between intestinal ALP-positive lesions and indicated considerable heterogeneity between intestinal subtypes.

6. Mechanisms of induction of intestinal metaplasia

The development of ALP-positive foci was increased by the administration of ranitidine, an H_2 receptor antagonist [13], and the administration of a crude stomach extract [7], pyloroplasty or pyloroplasty plus vagotomy [14]. On the other hand, intestinal metaplasia was decreased by cysteamine [13], which increases gastric acid secretion, administration of histamine [15] or the removal of the submandibular glands [15]. A close relationship between the fundic pH and ALP-positive foci was seen. The latter promptly disappeared because of their low resistance to acidic conditions [15]. Subtotal resection of the fundus increased the development of intestinal metaplasia induced by X-irradiation as assessed in terms of ALP-positive foci and total intestinal metaplasia [16]. The goblet cells appeared in the pylorus until 7 days of age, and then disappeared by the age of 14 days, which is concurrent with the decrease in the pH value and increase in the number of parietal cells [17]. Therefore, our working hypothesis is as follows: the elevation of the gastric juice pH due to the disappearance of the parietal cells is one of the principal factors responsible for the development of intestinal metaplasia, and they also provided evidence that intestinal metaplasia may be reversible.

Tissue differentiation in the gastrointestinal tract appeared to be unstable, and the esophagus can change to the glandular stomach or duodenum, glandular stomach to the intestine, small intestine to the glandular stomach, and large intestine to the small intestine under the influence of several GI tract diseases. The gastric tissue was transplanted into the duodenum [18]. The pepsinogen-positive chimeric gland with goblet cells appeared in the grafts. It was shown that stomach grafts newly differentiate into the intestine with goblet cells in the duodenum. In the duodenum, the microenvironment might thus be conducive to the development of metaplasia if this is associated with the increase in proliferation. The esophagus graft was transplanted into the glandular stomach or duodenum, which was newly differentiated into gastric or duodenal mucosa, respectively (Katayama and Watanabe, unpublished data). In conclusion, the gastrointestinal stem cells might newly differentiate into different environments in different gastrointestinal tracts.

7. Correlation between intestinal metaplasias and gastric tumors

The colonic mucosa transplanted into the gastric mucosa lacks susceptibility to N-methyl-N'-nitro-N-nitrosoguanidine (MNNG) [19] or to N-methylnitrosourea (MNU) given orally [20] but is sensitive to 1,2-dimethylhydrazine (DMH) carcinogenicity [21], whereas the normal gastric mucosa is not.

Regression analysis of gastric tumors against the frequency of intestinal metaplasia with or without Paneth cells per rat yielded a significant inverse relationship [22,23], suggesting that the development of the intestinal metaplasia and gastric tumors might be independent to the treatment with MNNG or MNU. Induction of intestinal metaplastic mucosa in the glandular stomach induced by X-rays was associated with a tendency for tumorigenesis in response to DMH [24,25], azoxymethane (AOM) or 2-amino-1-methyl-6-phenylimidazo[4.5-b]pyridine (PhIP) [unpublished data] in contrast to the non-susceptible normal gastric mucosa. Thus, the intestinal mucosal target cell is an important determinant of the action of colon carcinogens. Transplant experiments such as those reported here can be of assistance in clarifying the role of the macroenvironment in determining the risk of tumorigenesis.

The number of intestinal metaplasias with ALP-positive foci induced by X-rays in the Donryu rats was similarly decreased by the treatment with AOM, but that aberrant crypt-like foci appeared within some of the affected areas, with the appearance of cystic structures with the pyknotic nuclei that exhibited the binding of anti-8-hydroxyguanosine [26]. Thus, it would appear that areas of intestinal metaplasia with or without Paneth cells induced by X-irradiation might be susceptible to carcinogen damage, leading either to their deletion or to initiation, giving rise to tumors. This would be in line with the alternative possibility that the effects of irradiation and DMH and other colon carcinogens on glandular stomach epithelial cells might additively or synergistically cause carcinogenesis.

Thus, intestinal mucosal stem cell(s) would be expected to be susceptible to colon carcinogenesis, independent of the administration route or their location. With regard to our present finding of gastric carcinomas in animals receiving X-rays and colon carcinogens and that the induction of intestinal metaplasia with or without Paneth cells in the glandular stomach is associated with the susceptibility to tumorigenesis are of clear interest. Thus, the intestinal mucosal phenotype appears to be the important determinant of response to DMH-type carcinogens rather than the intestinal macroenvironment itself. The results are compatible with the conclusion that intestinal metaplasias are targets of DMH-type carcinogens in contrast to the counterpart normal gastric mucosa.

In summary, the presence of intestinal metaplasia with or without Paneth cells may increase the sensitivity to the induction of tumors by carcinogens of the DMH or AOM type but not the MNNG or MNU type. This route, however, is relatively minor compared to the main route of gastric carcinogenesis by carcinogens of the MNNG or MNU type acting on the normal glandular mucosa in the stomach. The protocol used in the present experiment may provide a new approach to help distinguish between the developmental events associated with intestinal metaplasia and gastric tumors.

References

[1] O.H. Jarvi, P. Lauren, On the role of heterotopias of the intestinal epithelium in the pathogenesis of gastric cancer, Acta Pathol. Microbiol. Scand., Suppl. 29 (1951) 26–43.
[2] W. Rubin, L.L. Ross, G.H. Jeffries, M.H. Slessenger, Intestinal metaplasia and heterotopia: a fine structural study, Lab. Invest. 15 (1966) 1024–1049.
[3] S.C. Ming, H. Goldman, D.C. Freiman, Intestinal metaplasia and histogenesis of carcinoma in human stomach: light and electron microscopic study, Cancer 20 (1967) 1418–1429.
[4] S. Nakamura, H. Sugano, K. Takagi, Carcinoma of the stomach in incipient phase: its histogenesis and histological appearances, Gann 59 (1968) 251–258.
[5] H. Watanabe, Pathological studies on intestinal metaplasia—in particular references to acid of gastric mucosa and gastric tumorigenesis, Proc. RINMB 29 (1988) 193–216 (in Japanese).
[6] H. Watanabe, Experimentally induced intestinal metaplasia in Wistar rats by X-ray irradiation, Gastroenterology 75 (1978) 796–799.
[7] H. Watanabe, I. Fujii, Y. Tedara, Induction of intestinal metaplasia in the rat gastric mucosa by local X-irradiation, Pathol., Res. Pract. 170 (1980) 104–114.
[8] H. Watanabe, M. Naito, K. Kawashima, A. Ito, Intestinal metaplasia induced by X-irradiation in different strains of rats, Acta Pathol. Jpn. 35 (1985) 841–847.
[9] H. Watanabe, N. Fujimoto, Y. Masaoka, M. Ohtaki, A. Ito, Strain differences in the induction of intestinal metaplasia by X-irradiation in rats, J. Gastroenterol. 32 (1997) 295–299.
[10] H. Lu, K. Shiraki, Y. Ishimura, M. Ohara, T. Uesaka, O. Katoh, H. Watanabe, The morphological changes of gastric mucosa in Mongolian gerbils by X-irradiation, Nagasaki Med. J. 75 (2000) 210–212.
[11] H. Watanabe, M. Naito, A. Ito, The effect of sex difference on induction of intestinal metaplasia in rats, Acta Pathol. Jpn. 34 (1984) 305–312.
[12] H. Watanabe, T. Okamoto, M. Matsuda, T. Takahashi, P.O. Ogundigie, A. Ito, Effects of sex hormones on induction of intestinal metaplasia by X-irradiation in rats, Acta Pathol. Jpn. 43 (1993) 456–463.
[13] H. Watanabe, M. Kamikawa, Y. Nakagawa, T. Takahashi, A. Ito, The effects of ranitidine and cysteamine on intestinal metaplasia induced by X-irradiation in rats, Acta Pathol. Jpn. 38 (1988) 1285–1296.
[14] I. Fujii, H. Watanabe, M. Naito, K. Kawashima, A. Ito, The induction of intestinal metaplasia in rats by pyloroplasty or pyloroplasty plus vagotomy, Pathol. Res. Pract. 180 (1985) 502–505.
[15] H. Watanabe, T. Okamoto, Y. Fudaba, P.O. Ogundigie, A. Ito, Influence of gastric pH modifiers on development of intestinal metaplasia induced by X-irradiation in rats, Jpn. J. Cancer Res. 84 (1993) 1037–1042.
[16] K. Kinoshita, H. Watanabe, Y. Ando, M. Katayama, H. Yamamoto, N. Hirano, S. Yoshikuni, T. Yamamoto, Effects of subtotal resection of the fundus on development of intestinal metaplasia induced by X-ray irradiation in Donryu rats, Pathol. Int. 50 (2000) 879–883.
[17] H. Watanabe, M. Naito, K. Kawashima, A. Ito, PH-related differentiation in the epithelia of the gastric mucosa of postnatal rats, Acta Pathol. Jpn. 35 (1985) 569–576.
[18] H. Watanabe, Y. Ando, T. Uesaka, Y. Ishimura, K. Shiraki, H. Lu, M. Ohara, S. Shoji, O. Katoh, Grafting of stomach tissue into the duodenum in F344 rats results in chimeric crypts and tumor development, J. Clin. Exp. Cancer Res. 19 (2000) 207–210.
[19] Y. Nakagawa, Y. Ando, N. Fujimoto, Y. Masaoka, M. Tanizaki, S. Shoji, O. Katoh, H. Watanabe, Colon tissue implanted into the glandular stomach in rats lacks susceptibility to N-methyl-N'-nitro-N-nitrosoguanidine (MNNG) carcinogenesis, Oncol. Rep. 4 (1997) 517–519.
[20] Y. Ando, T. Uesaka, S. Kido, N. Fujimoto, M. Tatematsu, Y. Ishimura, K. Shiraki, S. Hirata, K. Kuramoto, S. Shoji, O. Katoh, H. Watanabe, No susceptibility of colon tissue implanted into the glandular stomach of rats to N-nitroso-N-methylurea (MNU) carcinogenesis, Oncol. Rep. 5 (1998) 1373–1376.

[21] Y. Nakagawa, H. Watanabe, T. Takahashi, A. Ito, K. Dohi, Carcinogenicity of 1,2-dimethylhydrazine in colorectal tissue heterotopically transplanted into the glandular stomach of rat, Jpn. J. Cancer Res. 83 (1992) 4–30.

[22] H. Watanabe, A. Ito, Relationship between gastric tumorigenesis and intestinal metaplasia in rat given X-radiation and or N-methyl-N'-nitro-N-nitrosoguanidine, J. Natl. Cancer Inst. 76 (1986) 865–870.

[23] H. Watanabe, Y. Ando, K. Yamada, T. Okamoato, A. Ito, Lack of any positive effects of intestinal metaplasia on induction of gastric tumors in Wistar rats treated with N-methyl-N-nitrosourea in their drinking water, Jpn. J. Cancer Res. 85 (1994) 892–896.

[24] Y. Ando, H. Watanabe, M. Tatematsu, K. Hirano, C. Furihata, N. Fujimoto, T. Toge, A. Ito, Gastric tumorigenicity of 1,2-dimethylhydrazine on the background of gastric intestinal metaplasia induced by X-irradiation in CD(SD) rats, Jpn. J. Cancer Res. 87 (1996) 433–436.

[25] H. Watanabe, T. Uesaka, S. Kido, Y. Ishimura, K. Shiraki, K. Kuramoto, S. Hirata, S. Shoji, O. Katoh, N. Fujimoto, Gastric tumor induction by 1,2-dimethylhydrazine in Wistar rats with intestinal metaplasia caused by X-irradiation, Jpn. J. Cancer Res. 90 (1999) 1207–1211.

[26] H. Watanabe, N. Fujimoto, Y. Masaoka, M. Kurosumi, T. Oguri, T. Takahashi, S. Kido, S. Hirata, K. Kuramoto, S. Shoji, O. Katoh, Effects of azoxymethane on X-ray induced intestinal metaplasia in Donryu rats, Oncol. Rep. 5 (1998) 837–840.

Aberrant extracellular signaling induced by ionizing radiation and its role in carcinogenesis

Rhonda L. Henshall-Powell [a], Catherine C. Park [b], Mary Helen Barcellos-Hoff [a],*

[a] *Life Sciences Division, Lawrence Berkeley National Laboratory, 1 Cyclotron Road, Berkeley, CA 94720, USA*
[b] *University of California, San Francisco, CA 94143, USA*

Abstract

Multicellular organisms orchestrate the behavior of individual cells via extracellular signaling through the microenvironment. The lines of communication are diverse, ranging from the insoluble scaffold of the extracellular matrix, permeated with small, diffusible molecules of cytokines to the cell surface that is structured by adhesion receptors and gap junctions forming communication channels between cells. Ionizing radiation elicits rapid and persistent changes in extracellular signaling, as exemplified in our studies of the irradiated mammary gland by the rapid and persistent activation of transforming growth factor-$\beta1$. We have shown that such events can contribute to radiation's carcinogenic action in experiments in which nonirradiated, preoplastic mammary epithelial cells are transplanted to an irradiated stroma, in which significantly larger tumors arose more frequently. In recent studies, we analyzed the effect of radiation on extracellular signaling in human mammary cells using a three-dimensional culture model. Preliminary data indicate that the progeny of irradiated cells, i.e., survivors, display aberrant morphogenesis and cell–cell interactions, resulting in behaviors characteristic of malignancy in this model. We hypothesized that under certain conditions, radiation exposure prevents normal cell interactions, which in turn could predispose susceptible cells to genomic instability. © 2002 Elsevier Science B.V. All rights reserved.

Keywords: Transforming growth factor β; Mammary gland; Breast cancer; Carcinogenesis; Ionizing radiation; Microenvironment; Genomic instability

---

Abbreviations: TGF-β, transforming growth factor-$\beta1$; ECM, extracellular matrix; HMEC, Human mammary epithelial cells.
* Corresponding author. Tel.: +1-510-486-6371; fax: +1-510-486-4545.
E-mail address: mhbarcellos-hoff@lbl.gov (M.H. Barcellos-Hoff).

1. The role of the microenvironment

The behavior of individual cells is dictated by their interactions with each other via the microenvironment. Indeed, coordinated multicellular behavior is necessary for tissue function, which is orchestrated by extracellular signaling through the microenvironment [1]. The microenvironment, which consists of insoluble proteins in the extracellular matrix (ECM), soluble proteins like cytokines, and cell adhesion molecules (CAM) that link cells to the ECM and to each other, is essential for tissue-specific organization and differentiation [2,3].

In normal tissues, extracellular signaling is also critical for eliminating abnormal cells and suppressing neoplastic behavior (reviewed in Ref. [4]). Extracellular signaling controls cell migration, proliferation, and morphogenesis, all of which are disturbed during carcinogenesis. The disruption of cell adhesion has been postulated to contribute a rate-limiting step in neoplastic progression. Whether referred to as 'soil' or 'landscapers' for cancer, an intact, normal microenvironment is a critical barrier to neoplastic behavior. Disruption of ECM integrity, and thereby cell adhesion using transgenic manipulations can promote tumorigenesis [5]. Loss of E-cadherin, a cell-adhesion molecule crucial for epithelial cell interactions, leads to invasive cell behavior while restitution of E-cadherin to tumor cells impedes malignant behavior [6,7]. Nonmalignant human mammary epithelial cells (HMEC) treated with antibodies that alter ECM receptors known as integrins, exhibit disorganized growth characteristic of tumors. Conversely, manipulating integrins can revert breast cancer cell into normal mammary specific acinar organization. This normalization of cell interactions also reduces tumor formation in vivo [8]. These experiments suggest that appropriate microenvironment signaling can override malignant behavior even in tumor cells with an unstable genome. The importance of understanding extracellular control is further underscored by interferon-γ treatment for chronic myelogenous leukemia, which induces integrin expression that force cancer cells to reattach to the stroma [9].

We have previously proposed that the cell biology of irradiated tissues is indicative of a coordinated multicellular damage response program in which individual cell contributions are directed towards repair of the tissue [10]. However, as a function of dose or radiation quality, radiation-induced extracellular signaling can also disrupt multicellular communication. Our data, using murine and human models of breast cancer, suggest that such radiation-induced microenvironment remodeling can promote malignant progression in susceptible cell populations.

2. Radiation exposure leads to rapid and dynamic microenvironment remodeling

Over the last decade, we have demonstrated that radiation elicits a rapid and dynamic program of ECM remodeling. By comparing the composition ECM in irradiated mammary gland to that of liver and skin following high and low LET radiation exposures, we have concluded that the composition of irradiated microenvironment is a function of the tissue type, the dose, and the radiation quality [11,12]. In the mammary gland, ECM remodeling is mediated by the activation of a potent cytokine, transforming growth factor β1 (TGF-β).

TGF-β is produced as a latent complex that is secreted and requires extracellular activation that permits TGF-β to bind to ubiquitous receptors [13]. We defined antibodies that reveal activation in situ and showed that it is evidenced by the loss of the latent complex and unmasking of TGF-β [14]. Following radiation exposure, TGF-β activation is evident within an hour, persistent for at least a week, and detected following whole body doses of as little as 0.1 Gy [15]. As a consequence of its rapid activation, we demonstrated a novel mechanism of TGF-β activation via exposure to reactive oxygen species, which endows TGF-β with a redox sensor capability. As such, we proposed that TGF-β acts as an extracellular lynch pin released by radiation and other oxidative stressors, to orchestrate multicellular response to damage [16].

3. The irradiated microenvironment promotes neoplastic potential

To test the contribution of such alterations to the process of carcinogenesis, we created radiation chimeric mammary glands by taking advantage of the postnatal development of the mammary gland to surgically create a stroma without epithelium [17]. We then transplanted *unirradiated* mammary epithelial cells to an irradiated mammary stroma. The mammary cells we used were nontumorigenic by several criteria but harbor mutation in both alleles of the *p53* gene [18]. Infrequent small tumors that regress occur in sham-irradiated adult stroma. Nevertheless, when these cells were transplanted to irradiated (4 Gy) stroma, tumors formed in three-quarters of the transplants and the tumors were significantly larger. Furthermore, the effect of the irradiated stroma persisted up to 14 days after exposure. Since hemi-body irradiation resulted in tumors only on the irradiated side, we concluded that this effect was primarily due to altered stromal microenvironment.

Radiation-induced microenvironments constitute a new class of its carcinogenic action that affects the ability of cells to maintain appropriate communication between each other. As a consequence of disrupted communication, we hypothesized that the cells harboring neoplastic mutations are released from extracellular signals that suppress their proliferation in normal tissue [19].

4. Radiation induces a heritable HMEC phenotype that is characteristic of malignancy

To test whether radiation exposure alters cell communication, we used a model of nonmalignant HMEC cultured within an artificial ECM. Under these culture conditions, single cells proliferate as clones that organize into a structure typical of glandular epithelium, i.e., an acinus consisting of a single layer of polarized epithelial cells organized into a hollow sphere. Morphogenesis in these colonies is accompanied by appropriately localized proteins necessary for tissue structure and polarity. For example, E-cadherin is localized at the interface between cells, β1-integrin is baso-lateral and α6-integrin is basal [8]. Importantly, tumorigenic and nontumorigenic mammary epithelial cells, which are nearly indistinguishable when cultured as monolayers, are readily classified by their morphogenesis in three-dimensional ECM [20]. Tumor cells proliferate, fail to establish appropriate cell–cell and cell–ECM connections, and thus, form clumps. In contrast, nonmalignant mammary epithelial cells growth arrest after a short period of growth and form acini similar to those found in situ.

We used this tissue-specific morphogenesis to ask whether irradiated HMEC maintained appropriate microenvironment interactions. Cells were irradiated shortly after plating and TGF-β was added to some cultures to mimic the presence of an irradiated stroma. We observed that most colonies arising from cells treated with radiation and TGF-β showed pronounced morphological disorganization in comparison to colonies from sham controls or following single treatments (Barcellos-Hoff et al., unpublished data). Surprisingly, we also found that the number of cells per colony was increased in double treated specimens. Furthermore, using confocal microscopy and immunofluorescence, we observed that colonies from irradiated cells cultured in the presence of TGF-β showed a dramatic loss of E-cadherin, significantly increased β1-integrin, and decreased α6-integrin. A distinct collagen IV containing basement membrane was observed in all treatment groups, suggesting that the altered integrin expression was not due to the lack of appropriate ligand. Importantly, the phenotype occurred in almost all cells that survive irradiation, which clearly limits the role of mutational mechanisms. Together, these data suggested that colonies arising from irradiated cells exhibit a consistent phenotype consisting of inappropriate intercellular adhesion, deranged extracellular adhesion molecules, loss of gap junction proteins, and disorganized tissue-specific organization.

These observations indicate that radiation exposure of individual cells leads to a persistently altered phenotype in daughter cells that is characterized by the loss of critical controls imposed by the microenvironment to maintain tissue integrity. The resulting behavior is consistent with malignant progression. We have argued that microenvironment perturbations induced by irradiation are means of fostering neoplastic progression in susceptible (i.e., premalignant) cells [4]. Tlsty [21] has recently proposed that disturbances in cell adhesion can modulate pathways that control genomic stability. We predict that if such colonies arising from irradiated cells show increased genomic instability, it is a consequence of the lack of microenvironment control rather than as a direct result of DNA damage. Future studies will test whether the irradiated HMEC phenotype contributes to radiation-induced genomic instability.

5. Potential mechanisms underlying an irradiated phenotype

Since the disruption of extracellular interactions occurs in almost all colonies formed by cells that survive irradiation, the role of mutational mechanisms is clearly limited as the mechanism underlying this phenomenon. Phenotypic evolution is a more likely basis for the behavior of irradiated HMEC. Phenotype is driven by biochemical changes, due in part to extracellular signaling and epigenetic modulation of the genome. Both can lead to heritable phenotypes (as evidenced by the differentiation of more than 300 cell types from a human genome) or a reversible phenotype, such as in the 'activated' phenotypes that are observed in certain cells during inflammation or wound healing. If the mechanisms by which the irradiated phenotype is perpetuated involve extracellular signaling via soluble molecules or cell contact, irradiated cells will be able to influence unirradiated cells via a "bystander" mechanism. Radiation "bystander" effects have been demonstrated both in vitro and in vivo [22–24]. It is unclear whether bystander effects that promote neoplastic behavior are a function of direct cell contact, or mediated by soluble factors, or (more likely) both.

The relevance of studying epigenetic mechanisms in irradiated cells lies in their potency as harbingers of change: under certain circumstances, epigenetic modifications occur at high frequency in specific genes known to be players in the carcinogenic process. Epigenetic mechanisms encompass a wide range of events that affect how the genome is expressed but are not as well characterized as genomic sequence changes during cancer progression. DNA cytosine methylation in regions that are CpG-rich (CpG island) recruits methylated-DNA binding proteins that in turn interact with histone deacetylases and alter chromatin structure; each of these processes can act independently or in concert to favor inhibition of gene expression necessary for certain phenotypes. Little is known about mechanisms leading to aberrant methylation in epithelial cells and scarcely any data exist regarding methylation following radiation exposure. In addition to several Russian reports in the 1970s, one study in 1989 showed a dose-dependent decrease in 5-methylcytosine at 24, 48 and 72 h post-exposure to 0.5–10 Gy [25].

In tumors, aberrant methylation contributes to gene silencing and may contribute to the generation of heterogeneous phenotypes. Recently, the importance of epigenetic changes during carcinogenesis has gained attention [26]. Studies have demonstrated that hypermethylation of 14-3-3 sigma occurs in 91% of primary breast cancers and is strongly associated with the loss of gene expression in these tumors [27]. Also 50% of breast tumors exhibit E-cadherin epigenetic alterations [28]. The p16/INK4A and retinoblastoma are also among the genes that show changes in the methylation pattern during cancer progression. Romanov et al. [29] have shown that HMEC spontaneously bypass senescence and exhibit genomic instability which is accompanied by p16 methylation.

6. Conclusions

Our studies of the irradiated mammary gland and the progeny of irradiated human cells suggest that radiation exposure can promote malignant progression by pathways other than that of mutational mechanisms. Since the frequency of chromosome aberrations increases many cell generations after irradiation by an, as yet, unknown mechanism [30–32], we suggest that disruption of extracellular signaling, adhesion, and communication could precede, and may augment, destablization of the genome. Our hypothesis that genomic instability is as much a matter of the loss of a multicellular controls as the induction of cellular process [33]. Our future studies will concern the molecular mechanisms underlying these events.

Acknowledgements

Funding for this research was from the Specialized Center of Research and Training in Radiation Health from the National Aeronautics and Space Administration and from the

Low Dose Radiation Program in the Office of Biological and Environmental Research, Office of Energy Research of the US Department of Energy at Lawrence Berkeley National Laboratory.

References

[1] M.J. Bissell, M.H. Barcellos-Hoff, The influence of extracellular matrix on gene expression: is structure the message? J. Cell Sci. 8 (1987) 327–343.
[2] M.H. Barcellos-Hoff, M.J. Bissell, Mammary epithelial cells as a model for studies of the regulation of gene expression, in: K.S. Matlin, J.D. Valentich (Eds.), Functional Epithelial Cells in Culture, A.R. Liss, New York, 1989, pp. 399–433.
[3] M.H. Barcellos-Hoff, J. Aggeler, T.G. Ram, M.J. Bissell, Functional differentiation and alveolar morphogenesis of primary mammary epithelial cell cultures on reconstituted basement membrane, Development 105 (1989) 223–235.
[4] M.H. Barcellos-Hoff, It takes a tissue to make a tumor: epigenetics, cancer and the microenvironment, J. Mammary Gland Biol. Neoplasia 6 (2001) 213–221.
[5] C.J. Sympson, M.J. Bissell, Z. Werb, Mammary gland tumor formation in transgenic mice overexpressing stromelysin-1, Sem. Cancer Biol. 6 (1995) 159–163.
[6] K. Vleminckx, L.J. Vakaet, M. Mareel, W. Fiers, F. van Roy, Genetic manipulation of E-cadherin expression by epithelial tumor cells reveals an invasion suppressor role, Cell 66 (1) (1991) 107–119.
[7] J. Luo, D.M. Lubaroff, M.J. Hendrix, Suppression of prostate cancer invasive potential and matrix metalloproteinase activity by E-cadherin transfection, Cancer Res. 59 (15) (1999) 3552–3556.
[8] V.M. Weaver, O.W. Petersen, F. Wang, C.A. Larabell, P. Briand, C. Damsky, et al., Reversion of the malignant phenotype of human breast cells in three-dimensional culture and in vivo by integrin blocking antibodies, J. Cell Biol. 137 (1) (1997) 231–245.
[9] R. Bhatia, P.B. McGlave, C.M. Verfaillie, Treatment of marrow stroma with interferon-alpha restores normal beta 1 integrin-dependent adhesion of chronic myelogenous leukemia hematopoietic progenitors. Role of MIP-1 alpha, J. Clin. Invest. 96 (2) (1995) 931–939.
[10] M.H. Barcellos-Hoff, How do tissues respond to damage at the cellular level? The role of cytokines in irradiated tissues, Radiat. Res. 150 (5) (1998) S109–S120.
[11] E.J. Ernhart, HZE and proton induced microenvironment remodeling mediated by transforming growth factor-β1 [Doctoral]. Ft. Collins: Colorado State University; 1996.
[12] S.V. Costes, C.H. Streuli, M.H. Barcellos-Hoff, Quantitative image analysis of laminin immunoreactivity in 1 GeV/amu iron particle irradiated skin basement membrane, Radiat. Res. 154 (2000) 389–397.
[13] D.A. Lawrence, R. Pircher, P. Jullien, Conversion of a high molecular weight latent beta-TGF from chicken embryo fibroblasts into a low molecular weight active beta-TGF under acidic conditions, Biochem. Biophys. Res. Commun. 133 (1985) 1026–1034.
[14] M.H. Barcellos-Hoff, E.J. Ehrhart, M. Kalia, R. Jirtle, K. Flanders, M.L.-S. Tsang, Immunohistochemical detection of active TGF-β in situ using engineered tissue, Am. J. Pathol. 147 (1995) 1228–1237.
[15] E.J. Ehrhart, A. Carroll, P. Segarini, M.L.-S. Tsang, M.H. Barcellos-Hoff, Latent transforming growth factor-β activation in situ: quantitative and functional evidence following low dose irradiation, FASEB J. 11 (1997) 991–1002.
[16] M.H. Barcellos-Hoff, T.A. Dix, Redox-mediated activation of latent transforming growth factor-β1, Mol. Endocrinol. 10 (1996) 1077–1083.
[17] M.H. Barcellos-Hoff, S.A. Ravani, Irradiated mammary gland stroma promotes the expression of tumorigenic potential by unirradiated epithelial cells, Cancer Res. 60 (2000) 1254–1260.
[18] D.J. Jerry, D. Medina, J.S. Butel, p53 mutations in COMMA-D cells, In Vitro Cell. Dev. Biol. 30A (1994) 87–89.
[19] M.H. Barcellos-Hoff, The potential influence of radiation-induced microenvironments in neoplastic progression, J. Mammary Gland Bio. Neoplasia 3 (1998) 165–175.
[20] O.W. Petersen, L. Ronnov-Jessen, A.R. Howlett, M.J. Bissell, Interaction with basement membrane serves

to rapidly distinguish growth and differentiation pattern of normal and malignant human breast epithelial cells, Proc. Natl. Acad. Sci. U. S. A. 89 (1992) 9064–9068.

[21] T.D. Tlsty, Cell-adhesion-dependent influences on genomic instability and carcinogenesis, Curr. Opin. Cell Biol. 10 (5) (1998) 647–653.

[22] J.B. Little, Radiation carcinogenesis, Carcinogenesis 21 (3) (2000) 397–404.

[23] K.M. Prise, O.V. Belyakov, M. Folkard, B.D. Michael, Studies of bystander effects in human fibroblasts using a charged particle microbeam, Int. J. Radiat. Biol. 74 (6) (1998) 793–798.

[24] R. Ramesh, A.J. Marrogi, A. Munshi, C.N. Abboud, S.M. Freeman, In vivo analysis of the 'bystander effect': a cytokine cascade, Exp. Hematol. 24 (1996) 829–838.

[25] J.F. Kalinich, G.N. Catravas, S.L. Snyder, The effect of gamma radiation on DNA methylation, Radiat. Res. 117 (2) (1989) 185–197.

[26] L.M. Mielnicki, H.L. Asch, B.B. Asch, Genes, chromatin, and breast cancer: an epigenetic tale, J. Mammary Gland Biol. Neoplasia 6 (2001) 169–182.

[27] A.T. Ferguson, E. Evron, C.B. Umbricht, T.K. Pandita, T.A. Chang, H. Hermeking, et al., High frequency of hypermethylation at the 14-3-3 sigma locus leads to gene silencing in breast cancer, Proc. Natl. Acad. Sci. U. S. A. 97 (11) (2000) 6049–6054.

[28] J.R. Graff, E. Gabrielson, H. Fujii, S.B. Baylin, J.G. Herman, Methylation patterns of the E-cadherin 5′ CpG island are unstable and reflect the dynamic, heterogeneous loss of E-cadherin expression during metastatic progression, J. Biol. Chem. 275 (4) (2000) 2727–2732.

[29] S.R. Romanov, B.K. Kozakiewicz, C.R. Holst, M.R. Stampfer, L.M. Haupt, T.D. Tlsty, Normal human mammary epithelilal cells spontaneously escape senescence and acquire genomic changes, Nature 409 (2001) 633–637.

[30] M.A. Kadhim, S.A. Lorimore, K.M. Townsend, D.T. Goodhead, V.J. Buckle, E.G. Wright, Radiation-induced genomic instability: delayed cytogenetic aberrations and apoptosis in primary human bone marrow cells, Int. J. Radiat. Biol. 67 (3) (1995) 287–293.

[31] M.A. Kadhim, D.A. Macdonald, D.T. Goodhead, S.A. Lorimore, S.J. Marsden, E.G. Wright, Transmission of chromosomal instability after plutonium alpha-particle irradiation, [see comments], Nature 355 (6362) (1992) 738–740.

[32] C. Mothersill, M.A. Kadhim, S. O'Reilly, D. Papworth, S.J. Marsden, C.B. Seymour, et al., Dose- and time-response relationships for lethal mutations and chromosomal instability induced by ionizing radiation in an immortalized human keratinocyte cell line, Int. J. Radiat. Biol. 76 (6) (2000) 799–806.

[33] M.H. Barcellos-Hoff, A.L. Brooks, Extracellular signaling via the microenvironment: a hypothesis relating carcinogenesis, bystander effects and genomic instability, Radiat. Res. 156 (2001) 618–627.

Apoptosis induced in small intestinal crypts by low doses of radiation protects the epithelium from genotoxic damage

Christopher S. Potten

Epistem Ltd., The Incubator Building, Grafton Street, Manchester M13 9XX UK

Abstract

The ultimate stem cells in the small intestinal crypts divide about 1000 times in the lifetime of a mouse and through a series of transit dependent dividing cell generations produce about 2.5×10^{11} intestinal epithelial cells in the 3-year life span. In spite of the large number of stem cells in the small intestine and their rapid and extensive cell cycles, genetic defects leading to cancer or age-related deterioration are rare, indicating that the cells are well protected against these changes. Here, discussed are three ways in which protection is afforded. Firstly, it appears that the ultimate stem cells are protected against DNA replication-induced errors by the selective segregation at division of all the template strands of DNA for retention in the stem line as suggested by Cairns in 1975. Secondly, damage induced into the template strands by agents such as radiation is removed from the crypt by the altruistic apoptosis seen following low-dose radiation, as suggested by Potten in 1977. Thirdly, if an excessive number of stem cells are killed the crypt is protected by having a reserve population of radio-resistant potential stem cells, i.e. there is a hierarchy in the stem cell compartment. Together, these mechanisms provide an extraordinarily effective protection for these crucial cells. © 2002 Elsevier Science B.V. All rights reserved.

Keywords: Intestinal stem cells; DNA strand segregation; Apoptosis; Radiation response; Stem cell hierarchy; Protection of genome integrity

---

The crypts of the small intestinal epithelium are one of the most rapidly proliferating-tissues of the body. In the mouse each crypt contains about 250 cells, of which 150 cells divide twice a day. There are 7.5×10^5 crypts in the small intestine of the mouse, each producing about 300 cells per day; thus, in the 3-year life span of a mouse about 2.25×10^{11} cells are produced. This rapid turnover would be expected to make this tissue highly sensitive to cytotoxic agents such as radiation [1].

The cellular organisation in each crypt consists of four to six cell lineages each with its own lineage ancestor, self-maintaining, *ultimate stem cell*. Under steady state and unper-

0531-5131/02 © 2002 Elsevier Science B.V. All rights reserved.
PII: S0531-5131(01)00769-5

turbed conditions, the entire cell production from the crypt is ultimately dependent for the entire life of the animal on these stem cells. Thus, the production of 2.25×10^{11} cells depends on the survival and genetic integrity of the $3.0-4.5 \times 10^6$ small intestinal stem cells. These stem cells cycle once a day, i.e. have a cell cycle time twice as long as their dependent transit populations, which have in total about six generations. The stem cells thus divide about 1000 times during the 3-year lifespan of a laboratory mouse. As cells move through these transit generations, differentiation events may occur allowing the formation of the four discrete, differentiated cell lineages of the small intestine (Paneth cells, enteroendocrine cells, goblet cells and the predominant columnar enterocytes). It is believed that the first differentiation event does not occur until about the third generation in the transit lineage, implying that all cells below the third generation are undifferentiated cells indistinguishable in most ways from the ultimate lineage ancestor stem cell. In fact, if the ancestral stem cells are destroyed they can be repopulated from these undifferentiated, early lineage cells, which thus represent a *potential stem cell* population i.e. cells that are normally transitory and hence lost from the crypt but ones which can still function as stem cells if required [2,3].

It normally takes about 3 days for a cell to migrate from the upper crypt to the villus tip, from which it is shed at the end of its functional life span. It takes about 3 days for cells to move through the transit lineage, and hence the total life expectancy of cells born as a consequence of stem cell divisions is about 6 days in the mouse. This short life span and rapid migration means that the consequences of cytotoxic damage are rapidly expressed; for example, the destruction, or impairment of reproductive potential, of all actual and potential stem cells results in the loss of the cell lineages, the crypts and ultimately the villus mucosa, resulting in ulceration within a period of 3 to 6 days. The short life span of most of the crypt cells also means that these cells play little role in the gastrointestinal carcinogenesis sequence, which takes months in the mouse (decades in man) to accumulate the requisite sequence of genetic defects. The only cells that persist long enough to accumulate these errors are the permanent residents of the tissue—the ultimate stem cells. A truly remarkable fact about the small intestine is that in spite of the large number of cells and the enormous cell proliferation, this tissue rarely develops cancer. The adjacent large bowel is a common site for cancer development but the incidence in the small bowel in man is a hundredfold lower and in fact certain regions of the small bowel are almost entirely free of cancer. It is common to ask the question 'What is it that makes the large intestinal stem cells so prone to cancer development?' but less common to phrase the question 'How are the small intestinal stem cells so well protected?'

At present, no reliable markers are available to identify the ultimate stem cells, although an antibody to an RNA binding protein that developed from neural stem cells, Musashi-1 (Msi-1), does seem to be expressed in small intestinal early lineage cells [4]. It is also possible to label the DNA of these cells with precursors such as tritiated thymidine (^{3}HTdR), which is retained in the nucleus for long periods of time (label retaining cells, LRC) as a consequence of the slow cell cycle and a probable selective segregation of template DNA strands [5–7]. Such techniques may, in the future, permit more detailed studies of the biology of these crucial adult tissue stem cells and allow both the isolation of the cells and the genetic regulatory factors to be identified.

Even in the absence of these advances, the small intestinal epithelium is an attractive biological model for studying stem cell and transit cell behaviour because the cell lineages in the crypt are arranged in a precise fashion along the long axis of the crypt. Hence, by studying the characteristics and behaviour of cells at individual positions along the long axis of longitudinal sections through the crypt, one can study the behaviour of stem cells and their transit progeny [1–3].

If the intestinal epithelium is exposed to a cytotoxic agent such as radiation, one of the first detectable responses is the induction of apoptotic changes in some of the cells. The spatial distribution of these apoptotic cells along the crypt axis depends on the cytotoxic agent being used. Radiation, bleomycin, isopropyl–methane–sulphonate and adriamycin all target cells near the bottom of the crypt (cell positions [4–6]), i.e. cells early in the cell lineage [8,9]. Here, the focus will be on the effects of radiation on the small intestine.

The *radiation-induced apoptosis* is detectable within 2 to 3 h, reaches a peak between 3 to 6 h and declines over 24 h. The peak yield occurs over cell positions 4–6 but the distribution is quite broad with apoptoses being detectable up to about cell position 15 [10–12]. The development of such radiation-induced apoptosis is totally dependent on the *p53* gene, although wild-type p53 protein cannot be detected by immunohistochemistry in apoptotic cells. The apoptotic response is completely absent in *p53* knockout mice [13].

The most surprising feature of the radiation-induced apoptosis in the small intestine is its dose response relationship. Increases in the apoptotic yield can clearly be detected above the background level of spontaneous apoptosis, following doses as low as 1–5 cGy. The yield increases progressively with dose up to about 1 Gy at which point the response appears to saturate. There is little further increase in the yield of these early apoptotic cells when the dose is raised to 10 Gy [10–12]. This suggests that the crypt contains a discrete sub-population of proliferative cells, at cell positions 4–6, that are exquisitely radio-sensitive. In fact, these cells probably represent the most sensitive in vivo mammalian cells.

The sensitivity is such that a single hit anywhere on the entire DNA molecule of the cell triggers an altruistic apoptotic suicide, thus removing the damage and the damaged cell. This *p53*-dependent cell deletion process is activated in preference to any DNA repair mechanisms. In fact, the data suggest a complete inability to switch on DNA repair processes (the levels of apoptosis are independent of dose rate and the deduced survival curve for the susceptible cells is exponential to zero dose). The number of these apoptosis susceptible cells deduced from the plateau at dose saturation is about 4–6 per crypt. Thus, it can be concluded that since the apoptotic susceptible cells occur at the same cell positions in the crypt as the stem cells and with the same frequency, they are indeed the same cells and that they are using this altruistic apoptosis as a protective mechanism against DNA damage induced by agents such as radiation.

There are later waves of apoptosis that can be measured in the crypt following cytotoxic exposure. Some of these, notably the wave seen at about 24 h post exposure, are *p53*-independent and some of these waves are likely to be associated with the homeostatic mechanisms operating to restore the normal steady state balance between mitosis, apoptosis and differentiation.

The author has already alluded to the fact that there is a low level of naturally occurring, or *spontaneous apoptosis* in the crypts. This occurs at the stem cell position and is *p53*-independent since it is observed at the same levels in *p53* knockout mice. The frequency of this spontaneous apoptosis is very low when considered in relation to all the epithelial cells (about one apoptosis in every fifth crypt longitudinal section in the mouse). It occurs also in humans but probably at an even lower frequency. In the mouse the highest yield is at cell positions 4–6 and if the yield is expressed as a frequency of occurrence for cells at cell position 4, it suggests a spontaneous cell death rate of between 5% and 10%. Hence, it is concluded that this is also a stem cell-related phenomenon and it is hypothesised that it represents the homeostatic mechanisms operating to control the number of stem cells. An occasional symmetric division amongst the 4–6 stem cells is predicted from mathematical modelling [14] and this would result in an extra stem cell in a crypt.

Since each extra stem cell would produce a lineage of between 64 and 128 extra transit cells, this would distort the architecture of the crypt (generate a hyperplastic crypt). However, crypt size is remarkably constant and hence this hyperplasticity clearly does not occur, and it seems reasonable to conclude that any extra stem cells are rapidly detected and deleted by the spontaneous apoptosis mechanism.

In the adjacent large bowel, spontaneous apoptosis is virtually absent and radiation induced apoptosis occurs at a lower frequency, is not dose-saturated until about 6–8 Gy, and does not have any specificity for the stem cell position, which at this site is at the very base of the crypt (cell positions 1–2) [12]. It is thought that these differences are attributable to the expression of the cell survival (anti-apoptotic) gene *bcl-2* [15]. In the small intestine the *bcl-2* survival gene is not expressed. Thus, in the large bowel both the homeostatic stem cell regulation process and the apoptotic protective mechanism are compromised by the *bcl-2* gene. Thus, stem cell numbers may gradually drift upwards with the passage of time producing occasional hyperplastic colonic crypts and increasing the number of carcinogen target cells and thus cancer risk. Furthermore, stem cells incurring genotoxic damage are not deleted in the large bowel again as a consequence of the *bcl-2* gene but rather they activate repair pathways; such repair carries an element of risk of DNA damage, again raising the risk of cancer [16,17].

In the small intestine, following doses of 1 Gy or above, all the apoptotic susceptible cells are destroyed but all crypts survive, indicating that other more radio-resistant regenerative stem cells are present. Crypts are not destroyed until doses above about 8–9 Gy. Between and 9 and 16 Gy there is an exponential relationship between crypt survival and radiation dose [18,19]. Over this dose range surviving crypts are regenerated over a period of 3 days from single, surviving regenerative potential stem cells. The regeneration involves a re-establishment of the apoptotic-susceptible cell population [20] and the entire crypt lineages by a clonal expansion from the surviving cell. Hence, these cells are sometimes referred to as *clonogenic stem cells* and this assay represents a functional assessment of stem cell competence. Calculations based on experimental data suggest that the crypts may contain about 30 clonogenic cells [21]. One interpretation of the relationship between the apoptotic susceptible cells and the clonogenic cells is that these represent different parts of a stem cell hierarchy or lineage, with the day-to-day cell replacement being dependent on a few lineage ancestor stem cells, which protect their DNA from exogenous damage by the suicide mechanism and may protect themselves from replication-induced errors by selectively segregating the template DNA strands, which they consistently retain. If an occasional cell commits suicide, it is easily replaced from neighbouring lineage ancestor stem cells. If, however, all these stem cells are killed, a situation unlikely in nature but one which does occur in laboratory experiments and also possibly in some clinical radiotherapy situations, then cells in the next two to three generations, which remain undifferentiated, are capable of functioning as stem cells and regenerating the crypt. These cells are normally in transit through the crypt lineage but do not undergo differentiation until they reach the transition between the third and fourth generation. Once in the fourth dividing transit generation, these cells no longer are capable of crypt regeneration (a stem cell attribute) but are capable of several further rounds of division. The crypt is thus protected very effectively from high levels of damage by this hierarchical stem cell organisation.

Until recently, the ability to label the ultimate stem cells was lacking. However, it now appears that the early lineage cells (including the ultimate stem cells) express the RNA binding protein Musashi-1 (Msi-1) and this may prove useful for identification and isolation procedures [4].

Recently, we have been investigating whether or not the intestinal stem cells can have their DNA labelled in such a way that the label is retained in the cell for long periods of time, in much the same way as in epidermal keratinocyte stem cells (label retaining cells, LRC) [22–24]

It was suggested in 1975 by Cairns [5] that the crucial stem cells of the body would have evolved special protective mechanisms to conserve the integrity of their DNA. DNA replication represents one of the most hazardous stages, in terms of replication-induced errors, in the life cycle of the cell. A simple way of completely protecting against these errors would be, during asymmetric cell divisions, for the stem cell to selectively retain all the template strands and to discard the newly synthesised strands with their potential errors to the daughter cell, which is destined to be discarded from the tissue within a few days. If such a mechanism is operated, the template strands in stem cells could in principle be labelled when stem cells are making new stem cells (undergoing symmetric cell divisions), e.g. during gut development and post-irradiation regeneration, and this could provide an explanation for the label retention studies.

LRCs can be identified in the small intestinal crypt if repeated doses of tritiated thymidine are administered at critical stages during gut development or in the immediate post-irradiation crypt regeneration phase, i.e. at times when new stem cells are being made [6,7]. In both these situations, a small number of cells predominantly at cell position 4 appear to retain the tritiated thymidine label for long periods of time. Initially, it was thought, as it was for the skin experiments, that the label retention was the consequence of the slow cell cycle of the stem cells. However, this cannot be an adequate explanation because in both skin and gut, the stem cells should have been through many rounds of division based on current estimates of the stem cell cycle time. In the gut, label is apparently retained in spite of as many as 6–30 divisions. Four divisions would normally be expected to reduce the concentration of label to sub-threshold levels, assuming that the DNA is randomly segregated at division.

We have recently undertaken an experiment on the LRCs generated as indicated above, and based on the assumption that it is their template strands that are labelled and that they are selectively retaining these strands. LRCs were generated using ^{3}HTdR administration to either juvenile mice or post-irradiation during the regenerative phase, and then waiting for a suitable period of time to dilute the label from all other cell types. Bromodeoxyuridine (BrdUrd) was then repeatedly administered over a 48-h period to label the newly synthesised strands in all the cells in the crypt including the LRCs. We then followed the fate of the two labels with the passage of time. The BrdUrd labelling of the LRCs indicated that the LRCs were indeed passing through the cell cycle and, occasionally, label-retaining mitotic figures were seen. Following a dose of 1 Gy, some of the LRCs even underwent apoptosis. The ^{3}HTdR label in LRCs was retained for the 8–10 days of the experiment while the BrdUrd label in double-labelled cells was lost from the LRCs after a time equivalent to about two cell cycles (2–3 days) [7]. These results strongly suggest that the selective segregation of DNA, as suggested by Cairns, is indeed operating in the ultimate stem cells and this represents a third and powerful protective mechanism for maintaining the genetic integrity of the small intestinal crypt stem cells from DNA replication-induced errors.

The stem cells in the small intestinal crypts are thus extremely well protected against the occurrence of DNA errors, including those involved in carcinogenesis. Firstly, replication induced errors are prevented by the selective retention of all template DNA strands in the stem cell line. Secondly, randomly generated errors in the template strands (by agents such as radiation) are removed from the tissue by the altruistic apoptotic response. Finally, the tissue (crypt) itself is protected further by having a large reserve population of potential stem cells. The genetic regulation of these processes remain largely unknown, but *p53* and genes in the *bcl-2* family clearly play important roles. It is probable that different networks of genes act in different tissues to control various elements of these protective mechanisms. As a consequence of these processes in the small intestine, cancer rarely develops and the tissue continues to efficiently produce large numbers of cells for the lifetime of the animal, with only minor defects in the timing of the regenerative response in old age [25].

Acknowledgements

This work was supported by the Cancer Research Campaign, UK. I am grateful to Dr Catherine Booth for helpful comments, to my wife Sarah for her help in preparing the first draft of this manuscript, and to Christine Sutcliffe for her help with this and many other manuscripts.

References

[1] C.S. Potten, J.H. Hendry, Structure, function and proliferative organisation of mamallian gut, in: C.S. Potten, J.H. Hendry (Eds.), Radiation and Gut, Elsevier, 1995, pp. 1–31.
[2] C.S. Potten, C. Booth, M. Pritchard, The intestinal epithelial stem cell: the mucosal governor, Int. J. Exp. Pathol. 78 (1997) 219–243.
[3] C.S. Potten, Stem cells in gastrointestinal epithelium: numbers, characteristics and death, Philos. Trans. R. Soc. London, Ser. B 353 (1998) 821–830.
[4] C.S. Potten, C. Booth, G.L. Tudor, D. Booth, S.-I. Sakakibara, H. Okanol, Indentification of a putative stem cell marker, J. Cell Sci., (2002) Submitted for publication.
[5] J. Cairns, Mutation selection, and the natural history of cancer, Nature 255 (1975) 197–200.
[6] C.S. Potten, W.J. Hume, P. Reid, J. Cairns, The segregation of DNA in epithelial stem cells, Cell 15 (1978) 899–906.
[7] C.S. Potten, G. Owen, D. Booth, The ultimate intestinal stem cells protect the integrity of their genome by selective segregation of template DNA strands. Submitted for publication.

[8] K. Ijiri, C.S. Potten, Response of intestinal cells of differing topographical and hierarchical status to ten cytotoxic drugs and five sources of radiation, Br. J. Cancer 47 (1983) 175–185.
[9] K. Ijiri, C.S. Potten, Further studies on the response of intestinal crypt cells of different hierarchical status to cytotoxic drugs, Br. J. Cancer 55 (1987) 113–123.
[10] C.S. Potten, Extreme sensitivity of some intestinal crypt cells to χ and γ irradiation, Nature 269 (1977) 518–521.
[11] J.H. Hendry, C.S. Potten, C. Chadwick, M. Bianchi, Cell death (apoptosis) in the mouse small intestine after low doses: effects of dose-rate 14.7 MeV neutrons and 600 MeV (maximum energy) neutrons, Int. J. Radiat. Biol. 42 (1982) 611–620.
[12] C.S. Potten, H. Grant, The relationship between radiation-induced apoptosis and stem cells in the small and large intestines, Br. J. Cancer 78 (1998) 993–1003.
[13] A.J. Merritt, C.S. Potten, J.A. Hickman, C. Kemp, A. Balmain, P. Hall, D. Lane, The role of p53 paper in spontaneous and radiation-induced intestinal cell apoptosis in normal and p53 deficient mice, Cancer Res. 54 (1994) 614–617.
[14] M. Loeffler, T. Bratke, U. Paulus, Y.Q. Li, C.S. Potten, Clonality and life cycles of intestinal crypts explained by a state development stochastic model of epithelial stem cell organisation, J. Theor. Biol. 186 (1997) 41–54.
[15] A.J. Merritt, C.S. Potten, A.J.M. Watson, D.Y. Loh, J.A. Hickman, Differential expression of Bcl-2 in intestinal epithelia: correlation with attenuation of apoptosis in colonic crypts and the incidence of colonic neoplasia, J. Cell Sci. 108 (1995) 2261–2271.
[16] C.S. Potten, The significance of spontaneous and induced apoptosis in the gastrointestinal tract of mice, Cancer Metastasis Rev. 11 (1992) 179–195.
[17] C.S. Potten, Y.Q. Li, P.H. O'Connor, D.G. Winton, Target cells for the cytotoxic effects of carcinogens in the murine large bowel and a possible explanation for the differential cancer incidence in the intestine, Carcinogenesis 13 (1992) 2305–2312.
[18] H.R. Withers, M.M. Elkind, Microcolony survival assay for cells of mouse intestinal mucosa exposed to radiation, Br. J. Radiat. Biol. 17 (1970) 261–268.
[19] C.S. Potten, J.H. Hendry, The microcolony assay in mouse small intestine, in: C.S. Potten, J.H. Hendry (Eds.), Cell Clones: Manual of Mammalian Cell Techniques, Churchill Livingstone, Edinburgh, 1985, pp. 50–60.
[20] K. Ijiri, C.S. Potten, The re-establishment of hypersensitive cells in the crypts of irradiated mouse intestine, Int. J. Radiat. Biol. 46 (1984) 609–624.
[21] J.H. Hendry, S.A. Roberts, C.S. Potten, The clonogen content content of murine intestinal crypts: dependence on radiation dose used in its determination, Radiat. Res. 132 (1992) 115–119.
[22] J.R. Bickenbach, Identification and behaviour of label-retaining cells in oral mucosa and skin, J. Dent. Res. 60 (1981) 1611–1620.
[23] R.J. Morris, S.M. Fisher, T.J. Slaga, Evidence that the centrally and peripherally located cells in the murine epidermal proliferative unit are two distinct cell populations, J. Invest. Dermatol. 84 (1985) 277–281.
[24] F. Cotsarelis, R.R. Sun, R.M. Lavker, Label retaining cells reside in the bulge of the pilosebaceous unit: implication for follicular stem cells, hair cycle and skin carcinogenesis, Cell 61 (1990) 1329–1337.
[25] K. Martin, T.B.L. Kirkwood, C.S. Potten, Altered stem cell regeneration in irradiated intestinal crypts of senescent mice, J. Cell Sci. 111 (1998) 2297–2303.

Radiation-induced apoptosis and its role in tissue response

J.H. Hendry*

Experimental Radiation Oncology, Paterson Institute for Cancer Research, Christie Hospital NHS Trust, Wilmslow Road, Manchester, M20 4BX, UK

Abstract

A question pertinent to carcinogenesis after low doses and tissue response after high doses is whether the apoptotically sensitive cells in cell lineages are a part of the stem cell population or a daughter population, and whether mutations that affect stem cell response also affect the cell population susceptible to apoptosis. This has now been addressed by using mice null for each of several genes important for radiation response and cell survival. In wild-type mice, an acute dose of ≥ 1 Gy induces apoptosis by 4 h in ≤ 6 cells situated in the stem-cell zone of each small-intestinal crypt in mice. In mice null for the damage-processing gene *atm*, the level of apoptoses induced by 1 Gy was lower than in wild-types. The distribution of apoptoses within the crypt was unaffected. In contrast, the clonogens killed by much higher doses were three-fold more sensitive than in +/+ mice (microcolony assay). In mice null for the apoptosis-promoting gene *tp53*, the early wave of radiation-induced apoptosis was abolished. The crypts demonstrated a complex dose–survival curve, being more sensitive at lower doses and more resistant at higher doses, but not shifted totally in one direction or the other compared to +/+ animals. In mice null for the anti-apoptosis gene *bcl-2*, the early wave of induced apoptosis was similar to in the +/+, but the clonogens were slightly more sensitive. This first report of apoptosis (and clonogen radiosensitivity) in gut crypts of *atm* null mice indicates that radiation-induced apoptosis is both atm and p53 dependent. Also, the atm-null clonogens with markedly increased radiosensitivity either die via the second wave of (p53-independent) apoptosis and not the first wave, or by another mechanism such as premature differentiation or mitotic failure. The increased clonogen radiosensitivity in mice null for bcl-2 is compatible with its survival function, although bcl-2 has not been detected immunohistochemically in this site in +/+ mice. These data indicate the far greater importance of atm than two classical regulators of apoptosis, p53 and bcl-2, in the response of clonogens in the small intestine to high radiation doses. These studies show that the response of apoptotically susceptible cells at low dose, and clonogenic cells at high dose, can be differentially affected by particular mutations such as in atm. This suggests different modes of death of the two cell populations, and hence heterogeneity

---

* Tel.: +44-161-446-3123; fax: +44-161-446-3109.
E-mail address: Jhendry@picr.man.ac.uk (J.H. Hendry).

within precursor cell populations, which is a feature also of some other tissues. © 2002 Elsevier Science B.V. All rights reserved.

Keywords: Intestine; Apoptosis; Stem cells; Mutants; Irradiation

1. Introduction

Apoptosis is a non-inflammatory cellular depletion mechanism, which has been studied extensively [1]. It is vital in the development of organisms, allowing the correct tissue remodelling necessary in the sequence of developmental stages from zygote to adult. Apoptosis is also a component of normal tissue renewal. It allows abnormal cell types to be deleted during development and at certain stages of cellular differentiation and maturation, so preventing the amplification of abnormal cell types in cell production lineages. Particular cell and tissue types are susceptible to this form of cell death. Apart from the sub-populations in renewal tissues, which contributes a protective measure in the carcinogenic process, lymphocytes and thymocytes are also susceptible to induced apoptosis by external stimuli such as ionising radiation. These processes have been considered to reduce the risk of autoimmune disease and cancer [2], as well as the risk of transgenerational defects with regard to spermatogonial apoptosis [3].

Those cells that are susceptible to spontaneous apoptosis usually are also sensitive to radiation-induced apoptosis, and hence this provides a specific component to the overall tissue response. Such a component is evident in the duration of radiation-induced oligospermia after low-dose testicular irradiation. Radiation-induced apoptosis also forms the basis to the well-characterised early test for whole-body accidental irradiation, the blood lymphocyte count.

In the intestine, radiation-induced apoptosis has been studied extensively (eg. Ref. [4]). In the small intestine apoptosis is restricted largely to the stem-cell zone in the crypts, whereas in the large intestine it is more indiscriminate among the crypt-cell population. This suggests a better regulated system for cell deletion related to cell hierarchies in the small intestine compared to in the large intestine, which might contribute to the vastly lower incidence of cancers in the former site.

In the small intestine, the apoptotic response is highly radiosensitive, occurs within a few hours after low doses (≤ 1 Gy), is distributed around cell position 4 in the (mouse) crypt, is dose-rate independent to gamma-rays, and it shows an even higher sensitivity to densely ionising radiation compatible with a single track of ionisations being responsible for the induction of apoptosis in a single cell [5].

It remains unclear whether the apoptoses are induced in the stem cell population, in a transit daughter sub-population of cells, or both. An earlier publication [6] discussed two possibilities. The early apoptoses after low doses could represent either a hypersensitive component of the stem-cell population (in which case they would not be detected in the shape of the dose–response curve of the more resistant clonogens assayed after high doses), or they might be a few of the apparently large complement of radioresistant clonogens assayed after high radiation doses, and which died specifically by the early apoptotic mechanism. Since that time, there has been evidence that the number of clonogens, deduced from the irradiation responses, increased as size of the radiation dose increases [7,8]. This suggests either the presence of some artefact in the methodology, for example smaller clones resulting from higher doses giving greater levels of depletion than expected, or a phenomenon of recruitment of cells into the clonogen population. The latter might be associated with the greater overall cellular depletion after higher doses and the need for clonogen renewal and repopulation from as many survivors as possible to regenerate the depleted epithelium.

In recent years, various genetically modified strains of mice have become available, where a particular gene has been deleted, and yet there are still viable offspring. The role of that gene in radiation responses can then be tested by comparing the responses in the wild-type versus the knockout strain. Some of these null mutants for genes involved in specific apoptotic processes or the overall cellular radiosensitivity phenotype have provided further evidence regarding the relationship between apoptotic and clonogenic cells in the intestinal crypt.

2. Mouse null mutants

The p53 knockout mouse was one of the first null mutants to be studied in detail regarding radiation responses. Compared to the wild-type parental strain, the low spontaneous apoptotic rate is unaffected, but the induced rate following low doses of irradiation (1–5 Gy) is abolished [9,10]. The clonogen survival curve after higher doses (10–40 Gy at low dose-rate) tends to show higher survival levels after the highest doses used, but not after lowest, and the overall curve is not shifted totally in one direction or another [11]. The number of clonogens deduced from the back-extrapolate of the survival curve after low-dose-rate irradiation suggested that only a small proportion of the total clonogens dies by early apoptosis (Table 1) [5]. However, there is much uncertainty in the estimate of the number of clonogens in this case (e.g. the sort of variation that is observed between different series with the parental FVB strain mice, Table 1). Also, it is known that the deduced number is dose-dependent and smaller if deduced using only a lower range of dose, as described above.

In the large intestine there is evidence of some sensitisation of crypt clonogens in the p53 null mouse compared to the wild-type, possibly associated with reduced cell cycle delays after irradiation and less time for repair [12]. Regarding bcl-2, a gene product that

Table 1
Apoptosis levels and clonogen numbers in the null mutant mice compared to in the parental wild-type strains

		Apoptoses/crypt		Clonogens/crypt	
		@1 Gy	@8 Gy	n	α (Gy^{-1})
atm	−/−	2.1	5.6	12 ± 6	0.60 ± 0.10
FVB	+/+	3.5	6.3	13 ± 6	0.17 ± 0.02
p53	−/−	0.6	–	20 ± 33	0.12 ± 0.05
FVB	+/+	5.0	7.2	65 ± 44	0.23 ± 0.03
bcl2	−/−	5.2	–	13 ± 5	0.19 ± 0.02
C57	+/+	5.2	–	8 ± 3	0.14 ± 0.01

counteracts the effect of p53, in promoting survival rather than apoptosis, there is no difference in either the spontaneous or the 1Gy-induced level of apoptosis in small intestinal crypts compared to the wild-type mouse (Table 1) [13]. The clonogens are slightly more radiosensitive in the knockout mouse [14]. Both these observations are compatible with the known function of bcl-2 [15] and its low level (undetectable using immunohistochemistry) in this intestinal site [13]. It is more prevalent in the large intestine, specifically in the stem-cell zone [13].

The atm gene product is a good example of a protein involved in detection of DNA double-strand breaks, and in apoptosis and clonogenic survival after irradiation. In the atm null mouse, apoptotic radiosensitivity is lower than in the wild-type mouse (Fig. 1). The level of apoptoses induced by 1 Gy is about halved, although after 8 Gy the levels are similar (null vs. wild-type). This indicates that although apoptotic radiosensitivity is reduced, there is the same maximum number of susceptible cells (Table 1). In contrast, clonogenic radiosensitivity is markedly *increased* by about three-fold in terms of a dose-modifying factor (Fig. 2). The back-extrapolate (12–13) of the clonogen survival curves at low-dose-rate suggests that roughly only half of these (5.6–6.3) die by early apoptosis (Table 1) in *both* the null and wild-type strains. The remainder might die by other mechanisms, for example induced differentiation or mitotic failure. Indeed, in the atm nulls, virtually all the clonogens are killed by 8 Gy (Fig. 2), whereas in the wild-type the crypt surviving fraction is not yet below 1.0, showing that all crypts contain at least 1 surviving clonogen. Despite these differences, the number of early apoptoses per crypt does not increase above around 6 after 8 Gy (Table 1). The increased radiosensitivity of

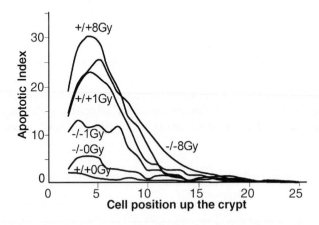

Fig. 1. Incidence of apoptotic cells as a function of cell postion up the crypt from the base in controls (0 Gy), and after 1 or 8 Gy. Apoptotic index = percentage of crypts with an apoptotic cell at each cell position. Values smoothed over three cell positions. Fifty crypts were scored per mouse, with two to five mice per group. Apoptoses per crypt (see text) calculated from the total number of apoptoses counted per crypt column multiplied by 2 (each crypt has two columns of cells in a longitudinal section) and divided by 0.6 (about 60% of apoptoses are counted in a single longitudinal section of a crypt). Unpublished data. +/+ = wild-type FVB strain mice, −/− = atm null FVB mice.

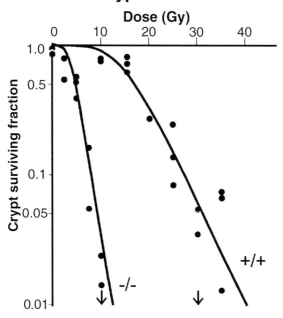

Fig. 2. Microcolones scored at day 3 after low-dose-rate irradiation (1 Gy gamma-rays per hour). Scores were crypt size-corrected to correct sampling frequencies. Arrows indicate observations off scale. Dose–response curves fitted using maximum likelihood techniques and two parameters (radiosensitivity quantified by alpha (Gy^{-1}) and initial clonogen content. Two to four mice per dose group. +/+ = wild-type FVB strain, −/− = atm null FVB mice.

the clonogens confirms the overall increase in gut radiosensitivity reported previously in these animals [16].

An alternative explanation is that the apoptotic cells are a more radiosensitive subpopulation *additional* to the more radioresistant clonogens. In this case a general shift upwards in the clonogen survival curve would be expected in the p53 null mice, if the apoptotically susceptible cells were also clonogenic. This was observed to some extent over limited dose ranges, but not as a consistent feature [11] (Table 1).

3. Conclusions

A comparison of apoptotic cell and clonogenic cell responses in intestinal crypts has given further evidence which relates to insights into their relationship.

(1) Spontaneous apoptosis in these null mutants appears to be at the same level or slightly higher than in the wild-types.

(2) The maximum induced number of early apoptoses (around 6 per crypt) is independent of these genotypes.

(3) Apoptotic sensitivity to 1 Gy is unaffected in the bcl-2 nulls, reduced in the atm nulls, and abrogated in the p53 nulls.

(4) The estimates of clonogen number are not sufficiently accurate to detect an increase in 6 cells, not dying by apoptosis and contributing to the clonogenic population in the p53 nulls. There was no general trend for an increase in crypt survival over the whole dose range which would be expected if cells not apoptosing contributed to the clonogen survival curve. Indeed in the original experiments there was evidence at the conventional assay time of 3 days for a *downward* shift in the clonogen survival curve, which recovered by the later assay time of day 4 [11]. This was interpreted as due to longer times necessary in the p53 nulls for degeneration and regeneration of damaged crypts lacking p53. This might also prevent some of the 'recruitment' occurring of the non-apoptotic precursor cells into the clonogen population, and hence prevent any upward shift in the clonogen survival curve.

(5) There are clear differences in radiosensitivity in the atm nulls regarding both apoptotic cells (*more resistant to low doses*) and clonogenic cells (*more sensitive to high doses*), but no evidence that the numbers of susceptible cells for either endpoint are changed. Hence, these null mutations have appeared to change the response phenotype, in a fairly dramatic way regarding atm, but these quantitative studies still have not clearly elucidated the relationship between the apoptotic cells and the clonogenic/stem cells. Together with additional evidence there is the suggestion that these early apoptotic cells are stem cells (as described by Potten in these Proceedings), which normally would be expected to be clonogenic. Also, that the clonogenic cell population includes stem-cell daughters that can function as regenerative cells in the right environment, in particular those such cells which survive higher cytotoxic doses.

References

[1] C.S. Potten (Ed.), Perspectives on Cell Death, Oxford Univ. Press, Oxford, New York, 1987, pp. 160–183.
[2] S. Kondo, Altruistic cell suicide in relation to radiation hormesis, Int. J. Radiat. Biol. 53 (1) (1988 Jan) 95–102.
[3] S. Kondo, Apoptotic repair of genotoxic tissue damage and the role of p53 gene, Mutat. Res. 402 (1–2) (1998 Jun 18) 311–319.
[4] C.S. Potten, Stem cells in gastrointestinal epithelium: numbers, characteristics and death, Philos. Trans. R. Soc. London, B. Biol. Sci. 353 (1370) (1998 Jun 29) 821–830.
[5] J.H. Hendry, C.S. Potten, Intestinal cell radiosensitivity: a comparison for cell death assayed by apoptosis or by a loss of clonogenicity, Int. J. Radiat. Biol. 42 (6) (1982 Dec) 621–628.
[6] J.H. Hendry, C.S. Potten, C. Chadwick, M. Bianchi, Cell death (apoptosis) in the mouse small intestine after low doses: effects of dose-rate, 14.7MeV neutrons, and 600 MeV (maximum energy) neutrons, Int. J. Radiat. Biol. 42 (6) (1982 Dec) 611–620.
[7] S.A. Roberts, J.H. Hendry, C.S. Potten, Deduction of the clonogen content of intestinal crypts: a direct comparison of two-dose and multiple-dose methodologies, Radiat. Res. 141 (3) (1995 Mar) 303–308.
[8] J.H. Hendry, S.A. Roberts, C.S. Potten, The clonogen content of murine intestinal crypts: dependence on radiation dose used in its determination, Radiat. Res. 132 (1) (1992 Oct) 115–119.
[9] A.J. Merritt, C.S. Potten, C.J. Kemp, J.A. Hickman, A. Balmain, D.P. Lane, P.A. Hall, The role of p53 in spontaneous and radiation-induced apoptosis in the gastrointestinal tract of normal and p53-deficient mice, Cancer Res. 54 (3) (1994 Feb 1) 614–617.
[10] A.R. Clarke, S. Gledhill, M.L. Hooper, C.C. Bird, A.H. Wyllie, p53 dependence of early apoptotic and proliferative responses within the mouse intestinal epithelium following gamma-irradiation, Oncogene 9 (6) (1994 Jun) 1767–1773.

[11] J.H. Hendry, D.A. Broadbent, S.A. Roberts, C.S. Potten, Effects of deficiency in p53 or bcl-2 on the sensitivity of clonogenic cells in the small intestine to low dose-rate irradiation, Int. J. Radiat. Biol. 76 (4) (2000 Apr) 559–565.

[12] J.H. Hendry, W.B. Cai, S.A. Roberts, C.S. Potten, p53 deficiency sensitizes clonogenic cells to irradiation in the large but not the small intestine, Radiat. Res. 148 (3) (1997 Sep) 254–259.

[13] A.J. Merritt, C.S. Potten, A.J. Watson, D.Y. Loh, K. Nakayama, K. Nakayama, J.A. Hickman, Differential expression of bcl-2 in intestinal epithelia. Correlation with attenuation of apoptosis in colonic crypts and the incidence of colonic neoplasia, J. Cell. Sci. 108 (Pt 6) (1995 Jun) 2261–2271.

[14] K.P. Hoyes, W.B. Cai, C.S. Potten, J.H. Hendry, Effect of bcl-2 deficiency on the radiation response of clonogenic cells in small and large intestine, bone marrow and testis, Int. J. Radiat. Biol. 76 (11) (2000 Nov) 1435–1442.

[15] S.J. Korsmeyer, Gene family and the regulation of programmed cell death, Cancer Res. 59 (1999 April) 1693s–1700s.

[16] C.H. Westphal, S. Rowan, C. Schmaltz, A. Elson, D.E. Fisher, P. Leder, atm and p53 cooperate in apoptosis and suppression of tumorigenesis, but not in resistance to acute radiation toxicity, Nat. Genet. 16 (4) (1997 Aug) 397–401.

Essential role of *p53* gene in apoptotic tissue repair for radiation-induced teratogenic injury

Toshiyuki Norimura*, Fumio Kato, Satoshi Nomoto

Department of Radiation Biology and Health, School of Medicine, University of Occupational and Environmental Health, Yahatanishi-ku, Kitakyushu 807-8555, Japan

Abstract

Early stage embryos of mice are radiosensitive to death, but the survivors are born without malformations; however, fetuses at the midgestational stage are highly susceptible to malformation at high, but not low doses of radiation. In order to elucidate the mechanisms of tissue repair of radiation-induced teratogenic injury, we compared the incidence of radiation-induced malformations and abortions in wild type *p53* (+/+) mice and *p53* (−/−) mice with the deficient *p53* gene. For the *p53* (+/+) mice, an X-ray dose of 2 Gy given at a high dose-rate (450 mGy/min) to the fetuses at 9.5 days of gestation was highly lethal and considerably teratogenic, whereas for the *p53* (−/−) mice the same treatment was only slightly lethal but highly teratogenic. However, when an equal dose of 2 Gy given at the same gestational period but at a 400-fold lower dose-rate (1.2 mGy/min), the dose was no longer teratogenic for the *p53* (+/+) mice, which are capable of *p53*-dependent apoptosis, whereas it remained teratogenic for the *p53* (−/−) mice, which are unable to carry out apoptosis. Hence, complete elimination of the teratogenic damage from the irradiated tissues requires the concerted cooperation of two mechanisms, *p53*-dependent apoptotic tissue repair as well as the well-known DNA repair. © 2002 Elsevier Science B.V. All rights reserved.

Keywords: p53-dependent apoptosis; Malformation; Radiation; Dose-rate dependency; Tissue repair

1. Background

Experimental studies with mice have established that irradiation during the pre-implantation period of the embryo (0.5 to 5 days of gestation) induces a high incidence of

---

* Corresponding author. Tel.: +81-93-691-7247; fax: +81-93-692-0559.
E-mail address: norimura@med.uoeh-u.ac.jp (T. Norimura).

prenatal death but virtually no anomalies in the survivors. After the implantation stage, however, the fetuses become progressively resistant to prenatal death, and the sensitivity to neonatal death or to gross anomalies at term reaches a peak around days 9 and 10 (organogenesis) [1,2]. Even at the stage most sensitive to the teratogenic effects of radiation, there are apparent dose rate effects and thresholds in teratogenesis [3]. This suggests that when DNA damage is produced by a small amount of radiation, it is efficiently eliminated by DNA repair. However, DNA repair is not perfect, and there must be defense mechanisms other than DNA repair. In order to elucidate the mechanisms of tissue repair of radiation-induced teratogenic injury, we compared the incidence of radiation-induced malformations and abortions in normal $p53$ (+/+) mice and $p53$ ($-/-$) mice with the deficient $p53$ gene, as the thymocytes and intestinal epithelial cells of the $p53$ ($-/-$) mice are much more radioresistant than those of the $p53$ (+/+) mice [4–7]. If there is a reciprocal relationship between radiation-induced killing and malformation during the early developmental stage, the $p53$ ($-/-$) mice would be expected to be extremely susceptible to radiation-induced malformation.

This paper provides experimental evidence that mouse fetuses have potent defense mechanisms against teratogenic injury caused by ionizing radiation, which function through $p53$-dependent apoptosis. When DNA repair functions efficiently, in concerted cooperation with $p53$-dependent apoptosis, there is a threshold dose rate for radiation-induced malformations.

2. Materials and methods

2.1. Transgenic mice

Mice carrying a disrupted, nonfunctional $p53$ gene, $p53$ ($-/-$), were derived by homologous recombination in an embryonic stem cell line from 129/SvJ mice, as described previously [8]. Mice with the heterozygous $p53$ (+/$-$) genotype were obtained by crossing the male $p53$ ($-/-$) mice with the female $p53$ (+/+) mice. The experiments were conducted according to the Guiding Principles for the Care and Use of Animals approved in 1987 by the Faculty Meeting of the University of Occupational and Environmental Health, Japan.

2.2. Examination of congenital malformations and deaths

The fertilized eggs at the two-cell stage were transplanted into pseudopregnant female MCH/ICR/jcl mice; 10 $p53$ ($-/-$) eggs into one side of the uterine horn of each recipient and 10 $p53$ (+/+) eggs into the other. The resultant pregnant mice were fed a laboratory diet and tap water ad libitum, and were exposed to 250 kVp X-rays or ^{137}Cs γ-rays at a high (450 mGy/min) or a low dose rate (1.2 mGy/min) on day 3.5 or 9.5–10.5 of gestation. On day 18.5 of gestation, the fetuses were removed by cesarean section and the number of implants, early deaths (deaths before the completion of the placenta, i.e., before day 9), late deaths (fetal deaths after day 10), living fetuses and the fetuses with phenotypic anomalies were recorded.

2.3. Measurements of apoptosis

The frequencies of cells dying by apoptosis were measured during or after irradiation. The pregnant mice were exposed to 0–3 Gy of X- or γ-rays on day 9.5 of gestation and killed after irradiation; the fetuses were then fixed and paraffin-embedded, and 4-μm sections were stained using the ApopTag in situ apoptosis detection kit (Oncor, USA; a modified TUNEL method [9]), which allows direct immunoperoxidase detection of digoxigenin-labeled DNA fragments in apoptotic cells.

3. Results

3.1. p53-dependent apoptosis suppresses radiation-induced teratogenesis

We compared the sensitivity of normal $p53$ (+/+) mice and $p53$ null $p53$ (−/−) mice to anomalies and death after X-irradiation with 2 Gy on day 9.5 of gestation. As can be seen in Fig. 1B, normal $p53$ (+/+) mice showed a 60% incidence of deaths and a 20% incidence of phenotypic anomalies, whereas the $p53$ (−/−) mice had a 7% incidence of deaths and a 70% incidence of anomalies [10]. Similarly, X-irradiation of the embryos at 3.5 days of gestation with 2 Gy was highly lethal but not teratogenic for the $p53$ (+/+) mice, whereas it was less lethal but highly teratogenic for the $p53$ (−/−) mice (Fig. 1A) [10]. This reciprocal relationship of radiosensitivity between the anomalies and embryonic lethality supports the notion that embryonic or fetal tissues have a $p53$-dependent 'guardian' of the tissue that aborts cells bearing radiation-induced teratogenic DNA damage. In fact, the frequency of cells dying from apoptosis after irradiation at 9.5 days

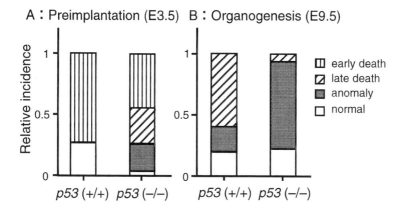

Fig. 1. Incidences of anomalies and deaths in fetuses of the $p53$ (+/+) and $p53$ (−/−) mice after X-irradiation with 2 Gy on day 3.5 or 9.5 of gestation. (A) X-irradiation of embryos at E3.5 induced no anomalies in the $p53$ (+/+) mice but nearly 100% incidence of anomalies in the living $p53$ (−/−) mice. (B) X-irradiation of the fetuses at E9.5, most of the $p53$ (+/+) fetuses died and few of the survivors malformed, whereas few of the $p53$ (−/−) fetuses died but there were many malformed survivors. Constructed from Ref. [10].

of gestation with 2 Gy markedly increased for the p53 (+/+) fetuses but did not increase for the p53 (−/−) fetuses [10].

Tissue repair is specific to multicellular organisms. Brash [11] proposed the concept of cellular proof reading for the p53-dependent apoptotic elimination of damage-bearing cells from irradiated tissues: cells, such as DNA polymerases, can erase their mistake. In this view, the damaged cells recognize their abnormality and protect both the tissue and the organism by committing suicide. In order to explain the complete repair of teratogenic injury, Kondo [12] proposed a cell-replacement repair model, in which the cells having induced teratogenic injury are deleted by suicide (apoptosis) and are then replaced by undamaged healthy cells.

3.2. Relationship between the incidence of radiation-induced malformations and the extent of p53-dependent apoptosis

After X-irradiation with 2 Gy on day 9.5 of gestation, the incidence of anomalies was 20%, 42% and 70% for the p53 (+/+), p53 (+/−) and p53 (−/−) mice, respectively

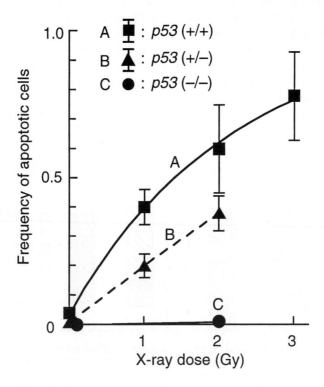

Fig. 2. Frequency of apoptotic cells in neural tubes 4 h after X-irradiation with the indicated doses, in (Curve A) p53 (+/+) mouse fetuses, (Curve B) p53 (+/−) mouse fetuses and (C) p53 (−/−) mouse fetuses on day 9.5 of gestation. Increases in the frequency of cells dying by apoptosis 4 h after X-irradiation with 2 Gy were 60%, 40% and 0% for p53 (+/+), p53 (+/−) and p53 (−/−) mice, respectively. Constructed from Refs. [10] and [13].

[10,13]. The increase in the frequency of cells dying from apoptosis in the neural tube and lumen 4 h after X-irradiation of the fetuses at 9.5 days with 2 Gy was, 60%, 40% and 0% for the p53 (+/+), p53 (+/−) and the p53 (−/−) mice, respectively (Fig. 2). After X-irradiation of the p53 (+/+) fetuses on 9.5 day of gestation with 3 Gy, the frequency of apoptotic cells increased rapidly to a maximum of 80% after 4 h, where it remained up until 8 h, and then returned almost to the control level after 48 h [13].

This means that in developing tissues of normal p53 (+/+) mice, the machinery of apoptosis is well prepared for a rapid and efficient removal of those cells that have failed to repair radiation-induced DNA damage. In fact, the cellular irregularities seen in the irradiated neural tubes at 4 h had returned to near normal after 48 h [13], i.e., a rapid healing best described as cell-replacement repair [12] had occurred. The higher susceptibility of the irradiated p53 (+/−) fetuses to malformation is related to a two-fold lower rate of apoptosis in the p53 (+/−) than in the p53 (+/+) fetuses. Furthermore, competent removal by apoptosis of the damaged cells from the irradiated tissues is impaired dramatically if one of the two wild type p53 alleles is lost.

3.3. Threshold effects in malformation after fetal irradiation in relation to dose-rate and p53-dependent apoptosis

Dose and dose-rate effectiveness is one of the most important subjects to estimate the risk of radiation exposure to humans. Nomura [3] reported that after fetal X-irradiation of normal mice with 1 Gy, the induced frequency of malformations was 25% after a single brief exposure at 540 mGy/min and only 5% after protracted exposure at 4.3 mGy/min, whereas severe combined immunodeficient (SCID) mice, which are defective of DNA double-strand break repair, showed an extremely high incidence of malformations, and such a dose-rate dependency was not observed. In order to obtain evidence that the p53 gene is indispensable in the reduction of the high teratogenic risk of radiation at a high dose rate to zero risk by lowering the dose rate, we compared the dose-rate dependence of

Table 1
Incidences of anomalies and deaths in p53 (+/+), p53 (+/−) and p53 (−/−) mice after fetal irradiation with 2 Gy at a high or a low dose-rate during days 9.5–10.5 of gestation (organogenesis)

Dose (Gy)	Dose-rate (mGy/min)	Genotype of p53	Placentas examined Number	Phenotypic anomalies Number	%[a]	Late deaths Number	%[b]
0		+/+	80	7	9	7	9
0		+/−	75	8	11	4	5
0		−/−	121	23	19	14	12
2	1.2	+/+	79	8	10	4	5
2	1.2	+/−	96	11	12	6	6
2	1.2	−/−	166	50	30	27	16
2	450	+/+	30	6	20	18	60
2	450	+/−	50	21	42	8	16
2	450	−/−	27	19	70	2	7

Data taken from Refs. [10,13,14].
[a] Incidence of anomalies in all embryos in completed placentas.
[b] Incidence of late deaths (after day 10 of gestation) among all embryos in completed placentas.

radiation-induced teratogenesis in wild type $p53$ (+/+), heterozygous $p53$ (+/−) and null $p53$ (−/−) mice.

When an X-ray dose of 2 Gy was given at a high dose rate (450 mGy/min) to the fetuses at 9.5 days of gestation, the $p53$ (+/+) mice were shown to be highly lethal and considerably teratogenic, whereas the $p53$ (−/−) mice were only slightly lethal but highly teratogenic (Table 1) [10,13]. However, when an equal dose of 2 Gy γ-rays given at the same 9.5–10.5 days of gestation but at a 400-fold lower dose rate (1.2 mGy/min), the dose was no longer teratogenic for the $p53$ (+/+) fetuses; and the malformation incidence of 10% after the protracted irradiation is not statistically different from the control level of 9% (Table 1) [14]. In contrast, the same protracted γ-irradiation with 2 Gy at 1.2 mGy/min remained teratogenic for the $p53$ (−/−) fetuses; and the malformation incidence of 30% for the irradiated $p53$ (−/−) fetuses is significantly higher than the control level of 19% (Table 1) [14]. Furthermore, the frequency of apoptotic cells in the neural tube after fetal γ-irradiation with 2 Gy at 1.2 mGy/min was markedly higher in the $p53$ (+/+) mice than in the control level, however, the frequency did not increase for the $p53$ (−/−) mice after irradiation [14]. These results suggest that the $p53$ gene is indispensable for a threshold effect in the risk of radiation at a low dose or dose rate.

4. Conclusion

The DNA damage-bearing cells are completely removed from the tissues after irradiation at low dose rates due to the concerted cooperation of DNA repair and $p53$-dependent apoptosis. Apoptotic cell death appears as a more rapidly operating mechanism by which cells with certain types of damage also from low dose rates are completely eliminated and may be responsible for a threshold for teratogenic risk of radiation. Proficient DNA repair and $p53$-dependent apoptosis are likely to be our most effective guardians against radiation-induced DNA damage in the embryonic or fetal tissues.

Acknowledgements

We wish to thank Drs. S. Kondo and M. Katsuki for their valuable suggestions and supplying the transgenic mice. This work was supported in part by Grants-in-Aid to T.N. for Scientific Research from the Ministry of Education, Science, Sports and Culture, Japan (06454641, 09480128).

References

[1] L.B. Russell, W.L. Russell, An analysis of the changing radiation response of the developing mouse embryo, J. Cell. Comp. Physiol. 43 (suppl. 1) (1954) 103–149.
[2] T. Nomura, High sensitivity of fertilized eggs to radiation and chemicals in mice: comparison with that of germ cells and embryos at organogenesis, Congenital Anomalies 24 (1984) 329–337.

[3] T. Nomura, Dose rate effectiveness and repair in radiation-induced mutagenesis, teratogenesis and carcinogenesis in mice, in: R.N. Sharan (Ed.), Trends in Radiation and Cancer Biology, Forschungszentrum Julich, Germany, 1998, pp. 149–155.
[4] S.W. Lowe, E.M. Schmitt, S.W. Smith, B.A. Osborne, T. Jacks, p53 is required for radiation-induced apoptosis in mouse thymocytes, Nature 362 (1993) 847–849.
[5] A.R. Clarke, C.A. Purdie, D.J. Harrison, R.G. Morris, C.C. Bird, M.L. Hooper, A.H. Wyllie, Thymocyte apoptosis induced by p53-dependent and independent pathways, Nature 362 (1993) 849–852.
[6] A.J. Merritt, C.J. Potten, C.J. Kemp, J.A. Hickman, A. Balmain, D.P. Lane, P.A. Hall, The role of p53 in spontaneous and radiation-induced apoptosis in the gastrointestinal tract of normal and p53-deficient mice, Cancer Res. 54 (1994) 614–617.
[7] K. Fujikawa, Y. Hasegawa, S. Matsuzawa, A. Fukunaga, T. Itoh, S. Kondo, Dose and dose-rate effects of X rays and fission neutrons on lymphocyte apoptosis in $p53(+/+)$ and $p53(-/-)$ mice, J. Radiat. Res. 41 (2000) 113–127.
[8] Y. Gondo, K. Nakamura, K. Nakao, T. Sasaoka, K. Ito, M. Kimura, M. Katsuki, Gene replacement of the p53 gene with the lacZ gene in mouse embryonic stem cells and mice by using two steps of homologous recombination, Biochem. Biophys. Res. Commun. 202 (1994) 830–837.
[9] Y. Gavrieli, Y. Sherman, S.A. Ben-Sasson, Identification of programmed cell death in situ via specific labeling of nuclear DNA fragmentation, J. Cell. Biol. 119 (1992) 493–501.
[10] T. Norimura, S. Nomoto, M. Katsuki, Y. Gondo, S. Kondo, p53-dependent apoptosis suppresses radiation-induced teratogenesis, Nat. Med. 2 (1996) 577–580.
[11] D.E. Brash, Cellular proof-reading, Nat. Med. 2 (1996) 525–526.
[12] S. Kondo, Health effects of low-level radiation, in: Osaka: Kinki University Press; Madison, WI: Medical Physics Publishing, 1993, pp. 73–92.
[13] S. Nomoto, A. Ootsuyama, Y. Shioyama, M. Katsuki, S. Kondo, T. Norimura, The high susceptibility of heterozygous $p53(+/-)$ mice to malformation after foetal irradiation is related to sub-competent apoptosis, Int. J. Radiat. Biol. 74 (1998) 419–429.
[14] F. Kato, A. Ootsuyama, S. Nomoto, S. Kondo, T. Norimura, Threshold effect for teratogenic risk of radiation depends on dose-rate and p53-dependent apoptosis, Int. J. Radiat. Biol. 77 (1) (2001) 13–19.

Susceptibility

Individual differences in chromosomal radiosensitivity: implications for radiogenic cancer

David Scott*

Department of Cancer Genetics, Paterson Institute for Cancer Research Campaign, Christie Hospital NHS Trust, Manchester M20 9BX, UK

Abstract

We have demonstrated interindividual differences in radiosensitivity in the normal population with chromosome breakage assays. In one such assay on G_2 cells, using the 90th percentile of normals as the cutoff value between a normal and a sensitive response, we found that patients with breast, colorectal and head and neck cancers had 40%, 30% and 30% sensitive cases, respectively. We propose that sensitive individuals carry germline mutations in low penetrance genes that predispose to common cancers via defects in processing DNA damage of the types induced by ionising radiation. Support for this hypothesis came from our demonstration of Mendelian segregation of chromosomal radiosensitivity in families of breast cancer cases. We speculate that 5–10% of the normal population carry such predisposing mutations and that, from studies of patients with severe reactions to radiotherapy, a subset of these will be sensitive to the deterministic effects of radiation. At present, there is no generally accepted, direct information on whether these putative gene carriers are at increased risk of radiogenic cancer. Indirect estimates suggest that if there is an enhanced risk, it will be less than the 10-fold increase proposed by the ICRP for carriers of mutations in highly penetrant tumour suppressor genes. © 2002 Elsevier Science B.V. All rights reserved.

Keywords: Chromosome aberrations; Ionising radiation; Lymphocytes; Cancer predisposition; Radiogenic cancer

Risk estimates for radiogenic cancer are derived from exposed populations that will include individuals of varying radiosensitivity. Some degree of enhanced chromosomal radiosensitivity has been observed in about 20 inherited cancer-prone disorders, but these comprise less than 1% of the population [1]. We have now used such assays on normal

---

* Present address: Yew Tree Cottage, Bow Green Rd., Bowdon, Cheshire, WA14 3LF, UK. Tel.: +44-161-941-2393; fax: +44-161-446-3109.
 E-mail address: dscott@picr.man.ac.uk (D. Scott).

healthy individuals and patients with various common cancers. Our primary aims were to investigate links with cancer predisposition and excessive normal tissue damage after radiotherapy, but our findings may also have implications for susceptibility to radiogenic cancer.

Lymphocytes were irradiated in either the G_2 or G_0 phases of the cell cycle. Over 200 healthy controls were G_2 tested and 100 G_0 tested in four different series [2–5]. From repeat assays on over 70 donors, we consistently found that intra-individual coefficients of variation (CV) were less than interindividual CVs. There was no influence of age or gender on radiosensitivity in either assay. In our G_2 studies, there was a skewing to the right in the sensitivity distribution, with about 10% "outliers" [2].

Using a cut-off value at the 90th percentile of controls to distinguish between a normal and a sensitive response, the proportions of G_2-sensitive breast [2], colorectal [4] and head and neck [5] cancer cases were approximately 40% (53/135), 30% (11/37) and 30% (13/42), respectively. Patients with cervical cancer or nonmalignant chronic disease gave values similar to controls [4]. Breast [2] and head and neck [5] cancer cases were also G_0-tested; the proportions of sensitive cases were 27% (35/130) and 35% (17/49), respectively, and there was no significant correlation between the results in the two assays.

For breast and colorectal cancers, there are known germline mutations that confer a high cancer risk. However, although some of these result in enhanced chromosomal radiosensitivity, their prevalence amongst all patients with these cancers is far too small to explain our findings [1]. We propose that, in addition to these highly penetrant, cancer-predisposing genes, there are other low penetrance genes that can be detected because of their influence on chromosomal radiosensitivity. There is good epidemiological evidence for the existence of such genes for breast and colorectal cancer [4]. Support for our hypothesis came from G_2 testing of blood relatives of breast cancer patients [3,6]. The results were consistent with the Mendelian segregation of one, or in some families, two genes conferring enhanced radiosensitivity. Preliminary results with the G_0 micronucleus assay on first-degree relatives of sensitive breast cancer cases also indicate heritability of radiosensitivity [7].

Our choice of the 90th percentile of controls as the dividing line between a normal and a sensitive response distinguished well between normal individuals and those carrying putative G_2 radiosensitising mutations [3,6]. In G_2 testing 74 clinically normal control individuals, Sanford and Parshad [8] found that four (5%) had an abnormally high response. We speculate that 5–10% of the normal population, detectable by virtue of their inherited G_2 sensitivity, have a modestly increased risk of developing certain common cancers. In addition, there may be a subpopulation who are G_0 sensitive and also cancer-prone. A possible example of a low penetrance, breast cancer-predisposing gene is *ATM* in ataxia–telangiectasia (A–T) heterozygotes, estimated to comprise up to 1.5% of the normal population and 7% of all breast cancer cases, although most estimates are considerably lower [9]. A–T heterozygotes are sensitive in both G_2 [10,11] and G_0 [12,13] assays.

Amongst our hypothesised cancer-predisposing genes, we suggest that there is a small but important subset conferring exaggerated sensitivity to the deterministic effects of ionising radiation. In a prospective study, we found that 9/123 (7%) breast cancer patients experienced severe acute reactions to radiotherapy. Seven of the nine (78%) had G_2 scores above our cutoff value, compared with 43/114 (38%) of normally reacting patients

($p=0.023$). Similar results were obtained for patients with severe late reactions, using the G_0 assay [13].

We have no direct information on whether enhanced chromosomal radiosensitivity might be a marker of increased risk of radiogenic cancer. Even for A–T homozygotes, with their extreme radiosensitivity, too few have lived long enough after radiotherapy to allow any estimate of such a risk. From a study of female A–T heterozygotes exposed to diagnostic irradiation, Swift et al. [14] concluded that they had an approximately 6-fold increase of breast cancer compared with nonexposed heterozygotes, but this interpretation has been criticised on several grounds [15].

Enhanced sensitivity to radiogenic cancer has been identified in only a few very rare inherited disorders characterised by a high incidence of spontaneous cancers attributed to germline mutations in tumour suppressor genes [1,15]. These include heritable retinoblastoma and the Li–Fraumeni syndrome, for which there is evidence of some degree of enhanced chromosomal radiosensitivity [1]. However, the primary mechanism responsible for their increased cancer risk is likely to be the high probability of inactivation or loss of their inherited wild-type allele [1,15].

Chromosomally radiosensitive individuals probably have some defect(s) in their ability to process DNA damage of the types induced by ionising radiation. This damage may arise from endogenous sources, such as reactive oxygen species, or from exogenous genotoxic exposures. The relative risk of cancer in mutation carriers compared with nonmutation carriers could therefore be similar for spontaneous and radiation-induced cancers. Athma et al. [16] estimated the spontaneous breast cancer risk in A–T heterozygotes as 3.8 times that in noncarriers, but others [17] have reported no increased risk in carriers. Estimates of risk of radiation-induced breast cancer in A–T heterozygotes are even more highly contentious [15]. Knight et al. [18] investigated cancer frequencies in relatives of individuals with a normal or an elevated G_2 response and reported more cancers in relatives of the latter group. The increases were 3.6- and 2.2-fold, respectively, in first- and second-degree relatives. However, these findings must be treated with caution because the ages of the relatives were unknown.

Another approach to estimating the magnitude of increased cancer risk (spontaneous or radiation induced) in the postulated 5–10% of the population with low penetrance mutations is to compare their mean level of G_2 chromosome damage with that of noncarriers. Our data would indicate a less than 2-fold difference, but Knight et al. [18] reported that G_2-sensitive normals had average yields 2–3 times greater than normally responding individuals.

The above tentative estimates of radiation risk for low penetrance damage-processing genes are lower than those for high penetrance tumour suppressor genes. On the basis of data from heritable retinoblastoma patients after radiotherapy and from studies of rodent strains carrying germline mutations of known tumour suppressor genes, the ICRP [15] and the NRPB [1] have made a judgement that the increase in individual radiation risk could be around 10-fold, although the limitations of the database on which this estimate is based are fully acknowledged. Using this estimate, they point out that the additional cancer risk in gene carriers, from environmental, occupational and medical diagnostic exposures, will be very small compared with their high spontaneous risk but that this additional risk will be more important after the higher exposures associated with radiotherapy. The same argument on the relative importance of low and high exposures is also used for carriers of

Table 1
Calculated lifetime risk[a] of fatal radiogenic breast cancer in women with germline mutations in high and low penetrance genes, compared with spontaneous risk

	Individual risk (%)		
	All individuals	Low penetrance carriers[b]	High penetrance carriers[c]
Spontaneous risk	4.5[d]	11.3	40.0[e]
Radiation risk (100 mSv)[f]	0.04	0.1	0.4
Radiation risk (2 Sv)[g]	0.8	2.0	8.0

[a] Normal female risk estimated at 0.4% Sv^{-1} [20].
[b] Assumed increase in radiation risk, 2.5-fold. The same increase was assumed for spontaneous risk.
[c] Assumed increase in radiation risk, 10-fold [15].
[d] Figure for the UK in 1997 [21].
[e] Nominal risk [15]. Approximates a 10-fold increase above the risk in whole population (see footnote d).
[f] Figure taken to represent an accumulated occupational exposure [15].
[g] Figure taken to represent dose to unaffected contralateral breast during radiotherapy [15].

mutations in low penetrance genes. Table 1 presents the calculated risk of fatal breast cancer, using values of 2.5- and 10-fold, respectively, to represent the increase in risk, above the average risk for the whole population, in carriers of low or high penetrance mutations. In spite of the likelihood that individual radiation risk from high penetrance genes will be greater than for those of low penetrance, the impact of the latter on the nonuniformity of population risk may be greater because of their higher prevalence.

These thoughts on the possible relevance of chromosomal radiosensitivity to the risk of radiogenic cancer will remain highly speculative until much more research has been undertaken. For example, more evidence is required of heritability of chromosomal radiosensitivity of the type we have obtained for families of breast cancer cases. A better understanding is needed of factors which determine experimental variability in these chromosome assays [compare Refs. [2] and [19]]. We need more information from humans and rodent models on whether or not known damage-processing genes that confer elevated chromosomal radiosensitivity also result in enhanced tumorigenic radiosensitivity. Patients with adverse reactions to radiotherapy should be investigated for risk of spontaneous and therapy-induced cancers. After further heritability studies, it may be possible to identify the radiosensitising genes through genetic linkage or candidate gene analysis. If the genes can be identified, their role in spontaneous and radiation-induced cancer risk could then be determined.

Acknowledgements

I would like to thank all my colleagues and collaborators, listed as authors in publications 2–7, 11–13 and 19, for their valuable contributions to this work. Funding was provided by the Cancer Research Campaign, the Christie Hospital Endowment Fund, the National Radiological Protection Board (UK) and the UK Coordinating Committee on Cancer Research.

References

[1] National Radiological Protection Board, Genetic heterogeneity in the population and its implications for radiation risk: report of an advisory group on ionising radiation, Doc. NRPB 10 (3) (1999) 1–47.
[2] D. Scott, J.B.P. Barber, A.R. Spreadborough, W. Burrill, S.A. Roberts, Increased chromosomal radiosensitivity in breast cancer patients: a comparison of two assays, Int. J. Radiat. Biol. 75 (1999) 1–10.
[3] S.A. Roberts, A.R. Spreadborough, B. Bulman, J.B.P. Barber, D.G.R. Evans, D. Scott, Heritability of cellular radiosensitivity: a marker of low penetrance predisposition genes in breast cancer? Am. J. Hum. Genet. 65 (1999) 784–794.
[4] K. Baria, C. Warren, S.A. Roberts, C.M. West, D. Scott, Chromosomal radiosensitivity as a marker of predisposition to common cancers? Br. J. Cancer 84 (2001) 892–896.
[5] R. Papworth, N. Slevin, S.A. Roberts, D. Scott, Sensitivity to radiation-induced chromosome damage may be a marker of genetic predisposition in young head and neck cancer patients, Br. J. Cancer 84 (2001) 776–782.
[6] D. Scott, Chromosomal radiosensitivity, cancer predisposition and response to radiotherapy, Strahlenther. Onkol. 176 (2000) 229–234.
[7] W. Burrill, J.B.P. Barber, S.A. Roberts, B. Bulman, D. Scott, Heritability of chromosomal radiosensitivity in breast cancer patients: a pilot study with the lymphocyte micronucleus assay, Int. J. Radiat. Biol. 76 (2000) 1617–1619.
[8] K.K. Sanford, R. Parshad, Detection of cancer-prone individuals using cytogenetic response to X-rays, in: G. Obe, A.T. Naterajan (Eds.), Chromosomal Aberrations: Basic and Applied Aspects, Springer-Verlag, Berlin, 1990, pp. 113–120.
[9] D.T. Bishop, J. Hopper, AT-ributable risks, Cancer Res. 15 (1997) 226.
[10] Y. Shiloh, R. Parshad, K.K. Sanford, G.M. Jones, Carrier detection in ataxia–telangiectasia, Lancet 1 (1986) 689–690.
[11] D. Scott, A. Spreadborough, E. Levine, S.A. Roberts, Genetic predisposition to in breast cancer, Lancet 344 (1994) 1444.
[12] D. Scott, Q. Hu, S.A. Roberts, Dose-rate sparing for micronucleus induction in lymphocytes of controls and ataxia–telangiectasia heterozygotes exposed to ^{60}Co γ-irradiation in vitro, Int. J. Radiat. Biol. 70 (1996) 521–527.
[13] J.B.P. Barber, W. Burrill, A.R. Spreadborough, E. Levine, A.E. Warren, C. Kiltie, S.A. Roberts, D. Scott, Relationship between in vitro chromosomal radiosensitivity of peripheral blood lymphocytes and the expression of normal tissue damage following radiotherapy for breast cancer, Radiother. Oncol. 55 (2000) 179–186.
[14] M. Swift, D. Morrell, R.B. Massey, C.I. Chase, Incidence of cancer in 161 families affected by ataxia–telangiectasia, N. Engl. J. Med. 325 (1991) 1831–1836.
[15] International Commission on Radiological Protection (ICRP) ICRP Publication 79 (1998) Genetic susceptibility to cancer, Annals ICRP 28 (1-2).
[16] P. Athma, R. Rappaport, M. Swift, Molecular genotyping shows that ataxia–telangiectasia heterozygotes are predisposed to breast cancer, Cancer Genet. Cytogenet. 92 (1996) 130–134.
[17] M.G. Fitzgerald, J.M. Bean, S.R. Hedge, H. Unsal, D.J. MacDonald, D.P. Harkin, D.M. Finkelstein, K.J. Isselbacher, D.A. Haber, Heterozygous *ATM* mutations do not contribute to early onset breast cancer, Nat. Genet. 15 (1997) 10–30.
[18] R.D. Knight, R. Parshad, F.M. Price, R.E. Tarone, K.K. Sanford, X-ray-induced chromatid damage in relation to DNA repair and cancer incidence in family members, Int. J. Cancer 54 (1993) 589–593.
[19] D. Scott, A.R. Spreadborough, L.A. Jones, S.A. Roberts, C.J. Moore, Chromosomal radiosensitivity in G_2-phase lymphocytes as an indicator of cancer predisposition, Radiat. Res. 145 (1996) 3–16.
[20] International Commission on Radiological Protection (ICRP) ICRP Publication 60 (1990) Recommendations of the International Commission on Radiological Protection, Annals ICRP 21 (13).
[21] Cancer Research Campaign, CRC CancerStats: Mortality—UK, 2000.

Susceptibility loci for radiation lymphomagenesis in mice

Nobuko Mori*, Masaaki Okumoto

Res. Inst. Advan. Sci. and Technol., Osaka Prefecture University, 1-2 Gakuen-cho, Sakai, Osaka 599-8570, Japan

Abstract

BALB/c and STS mice are susceptible and resistant, respectively, to radiation lymphomagenesis. Our previous study suggested that *Prkdc* encoding DNA-dependent protein kinase (DNA-PKcs) is a candidate for the lymphoma susceptibility locus on chromosome 16. The STS-derived segment with the wild-type *Prkdc* allele confers resistance in BALB/c, a variant defective in DNA-PKcs. The presence of a lymphoma susceptibility locus on chromosome 4 has also been suggested. To assess the effects of these loci on the susceptibility, we subjected chromosome 16 and chromosome 4 congenic lines by 4×1.1 and/or 1.7 Gy of X-irradiation. In the chromosome 16 congenic lines carrying the wild-type STS allele at *Prkdc*, lymphoma resistance was prominent after 4×1.1 Gy of X-irradiation, but modest at 4×1.7 Gy. Mice with the proximal half of STS-derived chromosome 4 showed resistance to the induction of lymphomas by the 4×1.7 Gy X-irradiation procedure. Thus, lymphoma susceptibility is controlled by more than one gene with different effects. © 2002 Elsevier Science B.V. All rights reserved.

Keywords: Radiation lymphomagenesis; Lymphoma susceptibility loci; Variation in DNA-PKcs; Mice

1. Background

A split dose of whole body X-irradiation efficiently induces lymphomas in susceptible strains of mice, such as BALB/cHeA (BALB/c) and B10, while strains such as STS/A (STS), MSM/Ms, and C3H/HeNirs are much more refractory to the induction [1–4]. The genetic basis underlying the difference in susceptibility to radiation lymphomagenesis is not well understood.

---

* Corresponding author. Tel.: +81-722-54-9838; fax: +81-722-54-9935.
E-mail address: morin@riast.osakafu-u.ac.jp (N. Mori).

Previously, we mapped an apoptosis susceptibility locus, radiation-induced apoptosis 1 (*Rapop1*), to chromosome 16 using the recombinant congenic CcS/Dem series, which contains a random set of 12.5% STS and 87.5% BALB/c genome [5]. The STS allele at *Rapop1* confers resistance to thymocyte apoptosis in the BALB/c background, since *Prkdc* encoding the catalytic subunit of DNA-dependent protein kinase (DNA-PKcs), a critical component of the DNA double-strand break (DSB) repair system and V(D)J recombination, was centered in the critical region for *Rapop1*. Furthermore, we sequenced the *Prkdc* cDNA from BALB/c and STS, and found variations in the DNA-PKcs between these strains. In the mice with the BALB/c-derived allele at *Prkdc*, the level of DNA-PKcs protein was significantly lower than that in mice with the STS-derived allele, though transcription was normal. As a consequence, BALB/c mice with the variant-type *Prkdc* allele exhibited diminished activity of DNA-PK [6]. In support of our findings, similar results are in other publications [7,8]. The severe combined immune deficiency (SCID) mutant is defective in DNA-PKcs and are severely impaired in DSB repair as well as V(D)J recombination [9]. The diminished activity of DNA-PK in BALB/c mice might be sufficient for normal V(D)J recombination, but insufficient for the repair of DSBs induced by radiation. A failure in rejoining the breaks might lead to a high sensitivity of BALB/c in apoptosis induced by radiation. Furthermore, we examined the susceptibility to lymphomagenesis induced by 1.1 Gy of X-irradiation using chromosome 16 congenic lines with the wild-type *Prkdc* allele against the BALB/c background, because the SCID mice exhibit a high sensitivity to lymphomagenesis by neonatal exposure to a low dose of ionizing radiation [10]. As a result, the STS-derived segment containing the wild-type *Prkdc* allele conferred resistance to lymphomagenesis [6]. Thus, our data strongly suggested *Prkdc* to be a candidate gene responsible for the susceptibility to apoptosis as well as lymphomagenesis.

STS mice are refractory to lymphomagenesis caused by 4×1.7 Gy of X-irradiation. The presence of a locus responsible for lymphoma resistance of STS mice, named *Lyr*, between *Tyrp1* (b) and *Ifna1* in the middle of chromosome 4 was suggested by Okumoto et al. [11], who used this procedure and a set of recombinant inbred mouse CXS/Hilgers strains whose parental strains were BALB/c and STS. To confirm that the *Lyr* locus is on chromosome 4 and better map its position, we constructed a series of congenic lines containing various STS-derived segments of chromosome 4 in the BALB/c background.

In the present study, we test the effect of different doses of X-rays on lymphomagenesis in the chromosome 16 congenic lines. Furthermore, we reconfirm the presence of *Lyr* on chromosome 4 using chromosome 4 congenics subjected to 4×1.7 Gy of X-irradiation, and discuss the implications for susceptibility to radiation lymphomagenesis.

2. Methods

2.1. Mice

C.S-R1 and C.S-R1L, congenic lines with STS-derived segments of chromosome 16 against the genetic background of BALB/cHeA (BALB/c), were established previously [6]. C.S-R1-4 is a subline of C.S-R1. Chromosome 4 congenic lines A, B, C, and D, which contain a variety of STS-derived segments of chromosome 4 on the BALB/c background,

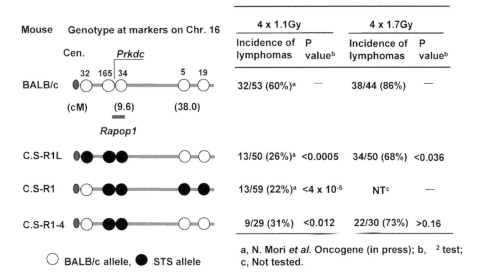

Fig. 1. Genotype of chromosome 16 congenic lines subjected to 1.1 or 1.7 Gy of X-irradiation four times and frequencies of lymphomas. Open and solid circles indicate genotype at micro satellite (Mit) markers. Marker names are abbreviated to numbers over the chromosome. Map positions of the markers relative to the centromere are given in centimorgans (cM) in parentheses. The location of *Prkdc* is indicated. The critical region for *Rapop1* is between D16Mit165 and D16Mit34 as indicated by a solid bar.

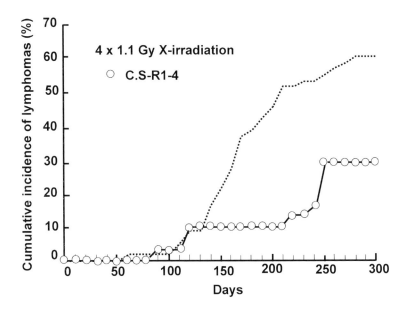

Fig. 2. Development of lymphomas in a chromosome 16 congenic line C.S-R1-4 subjected to 1.1 Gy of X-irradiation four times. Open circles represent the cumulative incidence of lymphomas in C.S-R1-4 mice. A broken line indicates development of lymphomas in BALB/c mice.

were constructed according to a modified method for speed congenic mice [12]. Briefly, chromosome 4 congenic lines were selected through genotyping of 70 micro satellite (Mit) markers distributed at intervals of approximately 20–25 centimorgans (cM) over all chromosomes during five backcrosses of (BALB/c × STS)F_1 mice to BALB/c. BALB/c and the congenic lines were maintained and bred at the conventional animal breeding facility of Osaka Prefecture University, Osaka, Japan. Only female mice were used for experiments.

2.2. Induction of lymphomas by X-irradiation

Four- to five-week-old mice were irradiated four times with 1.1 or 1.7 Gy of X-rays at weekly intervals with the X-ray generator Radioflex 350 (Rigaku Industrial, Osaka) equipped with a filter (0.3 mm copper plus 0.5 mm aluminum). The development of lymphomas was observed within 300 days of irradiation.

2.3. Statistics

The difference in the incidence of lymphomas between each congenic strain and the background strain BALB/c was analyzed by chi-square test or Fisher's exact probability test.

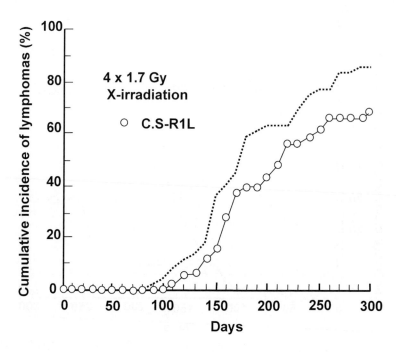

Fig. 3. Development of lymphomas in chromosome 16 congenics C.S-R1L and C.S-R1-4 subjected to 1.7 Gy of X-irradiation four times. Open circles represent cumulative incidences of lymphomas in C.S-R1L (upper) and C.S-R1-4 (lower). A broken line indicates development of lymphomas in BALB/c mice.

3. Results

3.1. Effects of different doses of X-irradiation on lymphoma induction in chromosome 16 congenics

In a previous study [6], we showed that two different chromosome 16 congenics C.S-R1L and C.S-R1 carrying the STS-derived wild-type *Prkdc* were relatively resistant to lymphomagenesis induced by the 4×1.1 Gy of X-irradiation as compared with their background, BALB/c. The genotype and frequency of lymphomas in these congenics are shown in Fig. 1. Additionally, we tested the line C.S-R1-4 derived from C.S-R1, which contains a diminished STS-derived segment covering the minimal region for *Rapop1*, as depicted in Fig. 1. The frequency of lymphomas in the C.S-R1-4 mice subjected to 4×1.1 Gy of X-irradiation was 31% (22/30), similar to those in other chromosome 16 congenics tested previously. The frequency was significantly lower than that in BALB/c mice. As shown in Fig. 2, no delay in the onset of lymphomas was detected in C.S-R1 mice or other chromosome 16 congenics (data not shown). Thus, the resistance to lymphomagenesis associated with the *Prkdc* wild-type allele was reproduced with 4×1.1 Gy of X-irradiation.

Since the frequency of lymphomas in STS mice subjected to 4×1.7 Gy of X-irradiation is less than 10% [2], we tested the induction in chromosome 16 congenic lines

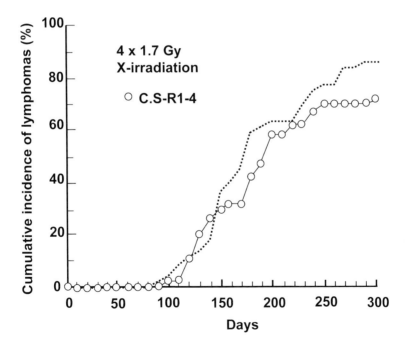

Fig. 4. Genotypes of chromosome 4 congenic lines subjected to 1.7 Gy of X-irradiation four times and frequencies of lymphomas. Open and solid circles indicate genotype at micro satellite (Mit) markers. Marker names are abbreviated to numbers over the bar. Map positions of the markers relative to the centromere are given in centimorgans (cM) in parentheses. The location of *Cdkn2a* is indicated. A solid bar indicates the *Lyr* region.

C.S-R1L and C.S-R1-4. The results are shown in Fig. 1. The frequency of lymphomas induced in BALB/c mice by this procedure was nearly 90% (38/44). On the other hand, the frequencies of lymphomas in C.S-R1L and C.S-R1-4 mice were 68% (34/50) and 73% (22/30), respectively. The cumulative incidence of lymphomas in the two congenics is plotted in Fig. 3. The results indicate that the resistance to lymphomagenesis of these congenics is rather modest or nonexistent at such doses, suggesting that other loci responsible for the resistance are present on other chromosomes.

3.2. Reconfirmation of the presence of Lyr, a locus responsible for lymphoma resistance, on STS-derived chromosome 4

To identify the gene responsible for the resistance to lymphomagenesis of STS mice, we constructed a series of chromosome 4 congenic lines and tested for lymphomagenesis at 4×1.7 Gy of X-irradiation. As shown in Fig. 4, the STS-derived segment distributed in the chromosome 4 congenics covered nearly all of chromosome 4. In lines A and C, frequencies of lymphomas were 50% (9/18) and 15% (2/13), respectively. Line B, whose STS-derived segment was included in the STS-derived part of line A, developed lymphomas at a lower incidence (4/7, 57%) similar to line A, though the difference was not significant. In line D with the STS-derived genome in the distal half of chromosome 4, the frequency of lymphomas was 91% (20/22), similar to that in BALB/c mice. Line C

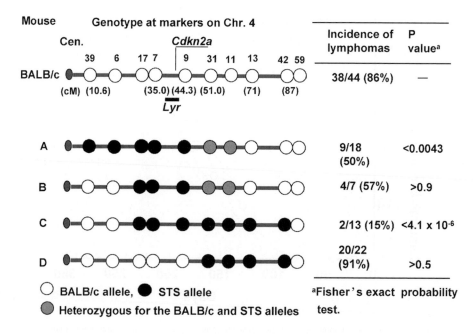

Fig. 5. Development of lymphomas in chromosome 4 congenic lines C and D, which were subjected to 1.1 Gy of X-irradiation four times. Open circles represent cumulative incidences of lymphomas in lines C (upper) and D (lower). A broken line indicates development of lymphomas in BALB/c mice.

showed a late onset as well as low frequency of lymphomas (Fig. 5, upper), while the development of lymphomas in line D exhibited a similar pattern to that in BALB/c (Fig. 5, lower). A delay in the onset of lymphomagenesis was also seen in lines A and B (data not shown). The results reconfirmed the presence of the *Lyr* locus responsible for lymphoma resistance in the proximal half of chromosome 4, where *Lyr* had originally been mapped.

4. Discussion

In the present study, we showed that the resistance to lymphomagenesis introduced by the STS-derived allele at the susceptibility locus on chromosome 16 in BALB/c mice was rather modest when the animals were exposed to 4×1.7 Gy of X-irradiation. At the same dose, however, an effect of the *Lyr* locus on chromosome 4 was detected. These results indicated that the resistance of STS mice is controlled by more than one gene. Of these, *Prkdc* is a candidate for the susceptibility locus on chromosome 16.

Proximal chromosome 4 contains *Cdkn2a* encoding a tumor suppressor, p16, as indicated in Fig. 4. Mutation and methylation of p16 are implicated in various tumors including lympomas [13,14]. Since p16 shows variation between BALB/c and STS, which influences the phosphorylation status of the Rb protein involved in cell cycle control [15], *Cdkn2a* encoding p16 might be a candidate for the susceptibility locus on chromosome 4. Preferential loss of heterozygosity (LOH) at markers on chromosome 4 in radiation-induced lymphomas in (BALB/c $\times$ STS)F_1 mice also suggests the involvement of genetic variation(s) in a tumor suppressor(s) in susceptibility to lymphomagenesis [16]. Because the region harboring the potential lymphoma resistance locus *Lyr* spans the proximal half of chromosome 4, further analysis is necessary to narrow down the region and identify the gene(s) responsible for the susceptibility.

5. Conclusions

The present study showed that susceptibility to radiation lymphomagenesis is controlled by more than one gene with different effects: lymphoma resistance conferred by the STS allele at the susceptibility locus on chromosomes 16 was rather modest in the lymphoma-sensitive BALB/c background, as compared with that given by the STS allele at the lymphoma susceptibility locus *Lyr* on chromosome 4. Although these loci have plausible candidates, *Prkdc* and *Cdkn2a* on chromosomes 16 and 4, respectively, further analysis is necessary to identify the genes responsible for the susceptibility to lymphomagenesis.

References

[1] M. Muto, T. Sado, I. Hayata, F. Nagasawa, H. Kamisaku, E. Kubo, Reconfirmation of indirect induction of radiogenic lymphomas using thymectomized, irradiated B10 mice grafted with neonatal thymuses from Thy 1 congenic donors, Cancer Res. 43 (1983) 3822–3827.

[2] M. Okumoto, R. Nishikawa, S. Imai, J. Hilgers, Resistance of STS/A mice to lymphoma induction by X-irradiation, J. Radiat. Res. (Tokyo) 30 (1989) 135–139.

[3] H. Kamisaku, S. Aizawa, M. Kitagawa, Y. Ikarashi, T. Sado, Limiting dilution analysis of T cell progenitors in the bone marrow of thymic lymphoma-susceptible B10 and -resistant C3H mice after fractionated whole body X-irradiation, Int. J. Radiat. Biol. 72 (1997) 191–199.

[4] M. Okumoto, N. Mori, N. Miyashita, K. Moriwaki, S. Imai, S. Haga, S. Hiroishi, Y. Takamori, K. Esaki, Radiation-induced lymphomas in MSM, (BALB/cHeA × MSM)F_1 and (BALB/cHeA × STS/A)F_1 hybrid mice, Exp. Anim. 44 (1995) 43–48.

[5] N. Mori, M. Okumoto, M.A. van der Valk, S. Imai, S. Haga, K. Esaki, A.A. Hart, P. Demant, Genetic dissection of susceptibility to radiation-induced apoptosis of thymocytes and mapping of *Rapop1*, a novel susceptibility gene, Genomics 25 (1995) 609–614.

[6] N. Mori, M. Matsumoto, M. Okumoto, N. Suzuki, J. Yamate, Variations in *Prkdc* encoding the catalytic subunit of DNA-dependent protein kinase (DNA-PKcs) and susceptibility to radiation-induced apoptosis and lymphomagenesis, Oncogene, in press.

[7] R. Okayasu, K. Suetomi, Y. Yu, A. Silver, J.S. Bedford, R. Cox, R.L. Ullrich, A deficiency in DNA repair and DNA-PKcs expression in the radiosensitive BALB/c mouse, Cancer Res. 60 (2000) 4342–4345.

[8] Y. Yu, R. Okayasu, M.M. Weil, A. Silver, M. McCarthy, R. Zabriskie, S. Long, R. Cox, R.L. Ullrich, Elevated breast cancer risk in irradiated BALB/c mice associates with unique functional polymorphism of the *Prkdc* (DNA-dependent protein kinase catalytic subunit) gene, Cancer Res. 61 (2001) 1820–1824.

[9] E.A. Hendrickson, X.Q. Qin, E.A. Bump, D.G. Schatz, M. Oettinger, D.T. Weaver, A link between double-strand break-related repair and V(D)J recombination: the SCID mutation, Proc. Natl. Acad. Sci. U. S. A. 88 (1991) 4061–4065.

[10] J.S. Danska, F. Pflumio, C.J. Williams, O. Huner, J.E. Dick, C.J. Guidos, Rescue of T cell-specific V(D)J recombination in SCID mice by DNA-damaging agents, Science 266 (1994) 450–455.

[11] M. Okumoto, R. Nishikawa, S. Imai, J. Hilgers, Genetic analysis of resistance to radiation lymphomagenesis with recombinant inbred strains of mice, Cancer Res. 50 (1990) 3848–3850.

[12] P.M. Visscher, Speed congenics: accelerated genome recovery using genetic markers, Genet. Res. 74 (1999) 81–85.

[13] M. Malumbres, I.P. Perez de Castro, J. Santos, B. Melendez, R. Mangues, M. Serrano, A. Pellicer, J. Fernandez-Piqueras, Inactivation of the cyclin-dependent kinase inhibitor p15INK4b by deletion and de novo methylation with independence of p16INK4a alterations in murine primary T cell lymphomas, Oncogene 14 (1997) 1361–1370.

[14] I.P. Perez de Castro, M. Malumbres, J. Santos, A. Pellicer, J. Fernandez-Piqueras, Cooperative alterations of Rb pathway regulators in mouse primary T cell lymphomas, Carcinogenesis 20 (1999) 1675–1682.

[15] S. Zhang, E.S. Ramsay, B.A. Mock, *Cdkn2a*, the cyclin-dependent kinase inhibitor encoding p16INK4a and p19ARF, is a candidate for the plasmacytoma susceptibility locus, *Pctr1*, Proc. Natl. Acad. Sci. U. S. A. 95 (1998) 2429–2434.

[16] M. Okumoto, Y.-G. Park, C.-W. Song, N. Mori, Frequent loss of heterozygosity on chromosomes 4, 12, and 19 in radiation-induced lymphomas in mice, Cancer Lett. 135 (1999) 223–228.

Age dependence of susceptibility for long-term effects of ionizing radiation

S. Sasaki*

*Research Center for Radiation Safety, National Institute of Radiological Sciences,
9-1 Anagawa 4-chome, Inage, Chiba 263-8555, Japan*

Abstract

This study was aimed at the elucidation of the characteristics of the long-term effects in mice irradiated during the intrauterine, neonatal and puberty period in comparison with those after exposure during the adult period. Female $B6C3F_1$ mice were irradiated during the intrauterine, neonatal, puberty, young adult or middle age adult period with Cs-137 gamma rays. All the mice were allowed to live out their entire life span under a specific pathogen-free condition, and the mean life span, age-specific mortality and lifetime incidences of the neoplasms were examined for each experimental group. Using life shortening as a comprehensive measure of long-term effects, the age dependence of susceptibility for long-term effects was analyzed. Mice of the neonatal and puberty periods were shown to be more susceptible to the induction of the overall long-term effects than the mice of the intrauterine and adult period. The temporal variation of the age-specific mortality was apparently influenced by the dose of radiation and the age at irradiation. The analysis of the age-associated change of susceptibility for the induction of neoplasms resulted in the conclusion that the age dependence of susceptibility is inherent for each type of neoplasm. Mice of the late fetal and neonatal period were shown to be highly susceptible to the induction of tumors of the endocrine glands. Statistically significant increase in incidence was detected for the induction of pituitary tumors and ovarian tumors after irradiation at 7 days of age with 10 cGy. © 2002 Elsevier Science B.V. All rights reserved.

Keywords: Ionizing radiation; Long-term effects; Age dependence; Carcinogenesis; Neonatal mice

1. Introduction

Age at the time of irradiation is one of the important biological variables influencing the long-term effects of ionizing radiation. Epidemiological studies indicated high suscepti-

---

* Tel.: +81-43-206-3162.
E-mail address: sasaki@nirs.go.jp (S. Sasaki).

bility of individuals during the intrauterine period [1–3] and childhood period [4–8] for radiation carcinogenesis. However, the risk assessment of exposure to ionizing radiation during the intrauterine and childhood period remains still obscure because the epidemiological data on lifetime incidences of neoplasms and temporal variation of risk are insufficient [9,10]. An experimental study on the relationship between the age at irradiation and the response of the animal to radiation is useful as a biological basis of risk assessment. In the first part of this report, results of the experiment using female B6C3F$_1$ mice on the influence of age at irradiation with gamma rays on life shortening, lifetime incidences of neoplasms and temporal variation of the age-specific mortality are described.

Female mice of the late intrauterine and neonatal period were shown to be highly susceptible to the induction of pituitary tumors and ovarian tumors in our previous studies [11–15]. These tumors developed in excess with a statistically significant difference after irradiation at day 0 with 48 cGy gamma rays [16]. An experimental study was designed to examine whether the increase in incidence can be detectable in mice irradiated with 10 cGy during the neonatal period.

2. Materials and methods

Two separate experiments were carried out. Females of the first generation hybrid mice between C57BL/6JNrs and C3H/HeNrs strain (B6C3F$_1$) were used in both experiments. After irradiation with gamma rays from Cs-137, the mice were allowed to live out their entire life span under a specific pathogen-free condition. Detailed methods for the maintenance of mice were described in previous reports [13,14]. Lifetime incidences of neoplasms were obtained based on autopsy and histological observation. Statistical examination was performed using Student's t-test and Fisher's direct probability method. Conditions of the gamma irradiation in each experiment were as follows.

2.1. Experiment on the age–response relationship

Ages at irradiation were day 17 of the late intrauterine period, that is 2 days before birth, day 0 of the neonatal period, day 35 of the puberty period, day 105 of the young adult period, and days 365 and 550 of the middle-aged adult period. The whole bodies of mice were irradiated with 190 cGy at dose rate of 87.0 cGy/min. The sample size of each irradiated and control group is shown in Table 1. The age-specific mortality was calculated for the age intervals of 50–250, 250–450, 450–550, 550–650, 650–750, 750–850, 850–950 and 950–1050 days, and expressed as number of deaths per 10,000 mouse/day (MD) at risk.

2.2. Experiment on the effects of low-dose irradiation during the neonatal period

Mice were irradiated at day 7 of the neonatal period with 10, 19, 48, 95, 190, 285, 380 or 475 cGy gamma rays at a dose rate of 80.2 cGy/min. The sample size of the lower dose group was larger than that in the higher dose group as summarized in Table 3. A simul-

Table 1
Mean life span, number of solid tumors and incidences of tumors in female B6C3F$_1$ mice irradiated at various ages with 190 cGy gamma rays

Age at irradiation	Number of mice	Mean life span (days)	Number of solid tumors	Pituitary tumors (%)	Ovarian tumors (%)	Liver tumors (%)	Harderian tumors (%)
Control	885	864.8	0.540	8.0	1.8	19.7	1.8
Day 17, IU[a]	93	831.2*	0.989***	31.2***	7.5***	22.6	1.1
Day 0	332	759.1***	1.369***	24.9***	30.3***	48.0***	0.3*
Day 35	80	752.8***	1.325***	8.8	38.8***	31.3**	17.5***
Day 105	81	788.7***	0.864***	7.4	24.7***	17.3	12.3***
Day 365	141	804.1***	0.709**	4.3	7.1***	22.7	1.4
Day 550	178	859.8	0.663*	9.0	6.7***	23.0	0.0*

[a] IU: intrauterine period.
* Statistical difference from the control: $P<0.05$.
** Statistical difference from the control: $P<0.01$.
*** Statistical difference from the control: $P<0.001$.

taneous control group was set up because this study was undertaken separately from previous experiments.

3. Results

3.1. Influence of the age at irradiation on the mean life span and lifetime incidence of the neoplasm

The mean life span, number of solid tumors per mouse, lifetime incidences of the pituitary, ovarian, liver and Harderian tumors in the control and irradiated groups are summarized in Table 1. The shortening of the mean life span was used as a comprehensive measure of radiation-induced long-term effects. It became evident that mice of the neonatal period and puberty period are highly susceptible to the life-shortening effect of gamma rays. The mean life span of the mice irradiated during the late intrauterine period was longer than that of the mice irradiated during the neonatal, puberty or young adult period. Mice of the late intrauterine period did not seem to be highly susceptible to the induction of the long-term effects by radiation. Similar findings were obtained in a previous experiment in which the mice were irradiated with 380 cGy gamma rays [17]. In a group irradiated at day 550 with 190 cGy, the mean life span was not different from that in the control group. Because the mean life span of these hybrid mice was 864.8 days, 550 days of age is the middle-aged adult period.

The increase in the mean number of solid tumors was used as an overall measure of susceptibility for the induction of solid tumors. Strictly speaking, the number of solid tumors in Table 1 implies the mean number of the types of solid tumors per mouse. For example, when three liver tumors and two lung tumors were found in a mouse, the number of solid tumors was counted as two in this case. The mean number of solid tumors increased with a statistically significant difference from the control in all of the irradiated groups examined in this study. Irradiation during the neonatal and puberty period was most effective to increase the number of solid tumors per mouse. The mean number of solid tumors increased sufficiently after irradiation during the late intrauterine period, but the effectiveness of irradiation during the late intrauterine period was significantly less than that during the neonatal and puberty period. Results of this experiment clearly showed that the overall susceptibility for the induction of solid tumors decreases with increasing age during the adulthood period.

The age dependence of susceptibility for the induction of a specific type of neoplastic disease is also given in Table 1. The increase in the incidence of pituitary tumors was observed after irradiation during the late intrauterine period and neonatal period. The highest incidence was found in mice irradiated during the late intrauterine period. Irradiation during the puberty period and adult period was not effective to induce pituitary tumors. The age dependence of susceptibility for ovarian tumorigenesis was revealed to be quite different from that for pituitary tumors. Ovarian tumors were developed in excess with a statistically significant difference in all the irradiated groups examined here. The exposure of mice to gamma rays during the neonatal, puberty and young adult period resulted in the development of ovarian tumors at high incidences. The highest incidence was found in a group irradiated during the puberty period. Irradiation during the late intrauterine period was evidently shown to be less effective than that during the neonatal period for the induction of ovarian tumors. The incidence of ovarian tumors after irradiation during the middle-aged adult period was lower than that in mice irradiated at young adulthood. Liver tumors developed at high incidence after irradiation during the neonatal period, and the increase in the incidence was also observed in mice irradiated during the puberty period. In contrast, the exposure of the adult mice did not result in the excess development of liver tumors. The increased incidence of liver tumors was not detected after irradiation during the late intrauterine period in this experiment, but enhanced liver tumorigenesis had been observed in the previous experiment in which the mice were irradiated with higher doses of gamma rays during the late fetal period [14]. Tumors of the Harderian gland were induced by irradiation during the puberty and young adult period, whereas the exposure during the late intrauterine and neonatal period was not effective in inducing this type of neoplasm. These findings led to a conclusion that the age dependence of susceptibility for radiation carcinogenesis is inherent to each type of neoplastic disease.

3.2. Influence of age at irradiation on the temporal variation of age-specific mortality

Temporal variation of the age-specific mortality in the control and irradiated groups is shown in Table 2. In the mice irradiated during the neonatal and puberty period, an increase in the age-specific mortality was remarkable and continued throughout their life. Continuous increase in the age-specific mortality was also observed after irradiation during the late intrauterine and young adult period. In a group irradiated at day 550 of the middle-aged adult period, however, a statistically significant increase in the age-specific mortality was not detected. The age-specific mortality of each interval of age was obviously dependent on the age at the time of irradiation.

Table 2
Temporal variation of age-specific mortality (number of deaths per 10,000 mice/day at risk) in mice irradiated at various ages with 190 cGy

Age (days)	Control	Age at irradiation (days)					
		17, IU[a]	0	35	105	365	550
50–250	0.06	0.00	0.30	0.63	0.00	0.00	0.00
250–450	0.28	0.54	1.37**	1.28	1.30	0.00	0.57
450–550	1.26	2.20	4.74***	9.44***	5.23*	0.71	1.72
550–650	4.57	5.62	13.3***	7.32	8.22	13.4***	7.18
650–750	13.0	19.1*	30.6***	39.6***	26.8**	33.5***	13.1
750–850	38.9	48.5*	82.3***	77.4**	60.6*	51.0*	38.9
850–950	68.2	86.3*	142***	86.7	92.4	78.6*	54.1*
950–1050	107	189*	227*	225	145	189*	115

[a] IU: intrauterine period.
* Statistical difference from the control: $P<0.05$.
** Statistical difference from the control: $P<0.01$.
*** Statistical difference from the control: $P<0.001$.

3.3. Effects of low-dose exposure during the neonatal period

The mean life span and incidences of thymic lymphomas, pituitary tumors and ovarian tumors in mice irradiated at 7 days of age are summarized in Table 3. The life-shortening effect of gamma irradiation at day 7 was slightly larger than that of irradiation at day 0 that was reported in a previous report [16]. The mean life span of the mice irradiated with 10 cGy was significantly shorter than that in the control group. The shape of the dose–response relationship for the induction of ovarian tumors was convex upward, and the highest incidence was observed in the group irradiated with 190 cGy. A statistically significant increase in the incidence of ovarian tumors was found after irradiation with 10 cGy gamma rays. The dose–response curve for the incidence of pituitary tumors was also convex upward. Pituitary tumors were developed at the highest incidence after irradiation with a higher dose than that for ovarian tumors. The incidence of pituitary tumors in mice irradiated with 10 cGy was significantly higher than that in the control group. It became evident that mice of 7 days of age are highly susceptible to the induction of ovarian tumors

Table 3
Mean life span and incidences of tumors in female B6C3F$_1$ mice irradiated at 7 days of age with gamma rays

Dose (cGy)	Number of mice	Mean life span (days ± S.E.)	Thymic lymphomas (% ± S.E.)	Pituitary tumors (% ± S.E.)	Ovarian tumors (% ± S.E.)
0	1003	871.1 ± 4.8	0.4 ± 0.2	8.1 ± 0.9	2.2 ± 0.5
10	1388	857.3 ± 4.2	0.5 ± 0.2	11.6 ± 0.9	14.7 ± 1.0
19	393	842.5 ± 6.4	0.5 ± 0.4	14.0 ± 1.8	23.4 ± 2.1
48	283	817.0 ± 7.1	0.0 ± 0.0	18.4 ± 2.3	41.7 ± 2.9
95	205	782.1 ± 8.4	0.5 ± 0.5	25.4 ± 3.0	44.4 ± 3.5
190	203	723.9 ± 11.4	2.0 ± 1.0	33.0 ± 3.3	47.3 ± 3.5
285	232	656.5 ± 11.2	4.7 ± 1.4	38.8 ± 3.2	37.9 ± 3.2
380	196	608.8 ± 13.5	7.7 ± 1.9	38.3 ± 3.5	30.6 ± 3.3
475	94	552.3 ± 17.2	9.6 ± 3.0	23.4 ± 4.4	19.1 ± 4.1

and pituitary tumors. In contrast, thymic lymphomas did not develop in excess after irradiation with low-dose gamma rays.

4. Discussion

Life-shortening and carcinogenic effects were not detected in previous studies in which mice of the young adult period were irradiated with 10 cGy or with lower doses of low LET radiation [18–21]. The present study showed that mice of the neonatal period are more susceptible than adult mice to the long-term effects of gamma rays. Irradiation of the neonatal mice with 10 cGy gamma rays resulted in the statistically significant shortening of mean life span and increase in the incidences of pituitary tumors and ovarian tumors. As shown in our previous reports, neonatal mice are also more susceptible than adult mice to the induction of liver tumors, lung tumors, bone tumors and thymic lymphomas [13–15]. It should be noticed that the developmental stage of neonatal mice corresponds to the late intrauterine period of human development. Therefore, the results of the present study suggest that individuals in the late intrauterine period are highly susceptible to the long-term effects of ionizing radiation in humans.

Although the presence of species difference in susceptibility to radiation carcinogenesis is evident, the age dependence of the susceptibility seems to be common beyond the difference of animal species. Epidemiological studies showed that individuals during the childhood period are susceptible to the induction of thyroid cancer by ionizing radiation [4,7]. Neoplastic and non-neoplastic diseases of the thyroid gland were observed in beagle dogs, which received gamma rays during the prenatal and postnatal development [22–24]. In an experimental study on thyroid tumorigenesis by radioactive iodine (I-131) using mice, administration during the late intrauterine period was more effective than that during the adult period [25]. For the induction of ovarian tumors, mice of the neonatal, puberty and young adult period are highly susceptible as shown in Table 1. The statistically significant increase in the incidence of ovarian tumors was detected among atomic bomb survivors in Hiroshima and Nagasaki [7]. Pituitary tumors were developed in excess as one of the intracranial tumors after exposure to ionizing radiation for treatment of skin hemangiomas during the infancy period [6]. As presented in this study and in previous reports, pituitary tumors were induced by irradiation during the late prenatal and early postnatal period in mice [11–16,26]. High susceptibility for the induction of the endocrine tumors during the intrauterine and juvenile period may be common in mammals including humans.

Liver tumors of hepatocellular type were induced by irradiation with ionizing radiation during the juvenile period in mice as shown in this report and in previous studies [12–16,27–29]. In contrast, hepatocellular tumors could not be induced in rats by radiation alone and by combination of radiation and partial hepatectomy [30–32]. Background incidence is high in mice and low in rats in general. Spontaneous mutation of *Ha-ras* gene was found in hepatocytes of mice [33]. This spontaneous mutation may be associated with the high susceptibility for radiation-induced hepatocarcinogenesis. When the rats were irradiated with X-rays after the treatment with chemical carcinogen, the incidence of hepatocellular tumors was remarkably higher than that in the group treated with chemical carcinogen alone [32]. In spite of the presence of these interesting findings, the mechanisms of species difference in susceptibility to radiation-induced hepatocarcinogenesis are not clear. The life span study of the atomic bomb survivors showed that the incidence of primary hepatocellular cancer was higher than the background incidence and that the highest incidence was observed in individuals exposed during the young adult period [7,34]. It may be possible that other agents are involved in hepatocarcinogenesis among atomic bomb survivors. Further studies are necessary to elucidate the mechanisms of species difference in susceptibility and age dependence of susceptibility for radiation-induced hepatocarcinogenesis.

In the age-attained model, which was proposed as a risk projection model, the increase in the age-specific mortality was assumed to be dependent on the dose of radiation, however, independent on the age at irradiation [9]. The age-specific mortality at each attained age was apparently dependent on the age at irradiation as shown in Table 2. The influence of age at irradiation on the temporal variation of the age-specific mortality may be common in mammals.

Acknowledgements

This study was included in the research project "Low-Dose Radiation Effects and Carcinogenesis" in the National Institute of Radiological Sciences. I am grateful to Dr. N. Fukuda for the helpful suggestion on the statistical analysis.

References

[1] A. Stewart, J. Webb, D. Hewitt, A survey of childhood malignancies, Br. Med. J. 1 (1958) 1495–1508.
[2] B. MacMahon, Prenatal X-ray exposure and childhood cancer, J. Natl. Cancer Inst. 28 (1962) 1173–1191.
[3] R. Doll, R. Wakeford, Risk of childhood cancer from fetal irradiation, Br. J. Radiol. 79 (1997) 130–139.
[4] E. Ron, B. Modan, D.L. Preston, E. Alfandary, M. Stovall, J.D. Boice, Thyroid neoplasia following low-dose radiation in childhood, Radiat. Res. 120 (1989) 516–531.
[5] B. Modan, A. Chetrut, E. Alfandary, L. Katz, Increased risk of breast cancer after low-dose irradiation, Lancet (1989) 629–631.
[6] P. Karlsson, E. Holmberg, M. Lundell, A. Mattsson, L.-E. Holm, A. Wallgren, Intracranial tumors after exposure to ionizing radiation during infancy: a pooled analysis of two Swedish cohorts of 28,008 infants with skin hemangioma, Radiat. Res. 150 (1998) 357–364.
[7] D.E. Thompson, K. Mabuchi, E. Ron, M. Soda, M. Tokunaga, S. Ochikubo, S. Sugimoto, T. Ikeda, M. Terasaki, S. Izumi, D.L. Preston, Cancer incidence in atomic bomb survivors: Part II. Solid tumors, 1958–1987, Radiat. Res. 137 (1994) S17–S67.
[8] D.A. Pierce, Y. Shimizu, D.L. Preston, M. Vaeth, K. Mabuchi, Studies of the mortality of atomic bomb survivors. Report 12, Part I. Cancer: 1950–1990, Radiat. Res. 146 (1996) 1–27.
[9] A.M. Kellerer, D. Barclay, Age dependences of in the modeling of radiation carcinogenesis, Radiat. Prot. Dosim. 41 (1992) 273–281.
[10] D.A. Pierce, M.L. Mendelsohn, A model for radiation-related cancer suggested by atomic bomb survivor data, Radiat. Res. 152 (1999) 642–654.
[11] S. Sasaki, T. Kasuga, F. Sato, N. Kawashima, Late effects of fetal mice X-irradiated at middle or late intrauterine stage, Gann 69 (1978) 167–177.
[12] S. Sasaki, T. Kasuga, Life shortening and carcinogenesis in mice irradiated neonatally with X-rays, Radiat. Res. 88 (1981) 313–325.

[13] S. Sasaki, T. Kasuga, Life shortening and carcinogenesis in mice irradiated at the perinatal period with gamma rays, in: R.C. Thompson, J.A. Mahaffey (Eds.), Life-Span Radiation Effects Studies in Animals: What Can They Tell Us, Department of Energy, Springfield, US, 1986, pp. 357–367.

[14] S. Sasaki, Influence of the age of mice at exposure to radiation on life shortening and carcinogenesis, J. Radiat. Res. 32 (Suppl. 2) (1991) 73–85.

[15] S. Sasaki, Age dependence of susceptibility to carcinogenesis by ionizing radiation in mice, Radiat. Environ. Biophys. 30 (1991) 205–207.

[16] S. Sasaki, N. Fukuda, Dose–response relationship for induction of solid tumors in female B6C3F$_1$ mice irradiated neonatally with a single dose of gamma rays, J. Radiat. Res. 40 (1999) 229–241.

[17] S. Sasaki, Consequences of prenatal irradiation in mice: carcinogenesis and CNS damage as a basis for human risks, in: E.M. Fielden, J.F. Fowler, H. Hendey, D. Scott (Eds.), Radiation Research, vol. 2. Taylor & Francis, London, 1987, pp. 652–657.

[18] J.B. Storer, L.J. Serrano, E.B. Darden, M.C. Jarnigan, R.L. Ullrich, Life shortening in RFM and BALB/c mice as a function of radiation quality, dose and dose rate, Radiat. Res. 78 (1979) 122–161.

[19] R.L. Ullrich, J.B. Storer, Influence of γ irradiation on the development of neoplastic disease in mice: I. Reticular tissue tumours, Radiat. Res. 80 (1979) 303–316.

[20] R.L. Ullrich, J.B. Storer, Influence of γ irradiation on the development of neoplastic diseases in mice: II. Solid tumors, Radiat. Res. 80 (1979) 317–324.

[21] V. Covelli, M. Coppola, V. Di Majo, S. Rebessi, B. Bassani, Tumor induction and life shortening in BC3F$_1$ female mice at low doses of fast neutrons and X-rays, Radiat. Res. 113 (1988) 362–374.

[22] S.A. Benjamin, W.J. Saunders, A.C. Lee, G.M. Angleton, C.H. Mallinckrodt, Non-neoplastic and neoplastic thyroid disease in beagles irradiated during prenatal and postnatal development, Radiat. Res. 147 (1997) 422–430.

[23] S.A. Benjamin, A.C. Lee, G.M. Angleton, W.J. Saunders, T.J. Keefe, C.H. Mallinckrodt, Mortality in beagles irradiated during prenatal and postnatal development: 1. Contribution of non-neoplastic diseases, Radiat. Res. 150 (1998) 316–329.

[24] S.A. Benjamin, A.C. Lee, G.M. Angleton, W.J. Saunders, T.J. Keefe, C.H. Mallinckrodt, Mortality in beagles irradiated during prenatal and postnatal development: II. Contribution of benign and malignant neoplasia, Radiat. Res. 150 (1998) 330–348.

[25] G. Walinder, A. Sjoden, Late effects of irradiation on the thyroid gland in mice: III. Comparison between irradiation of fetuses and adults, Acta Radiol.: Ther., Phys., Biol. 12 (1973) 201–208.

[26] Y. Nitta, K. Kamiya, K. Yokoro, Carcinogenic effect of in utero Cf-252 and Co-60 irradiation in C57BL/6N × C3H/HeF$_1$ (B6C3F$_1$) mice, J. Radiat. Res. 33 (1992) 319–333.

[27] S. Sasaki, T. Kasuga, F. Sato, N. Kawashima, Induction of hepatocellular tumor by X-ray irradiation at perinatal stage of mice, Gann 69 (1978) 451–452.

[28] V. Di Majo, M. Coppola, S. Rebessi, B. Bassani, T. Alati, A. Saran, C. Bangrazi, V. Covelli, Radiation-induced mouse liver neoplasms and hepatocyte survival, J. Natl. Cancer Inst. 77 (1986) 933–940.

[29] J.R. Maisin, G.B. Gerber, J. Vankerkom, A. Wamhersie, Survival and diseases in C57BL mice exposed to X-rays or 3.1-MeV neutrons at an age of 7 or 21 days, Radiat. Res. 146 (1996) 453–460.

[30] K. Christov, Liver cell proliferation and failure of X-radiation to produce hepatomas in rats, Radiat. Res. 74 (1978) 378–381.

[31] T. Kitagawa, K. Nomura, S. Sasaki, Induction by X-irradiation of adenosine triphosphatase-deficient islands in the rat liver and their characterization, Cancer Res. 45 (1985) 6075–6082.

[32] H. Mori, H. Iwata, Y. Morishita, Y. Mori, T. Ohno, T. Tanaka, S. Sasaki, Synergistic effect of radiation on N-2-fluorenylacetamide-induced hepatocarcinogenesis in male ACI/N rats, Jpn. J. Cancer Res. 81 (1990) 975–978.

[33] B.A. Moulds, J.I. Goodman, Spontaneous mutation at codon 61 of the *Ha-ras* gene in the nascent liver of B6C3F$_1$,C3H/He and C57BL/6 mice, Mutat. Res. 311 (1994) 1–7.

[34] J.E. Cologne, S. Tokuoka, G.W. Beebe, T. Fukuhara, K. Mabuchi, Effects of radiation on incidence of primary liver cancer among atomic bomb survivors, Radiat. Res. 152 (1999) 364–373.

Nutrition status and radiation-induced cancer in mice

Kazuko Yoshida [a,*], Yoko Hirabayashi [b], Toshihiko Sado [c], Tohru Inoue [b]

[a] National Institute of Radiological Sciences, 4-9-1 Anagawa, Inage-ku, Chiba 263-8555, Japan
[b] National Institute of Health Sciences, 1-18-1 Kamiyouga, Setagayaku, Tokyo 158, Japan
[c] Ohita University of Nursing and Health Sciences, Nozuhara, Ohita 870-120, Japan

Abstract

Dietary restriction, especially caloric restriction (CR), is a major carcinogenic modifier during experimental carcinogenesis. We attempted to examine the effects of CR on radiation-induced myeloid leukemia (MyL). The spontaneous incidence of MyL in C3H/He mice is 1.1%, and the incidence increases to 21.6% by X-ray irradiation. However, the incidence was decreased in the CR groups; it was 7.9% in RA (life-span restricted diet from the age of 6 weeks), 10.7% in RB (restricted diet after irradiation), 16.2% in RC (restricted from 6 weeks old until irradiation at 10 weeks old). The differences between both the RA and RB groups and the non-restricted group were statistically significant. The significantly fewer hematopoietic stem cells (HSC), potential target cells for radiation leukemia in the RA and RC at the time of radiation exposure, and the smaller cycling fraction of HSC under CR in the femur of restricted mice ($26 \pm 4.5\%$) than in the non-restricted group ($44 \pm 20\%$), were responsible for this underlying mechanism. CR contributes to the incidence reduction of leukemia at the initiation stage of leukemogenesis and, more significantly, to the reduction of MyL during the promotion stage of radiation leukemogenesis. © 2002 Elsevier Science B.V. All rights reserved.

Keywords: Myeloid leukemia; Calorie restriction; Hematopoietic stem cells; Cell cycle; Mice

1. Introduction

Restriction of dietary calories is a major modifier of experimental carcinogenesis, and is known to reduce the incidence of not only spontaneous neoplasms but also chemically induced cancers. We attempted to examine the effects of calorie restriction (CR) on radiation-induced myeloid leukemia (MyL) as an experimental model, because the incidence

* Corresponding author. Tel.: +81-43-206-3093; fax: +81-43-251-4582.
  E-mail address: yosida@nirs.go.jp (K. Yoshida).

of this disease has significantly increased among the survivors of the atomic bombs in Hiroshima and Nagasaki [1]. We clearly showed in our previous study that radiation-induced MyL was decreased with CR [2]. In the present study, we focused on whether CR modulated the initiation process or the promotion process, or both, during leukemogenesis, and the kinetics of hematopoietic stem cells (HSC) under CR were also analyzed to elucidate the underlying mechanism of the reduction. CR was designed with groups restricted from the age of 6 weeks until irradiation at 10 weeks (RC) and restricted after the time of irradiation (RB), and they were compared with the group restricted until natural death from the age of 6 weeks (RA) and the non-restricted group (NR). According to the theory of multi-step carcinogenesis, RC is thought to predict an effect on initiation, and RB on promotion.

2. Materials and methods

2.1. Mice and irradiation

Six-week-old male C3H/HeNirMs mice, bred at our institute, were used in the present study. Mice were exposed to 3 Gy of whole-body X-irradiation from a 200KV 20-mA source through a therapeutic X-ray irradiator (Shimadzu, Tokyo, Japan) filtered with 0.5 mm aluminum and 0.5 mm copper, at a dose rate of 0.614 Gy/min, and with a focus surface distance of 56 cm. All the mice of the irradiated groups were irradiated at the age of 10 weeks.

2.2. Diets

Details of the methods of CR and diets used are reported elsewhere [2]; briefly, the diets consisted of four different calorie-controlled regimens, that is, 60, 65, 70 and 95 kcal/week/mouse. Calorie intake was adjusted by controlling the amount of carbohydrate and dextrose, with an equal amount of other nutrients such as protein, lipid, vitamins and minerals.

2.3. Experimental procedure

Animals were divided into eight groups: (1) non-restricted diet with no irradiation (CNR) or with 3 Gy irradiation (3NR), (2) restricted diet group A with 3 Gy irradiation (3RA) or no irradiation (CRA), (3) restricted diet group B with 3 Gy irradiation (3RB) or with no irradiation (CRB), (4) restricted diet group C with 3 Gy irradiation (3RC) or with no irradiation (CRC). The non-restricted diet groups were fed a 95-kcal diet from the age of 6 weeks until natural death, that is, for their entire life spans. The restriction A (RA) groups were fed a 65-kcal diet between 6 and 10 weeks, followed by another calorie-restricted diet. The restriction B (RB) groups were fed a non-restricted diet between 6 and 10 weeks, followed by another calorie-restricted diet. Mice in the restriction A and B groups were started with other calorie restriction diets from the age of 10 weeks, that is, they were maintained as to their body weight at 25–27 g. This was controlled with appropriate combinations of the four original diets designed with graded doses of calories, as mice were necessarily fed with the minimum amount of calories required to maintain sufficient growth and development. The average caloric intake from the age of 10 weeks and thereafter was 75 kcal/week/mouse. The diet of the restriction C (RC) groups consisted of 65 kcal between 6 and 10 weeks of age, followed by a non-restricted diet (95 kcal).

2.4. Assay of HSC

Bone marrow or spleen cells were harvested from 10-week-old, both non-restricted and restricted, mice. The cells were analyzed for the presence of spleen colony forming cells (CFU-S) by Till and MuCulloch's method [3].

2.5. Size of cycling fraction of stem cells

Bromodeoxyuridine (BrdUrd) UV cytocide assay was performed for cell cycle analysis [4]. Briefly, both the non-restricted and restricted mice at the age of 50 weeks were implanted subcutaneously with an osmotic pump (Alza, Palo Alto, CA), which permits continuous infusion of BrdUrd at a dose rate of 2 mg/kg/h. Osmotic pumps were kept in mice for 5 days, and then bone marrow or spleen cells were harvested. The cells were exposed to a near-UV illuminator (Model UVP, Funakoshi, Tokyo, Japan) at a dose of 4000 J/m^2 with UV wavelengths longer than 300 nm and with a peak at 365 nm. The cells were then analyzed by CFU-S assay.

3. Results

3.1. Incidence of MyL

The incidence of MyL decreased with caloric restriction. The incidence of MyL in the non-irradiated groups CNR, CRA, CRB and CRC were 1.1%, 1.0%, 0% and 2.1%, respectively. When irradiated, the incidences increased up to 21.6% in 3NR, 7.9% in 3RA, 10.7% in 3RB and 16.2% in 3RC. Therefore, there were significant decreases from 3NR to 3RA and 3RB, although the one from 3NR to 3RC was not statistically significant.

3.2. Number of HSC at irradiation

The number of HSC, the possible target cells for radiation-induced leukemogenesis, was evaluated in both the restricted and non-restricted groups at the time of irradiation (10 weeks old). The number of CFU-S in the spleen decreased from 38 per 10^6 cells in the non-restricted group to 10 per 10^6 cells in the restricted group. The total number of CFU-S per spleen in the restricted group (0.93×10^3 per spleen) decreased to 10% compared with the non-restricted group (9.1×10^3 per spleen). The decrease in the number of CFU-S in bone marrow in the restricted group compared with the non-restricted group was less significant.

3.3. Size of cycling fraction of stem cells

The size of the CFU-S in the cell cycle in the femur significantly decreased statistically from 46% in the NR group to 26% in the RA and RB. In the spleen, the size of CFU-S also decreased from 31.4% in NR to 17.7% in RA and RB.

4. Conclusion

(1) Caloric restriction (CR) contributes to reducing the incidence of leukemia at an initial stage of leukemogenesis and, more significantly, to reducing the effect on the promotion process of leukemogenesis after irradiation.

(2) The number of hematopoietic stem cells (HSC) was significantly decreased in the restriction groups RC and RA, especially in the spleen, at the time of radiation exposure. Since the incidence of leukemia was greatly decreased in the RB group in which the number of stem cells was naturally similar to that in the NR group, and furthermore, the suppression of leukemia in RC was less significant, the decrease in the stem cell number at irradiation did not seem to be a major reason for the reduction of leukemia incidence by the restriction.

(3) The size of the cycling fraction of HSC in the restricted group was smaller than in the non-restriction group. The results described above imply that the HSC in the restriction groups proliferate at a much slower rate than those in NR mice do, and that the proliferation rate after irradiation may also contribute to the reduction seen in groups RA and RB.

Acknowledgements

The authors thank Ms. F. Watanabe, S. Wada, K. Nojima and M. Terada for their technical assistance. This work was supported by a special project grant for experimental studies on radiation health detriments and modifying factors from the Japan Science and Technology Agency.

References

[1] M. Ichimaru, M. Tomonaga, T. Amenomori, T. Matsuo, Atomic bomb and leukemia, Radiat. Res. 32 (1991) 162–167, suppl.
[2] K. Yoshida, T. Inoue, K. Nojima, Y. Hirabayashi, T. Sado, Calorie restriction reduces the incidence of myeloid leukemia induced by a single whole-body radiation in C3H/He mice, Proc. Natl. Acad. Sci. U. S. A. 94 (1997) 2615–2619.
[3] J.E. Till, E.A. McCulloch, A direct measurement of the radiation sensitivity of normal mouse bone marrow cells, Radiat. Res. 14 (1961) 213–222.
[4] Y. Hirabayashi, T. Matsumura, M. Matsuda, K. Kuramoto, K. Motoyoshi, K. Yoshida, H. Sasaki, T. Inoue, Cell kinetics of hematopoietic colony-forming units in spleen (CFU-S) in young and old mice, Mech. Ageing Dev. 101 (1998) 221–231.

Mutations in cervical cancer as predictive factors for radiotherapy

Khalida M.Y. Wani [a,b,c], N.G. Huilgol [b], T. Hongyo [a], H. Ryo [a], H. Nakajima [a], L.Y. Li [a], N. Chatterjee [b], C.K.K. Nair [c], T. Nomura [a,*]

[a] *Department of Radiation Biology and Medical Genetics, Graduate School of Medicine, Osaka University, Osaka 565-0871, Japan*
[b] *Division of Radiation Oncology, Nanavati Hospital and Medical Research Centre, Mumbai 400-056, India*
[c] *Biochemistry of Stress Response Section, Radiation Biology Division, B.A.R.C., Mumbai 400-085, India*

Abstract

We have investigated the role of the *p53*, *bak* and β-*catenin* genes in cervical carcinoma and assessed their predictive potential in patients treated with definitive radiotherapy. We examined 46 locally advanced cervical carcinoma (33 stage II cases and 13 stage III cases) punch biopsies taken prior to the initiation of radiotherapy, using a PCR-based assay followed by SSCP and direct sequencing. The *p53* gene was mutated in 19% (9/46) of the cases, *bak* mutations were found in 12% (5/42) of the cases, whereas β-*catenin* was altered in 17% (8/46) of the cases. There was a significant difference in stages II and III between the wild type and mutant *p53* patients ($p<0.01$). The results of this study suggest that the *bak* and β-*catenin* genes may play a role in the progression of cervical carcinoma. At present, we have found no significant difference in survival between the wild type and mutant patients for any of the genes analyzed. © 2002 Elsevier Science B.V. All rights reserved.

Keywords: p53; Cervical cancer; bak; β-catenin; Radiotherapy

1. Introduction

Cervical cancer represents the fifth most common neoplasm worldwide and radiotherapy remains the most important non-surgical treatment. Recent developments in cancer research have confirmed that carcinomas arise from accumulated genetic and epigenetic

---

* Corresponding author. Tel.: +81-6-6879-3811; fax: +81-6-6879-3819.
 E-mail address: tnomura@radbio.med.osaka-u.ac.jp (T. Nomura).

alterations in multiple genes. Some of those genes are likely to play crucial roles in the development of resistance to radiation.

The tumor suppressor gene *p53* regulates radiosensitivity in human malignancies after irradiation, giving rise to the hypothesis that tumors lacking functional *p53* are more radioresistant. Facing the complexities of genetic as well as epigenetic regulatory events, however, it seems unlikely that *p53* alone could serve as the single determinator for radiosensitivity. The β-*catenin* gene is involved in cell–cell adhesion. Mutational stabilization of this protein, however, causes it to function as an oncogene [1]. The *bak* gene, recently identified as a member of the Bcl$_2$ family of apoptosis regulatory genes, has been found to function as a potent inducer of apoptosis [2]. The purpose of our study was to determine the extent of *p53*, *bak* and β-*catenin* gene involvement in cervical cancer and to identify the correlation between mutations in these genes and patient outcome following radiotherapy.

2. Materials and methods

Punch biopsies were obtained from 46 proven cervical cancer patients prior to the initiation of radiotherapy. The *p53* exons 4–8, β-*catenin* exon 3, *bak* exons 3,4 and 6 were amplified by standard PCR, and subjected to non-radioactive SSCP analysis. SSCP bands with altered mobility were punched out and re-amplified. PCR products were purified and subjected to sequencing analysis. PCR–SSCP analysis and sequencing of the positive cases was repeated twice to rule out the possibility of PCR artifacts. Statistical analysis was done using the SPSS system (SPSS, Chicago).

3. Results and discussion

Nine tumors had missense mutations of the *p53* gene (19%). In this study, most of the mutations in *p53* (six cases) were found in exon 5, which encodes the DNA binding domain, mutations in exon 4 were detected in two cases including a 16 bp deletion and exon 8 was mutated in one case. There was a statistically significant difference in the frequency of *p53* mutations between stages II and III cases ($p < 0.01$). Five cases (12%) had missense mutations in the *bak* gene. Exon 3 was found to be mutated in two cases, and two cases showed mutations in exon 4, whereas exon 6 was mutated in one case. We report here the first known mutations to affect the functional domain of *bak*. Point mutations in exon 3 of the β-*catenin* gene were found in eight (17%) of the cases. Mutations affecting the highly conserved serine and threonine residues in exon 3 of the β-*catenin* gene, important for glutathione synthase kinase-3β phosphorylation, were relatively uncommon in our study. Two mutations affecting codon 41 by replacing a highly conserved threonine residue were found.

In conclusion, we have demonstrated for the first time that *bak* and β-*catenin* genes play a role in the progression of cervical cancer at-least in a subset of cases. Mutations in the *bak* gene were found more frequently in advanced stage tumors, suggesting that it is a late event in cervical carcinogenesis. There is no difference in overall survival, based on the ongoing follow-up studies of the patients, between patients with mutations and those without mutations in any of the genes analyzed.

Acknowledgements

This study was supported by the Research for the Future.

References

[1] W.S. Park, R.R. Oh, J.Y. Park, S.H. Lee, M.S. Shin, Y.S. Kim, S.Y. Kim, H.K. Lee, P.J. Kim, S.T. Oh, N.J. Yoo, J.Y. Lee, Frequent somatic mutations of the β-*catenin* gene in interstitial-type gastric cancer, Cancer Res. 59 (1999) 4257–4260.
[2] S. Kondo, Y. Shinomura, Y. Miyazaki, T. Kiyohara, S. Tsutsui, S. Kitamura, Y. Nagasawa, M. Nakahara, S. Kanayama, Y. Matsuzawa, Mutations of the *bak* gene in gastric and colorectal cancers, Cancer Res. 60 (2000) 4328–4330.

Lectin staining as a predictive test for radiosensitivity of oral cancers

P. Remani [a,*,1], V.N. Bhattathiri [b], L. Bindu [a], M. Krishnan Nair [b]

[a]*Division of Cancer Research, Regional Cancer Centre, Trivandrum 695 011, India*
[b]*Division of Radiotherapy, Regional Cancer Centre, Trivandrum 695 011, India*

Abstract

Owing to their specific carbohydrate binding properties, plant lectins have been used extensively as probes to study the surface architecture of transformed cells. The aim of our study was to evaluate the radiation-induced alterations in the lectin staining of tumour cells in order to see if there is any dose–effect relationship and if it has any predictive value regarding radiosensitivity. Twenty-eight patients with squamous cell carcinoma of oral cavity planned for radical radiotherapy were selected. Serial scrape smears were taken from each tumour before treatment and after the delivery of various fractions, usually 2 (7 Gy), 5 (17.5 Gy), 8 (28 Gy), and 11 (38.5 Gy). The smears were made to react with Jack Fruit Lectin (JFL) and stained with Diaminobenzedine. The nuclei were counterstained with Haematoxylin. A minimum of 100 cells from each sample was evaluated for the presence or absence of membrane staining. The results were expressed as the percentage of the total cells showing membrane staining. The mean percentage of the cells that showed lectin staining before treatment was 88.85 and after the delivery of various doses 7, 17.5, 28, and 38.5 Gy, respectively, they were 70.25 ($P<0.01$), 48.46 ($P<0.001$), 31.75 ($P<0.001$), and 19.53 ($P<0.001$). Since lectins bind to specific carbohydrate residues on the cell surface, the decrease in staining reflects the loss of these binding sites due to radiation-induced membrane damage. The different types of change and the treatment results indicate that this might prove to be a useful predictive test of radiosensitivity. © 2002 Elsevier Science B.V. All rights reserved.

Keywords: Oral cancer; Lectin staining; Predictive test; Radiosensitivity; Jack Fruit Lectin

---

[*] Corresponding author. Present address (from 1-6-2001 to 28-2-2002): Radiation Oncology Research Laboratory, Research Reactor Institute, Kyoto University, 590-0494, Osaka, Japan. Tel.: +81-724-51-2475; fax: +81-724-51-2627.
E-mail addresses: onokoji@rri.Kyoto-u.ac.jp, rcctvm@md2.vsnl.net.in (P. Remani).
[1] Tel.: +91-471-522337; fax: +91-471-447454.

1. Introduction

Lectins, carbohydrate-binding proteins or glycoproteins of defined specificity, have been extensively used in the study of cell surface characteristics [1]. Neoplastic transformation is accompanied by a variety of cell membrane changes and has been reflected by changes in lectin binding [2]. Variability in intrinsic radiosensitivity is an important factor that determines the control of cancers with radiotherapy [3]. One approach to predict radiosensitivity is to determine this directly by in vitro methods, such as tumour cell culturing and identification of the surviving fraction [4]. Another approach is to use radiation-induced cytomorphological changes of the cell membrane as markers of radiosensitivity. Lectins have proven to be valuable tools in the investigation of the molecular differences between tumour cells and normal cells by their ability to detect distinct changes in the cell surface glycoprotein. The cell membrane is a potential target for damage consequent to the preoxidation of lipids by radiation in it [5]. Since lectin staining by the cell membrane reflects its structural integrity, we studied the radiation-induced alterations in the lectin-staining pattern of tumour cells in order to see if there is any dose–effect relationship and if it has any predictive value regarding radiosensitivity.

2. Materials and methods

Twenty-eight patients with oral cavity tumours were included in the study. All of them had histologically proven squamous cell carcinoma and were treated with radiotherapy. Pretreatment smears were taken from the tumours of all patients. Subsequently, three to four smears were taken from each patient after the delivery of various fractions, usually after 2, 5, 8, and 11 fractions.

Scraping was done with previously wetted wooden spatula and the material was immediately spread on to a clean glass slide and fixed in a 3:1 solution of methanol and acetic acid. These were later made to react with Jack Fruit Lectin (JFL), which was isolated from the seeds of *Artocarpus integrifolia* and conjugated with horseradish peroxidase. Diaminobenzedene was used as the visualant. The nuclei were counterstained with Hematoxylin. All the preparations were stained at the same time and all the slides were evaluated by the same cytopathologist. A minimum of 100 cells from each sample were evaluated for the presence or absence of membrane staining. The results were expressed as the percentage of the total cells showing membrane staining.

3. Results

The percentage of cells showing lectin staining before treatment varied from 36% to 100% (mean 88.85%). With radiotherapy, the value decreased in general and this decrease was dose related. Table 1 shows the overall reduction in the proportion of lectin staining cells with radiation.

Regarding the rate of change in individual tumours, three patterns could be identified including the tumours, which had a moderate to high percentage of lectin staining cells

Table 1
Changes in lectin staining with radiation

Sampling occasion	Number of samples	Mean percentage of lectin staining cells (%)
Before treatment	28	88.85
After doses up to 17.5 Gy	18	48.46
After doses, 17.5 to 28 Gy	16	31.75
After doses, 28 to 38.5 Gy	14	19.53

before the treatment, which decreased to less than 10% with doses less than 17.5 Gy. Type 2 tumours also had an initial moderate to high percentage of lectin staining cells but the rate of decrease was slower. In type 3, the cells showed no significant change in the proportion of lectin staining even after doses of more than 28 Gy had been given.

4. Discussion

Radiation is known to damage the bilayer of lipids in the cell membrane by lipid peroxidation, resulting in membrane degradation and structural damage [6]. It is not known whether radiation produces any direct damage to the carbohydrate moieties on the cell surface. Since lipid peroxidation causes structural damage to the cell membrane, it is reasonable to presume that the damage results in the shedding of carbohydrate moieties to which the lectins bind and this effect will increase after higher doses of radiation.

The major method of cell lethality by radiation is mitotic death caused by DNA damage. However, it is recognized that radiation can cause direct damage to the cell membrane. It has been suggested that with lower doses, DNA lesions cause mitotic death, whereas with higher doses, as used in the therapeutic range, interphase death due to membrane damage occurs [7].

In the present study, the proportion of cells showing lectin staining before treatment varied from 36% to 100%, the mean being 88.85%. With radiation, the value decreased in general. Table 1 shows that there is a dose–effect relationship as shown by an overall reduction in the average percentage of cells showing lectin binding from 88.85% before the treatment to 19.53% after doses higher than 28 Gy.

The significance of the different patterns of changes in lectin staining induced by radiation and its clinical usefulness is not clear. In the present study, in 16 patients, the tumours showed either a slow or a rapid decrease in the proportion of lectin staining cells. Two of them did not complete treatment. Of the remaining 14 patients, 12 (85.7%) attained complete remission locally and 7 (50%) remained free of disease with a follow-up from 5 to 8 months. In contrast, none of the eight without decrease in lectin binding achieved local tumour control. The difference is not statistically significant, especially when the difference in tumour characteristics and treatment are considered. However, it suggests the distinct possibility that changes in the lectin binding doses reflect membrane sensitivity to radiation and that it might turn out to be a useful predictive tool for identifying the more radiosensitive or radioresistant tumours.

Acknowledgements

The financial support extended by the Indian Council of Medical Research is gratefully acknowledged.

References

[1] N. Sharon, Lectin receptors as lymphocyte surface markers, Adv. Immunol. 34 (1983) 213.
[2] M.E. Bramwell, H. Harris, An abnormal membrane glycoprotein associated with malignancy in a wide range of different tumours, Proc. R. Soc. London, Ser. B 201 (1978) 87–106.
[3] C.M.C. West, S.E. Davidson, R.D. Hunter, Evaluation of surviving fraction at 2 Gy as a potential prognostic factor for the radiotherapy of carcinoma cervix, Int. J. Radiat. Biol. 56 (1989) 761–765.
[4] R. Weichselbaum, M.A. Beckett, J.L. Schwartz, A. Dritschilo, Radioresistant tumour cells are present in head and neck carcinomas that recur after radiotherapy, Int. J. Radiat. Oncol., Biol., Phys. 15 (1988) 575–579.
[5] T. Alper, Cellular Radiobiology, Cambridge Univ. Press, Cambridge, 1979, pp. 221–224.
[6] B.B. Singh, Radiation induced molecular lesions in cells: a review, in: V.K. Jain, H.C Goel, W. Pohlit (Eds.), Recent Advances in Radiation Oncology, ICMR, New Delhi, 1990, pp. 33–41.
[7] A.W.T. Konings, Effect of heat and radiation on mammalian cells, Radiat. Phys. Chem. 30 (1987) 339–349.

Small dose pre-irradiation induced radioresistance and longevity after challenging irradiation in splenectomized C57BL/6 mice

K. Horie [a,*], K. Kubo [a,1], H. Kondo [b], M. Yonezawa [c]

[a] Department of Veterinary Radiology, Division of Veterinary Science, Graduate School of Agriculture and Biological Sciences, Osaka Prefecture University, 1-1 Gakuen-cho, Sakai, Osaka 599-8531, Japan
[b] Atomic Bomb Disease Institute, Nagasaki University School of Medicine, Nagasaki, Japan
[c] Research Institute for Advanced Science and Technology, Osaka Prefecture University, 1-2 Gakuen-cho, Sakai, Japan

Abstract

Pre-irradiation with 0.45 Gy of X-rays significantly decreased bone marrow death rate after challenging high-dose exposure to 6.75 Gy in splenectomized mice. However, the increment of the 30-day survival rate was smaller compared with that of the intact animals. The life span after day 30 of the control and pre-irradiated groups of mice was observed. The median survival time was 121 days for the control and 182 days for the pre-irradiated group. Statistical analyses by generalized Wilcoxon's rank sum test and log rank test showed that the pre-irradiation significantly elongated the life span of the high-dose irradiated mice. © 2002 Elsevier Science B.V. All rights reserved.

Keywords: Radioadaptive response; Life span; Small dose effects; Longevity; Survival rate

1. Introduction

We reported that pre-irradiation with 0.3–0.5 Gy induced radioresistance (decrease in bone marrow death rate after mid-lethal irradiation) in mice of both ICR [1,2] and C57BL/6 [3] strains 2 weeks later. Since the pre-irradiation of head was not necessary, the acquired radioresistance seemed to be induced by an adaptive response within blood-forming tissues. The spleen is an important organ for blood formation in mice. To know if the role of the spleen is essential for the induction of radioadaptive survival response, we examined whether the radioadaptive response is induced in splenectomized C57BL/6 mice or not.

---

* Corresponding author.
E-mail address: kiyohito@vet.osakafu-u.ac.jp (K. Horie).
[1] Tel./fax: +81-722-54-9497.

Effect of the pre-irradiation on life span of the two groups of mice, which survived bone marrow death, was also examined.

2. Materials and methods

Specific pathogen (especially *Pseudomonas aeruginosa*)-free 3-week-old C57BL/6N mice, purchased from Charles River Japan, were kept in a clean conventional environment. Acidified water, adjusted to pH 2.7 by adding HCl to tap water, was given to prevent contamination with bacteria. The animals were splenectomized at 4 weeks of age. When 6 weeks old, they were whole-body irradiated with 0.45 Gy of X-rays. Sham-irradiated control was run concurrently with the pre-irradiated group. Fourteen days later, the two groups of mice were again exposed to a mid-lethal dose with 6.75 Gy. The number of deaths that occurred within 30 days period was recorded. The difference within the 30-day survival rates of pre-irradiated and sham-irradiated control groups was examined statistically by the Chi-square test with Yates' correction. The life span was observed in the 30-day survived mice. The difference in the life span within the two groups was statistically examined with generalized Wilcoxon's rank sum test and the log rank test.

3. Results

The 30-day survival rate was significantly ($p < 0.001$) increased by the pre-irradiation in splenectomized mice; the 30-day survival rates with and without the priming irradiation were 88.4% (38/43) and 43.2%(19/44), respectively. The increment by the pre-irradiation in splenectomized mice (88.4–43.2% = 45.2%) was, however, smaller compared with that in the intact (nonsplenectomized) animals (78.0–8.0% = 70.0%) in the similar experimental conditions [3]. The difference (70.0–45.2% = 24.8%) may indicate the contribution of the spleen for the adaptive response.

Effect of pre-irradiation on the life span after the 30-day observation of bone marrow death was also examined. The median survival time was 121 (10–166) days after mid-lethal irradiation for the control and 182 (127–246) days for the pre-irradiated group. Statistical significance between the life span of the two groups was significant both in generalized Wilcoxon's rank sum test ($p = 0.0325$) and log rank test ($p = 0.0278$). The former test analyzes earlier phase in the death pattern, and the latter analyzes latter phase. Significance in the two statistical analyses shows that the pre-irradiation elongated the life span in both the earlier days and later days. The results indicate that the priming irradiation favored both short-term survival rate (30-day survival rate) and long-term survival time (life span) in splenectomized mice of C57BL/6N strain.

Acknowledgements

This work has been supported by a Grant-Aid for Scientific Research from the Central Research Institute Electric Power Industry Japan.

References

[1] M. Yonezawa, A. Takeda, J. Misonoh, Acquired radio-resistance after low dose X-irradiation in mice, J. Radiat. Res. 31 (1990) 256–262.
[2] M. Yonezawa, J. Misonoh, Y. Hosokawa, Two types of X-ray-induced radio-resistance in mice—presence of 4 dose ranges with distinct biological effects, Mutat. Res. 358 (1996) 237–243.
[3] J. Misonoh, M. Yonezawa, Dose ranges for radioadaptive response in mice—on the viewpoint of acquired radio-resistance after low dose irradiation, in: Health Effects of Low Dose Radiation, Challenges of 21st Century. Proceedings of the conference organized by British Nuclear Energy Society, 1997, pp. 196–200.

Suppression of X-ray-induced apoptosis by low dose pre-irradiation in the spleen of C57BL/6 mice

M. Yonezawa [a,*], A. Takahashi [b], K. Ohnishi [b], J. Misonoh [c], T. Ohnishi [b]

[a]*Division of Radiation Biology and Health Science, Research Institute for Advanced Science and Technology, Osaka Prefecture University, 1-2 Gakuen-cho, Sakai, Osaka 599-8570, Japan*
[b]*Department of Biology, Nara Medical University, Kashihara, Japan*
[c]*Abiko Research Laboratory, Central Institute of Electric Power Industry, Abiko, Japan*

Abstract

Pre-irradiation with 0.45 Gy of X-rays inhibited p53 accumulation determined by a Western blotting analysis in the spleen of C57BL/6 mice 5 h after challenging exposure to 3.0 Gy. Immunohistochemical analyses showed that the pre-irradiation decreased p53, Bax as well as apoptosis positive cell accumulation in the spleen on day 7 after 3.0 Gy. Recovery of spleen weight as well as endogenous CFUs after challenging exposure to 5.0 Gy was also stimulated by the pre-irradiation. These situations might favor recovery of hematopoietic function from radiation-induced acute injury, and might contribute to a decrease in bone-marrow death rate. © 2002 Elsevier Science B.V. All rights reserved.

Keywords: Radio adaptive response; Mouse survival; p53; Apoptosis; Endogenous CFUs

1. Introduction

Previously, we reported that pre-irradiation with 0.3–0.5 Gy induced radio-resistance (decrease in bone marrow death after mid-lethal irradiation) 2 weeks later in mice of both ICR [1] and C57BL/6 [2] strains. The induction of the radio-resistance did not need pre-irradiation of head, and thought to be induced by an adaptive response within blood-forming tissues. The spleen has been known to play an important role in hematopoiesis in mice. In the present paper, the molecular mechanism of radio-adaptive response was

---

* Corresponding author. Tel./fax: +81-722-54-9855.
E-mail address: yonezawa@riast.osakafu-u.ac.jp (M. Yonezawa).

examined in relation to signaling transduction to apoptosis in the spleen of the C57BL/6 mice. Effects of the pre-irradiation on the recovery of spleen weight and endogenous CFUs were preliminary examined.

2. Materials and methods

Specific pathogen (especially Pseudomonas bacteria)-free 4-week-old C57BL/6N mice, purchased from Charles River Japan, were kept in a clean-conventional environment. Acidified water, adjusted to pH about 2.7 by adding HCl to tap water, was given to prevent contamination with bacteria. When 6 weeks old, they were whole-body irradiated with 0.45 Gy of X-rays (260 kV, HVL 0.9 mm Cu, 0.45 Gy/min). Sham-irradiated control was run concurrently with the pre-irradiated group. The challenging dose was given 14 days afterwards.

To estimate molecular basis for the radio-adaptive response, we applied a Western blotting and an immuno-histochemical analyses after the challenging irradiation with 3.0 Gy in four groups: 0.0 and 0.0 Gy (intact), 0.45 and 0.0 Gy (priming irradiation only), 0.0 and 3.0 Gy (challenging irradiation only) and 0.45 and 3.0 Gy (priming and challenging irradiations). In addition, short-term accumulation of p53 in the spleen, 5 h after the challenging exposure, was measured using the Western blotting method. One to three animals per group were examined for this purpose. The samples of several tissues were immediately chilled in liquid nitrogen, and kept at dry ice temperature until analysis. The tissues were then pulverized by freeze fracturing in liquid nitrogen three times, and suspended in RIPA buffer. The insoluble cell debris was removed by centrifugation. Protein concentration of the supernatant was quantified using protein assay reagent (Bio-Rad Laboratories, Hercules, CA). The samples (20 g protein each) were subjected to 10% (w/v) SDS-polyacrylamide gel electrophoresis (PAGE), then transferred to an Immobilon™ PVDF membrane (Millipore Bedford, MA). The membranes were incubated with an anti-p53 monoclonal anti-body (Pab 421, which detects both wild type and mutant p53 in mammalian cells, Oncogene Science, Uniondale), and treated with a horseradish peroxidase-conjugated anti-mouse IgG anti-body (Zymed Labs., San Francisco, CA). The sensitivity of the visualization of the bands was enhanced using the BLAST blotting amplification system (Dupont/Biotechnology System NEN Research Products, Boston, MA). For confirmation of p53 band position, we used p53-deficient mouse embryonal fibroblast cells, MT158-8 (provided by Dr. O. Niwa, Kyoto Univ., Kyoto, Japan) as a negative control and protein size markers (Amersham Life Science, Buckinghamshire, UK). The densities of the bands were measured using a personal image analyzer LA-555D (PIAS, Osaka, Japan).

Long-term accumulation of p53, Bax and apoptosis positive cells in the spleen, 7 days after the challenging irradiation, was examined by an immuno-histological study. The p53 and Bax proteins on formalin-fixed paraffin-embedded sections were stained by the avidin–biotin peroxidase complex method using a HISTOFINE SAB-PO(R) Kit (Nichirei, Tokyo). The sections were treated with polyclonal anti-p53 antibody (CM1, Novocastra Lab., Newcastle, UK) or Bax antibody (Ab-1, Oncogene Research Products, Cambridge, MA), and were incubated with secondary antibody and the avidin–biotin peroxidase complex, stained with diaminobenzidine, and were counterstained with Harris' hematoxylin. Apoptotic cells in the sections were detected by staining with ApopTag in situ Detection Kit® (Intergen, NY). Three animals were used per group for this study, and 300 cells were counted in a field under microscope.

For the measurement of spleen weight and endogenous CFUs after challenging irradiation, mice were pre-irradiated at 8 weeks of age. They were killed by cervical dislocation on days 11, 12, 13 and 15 after the challenging irradiation with 5.0 Gy. The spleen was weighted, and fixed in a Bouin's solution. Colonies on the surface of the spleen were counted as endogenous CFUs. Five animals were used per point.

3. Results

Pre-irradiation with 0.45 Gy itself did not affect p53 accumulation in the spleen 14 days later, at the time just before challenging irradiation. Challenging irradiation with 3 Gy resulted in a significant increase (about 1.5 times) in p53 accumulation observed 5 h later, but the pre-irradiation completely inhibited the accumulation (Fig. 1). The short-term

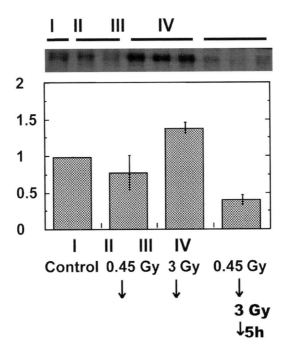

Fig. 1. Pre-irradiation with 0.45 Gy 14 days prior to the challenging exposure to 3.0 Gy prevented accumulation of p53, observed by a Western blotting analysis, in the spleen cells of C57BL/6 mice 5 h after the challenging irradiation. The numbers of animals used were 3, 3, 3 and 4 for 2 weeks after sham-irradiation (I), 2 weeks after pre-irradiation (II), 5 h after challenging irradiation (III), and 5 h after challenging irradiation 2 weeks after pre-irradiation (IV), respectively.

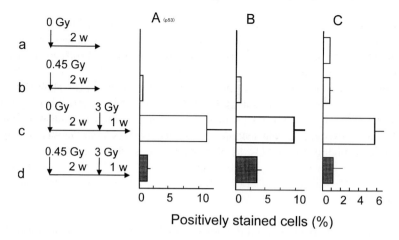

Fig. 2. Pre-irradiation with 0.45 Gy decreased p53 (A), Bax (B) and apoptosis (C) positive cells in the spleen 1 week after the challenging irradiation with 3 Gy in C57BL/6 mice, determined by immuno-histochemical method. Pre-irradiation was carried out 14 days before the challenging irradiation.

accumulation of p53 after challenging irradiation was not significantly affected by the pre-irradiation in other tissues, namely the brain, thymus, lung, small intestine and bone marrow.

Immuno-histological analysis showed that the pre-irradiation itself did not give any change in p53, Bax and apoptosis positive splenic cells 14 days later. Challenging exposure with 3 Gy itself accumulated p53, Bax and apoptosis positive cells in the spleen on day 7, but the pre-irradiation decreased induction of all p53, Bax and apoptosis positive cells (Fig. 2). Since Bax protein has been recognized to direct cell to apoptosis [3], these results seem to show that the pre-irradiation diminished signal transduction to apoptosis after the challenging irradiation. Significant effects of the pre-irradiation were not observed from measurements 3 days after the challenging irradiation (data not shown).

Recovery of spleen weight as well as endogenous CFUs (endo-CFUs) after challenging irradiation with 5.0 Gy was significantly enhanced by the pre-irradiation as shown in Table 1.

Table 1
Effect of pre-irradiation with 0.45 Gy on the recovery of the spleen weight and endogenous CFUs in C57BL/6 mice after challenging exposure to 5.0 Gy

	On day 11		On day 12		On day 13	
	Control	Pre-irradiated	Control	Pre-irradiated	Control	Pre-irradiated
Spleen weight (mg)	23.7 ± 2.3	25.5 ± 5.1	25.7 ± 4.9	30.2 ± 4.5**	23.6 ± 3.5	36.7 ± 6.0**
Endo-CFUs	5.0 ± 2.1	13.0 ± 6.6*	6.2 ± 2.7	33.0 ± 4.7***	6.4 ± 3.0	49.8 ± 12.0***

Expressed as average ± SD.
* $p < 0.05$.
** $p < 0.01$.
*** $p < 0.001$ (t-test).

4. Conclusion

Several experiments were carried out on the spleen, an important hematopoietic organ in mice, and we obtained several data to presume the molecular and biological mechanisms for the radio-adaptive response in C57BL/6 mice. Pre-irradiation with 0.45 Gy 14 days prior to the challenging irradiation with 3.0 Gy of X-rays prevented accumulation of p53 in the spleen measured 5 h after the challenging exposure. Inhibition by the pre-irradiation of p53, Bax and apoptosis positive cell accumulation in the spleen 7 days after the challenging exposure was also observed. This situation might favor recovery of hematopoietic function from radiation-induced acute injury. Stimulated recovery of splenic weight as well as endogenous CFUs, might contribute to a decrease in bone-marrow death rate.

References

[1] M. Yonezawa, J. Misonoh, Y. Hosokawa, Two types of X-ray-induced radio-resistance in mice. Presence of 4 dose ranges with distinct biological effects, Mutat. Res. 358 (1996) 237–243.
[2] J. Misonoh, M. Yonezawa, Dose ranges for radioadaptive response in mice—on the viewpoint of acquired radio-resistance after low dose irradiation, Health Effects of Low Dose Radiation, Challenges of 21st Century, Proceedings of the Conference Organized by British Nuclear Energy Society (1997) 196–200.
[3] X. Wang, T. Ohnishi, p53-dependent signal transduction induced by stress, J. Radiat. Res. 38 (1997) 179–194.

Effect of pre-irradiation of mice whole-body with X-ray on radiation-induced killing, induction of splenic lymphocyte apoptosis, and expression of mutated Ca^{2+} channel α_{1A} subunit

Kazuhiko Sawada [a,*], Akihisa Takahashi [b], Takeo Ohnishi [b], Hiromi Sakata-Haga [a], Yoshihiro Fukui [a]

[a]*Department of Anatomy, University of Tokushima School of Medicine, Tokushima 770-8503, Japan*
[b]*Department of Biology, Nara Medical University, Nara 634-8521, Japan*

Abstract

This study examined radiosensitive and radioadaptive responses following whole-body irradiation in rolling mouse Nagoya (RMN) which carries the mutation in the gene coding for the α_{1A} subunit of P/Q-type Ca^{2+} channel. Adult RMN and littermate controls (C3Hf/Nga) were subjected to a single exposure of X-irradiation at doses of 4, 5, and 6.75 Gy. RMN were slightly radiosensitive as compared to the control mice. All RMN died on day 8 after X-irradiation at 6.75 Gy, and all controls on day 13. In both RMN and controls, Bax- and TUNEL-positive lymphocytes in the splenic white pulp increased 12 h after irradiation at 3 Gy. Inductions of Bax- and TUNEL-positive lymphocytes were apparently suppressed 2 weeks following pre-irradiation with a low dose (0.45 Gy). In RMN, expression of the mutated Ca^{2+} channel α_{1A} subunit was not found in the lymphocytes and was not induced by the irradiation. These results suggested that the mutation in P/Q-type Ca^{2+} channel might modify the radiosensitivity. This may mean that the radioadaptive response results in the depression of apoptosis in splenic white pulp lymphocytes, but not in the expression of the mutated Ca^{2+} channel. © 2002 Elsevier Science B.V. All rights reserved.

Keywords: Bax; Apoptosis; Ca^{2+} channel; Adaptive response; Rolling mouse Nagoya

---

[*] Corresponding author. Tel.: +81-88-633-7052; fax: +81-88-633-7053.
E-mail address: sawada@basic.med.tokushima-u.ac.jp (K. Sawada).

1. Introduction

Rolling mouse Nagoya (RMN) is an ataxic mutant mouse that carries the mutation in the gene coding for the α_{1A} subunit of P/Q-type Ca^{2+} channel [1]. In this mutant, P-type Ca^{2+} current was selectively reduced [1] and some morphological and functional abnormalities in the central nervous system have been reported [2]. Using RMN, we examined the role of the P/Q-type Ca^{2+} channel in radiosensitivity and the radioadaptative response induced by pre-irradiation with a low dose.

2. Materials and methods

Adult male RMN and the controls (C3Hf/Nga) were subjected to a single exposure of X-irradiation at doses of 4, 5, and 6.75 Gy. Survival of each animal was monitored daily for 30 days following the challenge irradiation. The irradiation schedules and procedures for analysis of radioadaptive response of the splenic lymphocytes were performed according to the previous report [3]. The mice were irradiated by X-ray at 3 Gy, 2 weeks after irradiation with a low dose (0.45 Gy). Spleens were removed 12 h after the challenge irradiation and were fixed in Bouin's solution without acetic acid, embedded in paraffin, sectioned at 5 μm, and processed for Bax and the α_{1A} subunit immunohistochemistry and TUNEL staining.

3. Results

After irradiation at 4, 5, and 6.75 Gy, some RMN and the controls were dying and increasing the irradiation doses shortened the duration of survival. RMN were slightly radiosensitive compared to the control mice and begin to die 3 days following irradiation (Fig. 1). All RMN died on day 8 after X-irradiation at 6.75 Gy, and all the controls on day 13 (Fig. 1).

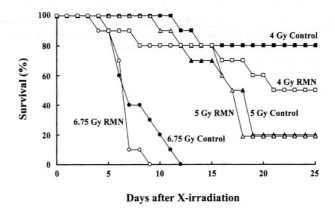

Fig. 1. Radiation-induced killing following whole-body irradiation.

Bax- and TUNEL-positive lymphocytes in the white pulp increased 12 h after irradiation at 3 Gy in both RMN and the controls. However, induction of Bax- and TUNEL-positive lymphocytes was apparently suppressed by pre-irradiation with a low dose. There were no differences in the frequencies of Bax- and TUNEL-positive lymphocytes between RMN and the controls in all irradiation groups. On the other hand, mutated Ca^{2+} channel α_{1A} subunit immunoreactivity was found in the cerebellum but not in the white pulp lymphocytes of RMN. Their expression was not induced by irradiation with or without pre-irradiation.

4. Conclusions

The results suggested that the altered function of the P/Q-type Ca^{2+} channel might modify radiosensitivity, resulting in shortened survival periods in RMN. This may mean that the radioadaptive response results in the depression of apoptosis in splenic white pulp lymphocytes but not in the expression of the mutated Ca^{2+} channel.

Acknowledgements

Mutant mice were kindly provided by Dr. S. Oda, Graduate School of Bio-Agricultural Science, Nagoya University.

References

[1] Y. Mori, M. Wakamori, S. Oda, C.F. Fletcher, N. Sekiguchi, E. Mori, N.G. Copeland, N.A. Jenkins, K. Matsushita, Z. Matsuyama, K. Imoto, Reduced voltage sensitivity of activation of P/Q-type Ca^{2+} channels is associated with the ataxic mouse mutation *rolling Nagoya* (tg^{rol}), J. Neurosci. 20 (15) (2000) 5654–5662.
[2] K. Sawada, H. Haga, Y. Fukui, Ataxic mutant mice with defects in Ca^{2+} channel α_{1A} subunit gene: morphological and functional abnormalities in cerebellar cortical neurons, Cong. Anom. 40 (2) (2000) 99–107.
[3] A. Takahashi, K. Ohnishi, I. Asakawa, N. Kondo, H. Nakagawa, M. Yonezawa, A. Tachibana, H. Matsumoto, T. Ohnishi, Radiation response of apoptosis in C57BL/6N mouse spleen after whole-body irradiation, Int. J. Radiat. Biol. 77 (9) (2001) 939–945.

Elevation of antioxidants in the kidneys of mice by low-dose irradiation and its effect on Fe^{3+}–NTA-induced kidney damage

T. Nomura [a,*], K. Yamaoka [b], K. Sakai [a]

[a] *Low Dose Radiation Research Center, Central Research Institute of Electric Power Industry, 2-11-1 Iwado-kita, Komae, Tokyo 201-8511, Japan*
[b] *Okayama University Medical School, Okayama, Japan*

Abstract

We previously examined the effects of 0.5 Gy of γ- or X-irradiation on endogenous antioxidative materials, including glutathione, glutathione-related enzymes, superoxide dismutase and catalase in the liver, pancreas and brain of mice. An increase in the antioxidants was observed soon after the irradiation. We also demonstrated that the low-dose irradiation gave suppressive effects on reactive oxygen species (ROS)-related disease models such as acute hepatitis, type I diabetes and Parkinson's disease. These results suggested that the suppression or inhibition of the ROS-related diseases were made through the elevation of the antioxidants. We studied the effects of irradiation (0.5 Gy of γ-ray) reducing the renal oxidative damage in Fe^{3+} nitrilotri-acetate (NTA) injected mice. We examined transaminase activity, lipid peroxide level and those levels of antioxidants in the mouse. Both transaminase activities and lipid peroxide level accelerated decrease to normal level by the irradiation. The levels of total glutathione content and the activities of its related enzymes were slightly increased in the irradiated group. On the other hand, catalase (CAT) activity was significantly increased by the irradiation. These findings suggested that low-dose irradiation relieved functional disorder in the kidney of mice with ROS-related diseases, probably through enhancing antioxidant, in particular CAT activity. © 2002 Elsevier Science B.V. All rights reserved.

Keywords: Antioxidants; Low-dose irradiation; Endogenous antioxidative materials; Glutathione

1. Background

NTA is a synthetic aminopolycarboxylic acid that chelates metals to form water-soluble complexes. Fe^{3+}–NTA complex has elevated levels of lipid and DNA oxidation products

---

* Corresponding author. Tel.: +81-3-3480-2111; fax: +81-3-3480-3539.
E-mail address: NOMURA@CRIEPI.DENKEN.OR.JP (T. Nomura).

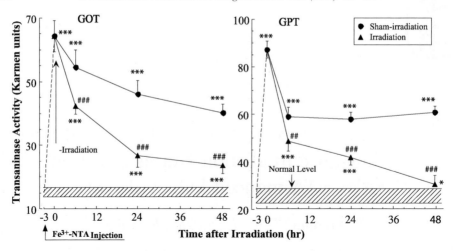

Fig. 1. Time-course changes of the GOT (left) and GPT (right) activities in the plasma of mice with Fe^{3+}–NTA pretreatment 3 h after 50 cGy γ-irradiation. * Significantly different when compared with normal group, and # significantly different when sham-irradiated group. Single mark is $p<0.05$, double is $p<0.001$, and triple is $p<0.005$, respectively.

[1]. It is thought that NTA toxicity is due to the binding of Fe^{3+} iron to form the complexes involved in oxidative damage.

We previously reported that 0.5 Gy of γ-rays increased the level of antioxidant substances such as superoxide dismutase, CAT and glutathione in various organs of normal mice, and lowered the lipid peroxide levels [2]. We have also shown the increase in antioxidant capacity in liver, pancreas and brain of normal mice or ROS-related disease models before or after low-dose irradiation (0.5 Gy of γ- or X-rays) [3,4]. These findings

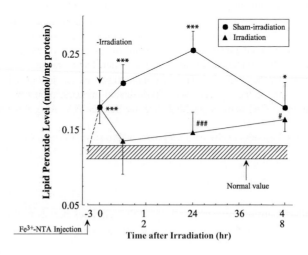

Fig. 2. Time-course of changes in lipid peroxide level of Fe^{3+}–NTA injected mouse kidneys after γ-irradiation. The symbols are described in Fig. 1.

lead us to the possibility that ROS-related disease might be inhibited or suppressed by the irradiation.

In this paper, we evaluate the effects of low-dose irradiation on kidney damage in Fe^{3+}-NTA treated mice.

2. Materials and methods

We used a female C57BL/6N mouse (Clea Japan, Tokyo) at 6 weeks of age. Whole body irradiation of mice was carried out with 0.5 Gy (1.17 Gy/min) of γ-rays from a ^{137}Cs source (Gammacell 40, Nordin International, Canada).

The renal toxicity induced by Fe^{3+}-NTA was conducted by the method [5]. Fe^{3+}-NTA (7.5 mg Fe/kg body weight) was given intraperitoneally. The irradiation of γ-rays was initiated 3 h after the injection of Fe^{3+}-NTA. There were four to six mice in each group. The activities of glutamic oxaloacetic transamirase (GOT) and glutamic pyruvic transamirase (GPT) in the serum, lipid peroxidation level and the levels of antioxidants in the kidneys were examined [4,6].

3. Results

As shown in Fig. 1, we demonstrated the time of change in the plasma GOT and GPT activities. Both activities are biochemical markers. The irradiation group rapidly returned to normal values as compared with the sham-irradiation group.

Lipid peroxide levels were greatly elevated by Fe^{3+}-NTA treatment (Fig. 2). The level of the irradiated group rapidly decreased after irradiation.

Table 1 shows the result of the total glutathione content and the activities of glutathione-related enzymes (GR, GPX and γ-GCS). The GR activity was significantly different in the irradiated group compared to the activity seen in the sham-irradiated group

Table 1
Time-course of changes in the activities of total glutathione and its related enzymes of mouse kidneys induced by Fe^{3+}-NTA after γ-irradiation

		Normal	0 h	6 h	24 h	48 h
GSH (nmol/mg protein)	Irradi	–	6.53 ± 0.29**	6.59 ± 0.25**	6.98 ± 0.25	7.61 ± 0.14*
	Sham	7.24 ± 0.20	6.53 ± 0.29**	6.70 ± 0.32*	6.80 ± 0.14	7.29 ± 0.38
GPX (units/mg protein)	Irradi	–	0.75 ± 0.03***	0.74 ± 0.04***	0.81 ± 0.06	0.96 ± 0.05#
	Sham	0.90 ± 0.05	0.75 ± 0.03***	0.72 ± 0.05***	0.74 ± 0.03***	0.85 ± 0.05
GR (units/mg protein)	Irradi	–	0.27 ± 0.03*	0.29 ± 0.01	0.30 ± 0.03	0.35 ± 0.02
	Sham	0.32 ± 0.02	0.27 ± 0.03*	0.28 ± 0.02	0.28 ± 0.01*	0.31 ± 0.02
γ-GCS (units/mg protein)	Irradi	–	1.68 ± 0.05***	1.68 ± 0.04***	1.82 ± 0.07***	2.05 ± 0.07
	Sham	2.00 ± 0.02	1.68 ± 0.05***	1.68 ± 0.05***	1.77 ± 0.07	1.97 ± 0.06

Abbreviations: GSH, glutathione; GPX, glutathione peroxidase; GR, glutathione reductase; γ-GCS, γ-glutamylcysteine synthetase; Irradi, irradiation; Sham, sham-irradiation.
The symbols are described in Fig. 1.

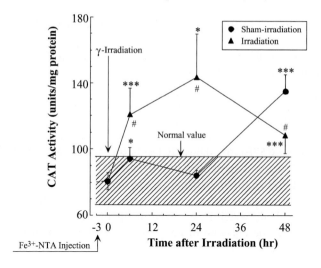

Fig. 3. Time-course of changes in CAT activity of mouse kidneys induced by Fe^{3+}–NTA after γ-irradiation. The symbols are described in Fig. 1.

at 48 h. Although the other antioxidants were not significantly different, there was a tendency that the level of the irradiated group was higher than the sham-irradiated group.

The result of the CAT activity is given in Fig. 3. The activity of the irradiated group significantly increased 24 h after irradiation.

4. Conclusion

In this paper, we study the effects of low-dose irradiation on kidney damage in Fe^{3+}–NTA treated mice. The irradiation inhibited renal oxidative damage as the result of lipid peroxidation and might accelerate the recovery from the damage. In the changing in GOT and GPT activities, these results reflected the renal damage. The result of the renal lipid peroxide is also reflected. The enzyme activity of CAT was extremely different from the irradiated group and the sham-irradiated group. The level of glutathione and the activities of its related enzymes, however, were not affected by the irradiation.

Glutathione is considered to act as a radical scavenger. CAT transforms H_2O_2 into H_2O as well as glutathione. In our paper, renal CAT activity changed in a similar fashion to the converted style of the level of lipid peroxide. This means that the CAT activity induced by the irradiation was protected by lipid oxidization. To find the relationship between the elevation of CAT activity and the level of glutathione and its related enzymes, the CAT activity might be easily inducible than the glutathione and its related enzymes to protect themselves against additional oxidative injury.

These findings suggest that low-dose irradiation relieved Fe^{3+}–NTA toxicity in the kidneys. The irradiation might be effective for the prevention and/or treatment of ROS-related diseases.

References

[1] T. Umehara, K. Sai, A. Takagi, R. Hasegawa, Y. Kurokawa, Cancer Lett. 54 (1990) 95–100.
[2] O. Matsuki, T. Nomura, S. Kojima, A. Kubodera, Radioisotopes 47 (1998) 291–299.
[3] S. Kojima, O. Matsuki, T. Nomura, N. Shimura, A. Kubodera, K. Yamaoka, H. Tanooka, H. Wakasugi, Y. Honda, S. Honda, T. Sasaki, Brain Res. 808 (1998) 262–269.
[4] T. Nomura, K. Yamaoka, Free Radical Biol. Med. 27 (1999) 1324–1333.
[5] M. Awai, Protein Nucl. Enzyme 33 (1988) 2844–2854.
[6] T. Nomura, K. Sakai, S. Kojima, M. Takahashi, K. Yamaoka, in: T. Yamada, C. Mothersill, B.D. Michael, C.S. Potten (Eds.), Biological Effects of Low Dose Radiation, Excepta Medica International Congress Series, vol. 1211, Elsevier, Netherlands, Amsterdam, 2000, pp. 101–106.

Suppressive effect of long-term low-dose rate gamma-irradiation on chemical carcinogenesis in mice

Kazuo Sakai*, Toshiyasu Iwasaki, Yuko Hoshi, Takaharu Nomura, Takeshi Oda, Kazuko Fujita, Takeshi Yamada, Hiroshi Tanooka

Low Dose Radiation Research Center, Central Research Institute of Electric Power Industry, 2-11-1 Iwado-kita, Komae, Tokyo 201-8511, Japan

Abstract

Female ICR mice, 6 weeks old, 35 in each group, were exposed to gamma-rays from a ^{137}Cs source in the long-term low-dose rate irradiation facility at the Central Research Institute of Electric Power Industry (CRIEPI). The dose rate was 2.6 (A), 0.96 (B), or 0.30 mGy/h (C). Thirty-five days later, the mice were injected in the groin with 0.5 mg of methylcholanthrene (MC) dissolved in olive oil and irradiation was continued. Cumulative tumor incidences after 216 days following MC injection were 89% in group A, 76% in group B, and 94% in group C. The one in the non-irradiated control group was 94%. The difference in the tumor incidence between the control and position B was statistically significant, indicating the suppressive effect of the low-dose rate irradiation on the process of MC-induced carcinogenesis with an optimum dose rate around 1 mGy/h. © 2002 Elsevier Science B.V. All rights reserved.

Keywords: Low-dose rate irradiation; Gamma-rays; Chronic irradiation; Suppression of carcinogenesis; Methylcholanthrene

1. Introduction

Low-dose ionizing radiation is known to stimulate certain biological activities, in vitro and in vivo, including (1) anti-oxidative capacity [1], (2) repair of damage to DNA [2], (3) the process of apoptosis [3,4], and (4) immune responses [5,6].

---

* Corresponding author. Tel.: +81-3-3480-2111x2572; fax: +81-3-3480-3113.
  E-mail address: kazsakai@criepi.denken.or.jp (K. Sakai).

When we consider the effects of low-dose radiation, cancer is one of the most important concerns. Carcinogenesis is supposedly initiated by the damage to DNA induced by various agents. Such DNA damage can cause normal cells to become malignant. These malignant cells, in turn, proliferate, overwhelming the normal regulation of cell growth.

Some of the responses to low-dose radiation described above can, when working together, suppress the process of carcinogenesis: the increased amount of anti-oxidants, together with the increased capacity of DNA repair reduce the amount of potentially carcinogenic DNA lesions, the stimulated apoptotic system removes the potentially malignant cells, and, finally, the activated immune system removes tumor cells.

We have examined the possible suppressive effect of low-dose rate radiation on carcinogenesis using our long-term low-dose rate irradiation facility at the Central Research Institute of Eletric Power Industry (CRIEPI).

2. Materials and methods

2.1. Experimental animals

Female ICR mice (Nihon Clea) were kept under the clean conventional condition in an irradiation room in the low-dose rate irradiation facility at CRIEPI. The irradiation room is 9 m in width, 12 m in depth, and 5 m in height. The mice were six weeks old at the beginning of the irradiation. All animals were maintained on a light schedule from 07:00 to 19:00 h, and given food and water ad libitum following the guidelines for animal experiments at CRIEPI.

2.2. Gamma-irradiation

Irradiation was carried out using a 370-GBq ^{137}Cs source built in our irradiation room. Groups of 35 mice were placed at a position 3, 5, or 10 m from the gamma-ray source. The dose rate at each point, indicated in Table 1, was estimated by a fluorescent glass dosimeter [7]. Irradiation was carried out continuously except for 1–2 h on weekdays for animal check and care.

2.3. Carcinogen treatment

In the present study, the mice were injected in the groin with 20-methylcholanthrene (MC) [8], which had been dissolved in olive oil, with each mouse receiving 0.1 ml. When a

Table 1
Position of the mice and the dose rates

Distance from the source (m)	Dose rate (mGy/h)
3	2.6
5	0.95
10	0.30

Table 2
Cumulative tumor incidence[a] after 216 days following MC injection in mice irradiated under different irradiation conditions

Non-irradiated control	0.30 mGy/h	0.95 mGy/h	2.6 mGy/h
94	94	76	89

[a] The cumulative tumor incidence is presented as the percentage of mice with tumors.

tumor reached the size of 10–15 mm in diameter, the mouse was sacrificed to examine the tissue section.

3. Results

3.1. Tumor incidence

The mice were irradiated at the respective dose rates for 35 days, then injected with MC. After the MC injection, irradiation was continued at the same dose rate. The first tumors appeared in each group approximately 60 days after the MC injection. There was little difference in the timing of tumor appearane among the groups. Tumor incidences 100, 150, 200, and 216 days after MC injection are summarized in Table 2.

As is shown in Table 2, the tumor incidence was the lowest in the group that was irradiated at a dose rate of 0.95 mGy/h.

After 216 days following the MC injection, when the observation was terminated, the difference in tumor incidence between the control mice and those irradiated at 0.96 mGy/h was statistically significant ($p < 0.05$), while the difference between the control group and 0.3 or 2.6 mGy/h group was not.

3.2. Spectrum of tumor type

Tissue examination revealed that 60% of the tumors were fibrosarcoma, and 40% squamous cell carcinoma. There was no difference in the spectrum among the groups of different irradiation conditions including the non-irradiated control group. No other tumor was found in the irradiated or non-irradiated mice.

4. Discussion

The results of the present study demonstrated the suppressive effects of low-dose rate irradiation on the process of carcinogenesis induced by MC in mice. The mechanism underlying the suppressive effect might be the increase in the anti-oxidative capacity, the increase in the DNA repair capability, the stimulation of apoptosis, or the stimulation of the immune function. Which, or which combination of these factors is involved remains for future study to elucidate.

It should be noted that the 0.95-mGy/h dose rate was not only more effective than the lower dose rate (0.30 mGy/h), but also more effective than the higher dose rate (2.6 mGy/

h). This observation suggests that there may be an "optimum" dose rate that suppresses the carcinogenic process.

This kind of so-called "window effect" has been reported for the induction of antioxidants [1], for the radiation adaptive responses, in which, supposedly, an increase in DNA repair capacity is involved, in cultured murine [10] and human [11] cells, for the stimulation of apoptosis [4], and for the stimulation of the immune function in mice [9]. These window effects might explain the presence of the optimum dose rate for the tumor suppression observed in the present study.

References

[1] S. Kojima, et al., Induction of mRNAs for glutathione synthesis-related proteins in mouse liver by low doses of gamma-rays, Biochim. Biophys. Acta 1381 (1998) 312–318.

[2] X.C. Le, et al., Inducible repair of thymine glycol detected by an ultrasensitive assay for DNA damage, Science 280 (1998) 1066–1069.

[3] S.P. Cregan, et al., Apoptosis and the adaptive response in human lymphocytes, Int. J. Radiat. Biol. 75 (1999) 1087–1094.

[4] K. Sakai, Effects of low-dose preirradiation on radiation-induced cell death in cultured mammalian cells, in: T. Yamada, C. Mothersill, B.D. Michael, C.S. Potten (Eds.), Biological Effects of Low Dose Radiation, Excerpta Medica International Congress Series 1211, Elsevier, Amsterdam, 2000, pp. 53–58.

[5] M. Nogami, et al., Mice chronically exposed to low dose ionizing radiation possess splenocytes with elevated levels of HSP70 mRNA, HSC70 and HSP72 and with an increased capacity to proliferate, Int. J. Radiat. Biol. 63 (1993) 775–783.

[6] M. Nogami, et al., T cells are the cellular target of the proliferation-augmenting effect of chronic low-dose ionizing radiation in mice, Radiat. Res. 139 (1994) 47–52.

[7] Y. Hoshi, et al., Application of a newly developed photoluminescence glass dosimeter for measuring the absorbed dose in individual mice exposed to low-dose rate 137Cs gamma-rays, J. Radiat. Res. 41 (2000) 129–137.

[8] H. Tanooka, et al., Dose response and growth rates of subcutaneous tumors induced with 3-methylcholanthrene in mice and timing of tumor origin, Cancer Res. 42 (1982) 4740–4743.

[9] S.Z. Liu, et al., Signal transduction in lymphocytes after low dose radiation, Chin. Med. J. 107 (1994) 431–436.

[10] M.S. Sasaki, Cytogenetic biomonitoring of human radiation exposures: possibilities, problems and pitfalls, J. Radiat. Res. 33 (1992) 44–53, Suppl.

[11] E.J. Broome, et al., Adaptation of human fibroblasts to radiation alters biases in DNA repair at the chromosomal level, Int. J. Radiat. Biol. 75 (1999) 681–690.

Possible role of elevation of glutathione in acquisition of enhanced immune function of mouse splenocytes exposed to low-dose γ-rays

Shuji Kojima [a,*], Shirane Matsumori [a], Hirokazu Ishida [a], Mereyuki Takahashi [a], Kiyonori Yamaoka [b]

[a]*Department of Radiochemistry, Faculty of Pharmaceutical Sciences, Science University of Tokyo, Research Institute for Biological Sciences, 2669 Yamazaki, Noda, Chiba 278-0022, Japan*
[b]*Okayama University Medical School, Shikata, Okayama 700-8558, Japan*

Abstract

The relation between induction of an increased glutathione level and the enhanced immune functions, including proliferative response and natural killer (NK) activity, of mouse splenocytes by a low dose of γ-rays was investigated. Glutathione level in mouse splenocytes significantly increased 2 h after whole-body γ-ray irradiation at 0.5 Gy, peaked at 4 h and thereafter, decreased almost to the initial (0 h) level within 12 h after irradiation. A significant enhancement of Con A-induced proliferation was recognized in the splenocytes from the whole-body-irradiated animals obtained at post-irradiation between 2 and 6 h. NK activity was also enhanced at the same time periods after the irradiation. Addition of glutathione to splenocytes obtained from normal mice enhanced both Con A-induced proliferative response and NK activity in a dose-dependent manner. These enhancements were completely blocked by buthionine sulfoximine, a specific inhibitor of the de novo pathway of glutathione synthesis. These results suggest that induction of endogenous glutathione synthesis in splenocytes after low-dose γ-ray irradiation is partially responsible for the appearance of enhanced immune function. © 2002 Elsevier Science B.V. All rights reserved.

Keywords: Glutathione; Immune function; Low-dose γ-rays; Adaptive response

1. Introduction

Low doses of ionizing radiation induce various effects, including radio-adaptive response, activation of immune function, stimulation of growth rate, and enhancement of resistance to high-dose radiation. These phenomena have been generally called

---

* Corresponding author. Tel/fax: +81-471-23-9755.
E-mail address: kjma@rs.noda.sut.ac.jp (S. Kojima).

"radiation hormesis", although little is yet known about the mechanisms involved [1]. In our previous studies, changes of endogenous reduced glutathione (GSH) were examined in mice exposed to whole-body γ-ray irradiation, and it was found that low doses of radiation significantly increased GSH levels in organs such as the liver, pancreas and brain [2–5]. GSH has direct or indirect roles in many biological processes, including protein and DNA synthesis, amino acid transport, activation of enzyme activities, activation of metabolism, and protection of cells from damage caused by reactive oxygen species (ROS) [6]. Most cell damage caused by ionizing radiation is also mediated by ROS generated from the interaction between radiation and water molecules in cells. Intracellular GSH scavenges these ROS and protects the cells from radiation toxicity. Since thiol-containing compounds, such as GSH, have been found to modulate mouse lymphocyte activities [6,7], we examined whether or not the increase of glutathione induced by small doses of γ-rays enhances the mitogenic response and the natural killer (NK) activity of mouse splenocytes, which was the most popular assay method for the estimation of immune function.

2. Methods

Male ICR strain mice, 7 weeks of age, were divided into irradiated and non-irradiated control groups. Irradiation was done with γ-rays from a ^{137}Cs source at a dose of 0.5 Gy (1.11 Gy/min). Glutathione content in the splenocytes was measured using a modified

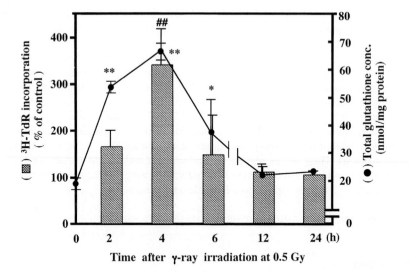

Fig. 1. Effect of low-dose whole-body γ-ray irradiation on the total glutathione (GSH+GSSG) levels and the proliferative response of mouse splenocytes to Con A. Splenocytes were obtained from male ICR mice after whole-body γ-ray irradiation at 0.5 Gy. The proliferative response of splenocytes was assessed in terms of ^{3}H-thymidine (^{3}H-TdR) incorporation into DNA of the cells stimulated by Con A (2 μg/ml). Results are represented as percentages (%) of the non-irradiated control at each time interval. Each point indicates the mean ± SD of four mice. *, ** Significantly different from the 0 h group at $p<0.05$ and 0.01, respectively. # Significantly different from the non-irradiated control group at $p<0.05$.

spectrophotometric technique. Enhancement of *Concanavalin* A (Con A)-induced proliferative response of the splenocytes after whole-body γ-ray irradiation was estimated from the ^{3}H-thymidine incorporation. NK activity was measured by ^{51}Cr release into the medium from the target cells (YAC cells).

3. Results

The relationship between the total glutathione levels and Con A-induced proliferative responses of splenocytes after γ-ray irradiation (0.5 Gy) is shown in Fig. 1. The total glutathione level increased 2 h after irradiation, reaching a maximum at around 4 h. Thereafter, the level reverted to the 0 h value by 12 h post-irradiation. Con A-induced proliferative response of the splenocytes increased between 2 and 6 h post-irradiation. A significantly increased response was observed in the splenocytes obtained at 4 h post-irradiation. The elevation observed at 4 h post-γ-ray irradiation was completely abolished by pre-treatment with BSO, a specific inhibitor of the de novo pathway of glutathione synthesis.

NK activity of the splenocytes after whole-body γ-ray irradiation with 0.5 Gy was assayed as a function of time at the constant effector cells/target cells ratio of 50:1. The activity increased significantly between 4 and 6 h post-irradiation, and returned to the non-

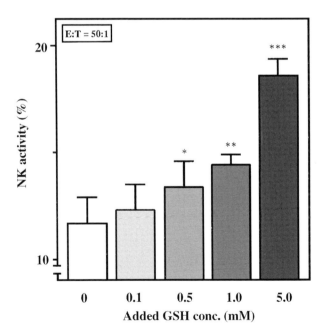

Fig. 2. Effect of exogenously added GSH on NK activity of mouse splenocytes. Splenocytes were obtained from normal ICR mice and incubated with GSH (0–5 mM) for 3 h and subjected to NK assay. Each point indicates the mean ± SD of four independent assays. *, **, *** Significantly different from the 0 mM group at $p<0.05$, $p<0.01$, and $p<0.005$, respectively.

irradiated control value by 24-h post-irradiation. Exogenously added GSH, clearly elevated the NK activity in a dose-dependent manner as well (Fig. 2).

4. Discussion

Glutathione is found at high concentrations in most living organisms, and possesses multifunctional roles as noted in the Introduction. We have recently found that low doses of radiation, unlike higher doses, do not always cause a decrease of cellular GSH level, but can increase the glutathione level. This phenomenon is considered to be an adaptive response. Here, we examined the relationship between the induction of glutathione and immune functions, such as Con A-induced proliferative response and NK activity, of splenocytes in mice after a lower dose of γ-ray irradiation.

The glutathione level transiently increased soon after the irradiation and reached a maximum at around 4 h post-treatment with γ-ray (0.5 Gy). A significantly high splenocyte proliferative response was also recognized at 4 h post-irradiation. The time at which the response reached the maximum coincided well with that of maximum total glutathione levels of the splenocytes in the γ-ray-irradiated mice. NK activity was also enhanced at the same time periods after the irradiation. Addition of glutathione to splenocytes obtained from normal mice enhanced both Con A-induced proliferative response and the NK activity in a dose-dependent manner. These enhancements were both completely blocked by buthionine sulfoximine, a specific inhibitor of the de novo pathway of glutathione synthesis.

These results indicate that the induction of endogenous glutathione synthesis in splenocytes after low-dose γ-ray irradiation is partially responsible for the appearance of enhanced immune function, and may give some help in understanding the mechanisms of carcinostatic effects induced by low doses of ionizing radiation as previously reported by other authors.

References

[1] T.D. Luckey, Radiation Hormesis, CRC Press, Boca Raton, 1991.
[2] S. Kojima, O. Matsuki, I. Kinishita, T. Gonzalet-Valdes, N. Shumura, A. Kubodera, Does small dose γ-ray irradiation induce endogenous antioxidant potential in vivo? Biol. Pharm. Bull. 20 (1997) 601–604.
[3] S. Kojima, O. Matsuki, T. Nomura, A. Kubodera, Y. Honda, S. Honda, H. Tanooka, H. Wakasugi, K. Yamaoka, Induction of mRNAs for glutathione-related proteins in mouse liver by small doses of γ-rays, Biochim. Biophys. Acta 1381 (1998) 312–318.
[4] S. Kojima, K. Teshima, K. Yamaoka, Mechanisms involved in the elevation of glutathione in RAW 264.7 cells exposed to low doses of γ-rays, Anticancer Res. 20 (2000) 1589–1594.
[5] K. Teshima, A. Yamamoto, K. Yamaoka, S. Kojima, Involvement of calcium ion in elevation of mRNA for γ-glutamyl cysteine sythetase (γ-GCS) induced by low-dose γ-rays, Int. J. Radiat. Biol. 76 (2000) 1631–1639.
[6] L. Kraus, M.A. Gougerot-Pocidalo, P. Lacombe, J.J. Pocidalo, Depression of Con A proliferative response of immune cells by in vitro hyperoxic exposure-protective effects of thiol compounds, Int. J. Immunopharmacol. 7 (1985) 753–760.
[7] M. De La Fuente, M. Dolores-Ferrández, M. Del Rio, M. Sol-Burgos, J. Miquel, Enhancement of leukocyte function in aged mice supplemented with the antioxidant thioproline, Mech. Ageing Dev. 104 (1998) 213–225.

Radiation protection effect on Hatakeshimeji (lyophyllum decastes sing)

Yeunhwa Gu [a,*], Yuuichi Ukawa [b], Sangrae Park [a], Ikukatsu Suzuki [a], Kenichi Bamen [a], Toshihiro Iwasa [a]

[a] Department of Radiological Technology, Suzuka University of Medical Science, 1001-1 Kishiokacho, Suzuka City, Mie 510-0293, Japan
[b] Research Center Eishogen Co., LTD. Fukaya, Saitama 366-0815, Japan

1. Background

We studied the effects of Hatakeshimeji (lyophyllum decastes sing) toward the side effect reduction of the radiation in the mice after an X-ray irradiated in the mice and the immunity reinforcement. An experimental group was divided into a medication group as the control group (no X-ray, and only saline is given), Hatakeshimeji (lyophyllum decastes sing) extract medication group, only X-ray irradiation group (X-ray irradiation, only saline medication) and X-ray irradiation plus Hatakeshimeji (lyophyllum decastes sing) extract group. Together in one group, 10 male mice, 10 female mice were used. For 2 weeks, stomach zoning was done every other day, and a sutra mouth gave 250 mg/kg to the sample that dissolved in the saline. The whole body was irradiated. The amount of total irradiation X-ray was 2 Gy. Statistical treatment was conducted according to the t-official approval of Student. As for the number of the leukocytes, it decreased to some degree in the only X-ray irradiation group and the X-ray plus Hatakeshimeji (lyophyllum decastes sing) medication group after the X-ray irradiation 3 h, after 15 days. Moreover, the irradiation ratio was changed for 3 days rest with the mice after 30 days in the X-ray plus Hatakeshimeji (lyophyllum decastes sing) medication group and for the only X-ray irradiation group, and the number of leukocytes showed an increase in comparison with the X-ray irradiation group. The decrease in the number of the leukocytes due to the X-ray irradiation was eased by the medication of Hatakeshimeji (lyophyllum decastes sing). It is thought that there was an effect, which hastened the recovery of the number of the leukocytes. As for the number of lymphocytes, it decreased to some degree in the only X-ray irradiation group and in the X-ray plus Hatakeshimeji (lyophyllum decastes sing) medication group after the X-ray irradiation 3 h, after 15 days. Moreover, a change was

---

* Corresponding author.

found with the mice after 30 days of irradiation in the X-ray plus Hatakeshimeji (lyophyllum decastes sing). In the medication group and the only X-ray irradiation group, the number of lymphocytes was increased in comparison with the X-ray irradiation group.

2. Methods

The experimental group consists of the control group (only a physiology solution of salt is given), the lyophyllum decastes sing medication group, the only X-ray irradiation group and the X-ray irradiation plus lyophyllum decastes sing medication group. The mouse was bred for 1 month with a sutra mouth giving a sample. Blood was taken after 3, 12 h, 1, 3, 7, 15, and 30 days from the tail vein of the mice. It was administrated the day before the sample medication started and the day before X-ray irradiation. The measurement of the number of corpuscles was done. An X-ray irradiation condition irradiates a limited part at the line quantity rate 0.35 Gy/min, and the amount of total X-ray irradiation was 2 Gy. The statistical significance treatment was conducted according to the t-official approval of Student.

3. Results

Experimental studies with mice have established that the cells in the bone marrow are the ones to administer immunity, such as leukocyte, lymphocyte and monocyte, when it is X-ray damage. This means that the cellular number decreases. As leukocyte decreases, the symptoms such as severity of the immunity were also decreased. This time, the effect that controlled a decrease such as a leukocyte due to the X-ray irradiation, a lymphocyte, or a monocyte was admitted by the medication of Hatakeshimeji (lyophyllum decastes sing). This demonstrates with the mice that the effect enhanced the immunity recovery in radiotherapy, and the prevention of a side effect of the radiation. Furthermore, an increase in the effect on immunity function recovery at the time when the immunity power declines caused by the anti-cancer medicine medication and the effect on an anti-tumor by the anti-cancer medicine, prevention of a side effect by the anti-cancer medicine, and so on can be expected.

A continuous change in the number of the leukocytes is shown in Fig. 1. The number of leukocytes was decreased to some degree in the only X-ray irradiation group and the X-ray plus lyophyllum decastes sing medication group after the X-ray irradiation, for 3 h after 15 days. Moreover, the irradiation ratio was changed for 3 days rest with the mice after 30 days in the X-ray plus lyophyllum decastes sing medication group and for the only X-ray irradiation group, and the number of leukocytes was increased in comparison with the X-ray irradiation group. The decrease in the number of leukocytes due to the X-ray irradiation was eased by the medication of lyophyllum decastes sing. It is thought that there was an effect, which hastened the recovery of the number of leukocytes. A continuous change in the number of lymphocytes is shown in Fig. 2. The number of lymphocytes was decreased to some degree in the only X-ray irradiation group and the X-ray plus lyophyllum decastes sing medication group after the X-ray irradiation, for 3 h

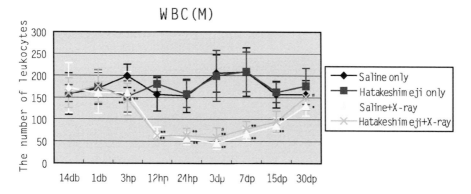

Fig. 1. The change in the number of leukocytes in the blood taken from the tail vein of whole body irradiated mice. Each linegram represents the mean value ± S.D. for five female mice leukocytes.

after 15 days. Moreover, a change was found in the mice after 30 days of irradiation in the X-ray plus lyophyllum decastes sing medication group and for the only X-ray irradiation group, and the number of lymphocytes was increased in comparison with the X-ray irradiation group. When it declined, it was observed in the lyophyllum decastes sing medication group as well. It was after irradiation for 30 days by the medication of lyophyllum decastes sing, that the effect on immunity function recovery promoted by lyophyllum decastes sing resulted in the normal value. A continuous change in the number of monocytes is shown in Fig. 3. The number of monocytes decreased to some degree in the same way as the number of lymphocytes in the only X-ray irradiation group and the X-ray plus lyophyllum decastes sing medication group after the X-ray irradiation, 3 h after 15 days. The irradiation ratio was changed for 7 days rest with the mice after 30 days in the X-ray plus lyophyllum decastes sing medication group and for the only X-ray irradiation group, and the number of monocytes had increased in the female mouse. In the monocytes,

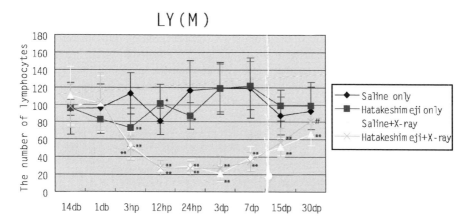

Fig. 2. The change in the number of lymphocytes in the blood taken from the tail vein of whole body irradiated mice. Each linegram represents the mean value ± S.D. for five female mice lymphocytes.

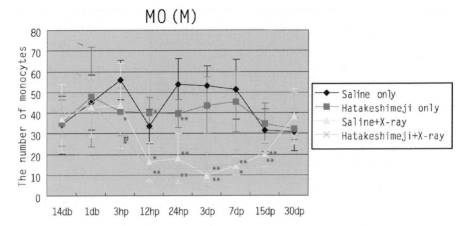

Fig. 3. The change in the number of monocytes in the blood taken from the tail vein of whole body irradiated mice. Each linegram represents the mean value ± S.D. for five female mice monocytes.

we presumed that the activation of the macrophages is further promoted by differentiating macrophages during the X-ray irradiation at the latter half using the medication of lyophyllum decastes sing.

4. Conclusions

Damage occurred to the marrow, which contains the cells to administer immunity, such as a leukocyte, lymphocyte, monocyte, when an X-ray is irradiated. This means that the cellular number decreases [1–3]. This develops into leukocyte decrease syndrome, with symptoms such as reduced immunity in severe cases. This time, the effect that controlled a decrease such as a leukocyte due to the X-ray irradiation, a lymphocyte, or a monocyte was admitted by the medication of lyophyllum decastes sing. It demonstrates with the mice that the effect enhanced the immunity recovery during radiotherapy and the prevention of side effects caused by radiation [4–7]. Furthermore, an increase in the effect on immunity function recovery at the time when the immunity function declines because of the anti-cancer medicine medication, and the effect on an anti-tumor by the anti-cancer medicine, prevention of a side effect by the anti-cancer medicine, and so on can be expected [8].

References

[1] Y. Fujimiya, Y. Suzuki, R. Katakura, T. Ebina, Tumor-specific cytocidal and immunopotentiating effects of relatively low molecular weight products derived from the basidiomycete, *Agaricus blazei* Murill, Anticancer Res. 19 (1999) 113–118.
[2] T. Ebina, Y. Fujimiya, Antitumor effect of a peptide-glucan preparation extracted from *Agaricus blazei* in a double-grafted tumor system in mice, Biotherapy 11 (1998) 259–265.

[3] Y. Fujimiya, Y. Suzuki, K. Oshiman, H. Kobori, K. Moriguchi, H. Nakashima, Y. Matumoto, S. Takahara, T. Ebina, R. Katakura, Selective tumoricidal effect of soluble proteoglucan extracted from the basidiomycete, *Agaricus blazei* Murill, mediated via natural killer cell activation and apoptosis, Cancer Immunol. Immunother. 46 (1998) 147–159.

[4] C. Tokuda, Y. Miyama, H. Takeda, H. Fukui, Antitumor activity of 1-hexylcarbamoyluracil which is carmofur substituted fluorine for hydrogen, Gan To Kagaku Ryoho 27 (2000) 1065–1067.

[5] H. Kamei, Y. Hashimoto, T. Koide, T. Kojima, M. Hasegawa, T. Umeda, Direct tumor growth suppressive effect of melanoidin extracted from immunomodulator—PSK, Cancer Biother. Radiopharm. 12 (1997) 341–344.

[6] D. Cohen-Aloro, O. Merimsky, S. Bar-Yehuda, B. Klein, S. Kayzer, P. Fishman, Mononuclear cells release low molecular weight factors with anti-cancer activity: a lower level of production by cells of cancer patients, Int. J. Oncol. 12 (1998) 921–925.

[7] H. Kamei, T. Koide, T. Kojimam, M. Hasegawa, K. Terabe, I. Umeda, Y. Hashimoto, Flavonoid-mediated tumor growth suppression demonstrated by in vivo study, Cancer Biother. Radiopharm. 11 (1996) 193–196.

[8] J. Beuth, J.M. Schierholz, G. Mayer, Immunomodulating and antimetastatic activity of thymic peptides in BALB/c mice, Anticancer Res. 19 (1999) 2993–2995.

General discussion

General discussion: molecular mechanisms of radiation carcinogenesis and their implications in radiation policy

(Chaired by T. Sugahara and John Little)

Dr. Tsutomu Sugahara

Good morning ladies and gentlemen, we shall now start our final session of this symposium. This session is very important as the title of the symposium indicates; 'Scientific Basis of Risk Assessment and the Mechanism of Radiation Carcinogenesis.' Concerning the mechanism of radiation carcinogenesis, we have had a three-day discussion from various aspects, and in this final session we would like to concentrate on the scientific basis of risk assessment. I hope our senior scientists will give us their own ideas about what we should be dealing with this problem. May I introduce the first speaker, Professor Streffer from Germany. He is a member of ICRP and I don't need much more introduction for him as he is very well known.

Dr. Christian Streffer

'Ohayo gozaimasu'! Thank you very much, Dr. Sugahara, for this introduction and for the invitation. As I have the privilege to be the first foreigner this morning, I would like to thank you all for the excellent party which we had yesterday evening. It was typical Japanese hospitality, always a great hospitality, and there is no other country where you can find this in that way. So thank you very much. Everybody enjoyed it—anybody who did not enjoy it, he doesn't deserve to come here!

Well, I think we are all aware about my first transparency. It shows the classical two types of dose–effect relationships for deterministic radiation effects with a threshold and for stochastic effects without a threshold. For regulation purposes, we are using a linear dose–effect relationship in the low dose range. We all know, as we have heard... during this conference, that we have measurements for health effects in the dose range of 100 mSv and above. I think we have very good data in this dose range, at least for acute radiation. It is certainly different for chronic irradiation, and ICRP as you all know, uses a reduction factor of 2, which has some uncertainty. But below 100 mSv, we certainly can only extrapolate cf. for radiation risk. And there is...a high degree of uncertainty in this low dose range. We have a number of data on cellular and molecular effects—what we have heard during this conference—and I may come back, perhaps, to one or another point. And of course there is a lot of discussion about the existence or non-existence of a threshold dose, also for radiation-induced cancer. When I have to give a talk in Germany, I always have to defend linear dose–effect curves, as there are quite a number of people telling me this is an underestimate of risk. But here, in the scientific community, there are many scientists who claim, assume and accept that this is an overestimation. It is certainly an open question, and I think we have to keep it for the moment as an open question.

What are the arguments? The arguments are: firstly, epidemiological studies do not show increased cancer rates in the...dose range below 100 mSv [1]. And if you are a pure and very solid scientist, you would say on first sight, 'anything I cannot measure is not there.' But certainly there are some indications and measurements, cf. chromosomal aberrations, of effects below 100 mSv with a dose–effect curve without a threshold dose (UNSCEAR, 2000), although we do not know how much impact this has on health effects. The second point is that DNA repair reduces radiation-induced lesions in DNA very efficiently, and there is not doubt that this is certainly the case. However, one has to consider—and I think we all know—that not all DNA damage is repaired even in the low dose range and if we have high LET radiation, like neutrons, we have considerable radiation damage which is not repaired, cf. complex double strand breaks in DNA. Or at least, there is less repair than with low LET radiation...where repair is very rapid, which has been shown with many cell systems cf. with the comet assay [2]. Further, we have heard during the conference, frequently, that there are individuals with a high susceptibility, and in these individuals again, repair is much less than it is in normal persons [3,4]. Again, this has been shown with the comet assay. Here, you see data measured in lymphocytes from an AT patient and a second patient who was treated in regular radiotherapy and who developed severe side effects (Fig. 1). The genetic predisposition has to be considered seriously in this connection and there are some sensitive individuals for whom the genetic background has not be defined as David Scott has explained yesterday. So there is certainly quite a bit of radiation damage which cannot be repaired easily.

The next point which is brought forward is adaptive response. It is generally assumed that repair is induced by low doses of radiation and therefore cells become more resistant [5]. However, everybody who has worked in this field knows that this adaptive response is dependent on many factors. I cannot discuss all these factors here, but you know the adapting dose is important, the interval between the adapting dose and the challenging dose is important, cell proliferation is important, and again, genetic disposition is apparently important. There are individuals where you do not see adaptive response; under the same conditions, you will see adaptive response in other persons. And there are also cell systems where you do not see adaptive response, for instance, during embryological development of the very early stages, where nobody has shown up to now that adaptive response occurs [6]. So I want to say adaptive response is certainly possible, but it is bound to many well-defined conditions. It is a narrow window and it is not a general principle.

The next point is apoptosis. These processes remove damaged cells including neoplastic ones. However, again we have a very wide variety of apoptosis in humans. You see here data from Crompton and co-worker (1999), where the apoptosis was measured in

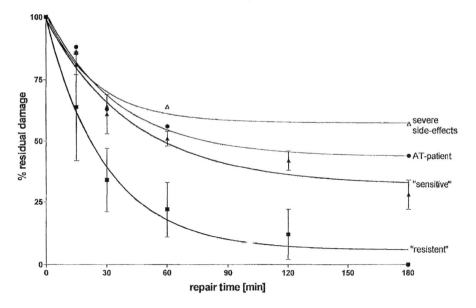

Fig. 1. Repair kinetics in human lymphocytes measured by the comet assay after irradiation with 2 Gy X-rays in vitro. Between "resistant" and "sensitive" one finds the range of normal healthy persons, "severe side effects" represents a patient with severe side effects during radiotherapy for esophagus carcinoma.

lymphocytes of cancer patients. There are some patients who have a higher radiosensitivity with less than normal apoptotic activity. The apoptotic activity in lymphocytes from AT patients again is very low (Crompton et al., 1999). So the apoptotic activity in individuals is very different, and high susceptibility to radiation is apparently accompanied with low apoptotic activity.

I think we all agree that carcinogenesis is a multiple-step process. We have several steps following each other—it is a sequence of events with mutations (UNSCEAR, 2000). We know that it takes a very long time from the first mutation up to the development of a metastatic carcinoma after irradiation and we know comparatively little about the processes between initiation and the clinical appearance of a cancer. We know quite a bit about the first steps (initiation, cell transformation) and we are discussing these processes. And, as it is sometimes brought forward, the probability of cancer would be very small if these several mutations were to take place independently. In this respect, probably genomic instability is very important [7,8] and I want to discuss this phenomenon somewhat, especially on experiments which we have done in our laboratory [9].

I want to stress that genomic instability is not a typical characteristic process after radiation damage. Genomic instability is observed without any irradiation and it occurs already during the first developmental divisions from the one-cell embryo to two cells from 2-cells to 4-cells and so on. You see how chromosomal aberrations increase already during these very early developmental steps (Table 1). In further experiments, we have irradiated the zygote and we let develop the embryo in vivo, the radiation dose was between 100 mGy and 1 or 2 Gy fast neutrons or X-rays. These embryos developed to fetuses and then we took fibroblasts from the fetuses and looked for chromosomal

Table 1
"Spontaneous" chromosomal aberrations (CA) in preimplantation mouse embryos

Development stage (division of blastom.)	CA per 100 metaph.
1-Cell → 2-Cell	2.3
2-Cell → 4-Cell	4.2
4-Cell → 8-Cell	7.7

aberrations, the frequency of chromosomal aberrations increased after irradiation in comparison to the control (Table 2). In this case, the irradiation had taken place many cell generations before—around 30 cell generations—before the determination of the chromosomal aberrations [9]. Therefore, these aberrations can not be directly induced by radiation. The effects developed later and this is due to genomic instability. This genomic instability can even be transmitted to the next mouse generation [10]. The effect is certainly smaller, but there is still some significant radiation effect (Table 3). The animals had been irradiated in the generation before, in the zygote stage. The animals from the irradiated zygotes were grown up and females were mated with unirradiated males and the embryos as well as fetuses resulting form the matings were studied. And it is in the next generation where we also see an increase of genomic instability. Like us, others have shown that genomic instability develops not only in vitro, but also in vivo and this has also been discussed during this meeting here.

So I think today it is quite clear that genomic instability develops also in vivo, and I think that we have some indication that this also happens in men. We have looked into uranium miners, for...the number of micronuclei as well as the number of micronuclei with centromeres. The micronucleated lymphocytes were studied where the centromeres have been stained by FISH technique with DNA probes for centromeres [11] (Streffer, 1998). Thus, we can see whether or not these micronuclei have a centromere and whether the micronucleus originates from an acentric fragment or from a whole chromosome. And if you do this with lymphocytes of uranium miners, we see that in healthy persons there is a very high percentage of micronuclei which have a centromere. This number is always found between 70% and 80%, the average is 76%. In uranium miners without lung carcinomas, it is around 62% and in uranium miners with lung carcinomas it is 55%. These numbers are significantly different from each other, as the individual variability is very low for this phenomenon (Table 4) [11,12]. And I would like to remind you that these are the lymphocytes of individuals who have been exposed to radon and uranium several decades before the measurements. The exposure was in the 40s and 50s and the measurements were done at the end of the 90s or just last year. So this can only be explained by a development of genomic instability, and if you look for the chromosomal aberration, you will see again that uranium miners have a very high increase of

Table 2
Chromosome aberration in fibroblasts from fetuses (19 days, p.c.) of mice after X-ray irradiation of zygots (1 h, p.c.) (number of aberrations/number of metaph. in %)

Mouse strain	Contr.	1 Gy	2 Gy
C 57 BL	22/795 (2.8%)	136/626 (21.7%)	109/400 (27.5%)
HLG	29/400 (7.3%)	48/400 (12.0%)	56/322 (17.4%)

Table 3
Teratogenic analysis. Teratogenic effects in the first and second generation after X-irradiation of HLG/Zte mice with 1 Gy

	Irradiated group		Controls (%)
	First generation (%)	Second generation (%)	
Gastroschisis	15.7	6.5	3.5
Early resorptions	64.8	18	9.8
Late resorptions	2.7	2.4	0
Sterile individuals	–	62	34

chromosomal aberrations where the exposure was a long time ago [11]. So, this is a development of genomic instability in humans in vivo, and therefore, certainly these processes can accelerate the development of carcinogenesis.

In order to have a dose–effect curve without a threshold, it is necessary that only one cell is involved in the initial process. If an interaction between cells has to take place during the early events, then certainly we will have a dose–effect relation with a threshold. Thus, the dose–effect curve after irradiation of the zygote for the development of malformations has no threshold. When the preimplantation embryos were irradiated at the stage of 2-, 4- and 8-cell embryos, the dose–effect curve for malformations has a threshold, however [13]. So this is a typical example where it can be shown for the same effect (malformation) dose–effect curves with a threshold in case of a multi-cellular system and without a threshold in the case of a unicellular system.

I have only discussed the part of the dose effect in the lower dose range below 100 mSv, but we still cannot rule out a dose effect without a threshold—and I don't think we should rule out this in the present situation. But certainly, there may also be the possibility for a threshold for carcinogenesis. It still remains an open question. For reasons of precaution in radioprotection it is, however, reasonable to use the concept of "no threshold dose" for carcinogenic risk.

As for molecular effects, we have heard during this meeting that there are effects of in the dose range of 1 cGy. This was very interesting, but this did not imply that the effect is connected to health problems. At least, I did not hear anybody talking about health effects. The researchers were talking about gene expression, they were talking about certain cellular changes but nobody talked about health effects. So I think the question is still open but it becomes narrower before we get more data in the low dose effect. And if I may use

Table 4
Number of micronuclei per binucleated cell (MN/BNC) (A) and MN with centromers (B) in the lymphocytes of 5 healthy persons (H.P.), 14 healthy uranium miners (U.M.) and 14 uranium miners with bronchial cancer (U.M.C.)

		H.P.	U.M.	U.M.C.
(A) MN/BNC	Average value	18.8	21.4	42.4
	S.D.	3.9	10.2	39.3
	P (t-test)	–	n.s.	n.s.
(B) MN with centromers	Average value	76.4	62.1	54.9
	S.D.	3.1	3.5	3.1
	P (t-test)	–	<0.0001	<0.0001

my last transparency, I would like to show a picture which was used yesterday evening by Dudley Goodhead in another way. A very famous German philosopher, Karl Jaspers, said, 'Science is always moving, it is always on the way. It is always in front of the gate. It enters through a gate, but when it has passed the gate there is another gate.' And I think this is very nicely symbolized by these Japanese kanji; science—kagaku; gakumon. In Tokyo University, there is a big gate where you have to go through, I think this is again a symbol of science in this country. You can open—akeru—a gate. And then, you see through the gate some sunshine—aida—in between, you see some sunshine—hi—and then there is another gate, but this is closed. And I think we open steadily a gate, yes, we open steadily a gate and we come to another gate, and with fascination open this gate again. Will we open the last gate and when? So thank you very much.

Chinese character	学問	門	開	間	日	閉
Pronounciation	gakumon	mon	akeru	aida	hi	tojiru
Meaning	science	gate	open	between	sunshine	close

Dr. John B. Little

It was a very nice introduction to the session. I would next like to introduce Dr. Dudley Goodhead from the MRC unit in Harwell. Dudley, like I think all of our speakers today, also requires no introduction.

Dr. Dudley T. Goodhead

I should like first to state my deepest thanks to the organizers for inviting me here and giving me the opportunity to enjoy the symposium.

This diagram (Fig. 1) illustrates the basic problem under consideration. Epidemiological data show very clearly that there is excess cancer risk at some doses of radiation. Such data are generally at quite high doses—even 0.1 Sv is high in terms of most human exposures. What we really need to know is what is happening down at the very low doses that are of practical relevance in most of radiation protection. Different people have quite strong

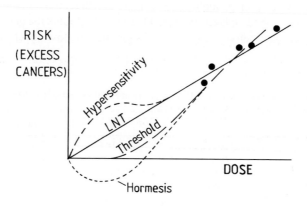

Fig. 1.

arguments for drawing these lines of extrapolation to low doses in these different ways (Fig. 1). There are tangible arguments to support each of these types of extrapolation. The most used extrapolation is the one recommended by ICRP, essentially linear with no threshold, and usually with a dose rate reduction factor also included. We know that, following irradiation there can be a raised risk of cancer, but there is a great 'black box' of uncertainty about the mechanisms that govern this outcome. None of us have adequate knowledge as to the detailed mechanisms inside the black box. The epidemiologists jump over the box by empirically modelling the observed cancer data, but then they have to make extrapolations as to what the risks might be at low doses; these are implicitly dependent on guesses as to what is happening in the box. Others apply animal models, or study cellular systems such as chromosome aberrations, to add to understanding of some of the processes in the box, and attempts are in progress to built quantitative mechanistic models to incorporate some of what are believed to be the relevant processes. There is no doubt that within the box there is a complex chain of interlinked events and processes, but for risk estimation simplifying assumptions and choices must be made, sometimes with weak foundations.

What we do know, of course, is that 'tracks' of ionizations from charged particles start off the process at the time of irradiation and somehow thereby enhance the probability of cancer arising. Additionally, we know that the details of the tracks themselves are important, because if we change the nature of the tracks, with different qualities of radiation, or we change the time course of delivery, then that changes the probability of the outcome. Three key questions in respect of extrapolation for practical application are from high dose to low dose, from high to low dose rate and to other qualities of radiation. The nature and capabilities of tracks are central to these questions, because the radiation tracks start and strongly influence the process. These tracks are highly structured at the microscopic levels of atoms, molecules and cells. The patterns of ionizations along the paths of the charged particles are highly stochastic; they contain isolated ionizations and a great diversity of clusters of ionizations, in varying proportions depending on the type of radiation. What we need for practical risk estimation, is to extrapolate from the cancer effects following hundreds or thousands of electron tracks through each cell in the body, such as for the A-bomb survivors, to a single track through a cell. That is what is typically received per cell per year from natural background radiation—a single isolated electron track through each cell in the body on average. Even in high background areas such as some parts of India or Iran, there might only be on average one track per cell per week or every few weeks. So the tracks are mostly well isolated in space and time and we need, therefore, to understand what can result biologically from a single track. The extrapolation needs to be from situations where all cells in the tissue have each received hundreds of tracks (the A-bomb situation). This is surely an enormously long extrapolation to undertake without clear mechanistic guidance.

The standard paradigm that implicitly underlies the ICRP thinking, is that cancer results from a multistage process, developing from a single cell via multiple changes, and that radiation can cause one (usually) or more or these changes by inducing a chromosome aberration (or some other mutation) as an immediate consequence of the irradiation. Hence, it is implicitly assumed that the dose response for cancer induction largely follows the dose response for chromosome aberrations or mutations, with linear and dose-squared components. Whether or not these assumptions are clearly acknowledged, they strongly underlie the thinking of ICRP, UNSCEAR and many other committees that have made extrapolations of risk from cancer data for radiation. Fig. 2 illustrates these general concepts for low-LET radiation. The linear component is conventionally ascribed to single tracks and therefore assumed to be independent of dose-rate, while the curvature is put down to multiple-track effects. The high-LET radiation response is dominated by the high effectiveness of the single tracks. The concepts illustrated in Fig. 2 are strongly based on information for induction of chromosome aberrations, apart from some apparent saturation effects at the highest doses. If one scans quickly through published official reports, one sees similar diagrams in, for example, ICRP Report 60, NCRP Report 104, UNSCEAR (1993) and NRPB (1997). The latter report was to estimate the relative biological effectiveness of neutrons for cancers, to guide the choice of weighting factors to use under specific circumstances instead of the radiation weighting factor, or quality factor: to do this the authors essentially went straight back to data on lymphocyte chromosome aberrations and assumed that these set the rules by which they should analyse all available data to obtain relevant weighting factors that would be appropriate for cancer risk at low doses. In such ways, chromosome aberrations have been given great importance in the concepts that underlie extrapolations of cancer risk.

If this chromosome aberration approach to risk is accepted, then a key question becomes whether or not one particle, a single electron track, can cause an aberration, or other type of mutation. If one electron can indeed do so, then it alone can contribute to the cancer process and there should be no threshold, consistent with ICRP thinking. Conversely, if one electron cannot create an aberration, that is, if some combination of electrons are required, then there would be a dose threshold below which there is no risk. So, let us consider the electron tracks in more detail. As we irradiate cells in the body with gamma rays, X rays or other low-LET radiations, electron tracks pass through the cells and these tracks produce sparse ionizations and also clusters of ionizations that are due mostly

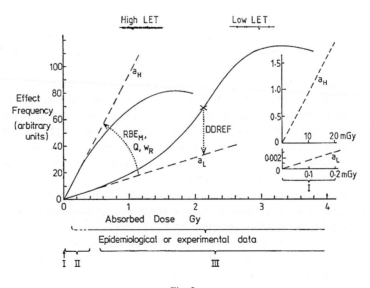

Fig. 2.

to the low-energy secondary electrons. A single track can produce a diversity of different patterns of ionizations. We know, of course, that a cluster of ionizations can produce a double-strand break in the DNA. For example, it may produce one or more direct ionizations in the DNA, each of which is capable of breaking a strand, and there may also be one or more OH radicals nearby from ionizations in the water and any of these can by chance diffuse to the DNA and cause breakage. Thus, there can be a double-strand break from various combinations of direct and indirect ionizations. Because this can result from a single electron track, it can occur at any dose, however small. If we look in detail at a section of an electron track on the scale of DNA, with a diameter of 2 nm, we see that the patterns of ionizations are quite likely to show clusters over dimensions of a few nanometrres, especially from the low-energy secondary electrons. It is very clear from detailed theoretical analyses of electron tracks, that within low-energy electron tracks there is much clustering of ionizations. If this occurs in or very near to DNA, a double-strand break is quite likely to result. Additionally, however, an ionization cluster is quite likely to produce also additional breaks in the immediate vicinity. In such a case, the result may well be not just a simple double-strand break, but three, four or even more breaks and possibly also a few base damages. This would be a rather severe complex clustered damage in the DNA, all as a result of a single electron track. It is not difficult to imagine that when the cell tries to repair clustered DNA damage, this is likely to result in some loss or distortion of coding information or, potentially, a visible chromosome rearrangement. So even a single track is able to produce clustered damage that leads to an aberration or mutation. Similar considerations of the track structure of high LET radiations, show that a single track, say of an alpha particle, can very readily produce clustered damage if it passes through or very close to DNA; sometimes these are even more severe than can be produced by an electron.

Detailed track structure analyses show that clustered DNA damage is a very special feature of all types of ionizing radiation. This makes radiation very different from the natural oxidative processes that occur in living cells. Therefore, we need to be very careful in making comparisons between the expected repair of normal oxidative damage, mostly single-base damage, and the repair of ionizing-radiation damage. Certainly, ionizing radiation will produce some simple damage, similar to oxidative damage, but radiation will also produce a high proportion of these clustered damages for which the repair may be much more problematic. A number of times during this symposium, people appeared to make the erroneous assumption that oxidative damage in general is the same as radiation damage. The clustering of ionizations within a track sets radiation apart from most other DNA-damaging agents, and this occurs even at the very lowest doses. This can produce a DNA double-strand break, and more complex clustered damage, all from a single particle, even a single electron in low-LET radiation. We can say quite convincingly from present theoretical and experimental evidence that even a single electron can produce a mutation or a complete chromosome aberration. From detailed theoretical Monte Carlo analyses of large numbers of radiation tracks and their likely chemical reactions with DNA, we can make meaningful quantitative estimates of the damage spectrum for different types of radiation. The DNA damage can then be classified into single-strand breaks, simple double-strand breaks and complex double-strand breaks with one or more additional breaks. Such analyses have shown that about 20–30% of the double-strand breaks from low-LET radiation have additional breaks associated with them, making complex lesions of three, four or more breaks. If we include base damage, then this proportion of complex double-strand breaks rises to about 50%. Of course, there is much more complex damage from high-LET alpha particles, including up to more than 90% of the double-strand breaks.

Experiments have shown that the low-energy-electron tracks efficiently produce chromosome aberrations and that the yield of simple exchange aberrations increases with a strongly linear dose dependence, implying that a single track produces an aberration. In summary, a single electron can produce clustered damage in DNA, can produce an aberration or a mutation, and therefore is assumed to contribute to cancer initiation, or another of the mutation steps in multistage carcinogenesis. On this basis one would expect no threshold for radiation-induced cancer risk because a single electron has the capability, albeit with small probability, of contributing to a cancer development. Every electron carries a risk, although it is very small. This is the implicit thinking behind much of the ICRP type of approach, following the standard paradigm of radiation carcinogenesis.

It is known, however, that radiation can act also in a wide variety of other ways. We have seen and heard many of these described during this week. They include effects by signalling pathways, activation of genes, adaptive responses, effects on bystander cells and induced genomic instability. These all question one or another aspect of the standard paradigm of radiation carcinogenesis. The assumption of no absolute threshold should remain secure if a single track can produce a mutation which contributes to cancer, but the assumed linearity at low and moderate doses would be very much questioned by all of these additional processes that take place. It was very clear from many of the talks at the meeting that cells do not act solely as isolated individuals. They act as a community in a tissue microenvironment; the experience of one cell can influence the state and the response of its neighbours (Fig. 3). If multiple-cell effects are considered able to influence the probabilities of carcinogenic outcomes, then expectations for linearity of dose response become more problematic. The exposed groups of A-bomb survivors had hundreds or

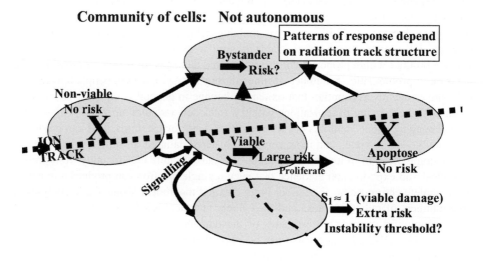

Fig. 3.

thousands of tracks through each cell in the tissue, but we are now trying to extrapolate to the situation where a single isolated electron passes through a cell and only a very small proportion of other cells in the tissue receive an electron track even in the same week. How can we extrapolate reliably between these vastly different situations? The assumption of linearity may be questioned by tissue factors.

I suggest, however, that a much bigger question is the one that Christian Streffer raised, and that is whether the conventional paradigm of radiation carcinogenesis is itself secure. Consider the multistage process from a normal cell to a tumour cell, via many mutational changes (Fig. 4). The conventional paradigm is that radiation acts directly on one of these stages by inducing a mutation. That is, it acts like a simple, instant mutagen, and therefore the above single-track expectations apply. But if radiation induces instability in the cell and its progeny, then it can enhance the probability of, not just one, but many of these multiple mutations. And so we have quite a different process, that not only increases the probability of one mutational step, instantaneously, but instead permanently changes the state of the cell so that for evermore the mutation rate is raised. This then brings us to the very important question: can radiation-induced instability be a cause of radiation cancer? If so, is this the major route to radiation-induced cancer? Now, it seems to me that throughout much of this meeting, this was assumed to be the case. To my knowledge, however, this crucial role for induced instability in radiation carcinogenesis has not been formally proven; I recognise that some leading ICRP people that I talk to are extremely resistant to accepting the concept that radiation-induced instability might contribute to cancer at all. Can it really contribute, and if it does, is this the main mechanism for radiation cancer? I believe that these are really the key questions. Then, if instability really does act in this way, can a *single* electron induce instability. At present we seem quite far away from being able to argue convincingly, one way or the other that a single electron has this capability. If a single electron cannot induce instability, then there should be a threshold dose for low-LET radiation below which there is no such risk. If the cancers in the A-bomb survivors arose through induced instability (from many hundreds of electrons per cell in a very short time interval) and a single electron does not have this capability, then there could be a threshold for low-LET radiation and application of a linear extrapolation from higher doses would be misleading.

By contrast, there is already much data to show that there is not a dose threshold for high-LET alpha-particle irradiation to induce instability (or almost any other cellular

TWO PARADIGMS FOR RADIATION TUMOURIGENESIS:

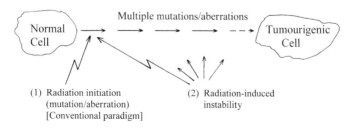

Fig. 4.

effect). Why might there such differences with radiation quality? Is it because of the spatial properties of the high-LET tracks, giving more severe clustered damage? Is it because of the large total amount of damage to the cell in an instant of time as the high-LET track passes through? Is it this temporal property of the track, or the microscopic spatial properties? Or is it simply that the traversed cell always receives a high dose from the single track? We do not yet know the answers to these questions for induced instability because the mechanism is not yet understood.

The instability paradigm offers a fundamentally different mechanism to radiation carcinogenesis and this could very seriously question the applicability of the conventional paradigm. There could well be a threshold at low doses of low-LET radiation. Or it could be that both mechanisms contribute to cancer. At present, it seems that there is insufficient information to be able to say definitively that radiation-induced instability does lead to cancer. But, conversely, there is also insufficient information to say definitively that a single instant mutation or aberration is the main contributor to radiation-induced cancer. So neither the instability nor the conventional paradigm is proven.

The practical implications of the instability paradigm, if true, are also not yet clear, so we can only speculate. I shall end with two brief speculations; I emphasize that these are speculations only, because there are not the data to verify them and we do not have sufficient knowledge of the biological mechanisms to allow us to make reliable predictions from theoretical analyses of the radiation tracks and the relevant molecular damage. Let us assume, however, that radiation-induced instability is the major route to radiation carcinogenesis, and that a single electron is unable to induce instability in a cell. Then, of course, there would be a dose threshold of no risk: the probability of a cell being induced with instability at low doses would be zero at low doses, and would rise with increasing dose (number of tracks) until a plateau is reached when the cell has been fully induced. The dose response per cell might be as sketched in Fig. 5, with no risk at low doses. In this case, linear extrapolation from higher doses would overestimate the risk at low doses. Conversely, consider the case where a single electron through a cell *can* induce instability, with a probability that increases with the number of electrons until a plateau is again reached. Then extrapolating back linearly from high doses (many hundreds of electrons, on the plateau) would *under*estimate the risk at low doses because a single electron is proportionately more effective than a hundreds of electrons.

By contrast, for high-LET radiation, we know that a single alpha particle can induce instability, so however low the overall dose to the tissue, an individual cell within it that is struck by an alpha particle can become induced to instability (Fig. 6). Of course, the number of cells that are struck increases with dose, so the overall risk to the tissue will increase with dose in a more or less linear way until all the relevant cells have been induced by direct traversals or by bystander effects. If there is a threshold for low-LET radiation, then in this case the RBE for high-LET radiation would tend to infinity at low doses, because the low-LET risk is zero. This might very seriously compromise the attempts to apply simple radiation weighting factors (or quality factors) to assess low-dose risks from different radiation types.

I end with the suggestion that the role of instability in radiation cancer is the most crucial open question underlying how we should best estimate radiation risk at low doses. Can a single electron, or can even a few electrons, induce instability? We know

IMPLICATIONS FOR RAD PROT? SPECULATION

<u>Assume</u> induced instability is a major mechanism in radiation carcinogenesis

<u>For Low-LET radiation</u>:

<u>Possible</u> threshold <u>if</u> level of simultaneous damage is needed to trigger instability

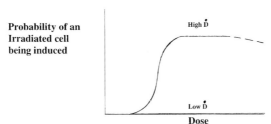

Then: Risk $_{D, \dot{D} \to 0} \to 0$

DDREF $\to \infty$

Fig. 5.

that a single high-LET track can. But most crucially, is induced instability the main route to radiation-induced cancer, or is it a significant contributory route, or is it not a route at all?

IMPLICATIONS FOR RAD PROT? SPECULATION

<u>Assume</u> induced instability is a major mechanism in radiation carcinogenesis

<u>For High-LET radiation</u>:

<u>No</u> threshold − single track is sufficient

Then: RBE $_{D, \dot{D} \to 0} \to \infty$

Fig. 6.

Dr. Hiromichi Matsudaira

RERF has reported (1996) the excess cancer mortality for period 1950–1990 among 50, 113 survivors exposed to above 5 mSv with average dose of about 250 mSv. Total deaths were 173 for leukemia and 4565 for solid cancer, respectively. Excess leukemia deaths were 78/173 (45%) and those of solid cancer, 376/4565 (8%). So, spontaneous leukemia deaths become 95/173 (55%) and those of solid cancer, 4189/4565 (92%). This implies that even in A Bomb survivors with known doses, the death from spontaneous malignancy is more important than that induced by radiation. Another point is to what level should we reduce the estimated excess death in radiation protection by restricting the exposure. Answer may be different even among experts.

We, ordinary Japanese, are exposed to about 4 mSv/year, including 1 mSv of natural background, 0.5 mSv of indoor radon, and 2–3 mSv of medical X-rays. So, during the total lifetime of 75 (male) and 80 years (female), we receive about 300 mSv, the amount of radiation which is comparable to the average of A-bomb survivors mentioned above. Yet, we live out our natural life.

Low dose and low dose-rate data are presented and discussed by many authors in this Symposium. As a Japanese participant, I would like to refer, in particular, to the results obtained in Japanese institutions. Pierce and Preston in RERF (*Radiat. Res.* **154** (2000) 178–186) have shown a significant cancer risk (incidence) below a dose of 100 mSv in A-bomb survivors. Similarly, Sasaki (NIRS) has shown a significant increase in endocrine tumors at a dose of 100 mGy given in 7-day-old SPF B6C3F1 female mice.

It has been claimed that low dose or low dose-rate radiation may lengthen the life span of humans or of mice. But recent data of Cologne and Preston (*Lancet* **356** (2000) 303–307) do not show any evidence of life-lengthening in A-bomb survivors. Similarly, the data of Sasaki suggest a significant decrease in the mean life span of B6C3F1 female 7-day-old mice irradiated with 100 mGy of gamma rays.

Sato, Otsu and their group at IES in Aomori have been studying the effects of low-dose rate gamma rays in SPF B6C3F1 mice of both sexes. Irradiation was initiated at an age of 8 weeks and continued for 400 days, with dose rates of 20, 1 and 0.05 mGy/day. A clear evidence of life shortening was obtained following irradiation of 20 mGy/day with a total dose of 8 Gy for both sexes, due possibly to the occurrence of malignant tumors. Irradiation of 1 mGy/day with a total dose of 400 mGy may shorten slightly the life span of female mice, although the exact analysis is not yet complete at the present time.

A point will be raised for dose and dose-rate effectiveness factor, DDREF. Usually values for DDREF are obtained, particularly for radiation protection purposes from human data, by the analysis of linear-quadratic dose–response relationship, taking so-called α/β ratio as a measure. A large value of DDREF is not expected for a system which shows apparently linear dose–response relationship for the acute effects, such as solid cancer data for A-bomb survivors. However, Tobari et al. (*Mutat. Res.* **201** (1988) 81–87) have found a value of 10 for translocation induction in spermatogonia of crab-eating monkey (*Macaca fascicularis*) by comparing the effects of acute (0.25 Gy/min) and chronic (1.08 mGy/h) gamma rays. This point may need a careful examination in future.

In our association, we have been examining since 1990 the causes of death for about 176,000 male workers in nuclear industries with known radiation doses. In one study, about 120,000 workers who were alive at the beginning of 1995 were followed prospectively for 4.5 years. Their average age was about 51 years and average cumulative dose, 15 mSv. The rates of various causes of death were compared with those of age-adjusted Japanese male (external comparison) and among those of separate dose groups (internal comparison), with or without taking the latency of 2 years for leukemia and 10 years for solid cancers.

No healthy worker's effects were found for cancer death. A positive relationship with dose was found for all cancers and some digestive tract cancers, particularly for esophageal cancer, on trend test. In a separate study, the effects of confounding factors were examined in about 50,000 workers who responded to the questionnaires (more than 80%). Positive trend with dose was found for heavy drinking, etc., and negative trend, for mass examination of upper digestive tract. These results indicate the importance of a close examination of life style factors for epidemiological studies of radiation workers whose exposure tends to become smaller in recent years.

Radiation protection policy should be hopefully based on the knowledge of low and low-dose rate radiation. Since the protection involves various types and qualities of exposure for various peoples with different ages and radiosensitivity, and no definite answer is not yet available from molecular studies as to the shape of dose–response relationship for radiogenic cancer induction, the adoption of linear non-threshold hypothesis is warranted as a precautional measure.

ICRP is discussing the introduction of new concept into the recommendations to be issued around 2003–2005, based on the importance of the control of individual dose, and taking the multiples of natural background radiation as basis of action levels for workers, members of the public and medical exposure.

The concept of collective dose for indiscriminately low doses and long times in future may be discarded.

Personally, I feel, in addition to the natural and medical exposure which ranges 3–4 mSv/year in average, a limit or action level of 100 mSv in 5 years for workers and 1/10 of that for members of the public may be appropriate for radiation protection, although radiations at these levels may not show detectable cancer risks.

Thank you very much.

Dr. John B. Little

Thank you very much, Dr. Matsudaira, for giving us a very complementary point of view to the previous two speakers. I next have the pleasure to introduce Lowell Ralston from the Environmental Protection Agency in the USA, from Washington. Lowell...

Dr. Lowell G. Ralston

Thank you Mr. Chairman, and good morning everyone.

Someone once said that when you get to this point in a meeting, you've probably heard it all...but, unfortunately, you haven't heard it from everyone. With this in mind, I would like to share with you a slightly different viewpoint, a regulator's perspective, on how mechanistic approaches to the study of cancer may affect radiation protection practices and policies in the future. I will begin with an overview of EPA's current risk assessment approach and then provide examples of a few intriguing studies that may ultimately alter our risk estimates. Finally, I will comment briefly on the implications of these new studies with respect to future risk assessment and protection approaches.

As it has done for the past three decades, EPA continues to be one of a handful of federal regulators in the United States charged with radiation protection. Through Executive Order and Congressional legislation, EPA has the authority and responsibility to protect human health and the environment from uncontrolled releases of radioactivity and unnecessary exposures to ionizing radiation. Under this authority, EPA conducts its mission by providing guidelines, standards, and policies for a wide range of environmental and occupational exposures. The Agency also develops methods for estimating radiation doses and lifetime excess cancer risks.

Radiation risk assessment is central to EPA's programs. As Fig. 1 illustrates, this is a very complex and uncertain process, involving more than just dose–response relationships. In fact, risk assessment consists of many steps, including the characterization of radioactive sources, the transport of radionuclides through the environment, and the identification of internal and external exposure pathways for chronic and acute scenarios. These assessments often involve multiple radionuclides and, sometimes, hazardous chemical contaminants as well. Models are commonly used to estimate doses and risks to individuals from radioactive sources or, when run in reverse, to calculate activity concentrations for specific radio-nuclides in various environmental media corresponding to target risk limits.

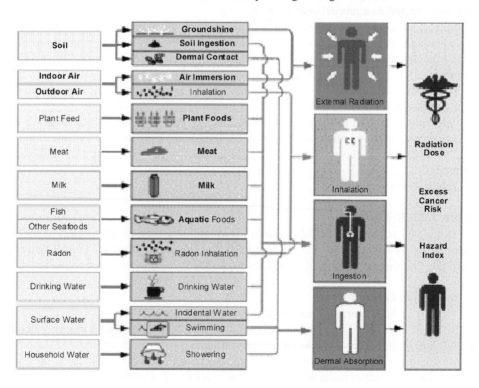

Fig. 1. Exposure pathways in radiation risk assessment. (Source: USDOE at http://web.ead.anl.gov/resrad/baseline.html.)

To assist risk assessors, EPA develops radionuclide-specific cancer risk conversion factors, or risk coefficients, for ingestion, inhalation, and external exposures. In developing risk coefficients for internal exposures, we consider a number of factors, as illustrated in Fig. 2. For example, we assume constant concentrations of radionuclides in environmental media and adjust for age- and gender-dependent ingestion and inhalation rates. To calculate temporal radioactivity concentrations in specific tissues after uptake, we use element-specific biokinetic models that account for the distribution, retention, and excretion of each nuclide in the body over time. We couple these tissue-specific activity concentrations with the radiological properties of each nuclide to arrive at estimates of tissue-specific absorbed dose rates. Finally, we combine the dose estimates with tissue-, age-, and gender-specific risk per unit dose estimates to obtain estimates of the total risk to those tissues, and to the total body.

Fig. 3 shows the age- and gender-dependent usage rate curves EPA employs to adjust its risk estimates. For internal exposures, we use ICRP's current physiologically based biokinetic and dosimetric models, correcting for age-dependent absorption of activity from the gut to blood, age-dependent organ masses, and specific absorbed fractions. Of course, if you don't live long enough, you can't contract or die from cancer, so we also correct for all competing causes of death using a life-table approach. To do this, we use recent estimates for total mortality rates for the U.S. population, as well as for cancer mortality rates for the same population for the same period of time.

For each cancer site, EPA applies age- and gender-specific cancer risk models. As shown in Fig. 4, below 20 cGy, we extrapolate linearly without a threshold. For chronic low dose, low-LET radiation exposures, we decrease our risk estimates using a dose and dose rate effectiveness factor (DDREF) of 2 for all sites, except breast. For high-LET exposures, we increase risk estimates using a relative biological effectiveness factor (RBE) of 20 for all sites, except breast and leukemia. We implement our models using a computer code called DCAL (*Dose Cal*culation) developed for us by Oak Ridge National Laboratory. With it, we have tabulated risk coefficients for about 800 radionuclides in our Federal Guidance Report No. 13, available electronically at our web site at www.epa.gov/radiation.

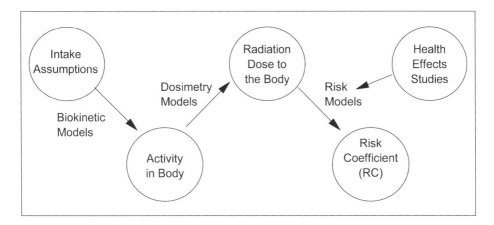

Fig. 2. Components of EPA radionuclide risk coefficients.

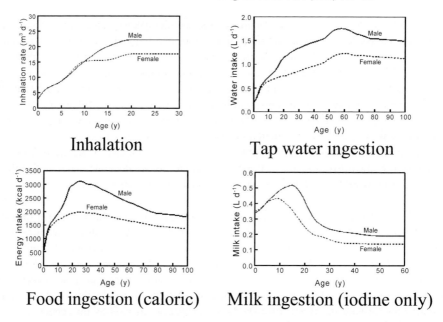

Fig. 3. Age- and gender-dependent usage rates.

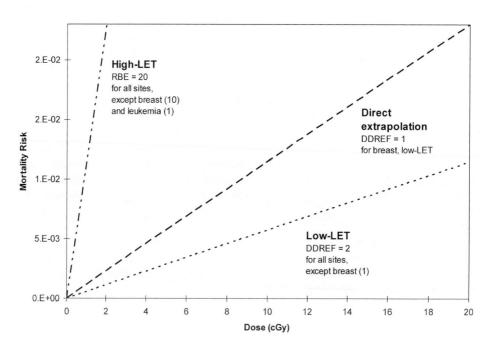

Fig. 4. Adjustments to low- and high-LET risk estimates.

EPA is aware that there are large uncertainties attached to its risk estimates. To address this issue, we have begun to quantify the uncertainties for many of our model inputs, some of which I've already talked about. Of these, it seems to be the shape of the dose–response curve below 10 cGy that has captured most people's attention. This is understandable since most occupational and environmental exposures occur in this region, and 10 cGy might be thought of as a nominal detection limit for our current epidemiological data and approaches. So the question becomes: How are we going to reduce the uncertainties in our model and improve our confidence in our risk estimates?

To quote a famous Western philosopher, "The secret to finding something is knowing where it is" [Tigger, Winnie the Pooh]. As proof of this principle, consider the example of the treatment of chronic myeloid leukemia (CML) using a new drug called Gleevec [1]. As many of you know, CML is the result of a reciprocal translocation between chromosomes 9 and 22, resulting in what's called the Philadelphia chromosome. This abnormal chromosome encodes for a fusion protein, bcr–abl, that conveys a proliferative growth advantage to immature white blood cells. By understanding the mechanisms involved, researchers developed Gleevec, a kinase inhibitor, to block activation of this protein by phosphorylation. As a result, the investigators were able to essentially shut down the unregulated growth of white cells, and thereby provide almost complete remission with little or no side effects in the majority of patients taking the drug. And so, in this example, the secret to finding a cure for CML was, in a real sense, knowing where it was—that is, in understanding the mechanisms involved. With this example in mind, we can ask: Can mechanistic approaches to the study of radiation carcinogenesis improve risk assessment?

EPA believes that traditional approaches to cancer risk assessment alone cannot provide complete answers to all radiation protection questions. Therefore, we now look to new mechanistic approaches made possible by recent advances in biotechnology. These approaches make sense because we know that cancer involves alterations in cellular DNA and changes in gene expression in response to endogenous and exogenous damage. Among the recent advances, the Human Genome Project and derivative programs have provided us with sequence information, the positions of certain genes along the chromosomes, the identities of several tumor suppressor genes and oncogenes, and protein expression data. With this information, along with our current understanding of the cell cycle, of signal transduction pathways, and of DNA damage and repair pathways, we now have maps and markers to help guide us through the uncharted territory ahead. And, as you've heard from many of the speakers during this symposium, we have powerful new tools, including FISH (fluorescence in situ hybridization) for cytogenetic studies and DNA microarrays for gene expression studies. These tools allow us to ask and answer questions previously unanswerable, and offer improvements in the sensitivity and specificity of measurements, thereby improving signal-to-noise ratios for the detection of early events in carcinogenesis. In addition, we have a core group of dedicated, talented, and hard-working scientists—all of you in the audience—who are working together in creative new ways to solve these pressing problems. Moreover, in the United States, where there is renewed interest in nuclear power and nuclear waste issues, there is funding available for research. So the take home message is: It is a very good time to be doing this kind of work. But what do we hope to learn from mechanistic approaches to the study of radiation carcinogenesis?

Obviously, we would like to learn everything we can, seeking answers along the way to some of the important questions for radiation protection purposes. For example, why do some individuals develop cancers and others don't when given seemingly equal doses of radiation? Why do different tissues and different cells respond differentially? Are there limits to radiation bioeffects, and if so, where are those limits? Can we use alterations in DNA and changes in gene expression as biomarkers or fingerprints for ionizing radiation? Can we apply our understanding of these alterations and their mechanisms for early detection and treatment of some kinds of cancer, as I alluded to in the example for Gleevec? Certainly, there are many other questions we would like to answer, but what have we learned so far?

We know from your collective research, both theoretically and experimentally, especially from the recent microbeam studies, that, as Dr. Goodhead said, a single track of ionizing radiation of both types can cause a wide spectrum of damage in cellular DNA, from simple base damage and single strand breaks to double strand breaks and more complex damage. Studies have shown that these lesions occur both in targeted cells, through nuclear and cytoplasmic irradiation, and in non-targeted, bystander cells. In evaluating the biological significance of these damages, we now know that clustered damages are very important and possibly unique to ionizing radiation; these are particularly complex lesions that are difficult to repair, especially by the process of non-homologous end joining. We also know that in those cells surviving this kind of damage, incorrect repair can lead to permanent aberrations and mutations. These types of alterations have been shown to increase linearly without a threshold, at least down to few cGy, and are biologically relevant with respect to carcinogenesis in people.

Understandably, not everyone agrees with EPA's low dose radiation risk estimates. Some believe that the risk estimates should be lower than what EPA projects them to be, while others believe they should be higher. We have heard from previous speakers that there may be new evidence to support both arguments. For example, bystander and inverse dose rate effects may argue for increased risks, whereas threshold models and adaptive responses may suggest lower risks or no effects at low doses. Increases or decreases in DDREF and RBE would cause corresponding increases or decreases in risk estimates. So it is extremely important that we learn as much as we can about these phenomena and the magnitude of effects they convey. I am going to touch on each of these briefly, since a number of speakers have already covered them in detail.

With respect to radiation-induced bystander effects, I would like to thank Dr. Little and Dr. Hei for doing such a wonderful job summarizing their respective research concerning this phenomenon; both are recognized pioneers in their fields. As they point out, several investigators have looked at bystander effects for a variety of end points after low- and high-LET irradiation. Brenner, Little and Sachs [2] recently proposed a quantitative model for the bystander phenomenon based primarily on the single-cell/single-particle microbeam experiments of Sawant and coworkers [3], as well as on studies of bystander effects by other investigators. As illustrated by the curve labeled "Total" in Fig. 5, their model predicts that alpha-particle-induced in vitro oncogenic transformation is non-linear below about 40 cGy, due to a combination of "direct" and indirect, "bystander" effects. According to the authors, "direct" effects arise when the nucleus of every cell in the population is irradiated with exactly one or more than one alpha particle, taking the form

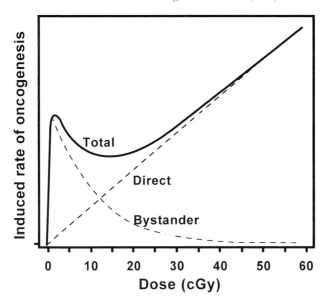

Fig. 5. Predicted contribution of bystander effects to radiation oncogenesis. (Source: D.J. Brenner, J.B. Little, R.K. Sachs, *Radiat. Res.* **155** (2001) 402–408.)

of a linear dose–response curve. "Bystander" effects are evident when a small fraction of the cells in the population are irradiated with exactly one or more than alpha particle. In fact, the model predicts that bystander effects act in a binary (all or none) fashion, predominating at the lowest doses, i.e., one or two alpha particles, and saturating as more and more cells are hit. The authors concluded that, "if [their] postulated mechanisms are applicable in vivo, the consequences for low-dose risk estimation might be major."

Dr. Kiefer did an excellent job of summarizing his work on inverse radiation dose rate effects. Fig. 6 reproduces the results of a study by Vilenchik and Knudson [4], who looked at data from several investigators, including Kiefer, Little, and Evans, and found this wonderful parabolic (U-shaped) dose rate–response curve. Specifically, they looked at mutations in both somatic and germ-line cells in mice, and found a region between 0.1 and 1 cGy/min where there was essentially error-free DNA repair; they named it the minimum mutagenic dose rate region or MMDR. In this range, they discovered that 1 cGy of low-LET radiation produced as much damage as a cell sees in 1 min from endogenous sources of oxidative damage. These were simple damages such as base damages and single strand breaks. They postulated that irradiation in this dose rate region increases the damage rate by 10–100%, and that the cell repairs itself with few, if any, errors. Based on this observation, they concluded that the cell was exquisitely tuned to this damage frequency. However, they speculated that below about 0.1 cGy/min, the cell doesn't detect the damage signal, resulting in the inverse dose rate response, whereas above 1 cGy/min, the cell cannot keep up with the damage and eventually fails, yielding a linear increase in mutation rate with increasing dose rate. They further opined that these observations explained, in part, adaptive responses, since many studies deliver priming or adapting doses within MMDR. However, they questioned the significance of the MMDR with respect to environmental exposures, pointing out that the

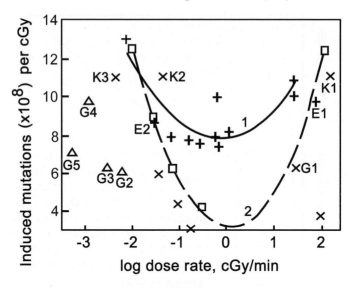

Fig. 6. Inverse radiation dose-rate effects. (Source: M.M. Vilenchik, A.G. Knudson Jr., *PNAS* **97** (2000) 5381–5386.)

natural background radiation dose rate from low-LET radiation (10:R per hour) is about a million times smaller than the MMDR.

As I mentioned, EPA applies a DDREF below 20 cGy to reduce its low dose, low-LET risk estimates. Recently, Sorensen and coworkers [5] studied in vivo dose rate effects by looking at translocations in mice following acute, chronic, and fractionated gamma irradiation at cumulative doses between 0 and 350 cGy. As shown in Fig. 7, they observed a linear-quadratic curve for acute exposures, and a lesser, linear dose–response curve for chronic exposures. With these curves, they calculated a DDREF of 14 at 350 cGy and a DDREF of 3 at 50 cGy. I remind you that EPA currently uses ICRP's recommended value of 2 at 20 cGy, which appears to be in good agreement with their findings. Of course, other experimental systems and damage end points may lead to different values for DDREF.

Threshold models suggest lower or no risks at low doses and low dose rates. For example, Roland [6] studied bone cancer mortality in female radium dial painters, and as shown in the table below, he found no excess cases of osteosarcoma below about 1000 cGy. Based on this observation, Rowland postulated a dose threshold for bone cancer induction. The table also shows EPA's estimates of the expected number of cases for the corresponding dose ranges. Based on a comparison of observed and expected values, we find that our estimates of less one bone cancer case in each of the dose ranges below 1,000 cGy are consistent with the observations of no cases in these ranges. This finding may provide an alternative explanation to Rowland's postulated threshold. Unfortunately, there are simply too few cases in this cohort to arrive at any firm conclusions using current epidemiological techniques. To address questions like this, we need a better understanding of the cellular/molecular events and mechanisms involved in radiation-induced bone cancer. Preliminary results from several alpha-particle microbeam experiments designed to illuminate these mechanisms have demonstrated that the traversal of the nucleus of a single cell by a single alpha particle, the lowest dose possible, increases oncogenic transformation in vitro. Moreover, a few investigators have observed similar effects in non-targeted bystander cells. Some have suggested that these results provide plausible mechanistic arguments against thresholds. Certainly, all agree that the answer to the question of whether or not thresholds exist has important implications for radiation protection. And, if EPA were to set thresholds for certain cancers, given that people vary widely and unpredictably in their response to irradiation, where would we draw the lines?

Observed versus expected bone sarcomas in female radium dial painters

Dose range (cGy)	Cases	Observed (Roland 1997)	Expected (EPA*)
>10,000	26	13	>3
5000–10,000	38	17	5
2500–5000	47	15	3
1000–2500	49	1	1
500–1000	52	0	0.6
250–500	57	0	0.3
100–250	106	0	0.3
50–100	96	0	0.1
25–50	141	0	0.1
10–25	205	0	0.1
5–10	167	0	0.02
2.5–5	80	0	0.01
<2.5	466	0	<0.01

*Calculated using DCAL by N. Nelson and C. Nelson, EPA (unpublished).

Like threshold models, adaptive responses fall into another category of phenomena that may support lower risks at low doses and low dose rates. Fig. 8 shows the results of an experiment by Azzam and coworkers [7] who measured the rate of induced malignant transformation in groups of control and irradiated C3H 10T1/2 mouse embryo cells, preceded or not by an adapting dose. They found that the transformation rate in the treatment group pre-exposed to an adapting dose of 100 mGy before a challenging dose of 4 Gy was approximately 2.5 times lower than the transformation rate observed in the group exposed only to an acute dose of 4 Gy. Their approach and findings are consistent with those found in other adaptive response experiments, and serve as a useful example for illustrating a few key points about these types of studies. First, radiation-induced adaptive responses appear to reduce the magnitudes or rates of the biological endpoints of interest, as evidenced by the fact that the non-exposed control group always shows less of an effect than the exposure group receiving the adapting dose. But, in the studies we are aware of, adaptive responses do not entirely eliminate adverse effects. Second, most such studies are either too short in duration or not designed specifically to investigate possible late-term effects caused by undetected complex lesions. Third, the relevance of adapting responses with respect to environment exposures is highly questionable because, as I discussed earlier, adapting doses are often delivered at a rate (0.1–1 cGy/min; the MMDR region) orders of magnitude higher than the natural low-LET background exposure rate. And finally, in most adaptive response studies, researchers typically find a reduction in the magnitude or rate of the endpoint of interest by a factor of 3 or less.

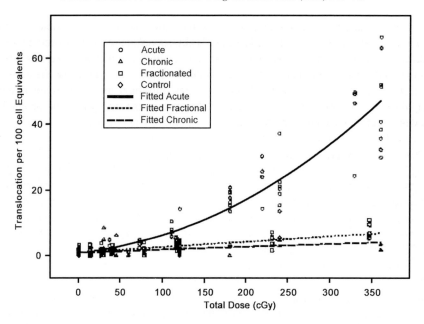

Fig. 7. In vivo dose rate effects and estimated DDREFs in mice following chronic gamma radiation. (Source: K.J. Sorensen, L.A. Zetterberg, D.O. Nelson, J. Grawe, J.D. Tucker, *Mutat. Res.* **457** (2000) 125–136.)

In reviewing these phenomena, we now appreciate that several opposing forces may affect the shape of the dose–response curve, and hence our risk estimates, below 10 cGy, and that all of them, or none of them, may be operational at any given point in time. To

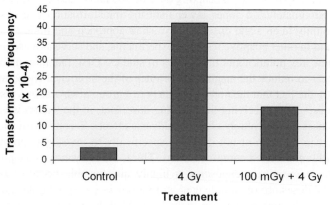

Fig. 8. Radiation-induced adaptive response in mouse embryo cells. (Source: E.I. Azzam, G.P. Raaphorst, R.E.J. Mitchel, *Radiat. Res.* **138** (1996) S28–S31.)

date, most of these phenomena have been observed only in vitro, and we clearly need to see if these same effects are active in vivo. And while all of these phenomena are very intriguing and worthy of investigation in their own right, we must remember that our ultimate goal is to use the knowledge gained this research to improve our understanding and assessment of human health risk from ionizing radiation. We need to find ways to extrapolate results obtained from cell systems and transgenic animals to man. I am confident that, collectively, we will find a way to meet these goals.

Meanwhile, EPA will continue to review and analyze the data you produce. We do not do this by ourselves; we often look to our Science Advisory Board for crucial consultations. We sponsor critical assessments by nationally recognized organizations, such as our recent co-sponsorship of the NAS-BEIR 7 study, which is currently underway. We also look to the international community for radiation protection advice, limits, and models. In addition, we apply our own criteria and systematic approach to weighing the evidence as it comes in. When that weight of evidence shifts, we do make changes to our models and our protection policies, albeit a little slowly at times.

In closing, I remind you that EPA is responsible for protecting people of all ages and both sexes from uncontrolled releases of radioactivity and unnecessary occupational and environmental radiation exposures. Risk assessment is an integral part of radiation protection; it is a complex and uncertain process, involving more than dose–response relationships. EPA uses state-of-the-art models to develop radionuclide-specific cancer risk coefficients for risk assessors. To reduce the uncertainties in our models and increase confidence in our risk estimates, EPA is looking beyond traditional approaches to evolving, mechanistic approaches to the study of radiation carcinogenesis. We believe that these new approaches may provide us with a more comprehensive understanding of the cellular and molecular events and processes involved in radiation carcinogenesis. This knowledge may also help us to better answer questions important to radiation protection.

I would like to acknowledge and thank my coworkers at EPA and at Oak Ridge National Laboratory. To Doctor and President Sugahara, to Dr. Little, and to the organizing committee, with a special nod to Dr. Niwa, I would like to say 'domo arigato' for your kind invitation to come here to speak to you today. And to all of you, thank you for a very informative and productive meeting. Thank you very much.

Dr. Tsutomu Sugahara

O.K., shall we continue our session? The next speaker is Dr. Sasaki, who retired 2 years ago from the Kyoto University Radiation Biology Center, but now is still working actively in the field of radiation research, and is the Vice President of the International Association of Radiation Research. Dr. Sasaki, please.

Dr. Masao S. Sasaki

First of all, I have to thank the organizers for inviting me to this panel. I am now getting a little nervous because all the other members of this panel are very distinguished scientists with ICRP, ICRU and NCRP backgrounds. So I'm feeling just like a little cat catching a mouse in front of tigers! But, on this occasion, I'd just like to present my idea on so-called adaptive defense mechanism—this will be rather new word. You may be puzzled but I shall introduce the word 'plasticity' in the defense system. In this symposium, many papers have discussed about low dose or low dose–rate effect. So our current concern is placed on low dose and low dose–rate effect in human populations. However, the current radiation protection strategy is still largely based on the stochastic concept of the microdosimetry of energy deposition to the target cells. In addition to the induction of the damage and the repair process, there is still a serious or general paucity of the implementation of the high-order biological regulatory system. Here, I'd just like to focus on the adaptive defense system. This topic has been presented by Professor Streffer, and also in this symposium, many papers dealt with adaptive response, although Dr. Streffer emphasized its viability. At the moment, most of studies were dealing with in vitro events, so the major question is what is the bridging of these phenomena to the people receiving radiation. So I'd just like to go out of the laboratory and focus on the human populations who are exposed to the low dose or low dose–rate radiation. This idea was stimulated by the recent puzzling observations on chromosome aberrations— particularly the chromosome translocations which originate from the stem cells. The finding is that the persistent translocation which originates from the stem cell is refractory to the accumulation of dose or severely suppressed to occur. At this moment, all the authors of those papers were thinking about probably that will be due to the errors in the estimation of dose, or insufficiency of the accuracy of the recorded dose, but my interpretation is different. The issue may be explained when we look carefully at what is the chromosome aberration we are looking at—what is the origin of the aberration? Those results will lead you to a question that the currently used unified quantity— DDREF—cannot be applied equally to the low dose of acute radiation and the low dose-rate of prolonged exposure.

We have about 500 billion lymphocytes in our body, in which about 60% or 70% of those were T lymphocytes, and we are looking at the chromosomes in those cells. But the lymphocytes are produced by the hemetopoiesis, supplied, committed, matured in the thymus and other lymphatic tissues, and go into the bloodstream. Only 2% of the lymphocytes were circulating in the bloodstream, but they are exchanged by the redistribution from the lymphocyte pool. So they are just turning over, but the differentiated lymphocytes disappear with their mean half life—their estimated mean half life is about 3 years. They disappear, but their disappearance is replenished by the new production of the lymphocytes. Depending on the time, the lymphocytes in the circulating blood are gradually replaced by those produced from stem cells. Therefore, if people are irradiated with an acute dose, we would expect the same amount of the chromosome aberration in terminally differentiate cells and in the stem cells. But, if the people who are gradually exposed or receiving protracted exposure to radiation and then you look at an unstable aberrations, the aberrations will disappear according to the lymphocyte lifetime since they cannot survive the cell division. So the unstable aberrations are those formed only in a differentiated cell. In contrast, the stable aberrations can survive the cell division, and therefore, the stable aberrations should be accumulated with accumulation of dose. Therefore, we would expect the stable aberrations to increase with accumulation dose, but not so for the unstable aberrations. This has been clearly demonstrated in the A-bomb survivors where the stable aberrations represent the dose. Such is also the case for Chernobyl's heavily exposed victims examined 4 years afterwards. This encourages the use of stable aberrations for the dosimetry for the accumulated dose. However, it may be rather optimistic. We have studied more than 1000 Japanese radiation workers, where the unstable aberrations tended to increase with total dose, while not strictly. This is because some are lost during exposure by the lymphocytes' lifetime. Age might be another factor. After adjustment by age and the dose rate of individual persons, the unstable aberrations show gradual increase with the accumulation of dose. However, such is not the case for the stable aberrations. Similar observations have been reported by Dr. Wei and Dr. Sugahara, in the high background radiation area. The unstable aberration increased with accumulation dose, but stable aberrations do not. Such puzzling data may be understood if you consider the difference in the origin and the difference in the repair capacity to respond to the low dose.

The absence of clear dose response of residual translocation frequencies has been also noted in the Chernobyl clean-up workers, workers in Mayak nuclear industry and people living near the Semipalatinsk nuclear test site. The observations have been discussed in favor of the uncertainties in dosimetry. However, an alternative interpretation may be that the lack of clear dose-response is a consequence of adaptive response of the stem cells. Currently, there are experimental data indicating that the long-term sustention of adaptation in the cells pre-exposed to protracted low dose radiation. If the prime dose was delivered over a long time, the adaptive stage persists for a longer time, for instance at least 40 days. This resembles a cellular memory and leads me to postulate the presence of plasticity in the adaptive defense system. In such adaptive condition, the dose response may be severely suppressed. According to the current DDREF quantity, dose linear term should be constant. However, in the adaptation the dose-linear term itself is variable. The presence of adaptation and its plasticity in the defense system would require reconsideration of the concept of DDREF. The dose effectiveness factor (DEF) can only be applied to a single shot of a low dose radiation, but the dose rate effectiveness factor (DREF) is different. Thank you very much.

Dr. John B. Little

Thank you very much, Dr. Sasaki, for this very interesting analysis. You may remember, Dr. Sasaki, that you hosted my first visit here in Kyoto 22 years ago. Our last formal speaker in this session is very appropriately, Warren Sinclair. As you know, Warren has served on many of the important agencies that have considered the radiation risks and the effects of low doses of radiation exposure, and he's a very wise person who has thought about this problem for many years.

Dr. Warren K. Sinclair

(The Scientific Basis of Radiation Protection Policy)

Thank you very much, Dr. Little for your introduction. Dr. Sugahara, ladies and gentlemen, it is an honor to be the last speaker on such a program as we have been privileged to hear in the last 3 days and to have an opportunity to address radiation protection policy and in particular the scientific basis of that policy. In this brief discussion I intend to address *first* some new perspectives on the dosimetry of the A-bomb survivors, because of its importance both to the survivors here in Japan and to the basis of risk estimation for cancer induction used in radiation protection worldwide, *second* some observations on uncertainties in risk estimation and *third* some general remarks about this meeting and the status of radiation research as it relates to radiation protection.

Dosimetry of the A-bomb survivors.

At low doses, 0.1Gy or less, the magnitude of the risk of cancer induction is the principal driver of radiation protection policy. The most important source of information on this risk is from the studies of A-bomb survivors. Critical to these risk estimates are both the assessment of the cancers induced (which RERF evaluates) and the dose causing the observed effects, which Japanese and American dosimetry committees examine and specify. The National Academy of Sciences in the U.S. maintains a Committee for this purpose, which has recently published a report on the Status of the Dosimetry for the Radiation Effects Research Foundation (DS86) [1]. The members of the Committee who produced this report are H. Agnew, H.L. Beck, R.F. Christy, S.B. Clark, N.H. Harley, A.M. Kellerer, K.J. Kopecky, W.M. Lowder, A.M. Weinberg, R.W. Young, M. Zaider and W.K. Sinclair-Chairman, (see Ref. [1], page iv]) Some are well known to many of you. The contents of the report are discussed briefly below and Figs. 1–6 illustrate certain technical items from the chapters of the report.

Gamma-ray measurements.

Fig. 1 ([1], Fig. 2-10) shows for Hiroshima very good agreement between the DS86 calculations for the gamma ray doses (full line) and direct thermoluminescent (TLD) measurements (points). Making a plausible correction of 20% does not improve agree-

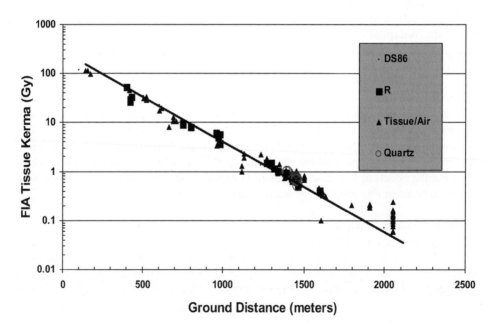

Fig. 1. Hiroshima TL measurements. Original reported units—roentgens, tissue, air, quartz—converted to FIA tissue kerma (Cullings, 2000).

ment, which is of the order of 10–20%. For Nagasaki (not shown, [1], Fig. 2-12), agreement for gamma rays is just as good although there are rather fewer TLD measurements. So there is very little concern about gamma rays because the calculated doses are well verified experimentally.

Neutron measurements.

With the neutrons that is not the case. Fig. 2 [2] shows for neutrons the measured to calculated ratio as a function of distance for thermal activation studies with ^{60}Co, ^{152}Eu, ^{36}Cl and updated to include fast neutron activation studies with ^{63}Ni (crosses on figure). The latter are very recent measurements and must be regarded as preliminary. In this plot of the ratio, agreement would mean a constant ratio of 1, but as can be seen the ratio rises with distance from 1 at about 800 m to about 10 (log scale) at about 1600 m. The recent ^{63}Ni measurements would fit better with a smaller ratio of about 3–5 at the same distance but are subject to revision.

Uncertainties in neutron measurements.

These neutron measurements are known to be subject to very many uncertainties and in this report these uncertainties have been evaluated. In Fig. 3 ([1], Fig. 3-5), the experimental points are shown again for Hiroshima with the uncertainties expressed as the 95% confidence limits (2 S.D.). It is apparent that there is the same trend of a higher ratio of measured to calculated with distance, as shown in Fig. 2 but the error bars (except perhaps for one point at 1800 m) now permit a ratio of 1 to be embraced by virtually all points. Thus, while it would be a very brave person who would say that *all* of the

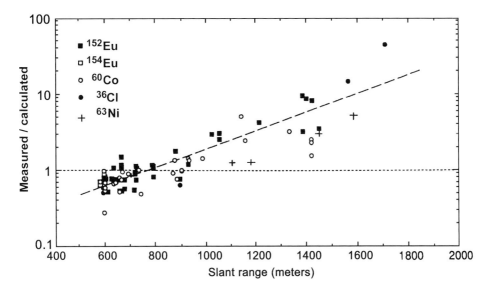

Fig. 2. Ratios of measured to calculated neutron activation in Hiroshima at various distances from the epicenter (slant range) (Straume et al., 1992).

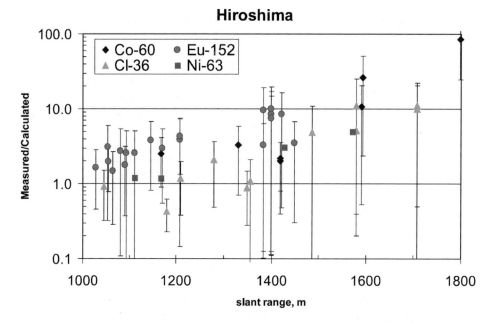

Fig. 3.

discrepancy between measured and calculated could be due to the uncertainties involved, it is, nevertheless, a possibility. At Nagasaki, on the other hand, calculations revised since 1986 and measurements agree well (i.e., ratio = about 1) ([1], Fig. 3-8).

Possibly, additional measurements with ^{63}Ni (of the fast neutrons that actually contribute to the dose) now in progress will clarify the situation with neutrons at Hiroshima.

Other physical entities.

Many other physical entities have been recalculated with greater precision since 1986. The Report ([1], Chap. 4) describes the re-estimates made from 1986 to 1993. For example, in 1986 (for DS86) neutron calculations were done with 27 energy groups and 20 angle bins whereas in 1993 more complete neutron recalculations were done with 174 neutron energy groups and a continuous angle distribution. These and other improved calculations are now available to incorporate in a revised DS86.

Biological dosimetry.

Again, since 1986, good progress has been made at RERF in biological dosimetry among the survivors. Several thousand have had persistent chromosome translocations assayed versus DS86 calculated doses, Fig. 4 ([1], Fig. 5-2) clearly shows more aberrations per unit dose for Hiroshima than for Nagasaki—about a factor of 2, just like the separate estimates of cancer risk for the two cities [3], even though in neither case is the factor of 2 *significantly* different. Measurements of the dose by electron spin resonance in teeth has also been accomplished for some 60 individuals with good agreement between

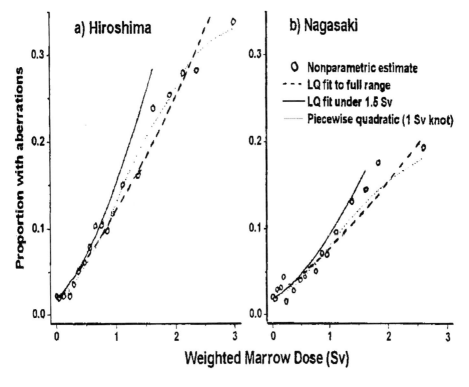

Fig. 4.

dose estimations from aberrations and from teeth ([1], Fig. 5-1). It is tempting to think that the extra neutrons in Hiroshima versus Nagasaki (a factor of 3 in DS86, perhaps higher experimentally) are responsible for the difference in effects between the two cities.

Implications for risk assessment.

In the chapter on implications for risk assessment ([1], Chap. 7) this possibility is explored in an illustrative rather than a definitive way. The chapter notes that if a higher RBE than 10 is used for neutrons, e.g. 20 or 50, which are not unreasonable values for the small neutron dose of ~ 11 mGy at 1Gy total dose to the bone marrow rather than to the colon, the fraction of the biological effect attributable to the neutrons becomes larger and may become significant even in DS86 as shown in Fig. 5 ([1], Fig. 7-2) (neutron contribution to the excess relative risk is ~ 20% at RBE = 20). The contribution of the neutrons may be even greater if the neutron component at Hiroshima is truly higher than DS86 calculates.

Recommendations of the report.

That question, the true magnitude of the neutrons at Hiroshima, still has to be resolved and the recommendations of the report, which concern chiefly the U.S. and Japanese working groups aiming at a revision of DS86, mainly address the neutron issue ([1], Chap. 8). Evidently, if the neutron component is responsible for a significant fraction of the risk, the risk due solely to gamma rays must be reduced. One of the recommendations of the

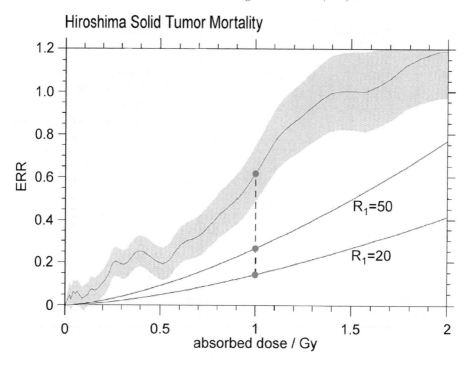

Fig. 5.

report is that if the gamma ray risk estimates from Hiroshima and Nagasaki data are to be used for other circumstances in which X- or gamma rays only (no neutrons) are the source of radiation exposure, allowance for the neutron contribution to the risk should first be examined (by RERF, by UNSCEAR, by ICRP or a BEIR Committee) before deciding upon risk estimates.

Conclusions of the report.

The report concludes that the gamma rays are the main contributor to the risk of induced cancer (98% or so of the absorbed dose at the organ level) and for these gamma rays calculation and measurement agree to within 10–20% and biological dosimetry confirms them to within 30–50%. The neutrons are still a problem and of uncertain magnitude since measurements and calculation do not seem to agree. It is hoped that these issues can be resolved by the U.S. and Japanese groups working to revise DS86, hopefully completing the work by 2002. The result is not expected to differ greatly from DS86 but there will be refinements and small changes are to be expected.

Appendices to the report.

The four appendices to the report include (Appendix A) a compendium of all measurements of gamma rays and neutrons at Hiroshima and Nagasaki known to the Committee and included in the RERF database.

Uncertainties in risk estimation.

Uncertainties in risk estimation have already been mentioned in this meeting, notably by Dr. Ralston. Uncertainties in the dosimetry are only one of five major components of the overall uncertainty in risk estimates derived from the A-bomb survivors. The others are epidemiological uncertainties (statistics, etc.), uncertainties in the transfer of risks between one population and another (the U.S. and Japan, for example), uncertainties in the projection of risk to a full lifetime risk (by age at exposure or attained age methods) and the question of dose and dose rate effectiveness corrections or factors to enable estimates derived from the high dose rates of the A bombs to be applied to the much lower dose rates more often encountered in other radiation protection circumstances, i.e. the DDREF.

These five major sources of uncertainty were combined by NCRP, together with an additional allowance for unknown uncertainty factors, to provide a measure of the overall uncertainties in the risk estimate for gamma rays derived from the A-bomb survivors which ranges from 1.2×10^{-2} Sv^{-1} (5% C.I.) to 8.8×10^{-2} Sv^{-1} for the nominal value of 5×10^{-2} Sv^{-1} [4] see Fig. 6. The most probable value of this distribution is about 3–4% Sv^{-1} rather than 5% Sv^{-1}. This assessment of uncertainty is supported by a further assessment by the U.S. EPA, which covers approximately the same range [5].

Each of the five major sources of uncertainty listed by NCRP [4] plays an important role in the overall assessment but the greatest single factor influencing the result is the choice of DDREF. The ICRP [6] used a DDREF of 2 in its 1991 recommendations, UNSCEAR 1993 [7] suggested up to 3, while NCRP used a wider range of values, 1–5, encompassing a number of different forms of possible dose response, in its evaluation of uncertainties [4]. Some values relate to individual circumstances such as lung cancer ~ 8 [8] or breast cancer ~ 1 [5].

US Population Lifetime Risk Coefficient
(Arrows Indicate 90% Confidence Interval)

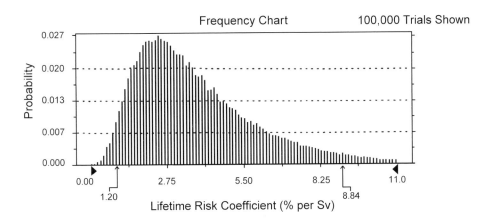

Fig. 6.

Values of DDREF lower than 1 reflect the possibility of supralinearity in the dose response at low doses some evidence for which exists in the A-bomb survivor data. Some authors have made an allowance for this possibility [9]. Overall however the linearity of the response for all cancers and for some individual cancers is impressive and significant excess of cancers is seen at low doses (0–0.1Gy) in the A-bomb survivor data [10].

Recent developments tend to modify the nominal value of 5×10^{-2} Sv^{-1} used by ICRP [6] and NCRP [11]. For example, in UNSCEAR 2000, an increased value 6×10^{-2} Sv^{-1}, would be better than 5×10^{-2} Sv^{-1}, but including an allowance for the neutron contribution at Hiroshima may reduce this 6×10^{-2} Sv^{-1} by 1 to 2×10^{-2} [1]. Using an attained age model for projection in preference to the age at exposure model may reduce the risk by a further factor of 1.6 or so. However, if incidence risks were to be preferred over mortality for stating nominal values for radiation protection, the risks would increase again by 1.6–1.7. Thus, the whole distribution given in Fig. 6, is not a static one but rather subject to re-evaluation as new information comes along. Eventually, it may be possible to narrow the range of uncertainties somewhat (for example—when the A-bomb survivor study is complete and projection to lifetime no longer a factor some 20 or so years from now). Genetic risk should also be included in the total detriment.

Radiation protection policy.

ICRP and NCRP have considered in the past [6,11] all the scientific evidence available to them before releasing recommendations on radiation protection. Judgement is inevitably involved in these recommendations and understandably not everyone agrees with ICRP and NCRP although most governments in the world follow ICRP and NCRP quite closely. ICRP and NCRP seek neither to exaggerate or to minimize the hazards of ionizing radiation. They believe that with proper respect for ionizing radiation hazards and observance of sensible procedures and dose limits, human beings can obtain the benefits of ionizing radiation in its widespread uses without incurring serious hazards for radiation workers or the public.

This meeting, Dr. Sugahara's symposia.

I want to say a few personal words about this meeting more generally. It is inevitable, of course, that I'd remember back to 1964 and the first time I came to Kyoto, when Dr. Sugahara's first symposium consisted of Dr. Mortimer Elkind and Dr. Robert Kallman and myself. All of the work in cell culture at that point was very new. We'd recently learned about the Elkind cellular recovery effect from Mort [12] and he and I [13] were exploring its applicability in many different cellular circumstances as well as in tissue to assist in understanding radiotherapy responses. We'd just started to learn also about cell cycle dependency—I was using synchronized mammalian cells to do this [14] and that was the main reason I was here. At that time, cell cultures looked tremendously promising for understanding many of the tissue responses that had already been established at least in outline.

Well, you can see from this meeting that we are still doing cell culture work, (e.g., Ref. [15] we are still doing more and more on tissue responses (e.g., Ref. [16]) and hopefully understanding them better. In fact, at this meeting, I would also say that some of the Japanese work on animal radiobiology that have been presented, show details and completeness that earlier works in other countries have not shown. I should mention, by way of example, the lifespan work in mice that Dr. Sasaki presented yesterday, in which even at 10 cGy he had a very definite decrease—statistically significant, he said—of lifespan. That's not the sort of thing one could've obtained out of earlier data where conflicting results sometimes suggested the possibility of stimulation and increased life span at low doses, an idea no longer supported by critical examination.

To me, this symposium has been a wonderful marker point in that respect, because I was talking to Dr. Niwa the other night about the future of radiation research; we were noting that the physicists had mainly sort of dropped out, and the chemists are no longer very interested in radiation biology. Radiation biology and medicine are pretty much left on their own amongst the four original disciplines that the Radiation Research Society in the U.S. once used to be proud of banding together. Now, we can see, however, that radiation physics is not actually over, because Dudley Goodhead has recently added dimensions to that in his own scientific work that I'm personally delighted to see, because they do contribute a lot to our understanding. And chemistry isn't over either, by any means, since chemistry plays a part in all molecular level science. Certainly cell culture and the tissue work are not over as we've seen them presented so well here, as indeed neither is radiation epidemiology and the important stimulating results from it that we did not have 40 years ago. But we do have now the addition in radiation research of the molecular dimension. Every speaker has, so far in some way, referred to that. How it will affect our risk estimates eventually? I don't know! Although initially it seems to add more confusion than clarity, I think there is no doubt at all that it will affect and eventually elucidate our understanding of the whole process of radiation damage and cancer induction. Looking a decade or two ahead—(I wish I could say I'll still be around to see it)—one can imagine that we will have a reasonably complete description of all the relevant events at the molecular level. To me that's one of the very promising things that has emerged from this meeting and other recent meetings like it and I am very grateful to all the speakers who have given me this feeling of confidence about radiation research in the future. Again, my gratitude, indeed the gratitude of all of us to Dr. Sugahara, for his 20th symposium in the field of radiation biology and the magnificent contributions that these symposia have represented. Thank you very much, Dr. Sugahara and your team, and to all the presenters in the symposium. A job very well done.

Dr. John B. Little

Thank you very much, Warren. If you would just remain on the podium and take a chair here, I'd like to ask all of our six participants if they would come up and sit up front here. I've counted, there are six chairs so we can enter the general discussion. This session, of course, was designed to finish the symposium with a discussion of current radiation policy, risk estimates and how the information we've heard over the last 3 days may help and aid us at this present time in terms of radiation policy. Particularly, how can we use—as Warren pointed out—the molecular data, the cellular data and the animal data. How we can use this information in developing a radiation policy and developing risk estimates. As Warren points out, there is still a great deal of work to do; there is certainly still empty spaces between these various experimental models right through to the epidemiology. But we have now our panelists up here and I would like to open the session to general discussion, questions, answers and any other problems that anyone would like to pose to our panelists, or even if the panelists would like to question each other—I think that would be just fine. Yes please, Jolyon and then Fred Burns.

Dr. Jolyon H. Hendry

Dudley, radiation tracks. You mentioned that a single alpha particle track can cause genomic instability and you asked the question, 'can a single electron track cause the same effect?' Now, with alpha particles the experiments are much easier because a single track gives you half a gray—500 mSv in a cell. And there are now methods available for giving a population of cells half a gray exactly to each cell. And even if you do that like people are starting doing now, a lot of the cells are killed, I suppose, and the ones that survive, not all of those develop genomic instability, only a proportion do. So even in those precise experiments that you can do, you can show that there is induced genomic instability but it isn't the same in every cell. On the other hand, with sparsely ionizing radiation, as you mentioned, at background levels of dose you are giving one track per cell per year at a mGy per year. And that is sparsely ionizing radiation over the cell versus the one single track. That is one fivehundredth of the alpha particle dose per year in one cell. Even if it's 10 times that high background or above, it's still a fiftieth of the alpha particle dose, so I've got a question for you which is, 'do you think that we will ever answer your question about whether a single sparsely ionizing track will cause genomic instability?' Because aren't we governed here still by statistics and dose–response curves and things, rather than one per cell exactly and an effect from that?

Dr. Dudley T. Goodhead

Yes, I mean, experimentally there would be ways to get down to delivering a single electron precisely to each cell in a population. The problem is, of course, detecting whether there is any effect or not and that might be a problem. But I think it's quite a challenge and I would remain optimistic that when we talk in 5 or 10 years time there will be a good indication of the answer to it. I think maybe it will come from understanding more of what the process is that allows the instability to be induced. What type of damage is it that induces the instability? Is it clustered damage? Is it the total amount of reactive oxygen species? But what is the type of damage that induces it? And then from that inferring what a single electron would do, and perhaps that's a slightly more likely way to get to the answer than giving direct observations on single electrons—although I'd still be hopeful that that might be possible.

Dr. John B. Little

I think it's rather interesting in the various studies of instability that we and others have done, that there doesn't seem to be much difference between the ability of X-rays or alpha particles to turn on the effect, and that the fraction of cells in which it's turned on is roughly similar.

Dr. Christian Streffer

Yes, I would like to underline this—our first data where we used X-rays. We have really a one-cell system. This is an isolated cell—the zygote—and there is no bystander effect, for the cell is surrounded by zona pellucida and there is no bystander effect possible under these conditions. And still it develops—genomic instability—and as I have shown, it is even transmitted to the next generation. Certainly it dilutes—the effect is smaller—but it is still there. The lowest dose with neutrons is 100 mGy; with X-rays—200, where we see a significant effect.

Dr. Frederic J. Burns

Fred Burns from New York. I basically want to follow up on that question that Dr. Hendry brought up. It's true that the human cancer cells in most solid tumors are extremely unstable cells that normally have something like modal chromosome number greater than 90, indeed, double the normal number so this is an extremely unstable configuration. And that radiation, especially high LET radiation, can produce some kind of instability in certain cells in culture. But it's not clear to me that that means the instabilities are in fact related to each other. There's one interesting observation, I believe, that is pretty general, that most malignant lesions have a benign precursor and, in fact, the benign lesion is pretty stable. So I don't know that there is any data suggesting that the benign lesion is unstable and therefore leads to the malignant one because of that instability, but I would think that would have to be a requirement. I guess this is a question for Dr. Goodhead.

Dr. Dudley T. Goodhead

Me again! There's some data that, for example, Mark Plumb...that's coming out shortly—Macdonald and others with Plumb—looking at the translocation or, at least, these non-clonal aberrations in radiation induced acute myeloid leukemia in the mouse. And there the average number of unstable aberrations—or at least non-clonal aberrations—is considerably greater in radiation induced AML than in spontaneous AML, and I think that's similar as being seen in some of the human studies. But if one then looks at the normal marrow in those same mice—not necessarily the leukemic mice, but the others after the radiation treatment—there is again a similar of instability induced in the normal tissue. So I think, along these lines, one would be suggesting that the instability is quite general and early, and feeds through to both the normal tissue and the tumor. Now, I'm not sure about your benign nodules. I'm not quite sure how that fits. Are you suggesting those don't have an instability? And, yes, I can't put that together with it, except to say it's a different system.

Dr. Amy Kronenberg

A couple of comments actually and maybe this will help Dudley a little bit. The first, of course, is that where several other people have tried to look at the issue of instability as a function of a level of exposure, there are a couple of conflicting reports for low LET radiation. Certainly, Bill Morgan would say that the risk is something, in his model system, on the order of 3% per Gy, so he shows a dose response. On the other hand, work from the Grezovski lab where fractionated exposures were compared with acute exposures, showed no sparing in that model system when you fractionated the dose over many days. The amount of instability that was seen using several different end points was the same as it was in the case of an acute exposure. Now that was two very different sorts of cellular models, but I think one of the other questions is going to be how are we going to think about instability when we have better coverage of the genomic response. The number of ways in which instability is being assessed at this point has been fairly limited to issues related to what you can see when you look down the microscope, looking at primarily chromosome aberrations which are a very wonderful end point. But with the genomic approach that many people are now able to apply since they have these unstable clones, trying to understand how the process of instability is controlled may give a different way to go back and look at the issue of instability at very low levels of exposure with greater precision. Of course, a large computing capacity will be required so I think we are going to need some more help from non-biologists to help us sort that out.

Dr. Martin F. Lavin

Thanks Jack. I wanted to change tack a little bit, but it could come back to Dudley, I guess, but it could be anyone else on the committee as well. And that was raising the issue of radiation versus oxidative damage, which he actually brought up and I think was referred to elsewhere as well. I think we'd all accept the uniqueness of radiation in causing complex damage, complex breaks in DNA and so on, and oxidative metabolism generating free radicals like that which give rise to single breaks. But I wanted to raise the special case, I guess, of AT which did come up in my talk during the week, and I referred to a number of examples from other people. And one I didn't refer to was some data from Yoshi Shilo's lab which isn't published yet, but where he's looked at a whole series of different genes using micro-ray technology and found that they're up-regulated in AT. And if you actually look at what happens when you expose a control cell to ionizing radiation with an equivalent of 4 Gy, the extent of up-regulation is about equivalent to what you see in an non-irradiated AT cell. So it's as though an AT cell has been exposed constantly to 4 Gy of radiation, I think that's the calculation, based on cells in culture, of course...sorry, not only cells in culture but cerebellum as well for mice. So I guess the point I wanted to raise was whether you can say that the type of oxidative damage that's occurring there isn't giving rise to double strand breaks in DNA, because, double strand breaks in DNA clearly are the lesion that exemplify AT. The question really is whether they arise because AT isn't around or whether they arise due to indirect effects on producing free radicals, which would give rise to breaks in DNA and be responsible for it. So I don't know whether Dudley or anyone else wanted to comment on the difference between oxidative damage and ionizing radiation damage, and especially the case in AT where you see this equivalent of 4 Gy there all the time.

Dr. Dudley T. Goodhead

Can I just make one quick comment and then perhaps somebody else should have a go. Radiation is a poor producer of oxidative damage and 4 Gy is a very high dose. If you delivered 4 Gy continuously to normal cells the effect would be an awful lot greater than you see in non-irradiated AT cells.

Dr. Martin F. Lavin

4 Gy is equivalent for induction of genes, I mean, that's what it's based on. You know, based on the level of gene induction.

Dr. Dudley T. Goodhead

Yes, but it's saying that whatever the radiation is doing is quite a small effect compared to the oxidative effects that you are talking about. Radiation is poor at producing these effects and 4 Gy, if delivered in normal cells, would be having a very much greater effect in other respects, I would suggest, because of the clustered damage. So I think I would see that argument as actually against suggesting that the oxidative damage is not the key process from the radiation.

Dr. Martin F. Lavin

So it would be generated elsewhere as a consequence of breaks within DNA, or something?

Dr. Dudley T. Goodhead

Clearly, double strand breaks must occur occasionally in normal cells by whether it's coincidences of oxidative damage, breaks or some of the repair processes from base damage and so on, and that's raised in AT cells. Yes, but I would still see that as a minor pathway as far as ionizing radiation is concerned.

Dr. Roy E. Shore

I'd like to change the topic a little bit. First, I'd like to congratulate the panel on a really excellent set of presentations—I thoroughly enjoyed them and learned from them. It seems to me that from what we've heard in these several days, it's exciting times for radiobiologists. A lot of relatively new concepts and new depth in pursuing them, things such as genomic instability, complex lesions, bystander effects, and stroma epithelial interactions. I find it to be very exciting, and yet at the same time, I kind of say "Well here are we, epidemiologists, sitting over in the corner and still counting dead bodies." And there's a real value to that, I don't want to diminish it. But I'm concerned about how do we get from here to there, how do some of these really exciting things that are happening in the lab get translated into ways that we can use them to define susceptibility to radiation effects or other possible things. And I'm interested in hearing what the members of the panel might have to say of avenues or of particular lines of research that look like they may have useful applications for us in epidemiology.

Dr. Lowell G. Ralston

It's a very good question, and certainly one of the thinking along those lines is to look for these unique damage patterns in chromosome abnormalities. Especially, what I've heard recently—inversions—which is something that doesn't happen as frequently, and so it's a more precise signal against a lower background. And certainly that's one way to pursue it, that we can't necessarily wait for a person to drop dead. They may be in the process of developing cancer and get run over by a bus. And so, we can look at these populations of individuals before they die or before they develop a frank tumor, which we can then count if they die from it. Perhaps we'd get a better understanding of when these processes actually occur. Of course, we've got this huge latent period of time where we have to wait for a term to occur, so rather than sit there and twiddle our thumbs waiting for it to show up we might actively and aggressively go seeking it on a molecular level.

Dr. Christian Streffer

Well, I certainly have not direct practical application. What we are trying, and I think we are trying together with the epidemiologists is to explain the mechanism of action of radiation. And only from these mechanisms we can see whether our extrapolations are correct or not. Both epidemiologists as well as the experimental biologists and physicists contribute each to this part. I think this was always the fascination of radiation research, that we had all the disciplines—as Warren Sinclair also pointed out—working together. Genomic instability, I think, is one of the important mechanisms for the development of cancer. And we had heard yesterday by the analysis of Dr. Mohrenweiser that there are many factors in carcinogenesis acting together, and certainly nobody will be able to pinpoint one single or, let's say, even 2 or 3 single factors because it's always a multifactorial plane. So in the individual case you certainly cannot say who will experience cancer and who will not, but which factors might be important and susceptibility, for instance, is certainly an important factor.

Dr. John B. Little

I could give you one example from the bystander effect at very low fluencies of alpha particles. That if one looks at very low fluencies of alpha particles, we are talking about 1% or 5% or fewer than the cells that are traversed by a particle in the population, and you find that the dose–response curve for the induction of mutations is super linear. There's a greater effect at lower doses than you would predict at higher doses, simply because now you're traversing one cell but you're producing mutations perhaps in four or five cells around it. So this doesn't mean that that is exactly what occurs in vivo in the case of low level exposure to residential radon, but it does give a model that one can test in vivo and in vitro.

Dr. Jurgen Kiefer

I'd like to come back to one of the first points mentioned by Dr. Streffer in his talk and this is repair. Repair, of course, is extremely important for understanding particularly low doses, and I think one has to keep in mind that the techniques, which are usually used for measuring repair, are not really techniques to measure repair. So what they do, they measure the disappearance of double strand breaks and this does not necessarily mean that they are correctly repaired. And also, the dose ranges where these techniques can be applied are far above what you use in normal cell work—so these are 10s of Gy which you use for the repair studies. So we really don't know whether the results we get are applicable to the lower dose range we are really interested in. As some of you know, there are now some approaches where you can measure fidelity of repair, and it has been shown that although there's always a complete disappearance of double strand breaks, at high doses there is about 50% of mis-repair. There is now done by Marcus Liebrich, experiments available by reducing the doses, and even at low doses in his experiment which is 5 Gy—which is not very low—there is still a substantial amount of mis-repair. So I think that is important, also, for the assessment of low doses.

Dr. Christian Streffer

I'm certainly aware about that. In such a talk you cannot explain the whole mechanisms and the whole pathways which are very complex. What I wanted to point out is that we have a very high efficiency of repair in our cells. Nobody would sit here if we would have no DNA repair, it would be a disaster with our DNA after each new synthesis of cell division; I think we all agree on that. What I wanted to point out is besides this high efficiency of repair, we have a very high variability again. There are individuals where repair is less, and there are radiation qualities where the repair is less than with others. I think this is very important for the evaluation of risk, that you have differences on an individual basis and also differences with radiation quality and with radiation conditions. And I think this again is the fascination of radiobiology or radiation research. I didn't go into complex DNA damages, as I knew that Dudley Goodhead is speaking after me and this is his domain!

Dr. Kazuo Sakai

Sakai from Tokyo. My question is a rather general one. So, we have learned a lot on achievements of the recent research at various levels; starting from molecular level, cellular level and tissue level and others, including epidemiology. But, as many of you mentioned, there are still gaps between them, and then the importance of bridging the gap is pointed out as a general comment. However, let me ask, what kind of specific study should be done to bridge the gap towards...to estimate the radiation risk in the human being—that is one thing. The other thing is that under the situation where we have started to realize the individual different predisposition to the radiation, for example, the cancer risk. We have started to know the mechanism and alike, and also on the other hand, if the radiation regulations are to be based on the individual level rather than the radiation source level, how should we involve this kind of information in the future radiation regulations? These are my two questions. Thank you.

Dr. Warren K. Sinclair

I think the proper study of mankind is man, is what somebody said a long time ago. And I think the epidemiologists, who might be feeling a little left out by the molecular trends, should not feel left out at all because the action is really with them. To point the way to those features of what they see in their studies, and people at RERF have been doing this. The features that they see in their studies that can be studied in the laboratory further and proven experimentally subsequently. So in the contest of molecular epidemiology, it's really the epidemiologist who is the leader and the finger pointing person for whatever end point he or she sees, as the one that deserves because of its peculiarities, because of its frequency, because of whatever—liver cancer, bone cancer, whatever. He sees the important features of that at the human level, and they can be studied further from that beginning, so to speak. So the epidemiologist is really at the start and at the beginning of the answer to your question.

Dr. Lowell G. Ralston

Yes, I just want to echo Warren's comment that epidemiologists are very, very important in this entire process and I'm sorry that they are feeling left in the corner right now, because they'll be back very shortly. I also want to say that while all of this seems like a free-for-all in trying to answer questions from different angles, I actually find that quite exciting. Because I think it's only by looking at this problem from many different angles that we'll get the information we need to form that critical jump in our understanding. So rather than try to focus it one way or the other by designing it as an experiment, I think we should leave it the way that it is right now with many people attacking it from different positions.

Dr. Tatsuo Matsuura

My name is Matsuura, I came from Tokyo. I have a fundamental comment as for the dosimetry of atomic bomb survivors. As for the estimation of those survivors, I believe that not only the acute dose but also the chronic dose of survivors should be taken into account. I believe the anomaly of the dose–response curve for the survivors who lived more than 2 km, as indicated in Dr. Preston's lecture on the first day and Dr. Matsudaira's later, shows it is not a direct acute dose, not due to the direct use, but due to, I believe, the acute dose. They must have worked...after the bombing they entered the city and worked in rescue or searching for their friends. So I reported this comment last year at the Hiroshima International Congress of Radiation Protection Association, as a poster session. So please read and give us some comments. Thank you very much.

Dr. John B. Little

I'd like to take the Chairman's prerogative and ask one last question, and I might direct this at Dr. Sasaki and Dr. Matsudaira, and that concerns the DDREF. I think there's a real question now; there's a task force of the ICRP committee 1, which is discussing this. How, given the current information we have now, should we deal with the DDREF? Should we try to assign it a specific value? Should we not use it and say we don't have enough information on it? I think this is an important question of policy for our estimation of risks at the present time.

Dr. Hiromichi Matsudaira

Very difficult questions. I cannot make precise comments, but my understanding is that people in Hiroshima tried three times in revising dosimetry—D57, D65, DS86 and then a new one. Every time...in the beginning, the people in RERF had many assistant nurses, and they went to each survivor; where he or she was at the time of bombing and what would be the shielding, and so on and so on. At present, I'm wondering if we have any possibility to change the doses for distance exposed people—I cannot say.

Dr. John B. Little

I think that the question that is in my mind is whether we can ascribe a DDREF a dose rate factor of 2, 4 or 5, or whether we should just, at this point, not use one at all until we have more of a clear idea of what it may be.

Dr. Warren K. Sinclair

We wrestled with this question at the time that it was done last time in 1990, and I can tell you, there were a lot of different opinions within the ICRP group itself. And in fact, we started with 3 and everything we did was actually based on 3 as the factor until committee 1, which was the, shall we say, final arbiter on the biology and so on, insisted that that was not good enough and they wanted 2. Well, 2 was the compromise I should say, between 1 and whatever else was on the table. I think ICRP has always indicated how arbitrary they think it is, because obviously there are biological situations where the dose rate factor is much greater than that, and others where it doesn't seem one applies. Possibly, the best thing to do is to leave it alone at this point. But I'm in the very fortunate position now, Jack, I'm a emeritus member of the ICRP, and it means that I get all the documents so I can see everything that's going on, but I don't have to say anything about it!

Dr. John B. Little

From the comfort of your home in California! I know there are several other questions, but I'm afraid because of the limitations of time and the next meeting, we have to proceed. So I'm going to turn the final microphone over to our Chairman and our distinguished host.

Dr. Tsutomu Sugahara

Thank you. Unfortunately, we have a time limit so I'm very sorry we that cannot continue our discussion furthermore. As a conclusion, my feeling about radiation protection principles is that we need a very simple one. But we biologists are much more interested in complex things, so it is only that we have a debate about this. But, for protection purposes at any time we must have a simple principle, so I hope the protection principle will be always watching what is going on in biology, and adapt it as far as possible. For this purpose in this symposium, for 3 days we discussed about very complex things, and at final, all is too short to summarize all of this discussion and to go to the one single principle. But I think this procedure would be very good and so I hope—I am too old—but the next generation will have another meeting in the near future and go to the consensus; all of the scientists to their protection of human beings. Thank you very much for your contribution. I am closing this symposium. Thank you very much for your attendance.

The chairpersons of the discussion session
John B. Little, Harvard School of Public Health, USA
Tsutomu Sugahara, Health Research Foundation, Japan
The speakers of the discussion session
Christian Streffer, University Essen, Germany
Dudley T. Goodhead, Medical Research Council, UK
Hiromichi Matsudaira, Radiation Effect Association, Japan
Lowell G. Ralston, U.S. Environmental Protection Agency, USA
Masao S. Sasaki, Kyoto University, Japan
Warren K. Sinclair, National Council on Radiation Protection, USA.

References

Dr. C. Streffer's presentation

[1] C. Streffer, Genetische Prädisposition und Strahlenempfindlichkeit bei normalen Geweben, Strahlenther. Onkol. 173 (1997) 462–468.
[2] W.-U. Müller, Th. Bauch, G. Stüben, H. Sack, C. Streffer, Radiation sensitivity of lymphocytes from healthy individuals and cancer patients as measured by the comet assay, Radiat. Environ. Biophys. 40 (2001) 83–89.

[3] C. Streffer, Threshold dose for carcinogenesis: what is the evidence? in: D.T. Goodhead, P. O'Neill, H.G. Menzel (Eds.), Microdosimetry. An Interdisciplinary Approach, The Royal Society for Chemistry, 1997, pp. 217–224.
[4] ICRP, ICRP Annals of the ICRP, Publication 79, Genetic Susceptibility to Cancer, Pergamon, 1998.
[5] A. Wojcik, C. Streffer, Adaptive response to ionizing radiation in mammalian cells: a review, Biol. Zent. Bl. 113 (1994) 417–434.
[6] W.-U. Müller, C. Streffer, F. Niedereichholz, Adaptive response in mouse embryos? Int. J. Radiat. Biol. 62 (1992) 169–175.
[7] J.B. Little, Radiation carcinogenesis, Carcinogenesis 21 (3) (2000) 397–404.
[8] C. Streffer, Genomic instability induced by ionizing radiation, Proceedings of the International Congress of the International Radiation Protection Association, May 14–19, 2000, Hiroshima, Japan, PS 1–3 (2000) pp. 1–5.
[9] S. Pampfer, C. Streffer, Increased chromosome aberration levels in cells from mouse fetuses after zygote X-irradiation, Int. J. Radiat. Biol. 55 (1989) 85–92.
[10] S. Pils, W.-U. Müller, C. Streffer, Lethal and teratogenic effects in two successive generations of the HLG mouse strain after radiation exposure of zygotes—association with genomic instability? Mutat. Res. 429 (1999) 85–92.
[11] A. Kryscio, W.-U. Müller, A. Wojcik, N. Kotschy, C. Streffer, A cytogenetic analysis of the long-term effect of uranium mining on peripheral lymphocytes using the micronucleus-centromere assay, Int. J. Radiat. Biol. (in press).
[12] C. Streffer, W.-U. Müller, A. Kryscio, W. Böcker, Micronuclei–biological indicator for retrospective dosimetry after exposure to ionizing radiation, Mutat. Res. 404 (1998) 101–105.
[13] W.-U. Müller, C. Streffer, S. Pampfer, The question of threshold doses for radiation damage: malformations induced by radiation exposure of unicellular or multicellular preimplantation stages of the mouse, Radiat. Environ. Biophys. 33 (1994) 63–68.

Dr. L.G. Ralston's presentation

[1] R. Skloot, A pill to cure cancer? Pop. Sci. (2001) 41–45 July.
[2] D.J. Brenner, J.B. Little, R.K. Sachs, The bystander effect in radiation oncogenesis: II. A quantitative model, Radiat. Res. 155 (2001) 402–408.
[3] S.G. Sawant, G. Randers-Pehrson, C.R. Geard, D.J. Brenner, E.J. Hall, The bystander effect in radiation oncogenesis: I. Transformation of C3H 10T1/2 cells in vitro can be initiated in the unirradiated neighbors of irradiated cells, Radiat. Res. 155 (2001) 397–401.
[4] M.M. Vilenchik, A.G. Knudson Jr., Inverse radiation dose-rate effects on somatic and germ-line mutations and DNA damage rates, PNAS 97 (2000) 5381–5386.
[5] K.J. Sorensen, L.A. Zetterberg, D.O. Nelson, J. Grawe, J.D. Tucker, The in vivo dose rate effect of chronic gamma radiation in mice: translocation and micronucleus analyses, Mutat. Res. 457 (2000) 125–136.
[6] R.E. Rowland, Bone sarcoma in humans induced by radium: a threshold response? Radioprotection 32 (1997) C1-331–C1-338.
[7] E.I. Azzam, G.P. Raaphorst, E.J. Mitchel, Radiation-induced adaptive response for protection against micronucleus formation in C3H 10T1/2 mouse embryo cells, Radiat. Res. 138 (1996) S28–S31.

Dr. W.K. Sinclair's presentation

[1] NAS/NRC, Status of the Dosimetry for the Radiation Effects Research Foundation (DS86). Report of the Committee on Dosimetry for RERF, National Academy Press, Washington, DC, 2001.
[2] T. Straume, S.D. Egbert, W.A. Woolson, R.C. Finkel, P.W. Kubik, H.E. Gove, P. Sharma, M. Hoshi, Neutron discrepancies in the DS86 Hiroshima Dosimetry System, Health Phys. 63 (1992) 421–426.
[3] D.A. Pierce, Y. Shimizu, D.L. Preston, M. Vaeth, K. Mabuchi, Studies of the mortality of the atomic bomb survivors report 12 part I cancer 1950–1990, Radiat. Res. 146 (1996) 1–27.
[4] NCRP, National Council on Radiation Protection and Measurements. Report 126, Uncertainties in Risk Estimates for Fatal Cancer Induction by Low LET Radiation NCRP, Bethesda, MD, 1997.

[5] EPA (U.S. Environmental Protection Agency), Estimating Radiogenic Cancer Risks, Addendum—Uncertainty Analysis, EPA Report 402-R-99-003, U.S. EPA, Washington, DC, 1999.
[6] ICRP (International Commission on Radiological Protection), The 1990 Recommendations of the ICRP, ICRP Publication 60 Ann. ICRP 21, no. 1–3, Pergamon, Oxford, 1991.
[7] UNSCEAR (United Nations Scientific Committee on the Effects of Atomic Radiation), Sources and Effects of Ionizing Radiation, United Nations, New York, 1993.
[8] G.R. Howe, Lung cancer mortality between 1950 and 1987 after exposure to fractionated moderate-dose-rate ionizing radiation in the Canadian fluoroscopy study and a comparison with lung cancer mortality in the atomic bomb survivors study, Radiat. Res. 142 (1995) 295–304.
[9] H.A. Grogan, W.K. Sinclair, P.G. Voillequé, Assessing Risks of Exposure to Plutonium, RAC Report no. 5-CDPHE-RFP-1998 Final (Rev. 2).
[10] D.A. Pierce, D.L. Preston, Radiation-related cancer risks at low doses among atomic bomb survivors, Radiat. Res. 154 (2000) 178–186.
[11] NCRP, National Council on Radiation Protection and Measurements, Report 116, Limitation of Exposure to Ionizing Radiation, NCRP, Bethesda, MD, 1993.
[12] M.M. Elkind, H.A. Sutton, Radiation response of mammalian cells grown in culture: I. Repair of X ray damage in surviving Chinese hamster cells, Radiat. Res. 13 (1960) 556–593.
[13] W.K. Sinclair, R.A. Morton, Recovery following X irradiation of synchronized Chinese hamster cells, Nature 203 (1964) 247–250.
[14] W.K. Sinclair, R.A. Morton, Variation in X ray response during the division cycle of partially synchronized chinese hamster cells in culture, Nature 199 (1963) 1158–1160.
[15] T.K. Hei, T.N. Zhou, M. Suzuki, G. Randers-Person, C.A. Waldren, E.J. Hall, The yin and yang of bystander versus adaptive response, Lessons from the Microbeam Studies, Proceedings of International Symposium on Radiation and Homeostasis, Kyoto, July, 2001.
[16] M.H. Barcellus-Hoff, Aberrant Extra cellular Signaling in the Progeny of Irradiated Human Cells, Proceedings of International Symposium on Radiation and Homeostasis, Kyoto, July, 2001.

Author index

Abe, K., 303
Abisheva, G., 47
Agematsu, K., 335
Andoh, T., 331
Angelini, S., 217
Antal, S., 119
Arase, Y., 335
Azzam, E., 229

Balásházy, I., 133, 341
Bamen, K., 495
Ban, S., 67
Barcellos-Hoff, M.H., 399
Baskar, R., 115
Bedford, J.S., 327
Bhattathiri, V.N., 463
Bindu, L., 463
Binks, K., 51
Bogdándi, N.E., 341
Botchway, S.W., 289
Bovornkitti, S., 31
Burlakova, E.B., 387
Burns, F.J., 175

Cameron, J.R., 35
Chatterjee, N., 459
Chipman, J.K., 289
Curtis, S.B., 283

Dám, A., 133, 341
Desaintes, C., 55
De Toledo, S.M., 229
Dikiy, N.P., 39

Edwards, G.O., 289
Egashira, A., 111
Enflo, A., 23

Falt, S., 217
Fujita, K., 487
Fujitaka, K., 323

Fukui, Y., 477
Fukumoto, M., 191, 221
Fukunishi, N., 317
Furusawa, Y., 295
Furuse, T., 101

Gajendiran, N., 383
Ghiassi-Nejad, M., 19, 35
Goloshchapov, A.N., 387
Gorski, A., 43
Goto, S., 317
Gu, Y., 495

Hall, E.J., 241
Hama-Inaba, H., 299
Hashiguchi, K., 331
Hatashita, M., 295
Hayashi, S., 295
Hayata, I., 299
Hazelton, W.D., 283
Hei, T.K., 241
Hendry, J.H., 415
Henshall-Powell, R.L., 399
Hergenhahn, M., 185
Hidvégi, E., 119
Hino, O., 163
Hirabayashi, Y., 455
Hirst, G.J., 289
Hiwasa, T., 335
Hofmann, W., 27, 133
Hollstein, M., 185
Hongyo, T., 115, 459
Horie, K., 467
Hoshi, M., 47, 127
Hoshi, Y., 487
Hossain, M., 123
Hou, S.-M., 217
Huilgol, N.G., 459
Husgafvel-Pursiainen, K., 217

Ikeda, M., 359
Ikura, T., 151
Ikushima, T., 19, 35, 331
Imai, T., 67
Inoue, T., 455
Ise, T., 355
Ishida, H., 491
Ishii-Ohba, H., 157
Ishikawa, Y., 191, 221
Ito, H., 335
Ivanov, V., 43
Iwamoto, K.S., 195
Iwasa, T., 495
Iwasaki, T., 487
Izumi, M., 317

Jin, Z.-H., 295

Kaikawa, T., 331
Kakinuma, S., 179
Kamiya, K., 47, 127, 151
Kannio, A., 217
Kano, E., 295
Karam, P.A., 35
Kashino, G., 347
Kato, F., 423
Katsura, Y., 179
Kellerer, A.M., 3
Khait, S., 43
Kiefer, J., 255
Kimura, A., 47, 67
Kitai, R., 295
Kobayashi, S., 157
Kodama, S., 237, 309, 313, 347, 351, 355, 363, 367, 371, 375
Kodama, Y., 143
Kohlpoth, M., 255
Koike, N., 151
Kojima, S., 75, 491
Kominami, R., 143
Kondo, H., 467
Konradov, A.A., 387
Kosugi-Okano, H., 143
Krishnan Nair, M., 463
Kubo, A., 179

Kubo, K., 467
Kubota, Y., 327
Kumada, T., 351
Kumagai, J., 351, 355
Kuntze, M., 255
Kuramoto, K., 67
Kurobe, T., 317

Land, C., 47
Lee, R., 323
Li, L., 221
Li, L.Y., 115, 459
Little, J.B., 229, 503
Liu, D., 221
Loomis, C., 175
Luebeck, E.G., 283
Lumniczky, K., 119
Luo, J.-L., 185

Majima, H., 179, 323
Maki, H., 111
Masuda, Y., 151
Matsuda, N., 359
Matsuki, A., 143
Matsumori, S., 491
Matsumoto, H., 295
McGeoghegan, D., 51
Medvedeva, E.P., 39
Meldrum, R.A., 289
Mifune, T., 75
Mishima, Y., 143
Misonoh, J., 471
Mita, K., 179
Mitani, H., 163
Miura, Y., 303
Miyakoda, M., 375
Miyazaki, T., 351, 355
Mohankumar, M., 383
Molochkina, E.M., 387
Momoi, H., 221
Moolgavkar, S.H., 283
Mori, N., 439
Mori, S., 75
Morimoto, S., 317
Morita, N., 359

Mortazavi, S.M.J., 19, 35
Muirhead, C.R., 83

Nagai, J., 179
Nagasawa, H., 229
Nair, C.K.K., 459
Nakajima, H., 115, 459
Nakajima, T., 299
Nakashima, E., 71
Nakatomi, S., 313
Namba, H., 201, 319
Neriishi, K., 71
Niroomand-rad, A., 35
Nishimura, M., 157, 179
Nitta, Y., 127
Niwa, O., 143
Noda, Y., 101
Nomoto, S., 423
Nomura, T., 105, 115, 459, 481, 487
Norimura, T., 423
Nyberg, F., 217

Ochiai, Y., 143
Oda, K., 67
Oda, T., 487
Ogiu, T., 157, 179
Ohira, C., 323
Ohnishi, K., 265, 471
Ohnishi, T., 265, 471, 477
Ohta, M., 379
Ohtsubo, T., 295
Ohyama, H., 299
Okaichi, K., 359
Okamura, H., 379
Okayasu, R., 327
Okumoto, M., 439
Okumura, Y., 359
Onishchenko, N.I., 39
Oshimura, M., 313, 347
Otsu, H., 101

Pálfalvi, J., 119
Pálfalvy, J., 341
Park, C.C., 399
Park, S., 495

Paunesku, T., 249
Pohl-Rüling, J., 27
Polonyi, I., 341
Potten, C.S., 407
Protić, M., 249

Rabes, H.M., 207
Randers-Pehrson, G., 241
Remani, P., 463
Rogounovitch, T., 319
Roy, N., 175
Ryo, H., 115, 459

Sado, T., 157, 179, 455
Saenko, V., 201, 319
Sáfrány, G., 119
Saito, Y., 143
Sakai, K., 481, 487
Sakata-Haga, H., 477
Sakaurai, J., 163
Sárdy, M.M., 341
Sasaki, S., 447
Sato, H., 327
Satyamitra, M., 123
Sawada, K., 477
Schneider, F., 119
Scott, D., 433
Shaikh, W., 289
Sharp, G.B., 59
Shibata, Y., 201
Shibuya, K., 75
Shimada, Y., 157, 179
Shimizu-Yoshida, Y., 319
Shinbo, T., 143
Shioura, H., 295
Shirasawa, H., 335
Shishkina, L.N., 387
Shore, R.E., 13, 175
Si, X.E., 115
Simmens, S.J., 137
Sugahara, T., 91, 503
Sugita, K., 75, 335
Sugiyama, K., 319
Sumii, M., 151
Suzuki, F., 157, 367

Suzuki, G., 67
Suzuki, I., 495
Suzuki, K., 237, 309, 313, 347, 355, 359, 363, 367, 371, 375
Suzuki, M., 241, 323, 371
Suzuki, N., 335
Suzuki, S., 303
Syaifudin, M., 115

Tachibana, A., 347
Takahashi, A., 265, 471, 477
Takahashi, H., 359
Takahashi, M., 151
Takahashi, M., 491
Takahashi, S., 327
Takahashi, Y., 143
Takamura, N., 201
Takeichi, N., 47
Tamaki, T., 313
Tanaka, H., 67
Tanaka, K., 383
Tanaka, T., 355
Tanizaki, Y., 75
Tanooka, H., 487
Teshima, J., 151
Teunen, D., 55
Tong, W.-M., 185
Treshchenkova, Yu.A., 387
Tsuji, H., 157
Tsuzuki, T., 111
Tsyb, A., 43

Ukai, H., 157
Ukawa, Y., 495
Uma Devi, P., 123
Urushibara, A., 313, 367

Wada, I., 191
Wakabayashi, Y., 143

Waldren, C.A., 241
Wang, B., 299
Wang, Z.-Q., 185
Wani, K.M.Y., 459
Watanabe, F., 157
Watanabe, H., 393
Watanabe, M., 237, 309, 313, 317, 347, 351, 355, 359, 363, 367, 371, 375
Wei, L.-X., 91
Wharton, C.W., 289
Woloschak, G.E., 249
Wright, E.G., 271

Xu, A., 241

Yamada, T., 487
Yamaguchi, C., 323
Yamaguchi, Y., 323
Yamaoka, K., 75, 481, 491
Yamashita, S., 201, 319
Yamauchi, K., 111, 313
Yamauchi, M., 363
Yasuda, H., 323
Yasuda, M., 379
Yatagai, F., 317
Yonezawa, M., 467, 471
Yoshida, K., 455
Yoshiyama, K., 111
Yukawa, O., 295, 299

Zabolotny, V.D., 39
Zhao, P., 175
Zhigitaev, T., 47
Zhou, H., 241
Zhumadilov, Z.H., 47
Zielinski, B., 185
Zook, B.C., 137

Keyword index

A-bomb 195
A-bomb survivors 3
Accumulation 363
Adaptive response 35, 241, 303, 477, 491
Age 71
Age dependence 447
Aging 303
Alpha particle 191
Ambient dose equivalent 3
Antimony 327
Antioxidants 481
α-particle 221
Apoptosis 265, 299, 407, 415, 471, 477
Arsenite 327
ATM 371
Atomic bomb 71

Background radiation 35
Bak 459
Bax 477
β-*catenin* 459
Brain cancer 137
Breast cancer 399
Bronchial carcinoma 133
Bystander effect 229, 241
Bystander effects 283, 289

Ca^{2+} channel 477
Calcium-deficient hydroxyapatite 379
Calorie restriction 455
Cancer 83, 175, 191, 317
Cancer incidence 51
Cancer predisposition 433
Carcinogenesis 55, 123, 283, 399, 447
Cell cycle 455
Cellular telephones 137
Centrosome 375
Cervical cancer 459
Chernobyl 127, 201, 207
Childhood cancer 13

Chromosomal instability 123
Chromosome 375
Chromosome aberration 383
Chromosome aberrations 433
Chromosome instability 313
Chromosome transfer 313
Chronic irradiation 487
Cohort 51
Correlation 39
CRAD 151

Delayed chromosome aberrations 367
Delayed effects 309
Deletion 331
Dentine 379
Deposition enhancement factors 133
Diagnostic radiation 13
DNA damage 309
DNA double-strand breaks repair 347
DNA-DSB 327
DNA-PK 265
DNA-PKcs 367
DNA repair 55
DNA strand segregation 407
Domestic radon 23
Dose and dose rate effectiveness factor (DDREF) 101
Dose rate 255
Dose-rate dependency 423
Dose rate effect 105

E6 335
Effective dose 3
EGCG 355
Eker rat 163
Elk-1 237
Enamel 379
Endogenous antioxidative materials 481
Endogenous CFUs 471
Epidemiology 55, 83
ERK 237

ESEEM 351
ESR 351
ESR dosimetry 379
Ethylnitrosourea 137

Fallout 47
FISH 383

Gamma irradiation 249
Gamma rays 255
Gamma-rays 487
Gap junction intercellular communication 289
Gap junction intracellular communication 229
Gastric cancer 393
Gene expression microarrays 249
Gene rearrangement 201, 207
Gene targeting 111, 185
Gene therapy 363
Genetic changes 191
Genetic instability 221
Genetic susceptibility 55
Genome destabilization 375
Genomic instability 255, 271, 399
Genotype/phenotype correlations 207
Glandular stomach 393
Glial cells 303
Glutathione 481, 491
γ-radiation 115

Haemopoietic stem cells 271
Heat shock 375
Heavy ion 323
Hematopoietic stem cells 455
Hepatitis C virus 59
Hepatocellular carcinoma 59
Hepatoma 151
High background area 27
High background radiation area 19
High-LET radiation 283
hSNK 317
Human 175
Human cells 355

Human organ/tissue 105
Human SCID cells 335

^{131}I 127
ICRP 35
Ikaros 179
Illegitimate recombination 347
Immune function 491
Inflammation 71
Intelligence 13
Internal exposure 127
Intestinal metaplasia 393
Intestinal stem cells 407
Intestine 415
Intrahepatic cholangiocarcinoma 221
Ionising radiation 433
Ionizing radiation 271, 359, 399, 447
Irradiated cells 295
Irradiation 415

Jack Fruit Lectin 463

Laser plasma 289
Lectin staining 463
Leukaemogenesis 271
Life span 323, 467
Lipid peroxide 75
Liver cancer 59
LNT 35
LOH 115, 179
Longevity 467
Long-lived radicals 351, 355
Long-term effects 447
Loss of heterozygosity 119
Low-density lipoprotein cholesterol 75
Low dose 105
Low-dose and low-dose rate exposure 101
Low dose fetal irradiation 123
Low-dose γ-rays 491
Low-dose irradiation 481
Low-dose neutrons 341
Low dose of X-ray 237
Low-dose radiation 323

Low-dose rate 265
Low-dose rate irradiation 487
Low doses 83
Low-level irradiation 387
Lung cancer 23, 27, 217
Lymphocytes 433
Lymphoma susceptibility loci 439

Macrophages 271
Malformation 423
Mammary gland 399
MAP kinase 237
MDS 67
Mechanistic modeling 55
Mental retardation 13
Metal 327
Methylcholanthrene 487
Mice 439, 455
Microenvironment 399
Micronuclei 67
Micro satellite 221
Microwaves 137
Mitochondrial DNA 331
Monoenergetic nuetrons 383
Mortality 51
Mouse 105, 331
Mouse survival 471
Mouse thymic lymphoma 143
M6P / IGF2r 195
Multistage models 283
Mutagenesis 229, 241
Mutants 415
Mutation 115, 151, 185, 255, 341, 347, 351, 355
Myeloid leukemia 101, 455

Natural radiation 19
Neonatal mice 447
NER gene 67
N-ethyl-*N*-nitrosourea 179
Neutrons 3, 383
Nitric oxide 295
Normal human fibroblasts 323
Notch1 157
NTRK1 207

Nuclear workers 51
Nucleotide excision repair 217
Nucleotide sanitization 111

Oncogenes 119, 157
Oral cancer 463
Oxidative DNA damage 111
Oxidative stress 229, 303
8-Oxoguanine 111

p53 185, 237, 265, 309, 363, 371, 375, 471
p53 195, 359, 459
Papillary thyroid carcinoma 207
p53-dependent apoptosis 423
Phosphorylation 371
PI3-kinase 359
p53 knockout 115
p53 mutation 217
Polo-like kinase 317
Polymorphism 143
Positional cloning 143
Predictive test 463
Prenatal radiation 13
Proliferating cell nuclear antigen (PCNA) 249
Promotion 283
Proteasome 363
Protection and therapy with antioxidants 387
Protection of genome integrity 407
Protein kinase C 299
Protein radical 351
Proteins 341
PTCH 175
Pulse train 289

Radiation 39, 51, 59, 83, 119, 175, 179, 191, 195, 207, 241, 303, 313, 317, 327, 355, 367, 423
Radiation carcinogenesis 101, 151, 163
Radiation-induced thymic lymphomas 157
Radiation instability 309
Radiation lymphomagenesis 439

Radiation protection 19
Radiation response 407
Radiation sensitivity 299
Radio-adaptive response 341
Radioadaptive response 467
Radio adaptive response 471
Radiofrequency 137
Radiogenic cancer 433
Radiogenic neoplasm 47
Radioresistance 295
Radiosensitivity 67, 359, 463
Radiotherapy 363, 459
Radon 23, 27
Radon inhalation 133
Radon therapy 75
Raf-1 299
Ramsar 35
Ras 157
Rat 127, 393
RBE 3, 383
Renal carcinogenesis 163
Repair 327
RET 207
Rev1 151
Risk coefficient 3
Risk estimation 23
Rolling mouse Nagoya 477
RT-PCR 67

SCID 105, 331
Scid mice 157
Scid mouse 367
Screening 47, 201
Senescence 371
Sensitivity to ionizing radiation 335
Sex 71
Signal transduction 201, 229
Simulated body fluid 379
Skin 175
Small dose effects 467
Smoking 23
Soft X-rays 289

Solid cancers 3
Spontaneous mutation 111
Stem cell hierarchy 407
Stem cells 415
Superoxide dismutase 75, 341
Suppression of carcinogenesis 487
Suppressor genes 119
Survival rate 467

T-cell leukemia 179
Telomerase activity 335
Telomere 323
Telomere shortening 371
Telomeric fusion 313
Telonomic instability 313, 367
Thermal effect 75
Thorotrast 191, 195, 221
Thymic murine lymphoma 299
Thyroid 47, 317
Thyroid cancer 39, 127, 201
Tissue repair 423
Transfection 335
Transforming growth factor β 399
Tsc2 gene 163
Tumorigenesis 115
Tumor induction 119
Tumor suppressor gene 143
Tumor susceptibility 143
Two-hit carcinogenesis 163

Variation in DNA-PKcs 439

Wasted mouse 249
Werner syndrome 347
Whole body 265
Wortmannin 359
WRN 347

X-irradiation 393
XPA 175
XPD genotype 217
X-rays 331